STRUCTURAL STEEL DESIGN

FIFTH EDITION

JACK C. McCORMAC
STEPHEN F. CSERNAK

Prentice Hall

Boston   Columbus   Indianapolis   New York   San Francisco   Upper Saddle River
Amsterdam   Cape Town   Dubai   London   Madrid   Milan   Munich   Paris   Montreal   Toronto
Delhi   Mexico City   Sao Paulo   Sydney   Hong Kong   Seoul   Singapore   Taipei   Tokyo

Vice President and Editorial Director, ECS: *Marcia J. Horton*
Executive Editor: *Holly Stark*
Editorial Assistant: *William Opaluch*
Vice President, Production: *Vince O'Brien*
Senior Managing Editor: *Scott Disanno*
Production Liaison: *Greg Dulles*
Production Editor: *Pavithra Jayapaul, TexTech International*
Operations Specialist: *Lisa McDowell*
Executive Marketing Manager: *Tim Galligan*
Market Assistant: *Jon Byrant*
Art Editor: *Greg Dulles*
Composition/Full-Service Project Management: *TexTech International*

The author and publisher of this book have used their best efforts in preparing this book. These efforts include the development, research, and testing of the theories and programs to determine their effectiveness. The author and publisher make no warranty of any kind, expressed or implied, with regard to these programs or the documentation contained in this book. The author and publisher shall not be liable in any event for incidental or consequential damages in connection with, or arising out of, the furnishing, performance, or use of these programs.

Library of Congress Cataloging-in-Publication Data

McCormac, Jack C.
   Structural steel design / Jack C. McCormac, Stephen F. Csernak. — 5th ed.
     p. cm.
   ISBN-13: 978-0-13-607948-4
   ISBN-10: 0-13-607948-2
1. Building, Iron and steel—Textbooks.   2. Steel, Structural—Textbooks.   I. Csernak, Stephen F.   II. Title.
   TA684.M25 2011
   624.1'821—dc22

                                2011010788

Prentice Hall
is an imprint of

www.pearsonhighered.com

10 9 8 7 6 5 4 3 2

ISBN-13: 978-0-13-607948-4
ISBN-10:   0-13-607948-2

Preface

This textbook has been prepared with the hope that its readers will, as have so many engineers in the past, become interested in structural steel design and want to maintain and increase their knowledge on the subject throughout their careers in the engineering and construction industries. The material was prepared primarily for an introductory course in the junior or senior year but the last several chapters may be used for a graduate class. The authors have assumed that the student has previously taken introductory courses in mechanics of materials and structural analysis.

The authors' major objective in preparing this new edition was to update the text to conform to both the American Institute of Steel Construction (AISC) 2010 Specification for Structural Steel Buildings and the 14th edition of the AISC Steel Construction Manual published in 2011.

WHAT'S NEW IN THIS EDITION

Several changes to the text were made to the textbook in this edition:

1. End of chapter Problems for Solution have been added for Chapter 1 of the textbook.
2. The load factors and load combinations defined in Chapter 2 of the textbook and used throughout the book in example problems and end of chapter problems for solution have been revised to meet those given in the ASCE 7-10 and Part 2 of the AISC Steel Construction Manual.
3. The classification of compression sections for local buckling defined in Chapter 5 of the textbook has been revised to the new definition given in Section B4.1 of the new AISC Specification. For compression, sections are now classified as *non-slender* element or *slender* element sections.
4. The AISC Specification provides several methods to deal with stability analysis and the design of beam-columns. In Chapter 7 of the textbook, the *Effective Length Method (ELM)* is still used, though a brief introduction to the *Direct Analysis Method (DM)* has been added. A more comprehensive discussion of the DM is reserved for Chapter 11 of the text.
5. In Chapter 11 of the textbook, both the *Direct Analysis Method* and the *Effective Length Method* are presented for the analysis and design of beam-columns. This is to address the fact that the presentation of the *Direct Analysis Method* was moved from an appendix to Chapter C of the new AISC Specification while the *Effective Length Method* moved from Chapter C to Appendix 7.
6. Most of the end of chapter *Problems for Solution* for Chapters 2 through 11 have been revised. For Chapters 12 through 18 about half the problems have been revised.
7. Various photos were updated throughout the textbook.

The authors would like to express appreciation to Dr. Bryant G. Nielson of Clemson University for his assistance in developing the changes to this manuscript and to Sara Elise Roberts, former Clemson University graduate student, for her assistance in the review of the end of chapter problems and their solutions. In addition, the American Institute of Steel Construction was very helpful in providing advance copies of the AISC Specification and Steel Construction Manual revisions. Finally, we would like to thank our families for their encouragement and support in the revising of the manuscript of this textbook.

We also thank the reviewers and users of the previous editions of this book for their suggestions, corrections, and criticisms. We welcome any comments on this edition.

Jack C. McCormac, P.E.
Stephen F. Csernak, P.E.

Contents

C H A P T E R   1

Introduction to Structural Steel Design

1.1 ADVANTAGES OF STEEL AS A STRUCTURAL MATERIAL

A person traveling in the United States might quite understandably decide that steel is the perfect structural material. He or she would see an endless number of steel bridges, buildings, towers, and other structures. After seeing these numerous steel structures, this traveler might be surprised to learn that steel was not economically made in the United States until late in the nineteenth century, and the first wide-flange beams were not rolled until 1908.

The assumption of the perfection of this metal, perhaps the most versatile of structural materials, would appear to be even more reasonable when its great strength, light weight, ease of fabrication, and many other desirable properties are considered. These and other advantages of structural steel are discussed in detail in the paragraphs that follow.

1.1.1 High Strength

The high strength of steel per unit of weight means that the weight of structures will be small. This fact is of great importance for long-span bridges, tall buildings, and structures situated on poor foundations.

1.1.2 Uniformity

The properties of steel do not change appreciably with time, as do those of a reinforced-concrete structure.

1.1.3 Elasticity

Steel behaves closer to design assumptions than most materials because it follows Hooke's law up to fairly high stresses. The moments of inertia of a steel structure can be accurately calculated, while the values obtained for a reinforced-concrete structure are rather indefinite.

Erection of steel joists. (Courtesy of Vulcraft.)

1.1.4 Permanence

Steel frames that are properly maintained will last indefinitely. Research on some of the newer steels indicates that under certain conditions no painting maintenance whatsoever will be required.

1.1.5 Ductility

The property of a material by which it can withstand extensive deformation without failure under high tensile stresses is its *ductility*. When a *mild* or *low-carbon* structural steel member is being tested in tension, a considerable reduction in cross section and a large amount of elongation will occur at the point of failure before the actual fracture occurs. A material that does not have this property is generally unacceptable and is probably hard and brittle, and it might break if subjected to a sudden shock.

In structural members under normal loads, high stress concentrations develop at various points. The ductile nature of the usual structural steels enables them to yield locally at those points, thus preventing premature failures. A further advantage of ductile structures is that when overloaded, their large deflections give visible evidence of impending failure (sometimes jokingly referred to as "running time").

1.1.6 Toughness

Structural steels are tough—that is, they have both strength and ductility. A steel member loaded until it has large deformations will still be able to withstand large forces. This is a very important characteristic, because it means that steel members can be subjected to large deformations during fabrication and erection without fracture—thus allowing them to be bent, hammered, and sheared, and to have holes punched in them without visible damage. The ability of a material to absorb energy in large amounts is called *toughness*.

1.1.7   Additions to Existing Structures

Steel structures are quite well suited to having additions made to them. New bays or even entire new wings can be added to existing steel frame buildings, and steel bridges may often be widened.

1.1.8   Miscellaneous

Several other important advantages of structural steel are as follows: (a) ability to be fastened together by several simple connection devices, including welds and bolts; (b) adaptation to prefabrication; (c) speed of erection; (d) ability to be rolled into a wide variety of sizes and shapes, as described in Section 1.4 of this chapter; (e) possible reuse after a structure is disassembled; and (f) scrap value, even though not reusable in its existing form. Steel is the ultimate recyclable material.

1.2   DISADVANTAGES OF STEEL AS A STRUCTURAL MATERIAL

In general, steel has the following disadvantages:

1.2.1   Corrosion

Most steels are susceptible to corrosion when freely exposed to air and water, and therefore must be painted periodically. The use of weathering steels, however, in suitable applications tends to eliminate this cost.

Though weathering steels can be quite effective in certain situations for limiting corrosion, there are many cases where their use is not feasible. In some of these situations, corrosion may be a real problem. For instance, corrosion-fatigue failures can occur where steel members are subject to cyclic stresses and corrosive environments. The fatigue strength of steel members can be appreciably reduced when the members are used in aggressive chemical environments and subject to cyclic loads.

The reader should note that steels are available in which copper is used as an anti-corrosion component. The copper is usually absorbed during the steelmaking process.

1.2.2   Fireproofing Costs

Although structural members are incombustible, their strength is tremendously reduced at temperatures commonly reached in fires when the other materials in a building burn. Many disastrous fires have occurred in empty buildings where the only fuel for the fires was the buildings themselves. Furthermore, steel is an excellent heat conductor—nonfireproofed steel members may transmit enough heat from a burning section or compartment of a building to ignite materials with which they are in contact in adjoining sections of the building. As a result, the steel frame of a building may have to be protected by materials with certain insulating characteristics, and the building may have to include a sprinkler system if it is to meet the building code requirements of the locality in question.

1.2.3   Susceptibility to Buckling

As the length and slenderness of a compression member is increased, its danger of buckling increases. For most structures, the use of steel columns is very economical because of their high strength-to-weight ratios. Occasionally, however, some additional steel is needed to stiffen them so they will not buckle. This tends to reduce their economy.

1.2.4   Fatigue

Another undesirable property of steel is that its strength may be reduced if it is subjected to a large number of stress reversals or even to a large number of variations of tensile stress. (Fatigue problems occur only when tension is involved.) The present practice is to reduce the estimations of strength of such members if it is anticipated that they will have more than a prescribed number of cycles of stress variation.

1.2.5   Brittle Fracture

Under certain conditions steel may lose its ductility, and brittle fracture may occur at places of stress concentration. Fatigue-type loadings and very low temperatures aggravate the situation. Triaxial stress conditions can also lead to brittle fracture.

1.3   EARLY USES OF IRON AND STEEL

Although the first metal used by human beings was probably some type of copper alloy such as bronze (made with copper, tin, and perhaps some other additives), the most important metal developments throughout history have occurred in the manufacture and use of iron and its famous alloy called steel. Today, iron and steel make up nearly 95 percent of all the tonnage of metal produced in the world.[1]

Despite diligent efforts for many decades, archaeologists have been unable to discover when iron was first used. They did find an iron dagger and an iron bracelet in the Great Pyramid in Egypt, which they claim had been there undisturbed for at least 5000 years. The use of iron has had a great influence on the course of civilization since the earliest times and may very well continue to do so in the centuries ahead. Since the beginning of the Iron Age in about 1000 BC, the progress of civilization in peace and war has been heavily dependent on what people have been able to make with iron. On many occasions its use has decidedly affected the outcome of military engagements. For instance, in 490 BC in Greece at the Battle of Marathon, the greatly outnumbered Athenians killed 6400 Persians and lost only 192 of their own men. Each of the victors wore 57 pounds of iron armor in the battle. (This was the battle from which the runner Pheidippides ran the approximately 25 miles to Athens and died while shouting news of the victory.) This victory supposedly saved Greek civilization for many years.

---

[1]American Iron and Steel Institute, *The Making of Steel* (Washington, DC, not dated), p. 6.

The mooring mast of the Empire State Building, New York City. (Courtesy of Getty Images/Hulton Archive Photos.)

According to the classic theory concerning the first production of iron in the world, there was once a great forest fire on Mount Ida in Ancient Troy (now Turkey) near the Aegean Sea. The land surface reportedly had a rich content of iron, and the heat of the fire is said to have produced a rather crude form of iron that could be hammered into various shapes. Many historians believe, however, that human beings first learned to use iron which fell to the earth in the form of meteorites. Frequently, the iron in meteorites is combined with nickel to produce a harder metal. Perhaps, early human beings were able to hammer and chip this material into crude tools and weapons.

Steel is defined as a combination of iron and a small amount of carbon, usually less than 1 percent. It also contains small percentages of some other elements. Although some steel has been made for at least 2000–3000 years, there was really no economical production method available until the middle of the nineteenth century.

The first steel almost certainly was obtained when the other elements necessary for producing it were accidentally present when iron was heated. As the years went by, steel probably was made by heating iron in contact with charcoal. The surface of the iron absorbed some carbon from the charcoal, which was then hammered into the hot iron. Repeating this process several times resulted in a case-hardened exterior of steel. In this way the famous swords of Toledo and Damascus were produced.

The first large volume process for producing steel was named after Sir Henry Bessemer of England. He received an English patent for his process in 1855, but his efforts to obtain a U.S. patent for the process in 1856 were unsuccessful, because it was shown that William Kelly of Eddyville, Kentucky, had made steel by the same process seven years before Bessemer applied for his English patent. Although Kelly was given the patent, the name Bessemer was used for the process.[2]

Kelly and Bessemer learned that a blast of air through molten iron burned out most of the impurities in the metal. Unfortunately, at the same time, the blow eliminated some desirable elements such as carbon and manganese. It was later learned that these needed elements could be restored by adding spiegeleisen, which is an alloy of iron, carbon, and manganese. It was further learned that the addition of limestone in the converter resulted in the removal of the phosphorus and most of the sulfur.

Before the Bessemer process was developed, steel was an expensive alloy used primarily for making knives, forks, spoons, and certain types of cutting tools. The Bessemer process reduced production costs by at least 80 percent and allowed, for the first time, production of large quantities of steel.

The Bessemer converter was commonly used in the United States until the beginning of the twentieth century, but since that time it has been replaced with better methods, such as the open-hearth process and the basic oxygen process.

As a result of the Bessemer process, structural carbon steel could be produced in quantity by 1870, and by 1890, steel had become the principal structural metal used in the United States.

Today, most of the structural steel shapes and plates produced in the United States are made by melting scrap steel. This scrap steel is obtained from junk cars and scrapped structural shapes, as well as from discarded refrigerators, motors, typewriters, bed springs, and other similar items. The molten steel is poured into molds that have approximately the final shapes of the members. The resulting sections, which are run through a series of rollers to squeeze them into their final shapes, have better surfaces and fewer internal or residual stresses than newly made steel.

The shapes may be further processed by cold rolling, by applying various coatings, and perhaps by the process of *annealing*. This is the process by which the steel is heated to an intermediate temperature range (say, 1300–1400°F), held at that temperature for several hours, and then allowed to slowly cool to room temperature. Annealing results in steel with less hardness and brittleness, but greater ductility.

The term **wrought iron** refers to iron with a very low carbon content (≤ 0.15 percent), while iron with a very high carbon content (≥ 2 percent) is referred to as **cast iron**. Steel falls in between cast iron and wrought iron and has carbon contents in the range of 0.15 percent to 1.7 percent (as described in Section 1.8 of this chapter).

---

[2]American Iron and Steel Institute, *Steel 76* (Washington, DC, 1976), pp. 5–11.

The first use of metal for a sizable structure occurred in England in Shropshire (about 140 miles northwest of London) in 1779, when cast iron was used for the construction of the 100-ft Coalbrookdale Arch Bridge over the River Severn. It is said that this bridge (which still stands) was a turning point in engineering history because it changed the course of the Industrial Revolution by introducing iron as a structural material. This iron was supposedly four times as strong as stone and thirty times as strong as wood.[3]

A number of other cast-iron bridges were constructed in the following decades, but soon after 1840 the more malleable wrought iron began to replace cast iron. The development of the Bessemer process and subsequent advances, such as the open-hearth process, permitted the manufacture of steel at competitive prices. This encouraged the beginning of the almost unbelievable developments of the last 120 years with structural steel.

1.4   STEEL SECTIONS

The first structural shapes made in the United States were angle irons rolled in 1819. I-shaped steel sections were first rolled in the United States in 1884, and the first skeleton frame structure (the Home Insurance Company Building in Chicago) was erected that same year. Credit for inventing the "skyscraper" is usually given to engineer William LeBaron Jenny, who planned the building, apparently during a bricklayers' strike. Prior to this time, tall buildings in the United States were constructed with load-bearing brick walls that were several feet thick.

For the exterior walls of the 10-story building, Jenny used cast-iron columns encased in brick. The beams for the lower six floors were made from wrought iron, while structural steel beams were used for the upper floors. The first building completely framed with structural steel was the second Rand-McNally building, completed in Chicago in 1890.

An important feature of the 985-ft wrought-iron Eiffel tower constructed in 1889 was the use of mechanically operated passenger elevators. The availability of these machines, along with Jenny's skeleton frame idea, led to the construction of thousands of high-rise buildings throughout the world during the last century.

During these early years, the various mills rolled their own individual shapes and published catalogs providing the dimensions, weight, and other properties of the shapes. In 1896, the Association of American Steel Manufacturers (now the American Iron and Steel Institute, or AISI) made the first efforts to standardize shapes. Today, nearly all structural shapes are standardized, though their exact dimensions may vary just a little from mill to mill.[4]

Structural steel can be economically rolled into a wide variety of shapes and sizes without appreciably changing its physical properties. Usually, the most desirable members are those with large moments of inertia in proportion to their areas. The **I**, **T**, and **C** shapes, so commonly used, fall into this class.

---

[3]M. H. Sawyer, "World's First Iron Bridge," *Civil Engineering* (New York: ASCE, December 1979), pp. 46–49.
[4]W. McGuire, *Steel Structures* (Englewood Cliffs, NJ: Prentice-Hall, 1968), pp. 19–21.

Pedestrian bridge for North Carolina Cancer Hospital, Chapel Hill, NC. (Courtesy of CMC South Carolina Steel.)

Steel sections are usually designated by the shapes of their cross sections. As examples, there are angles, tees, zees, and plates. It is necessary, however, to make a definite distinction between American standard beams (called *S beams*) and wide-flange beams (called *W beams*), as they are both I-shaped. The inner surface of the flange of a W section is either parallel to the outer surface or nearly so, with a maximum slope of 1 to 20 on the inner surface, depending on the manufacturer.

The S beams, which were the first beam sections rolled in America, have a slope on their inside flange surfaces of 1 to 6. It might be noted that the constant (or nearly constant) thickness of W-beam flanges compared with the tapered S-beam flanges may facilitate connections. Wide-flange beams comprise nearly 50 percent of the tonnage of structural steel shapes rolled today. The W and S sections are shown in Fig. 1.1, together with several other familiar steel sections. The uses of these various shapes will be discussed in detail in the chapters to follow.

Constant reference is made throughout this book to the 14th edition of the **Steel Construction Manual**, published by the American Institute of Steel Construction (AISC). This manual, which provides detailed information for structural steel shapes, is referred to hereafter as "the AISC Manual," "the Steel Manual," or simply, "the Manual." It is based on the 2010 **Specification for Structural Steel Buildings** (ANSI/AISC 360-10) (hereafter, "the AISC Specification"), published by the AISC on June 22, 2010.

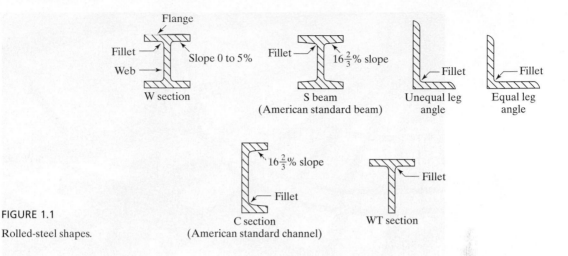

FIGURE 1.1

Rolled-steel shapes.

Structural shapes are identified by a certain system described in the Manual for use in drawings, specifications, and designs. This system is standardized so that all steel mills can use the same identification for purposes of ordering, billing, etc. In addition, so much work is handled today with computers and other automated equipment that it is necessary to have a letter-and-number system which can be printed out with a standard keyboard (as opposed to the old system where certain symbols were used for angles, channels, etc.). Examples of this identification system are as follows:

1. A W27 × 114 is a W section approximately 27 in deep, weighing 114 lb/ft.
2. An S12 × 35 is an S section 12 in deep, weighing 35 lb/ft.
3. An HP12 × 74 is a bearing pile section approximately 12 in deep, weighing 74 lb/ft. Bearing piles are made with the regular W rolls, but with thicker webs to provide better resistance to the impact of pile driving. The width and depth of these sections are approximately equal, and the flanges and webs have equal or almost equal thickness.
4. An M8 × 6.5 is a miscellaneous section 8 in deep, weighing 6.5 lb/ft. It is one of a group of doubly symmetrical H-shaped members that cannot by dimensions be classified as a W, S, or HP section, as the slope of their inner flanges is other than 16 2/3 percent.
5. A C10 × 30 is a channel 10 in deep, weighing 30 lb/ft.
6. An MC18 × 58 is a miscellaneous channel 18 in deep, weighing 58 lb/ft, which cannot be classified as a C shape because of its dimensions.
7. An HSS14 × 10 × 5/8 is a rectangular hollow structural section 14 in deep, 10 in wide, with a 5/8-in wall thickness. It weighs 93.10 lb/ft. Square and round HSS sections are also available.
8. An L6 × 6 × 1/2 is an equal leg angle, each leg being 6 in long and 1/2 in thick.

Roof framing for Glen Oaks School, Bellerose, NY. (Courtesy of CMC South Carolina Steel.)

9. A WT18 × 151 is a tee obtained by splitting a W36 × 302. This type of section is known as a structural tee.

10. Rectangular steel sections are classified as wide *plates* or narrow *bars*.

The only differences between bars and plates are their sizes and production procedures. Historically, flat pieces have been called bars if they are 8 in or less in width. They are plates if wider than 8 in. Tables 1-29, 2-3, and 2-5 in the AISC Manual provide information on bars and plates. Bar and plate thicknesses are usually specified to the nearest 1/16 in for thicknesses less than 3/8 in, to the nearest 1/8 in for thicknesses between 3/8 in and 1 in, and to the nearest 1/4 in for thicknesses greater than 1 in. A plate is usually designated by its thickness, width, and length, in that order; for example, a PL1/2 × 10 × 1 ft 4 in is 1/2 in thick, 10 in wide, and 16 in long. Actually, the term **plate** is almost universally used today, whether a member is fabricated from plate or bar stock. Sheet and strip are usually thinner than bars and plates.

The student should refer to the Steel Manual for information concerning other shapes. Detailed information on these and other sections will be presented herein as needed.

In Part 1 of the Manual, the dimensions and properties of W, S, C, and other shapes are tabulated. The dimensions of the members are given in decimal form (for the use of designers) and in fractions to the nearest sixteenth of an inch (for the use of craftsmen and steel detailers or drafters). Also provided for the use of designers are such items as moments of inertia, section moduli, radii of gyration, and other cross-sectional properties discussed later in this text.

There are variations present in any manufacturing process, and the steel industry is certainly no exception. As a result, the cross-sectional dimensions of steel members may vary somewhat from the values specified in the Manual. Maximum tolerances for the rolling of steel shapes are prescribed by the American Society for Testing and Materials (ASTM) A6 Specification and are presented in Tables 1-22 to 1-28 in the Manual. As a result, calculations can be made on the basis of the properties given in the Manual, regardless of the manufacturer.

Some steel sections listed in the Manual are available in the United States from only one or two steel producers and thus, on occasion, may be difficult to obtain promptly. Accordingly, when specifying sections, the designer would be wise to contact a steel fabricator for a list of sections readily available.

Through the years, there have been changes in the sizes of steel sections. For instance, there may be insufficient demand to continue rolling a certain shape; an existing shape may be dropped because a similar size, but more efficient, shape has been developed, and so forth. Occasionally, designers may need to know the properties of one of the discontinued shapes no longer listed in the latest edition of the Manual or in other tables normally available to them.

For example, it may be desired to add another floor to an existing building that was constructed with shapes no longer rolled. In 1953, the AISC published a book entitled *Iron and Steel Beams 1873 to 1952*, which provides a complete listing of iron and steel beams and their properties rolled in the United States during that period. An up-to-date edition of this book is now available. It is *AISC Design Guide 15* and covers properties of steel shapes produced from 1887 to 2000.[5] There will undoubtedly be many more shape changes in the future. For this reason, the wise structural designer should carefully preserve old editions of the Manual so as to have them available when the older information is needed.

1.5   METRIC UNITS

Almost all of the examples and homework problems presented in this book make use of U.S. customary units. The author, however, feels that today's designer must be able to perform his or her work in either customary or metric units.

The problem of working with metric units when performing structural steel design in the United States has almost been eliminated by the AISC. Almost all of their equations are written in a form applicable to both systems. In addition, the metric equivalents of the standard U.S. shapes are provided in Section 17 of the Manual. For instance, a W36 × 302 section is shown there as a W920 × 449, where the 920 is mm and the 449 is kg/m.

---

[5]R. L. Brockenbrough, *AISC Rehabilitation and Retrofit Guide: A Reference for Historic Shapes and Specifications* (Chicago, AISC, 2002).

Mariners, Ballpark, Seattle, WA. (Courtesy of Trade ARBED.)

1.6 COLD-FORMED LIGHT-GAGE STEEL SHAPES

In addition to the hot-rolled steel shapes discussed in the previous section, there are some cold-formed steel shapes. These are made by bending thin sheets of carbon or low-alloy steels into almost any desired cross section, such as the ones shown in Fig. 1.2.[6] These shapes—which may be used for light members in roofs, floors, and walls—vary in thickness from about 0.01 in up to about 0.25 in. The thinner shapes are most often used for some structural panels.

Though cold-working does reduce ductility somewhat, it causes some strength increases. Under certain conditions, design specifications will permit the use of these higher strengths.

Concrete floor slabs are very often cast on formed steel decks that serve as economical forms for the wet concrete and are left in place after the concrete hardens. Several types of decking are available, some of which are shown in Fig. 1.3. The sections

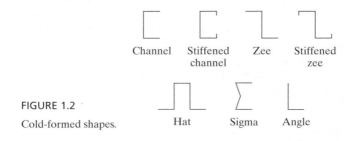

FIGURE 1.2
Cold-formed shapes.

Channel   Stiffened   Zee   Stiffened
           channel   zee

Hat   Sigma   Angle

---

[6]*Cold-Formed Steel Design Manual* (Washington, DC: American Iron and Steel Institute, 2002).

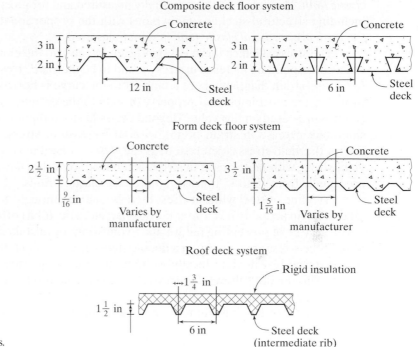

FIGURE 1.3

Some types of steel decks.

with the deeper cells have the useful feature that electrical and mechanical conduits can be placed in them. The use of steel decks for floor slabs is discussed in Chapter 16 of this text. There, composite construction is presented. With such construction steel beams are made composite with concrete slabs by providing for shear transfer between the two so they will act together as a unit.

1.7   STRESS–STRAIN RELATIONSHIPS IN STRUCTURAL STEEL

To understand the behavior of steel structures, an engineer must be familiar with the properties of steel. Stress–strain diagrams present valuable information necessary to understand how steel will behave in a given situation. Satisfactory steel design methods cannot be developed unless complete information is available concerning the stress–strain relationships of the material being used.

If a piece of ductile structural steel is subjected to a tensile force, it will begin to elongate. If the tensile force is increased at a constant rate, the amount of elongation will increase linearly within certain limits. In other words, elongation will double when the stress goes from 6000 to 12,000 psi (pounds per square inch). When the tensile stress reaches a value roughly equal to three-fourths of the ultimate strength of the steel, the elongation will begin to increase at a greater rate without a corresponding increase in the stress.

The largest stress for which Hooke's law applies, or the highest point on the linear portion of the stress–strain diagram, is called the *proportional limit*. The largest stress that a material can withstand without being permanently deformed is called the *elastic limit.* This value is seldom actually measured and for most engineering materials, including structural steel, is synonymous with the proportional limit. For this reason, the term *proportional elastic limit* is sometimes used.

The stress at which there is a significant increase in the elongation, or strain, without a corresponding increase in stress is said to be the *yield* stress. It is the first point on the stress–strain diagram where a tangent to the curve is horizontal. The yield stress is probably the most important property of steel to the designer, as so many design procedures are based on this value. Beyond the yield stress there is a range in which a considerable increase in strain occurs without increase in stress. The strain that occurs before the yield stress is referred to as the *elastic strain;* the strain that occurs after the yield stress, with no increase in stress, is referred to as the *plastic strain.* Plastic strains are usually from 10 to 15 times as large as the elastic strains.

Yielding of steel without stress increase may be thought to be a severe disadvantage, when in actuality it is a very useful characteristic. It has often performed the wonderful service of preventing failure due to omissions or mistakes on the designer's part. Should the stress at one point in a ductile steel structure reach the yield point, that part of the structure will yield locally without stress increase, thus preventing premature failure. This ductility allows the stresses in a steel structure to be redistributed. Another way of describing this phenomenon is to say that very high stresses caused by fabrication, erection, or loading will tend to equalize themselves. It might also be said that a

Erection of roof truss, North Charleston, SC. (Courtesy of CMC South Carolina Steel.)

steel structure has a reserve of plastic strain that enables it to resist overloads and sudden shocks. If it did not have this ability, it might suddenly fracture, like glass or other vitreous substances.

Following the plastic strain, there is a range in which additional stress is necessary to produce additional strain. This is called *strain-hardening*. This portion of the diagram is not too important to today's designer, because the strains are so large. A familiar stress–strain diagram for mild or low-carbon structural steel is shown in Fig. 1.4. Only the initial part of the curve is shown here because of the great deformation that occurs before failure. At failure in the mild steels, the total strains are from 150 to 200 times the elastic strains. The curve will actually continue up to its maximum stress value and then "tail off" before failure. A sharp reduction in the cross section of the member (called *necking*) takes place just before the member fractures.

The stress–strain curve of Fig. 1.4 is typical of the usual ductile structural steel and is assumed to be the same for members in tension or compression. (The compression members must be stocky, because slender compression members subjected to compression loads tend to buckle laterally, and their properties are greatly affected by the bending moments so produced.) The shape of the diagram varies with the speed of loading, the type of steel, and the temperature. One such variation is shown in the figure by the dotted line marked *upper yield*.

This shape stress–strain curve is the result when a mild steel has the load applied rapidly, while the *lower yield* is the case for slow loading.

Figure 1.5 shows typical stress–strain curves for several different yield stress steels.

You should note that the stress–strain diagrams of Figs. 1.4 and 1.5 were prepared for a mild steel at room temperature. During welding operations and during fires, structural steel members may be subjected to very high temperatures. Stress–strain diagrams prepared for steels with temperatures above 200°F will be more rounded and nonlinear and will not exhibit well-defined yield points. Steels

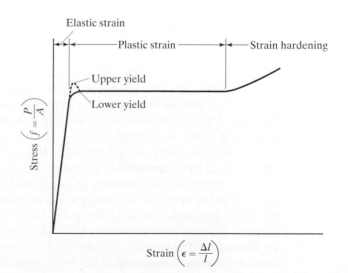

FIGURE 1.4

Typical stress–strain diagram for a mild or low-carbon structural steel at room temperature.

Puerta Europa, Madrid, Spain.
(Courtesy of Trade ARBED.)

(particularly those with high carbon contents) may actually increase a little in tensile strength as they are heated to a temperature of about 700°F. As temperatures are raised into the 800°-to-1000°F range, strengths are drastically reduced, and at 1200°F little strength is left.

Figure 1.6 shows the variation of yield strengths for several grades of steel as their temperatures are raised from room temperature up to 1800° to 1900°F. Temperatures of the magnitudes shown can easily be reached in steel members during fires, in localized areas of members when welding is being performed, in members in foundries over open flame, and so on.

When steel sections are cooled below 32°F, their strengths will increase a little, but they will have substantial reductions in ductility and toughness.

A very important property of a structure that has been stressed, but not beyond its yield point, is that it will return to its original length when the loads are removed. Should it be stressed beyond this point, it will return only part of the way back to its original position. This knowledge leads to the possibility of testing an existing structure by loading and unloading. If, after the loads are removed, the structure does not resume its original dimensions, it has been stressed beyond its yield point.

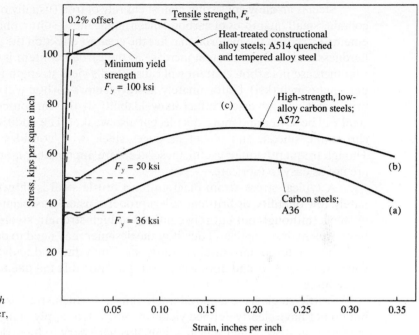

FIGURE 1.5

Typical stress–strain curves. (Based on a figure from Salmon C. G. and J. E. Johnson, *Steel Structures: Design and Behavior, Fourth Edition*. Upper Saddle River, NJ: Prentice Hall, 1996.)

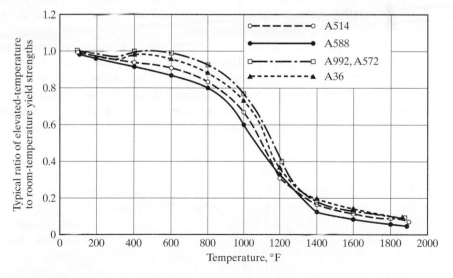

FIGURE 1.6

Effect of temperature on yield strengths.

Steel is an alloy consisting almost entirely of iron (usually over 98 percent). It also contains small quantities of carbon, silicon, manganese, sulfur, phosphorus, and other elements. Carbon is the material that has the greatest effect on the properties of steel. The hardness and strength of steel increase as the carbon content is increased. A 0.01 percent increase in carbon content will cause steel's yield strength to go up about 0.5 kips per square inch (ksi). Unfortunately, however, more carbon will cause steel to be more brittle and will adversely affect its weldability. If the carbon content is reduced, the steel will be softer and more ductile, but also weaker. The addition of such elements as chromium, silicon, and nickel produces steels with considerably higher strengths. Though frequently quite useful, these steels are appreciably more expensive and often are not as easy to fabricate.

A typical stress–strain diagram for a brittle steel is shown in Fig. 1.7. Unfortunately, low ductility, or brittleness, is a property usually associated with high strengths in steels (although not entirely confined to high-strength steels). As it is desirable to have both high strength and ductility, the designer may need to decide between the two extremes or to compromise. A brittle steel may fail suddenly and without warning when over-stressed, and during erection could possibly fail due to the shock of erection procedures.

Brittle steels have a considerable range in which stress is proportional to strain, but do not have clearly defined yield stresses. Yet, to apply many of the formulas given in structural steel design specifications, it is necessary to have definite yield stress values, regardless of whether the steels are ductile or brittle.

If a steel member is strained beyond its elastic limit and then unloaded, it will not return to a condition of zero strain. As it is unloaded, its stress–strain diagram will follow a new path (shown in Fig. 1.7 by the dotted line parallel to the initial straight line). The result is a permanent, or residual, strain.

The line representing the stress–strain ratio for quenched and tempered steels gradually varies from a straight line so that a distinct yield point is not available. For such steels the yield stress is usually defined as the stress at the point of unloading, which corresponds to some arbitrarily defined residual strain (0.002 being the common value). In other words, we increase the strain by a designated amount and from that point draw a line parallel to the straight-line portion of the stress–strain diagram, until

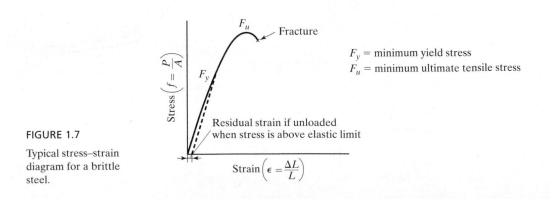

FIGURE 1.7

Typical stress–strain diagram for a brittle steel.

F_y = minimum yield stress
F_u = minimum ultimate tensile stress

the new line intersects the old. This intersection is the yield stress at that particular strain. If 0.002 is used, the intersection is usually referred to as the yield stress at 0.2 percent offset strain.

1.8   MODERN STRUCTURAL STEELS

The properties of steel can be greatly changed by varying the quantities of carbon present and by adding other elements such as silicon, nickel, manganese, and copper. A steel that has a significant amount of the latter elements is referred to as an *alloy steel*. Although these elements do have a great effect on the properties of steel, the actual quantities of carbon or other alloying elements are quite small. For instance, the carbon content of steel is almost always less than 0.5 percent by weight and is normally from 0.2 to 0.3 percent.

One-half of a 170-ft clear span roof truss for the Athletic and Convention Center, Lehigh University, Bethlehem, PA. (Courtesy of Bethlehem Steel Corporation.)

The chemistry of steel is extremely important in its effect on such properties of the steel as weldability, corrosion resistance, resistance to brittle fracture, and so on. The ASTM specifies the exact maximum percentages of carbon, manganese, silicon, etc., that are permissible for a number of structural steels. Although the physical and mechanical properties of steel sections are primarily determined by their chemical composition, they are also influenced to a certain degree by the rolling process and by their stress history and heat treatment.

In the past few decades, a structural carbon steel designated as A36 and having a minimum yield stress $F_y = 36$ ksi was the commonly used structural steel. More recently, however, most of the structural steel used in the United States is manufactured by melting scrap steel in electric furnaces. With this process, a 50 ksi steel, A992, can be produced and sold at almost the same price as 36 ksi steel.

The 50 ksi steels are the predominant ones in use today. In fact, some of the steel mills charge extra for W sections if they are to consist of A36 steel. On the other hand, 50 ksi angles have on occasion been rather difficult to obtain without special orders to the steel mills. As a result, A36 angles are still frequently used. In addition, 50 ksi plates may cost more than A36 steel.

In recent decades, the engineering and architecture professions have been continually requesting increasingly stronger steels—steels with more corrosion resistance, steels with better welding properties, and various other requirements. Research by the steel industry during this period has supplied several groups of new steels that satisfy many of the demands. Today there are quite a few structural steels designated by the ASTM and included in the AISC Specification.

Steel dome. (Courtesy of Trade ARBED.)

Structural steels are generally grouped into several major ASTM classifications: the carbon steels A36, A53, A500, A501, and A529; the high-strength low-alloy steels A572, A618, A913, and A992; and the corrosion-resistant high-strength low-alloy steels A242, A588, and A847. Considerable information is presented for each of these steels in Part 2 of the Manual. The sections that follow include a few general remarks about these steel classifications.

1.8.1   Carbon Steels

These steels have as their principal strengthening agents carefully controlled quantities of carbon and manganese. Carbon steels have their contents limited to the following maximum percentages: 1.7 percent carbon, 1.65 percent manganese, 0.60 percent silicon, and 0.60 percent copper. These steels are divided into four categories, depending on carbon percentages:

1. Low-carbon steel: < 0.15 percent.
2. Mild steel: 0.15 to 0.29 percent. (The structural carbon steels fall into this category.)
3. Medium-carbon steel: 0.30 to 0.59 percent.
4. High-carbon steel: 0.60 to 1.70 percent.

1.8.2   High-Strength Low-Alloy Steels

There are a large number of high-strength low-alloy steels, and they are included under several ASTM numbers. In addition to containing carbon and manganese, these steels owe their higher strengths and other properties to the addition of one or more alloying agents such as columbium, vanadium, chromium, silicon, copper, and nickel. Included are steels with yield stresses as low as 42 ksi and as high as 70 ksi. These steels generally have much greater atmospheric corrosion resistance than the carbon steels have.

The term *low-alloy* is used arbitrarily to describe steels for which the total of all the alloying elements does not exceed 5 percent of the total composition of the steel.

1.8.3   Atmospheric Corrosion-Resistant High-Strength Low-Alloy Structural Steels

When steels are alloyed with small percentages of copper, they become more corrosion-resistant. When exposed to the atmosphere, the surfaces of these steels oxidize and form a very tightly adherent film (sometimes referred to as a "tightly bound patina" or "a crust of rust"), which prevents further oxidation and thus eliminates the need for painting. After this process takes place (within 18 months to 3 years, depending on the type of exposure—rural, industrial, direct or indirect sunlight, etc.), the steel reaches a deep reddish-brown or black color.

Supposedly, the first steel of this type was developed in 1933 by the U.S. Steel Corporation to provide resistance to the severe corrosive conditions of railroad coal cars.

You can see the many uses that can be made of such a steel, particularly for structures with exposed members which are difficult to paint—bridges, electrical transmission towers, and others. This steel is not considered to be satisfactory for use where it is frequently subject to saltwater sprays or fogs, or continually submerged in water (fresh or salt) or the ground, or where there are severe corrosive industrial fumes. It is also not satisfactory in very dry areas, as in some western parts of the United States. For the patina to form, the steel must be subjected to a wetting and drying cycle. Otherwise, it will continue to look like unpainted steel.

Table 1.1 herein, which is Table 2-4 in the Steel Manual, lists the 12 ASTM steels mentioned earlier in this section, together with their specified minimum yield strengths (F_y) and their specified minimum tensile strengths (F_u). In addition, the right-hand columns of the table provide information regarding the availability of the shapes in the various grades of steels, as well as the preferred grade to use for each of them. The preferred steel in each case is shown with a black box.

You will note by the blackened boxes in the table that A36 is the preferred steel to be used for M, S, HP, C, MC, and L sections, while A992 is the preferred material for the most common shapes, the Ws. The lightly shaded boxes in the table refer to the shapes available in grades of steel other than the preferred grades. Before shapes are specified in these grades, the designer should check on their availability from the steel producers. Finally, the blank, or white, boxes indicate the grades of steel that are not available for certain shapes. Similar information is provided for plates and bars in Table 2-5 of the Steel Manual.

As stated previously, steels may be made stronger by the addition of special alloys. Another factor affecting steel strengths is thickness. The thinner steel is rolled, the stronger it becomes. Thicker members tend to be more brittle, and their slower cooling rates cause the steel to have a coarser microstructure.

Referring back to Table 1.1, you can see that several of the steels listed are available with different yield and tensile stresses with the same ASTM number. For instance, A572 shapes are available with 42, 50, 55, 60, and 65 ksi yield strengths. Next, reading the footnotes in Table 1.1, we note that Grades 60 and 65 steels have the footnote letter "e" by them. This footnote indicates that the only A572 shapes available with these strengths are those thinner ones which have flange thicknesses $\leq$ 2 inches. Similar situations are shown in the table for several other steels, including A992 and A242.

1.9   USES OF HIGH-STRENGTH STEELS

There are indeed other groups of high-strength steels, such as the ultra-high-strength steels that have yield strengths from 160 to 300 ksi. These steels have not been included in the Steel Manual because they have not been assigned ASTM numbers.

It is said that today more than 200 steels exist on the market that provide yield stresses in excess of 36 ksi. The steel industry is now experimenting with steels with yield stresses varying from 200 to 300 ksi, and this may be only the beginning. Many people in the steel industry feel that steels with 500 ksi yield strengths will be made available within a few years. The theoretical binding force between iron atoms has been estimated to be in excess of 4000 ksi.[7]

---

[7]"L. S. Beedle et al., *Structural Steel Design* (New York: Ronald Press, 1964), p. 44.

TABLE 1.1   Applicable ASTM Specifications for Various Structural Shapes

Steel Typ	ASTM Designatio		F_y Min. Yield Stress (ksi)	F_u Tensile Stress[a] (ksi)	W	M	S	HP	C	MC	L	Rect. (HSS)	Round (HSS)	Pipe
Carbon	A36		36	58–80[b]	▨	■	■	▨	■	■	■			
	A53 Gr. B		35	60										■
	A500	Gr. B	42	58								▨	■	
		Gr. B	46	58								■	▨	
		Gr. C	46	62								▨	▨	
		Gr. C	50	62								▨	▨	
	A501	Gr. A	36	58								▨	▨	
		Gr. B	50	70								▨	▨	
	A529[c]	Gr. 50	50	65–100	▨	▨			▨	▨	▨			
		Gr. 55	55	70–100	▨	▨			▨	▨	▨			
High-Strength Low-Alloy	A572	Gr. 42	42	60	▨	▨	▨	▨	▨	▨	▨			
		Gr. 50	50	65[d]	▨	▨	▨	■	▨	▨	▨			
		Gr. 55	55	55	▨	▨	▨	▨	▨	▨	▨			
		Gr. 60[e]	60	60	▨	▨	▨		▨	▨	▨			
		Gr. 65[e]	65	65	▨	▨	▨		▨	▨	▨			
	A618[f]	Gr. I & II	50[g]	70[g]								▨	▨	
		Gr. III	50	50								▨	▨	
	A913	50	50[h]	60[h]	▨									
		60	60	75	▨									
		65	65	80	▨									
		70	70	90	▨									
	A992		50	65[i]	■									
Corrosion Resistant High-Strength Low-Alloy	A242		42[j]	63[j]	▨				▨					
			46[k]	67[k]	▨				▨					
			50[l]	70[l]	▨				▨					
	A588		50	70	▨	▨	▨	▨	▨	▨	▨			
	A847		50	70								▨	▨	

■ = Preferred material specification
▨ = Other applicable material specification, the availability of which should be confirmed prior to specification
☐ = Material specification does not apply

a   Minimum unless a range is shown.
b   For shapes over 426 lb/ft, only the minimum of 58 ksi applies.
c   For shapes with a flange thickness less than or equal to 1½ in. only. To improve weldability, a maximum carbon equivalent can be specified (per ASTM Supplementary Requirement S78). If desired, maximum tensile stress of 90 ksi can be specified (per ASTM Supplementary Requirement S79).
d   If desired, maximum tensile stress of 70 ksi can be specified (per ASTM Supplementary Requirement S91).
e   For shapes with a flange thickness less than or equal to 2 in. only.
f   ASTM A618 can also be specified as corrosion-resistant; see ASTM A618.
g   Minimum applies for walls nominally ¾-in. thick and under. For wall thicknesses over ¾ in., F_y = 46 ksi and F_u = 67 ksi.
h   If desired, maximum yield stress of 65 ksi and maximum yield-to-tensile strength ratio of 0.85 can be specified (per ASTM Supplementary Requirement S75).
i   A maximum yield-to-tensile strength ratio of 0.85 and carbon equivalent formula are included as mandatory in ASTM A992.
j   For shapes with a flange thickness greater than 2 in. only.
k   For shapes with a flange thickness greater than 1½ in. and less than or equal to 2 in. only.
l   For shapes with a flange thickness less than or equal to 1½ in. only.

Although the prices of steels increase with increasing yield stresses, the percentage of price increase does not keep up with the percentage of yield-stress increase. The result is that the use of the stronger steels will quite frequently be economical for tension members, beams, and columns. Perhaps the greatest economy can be realized with tension members (particularly those without bolt holes). They may provide a great deal of savings for beams if deflections are not important or if deflections can be controlled (by methods described in later chapters). In addition, considerable economy can frequently be achieved with high-strength steels for short- and medium-length stocky columns. Another application that can provide considerable savings is hybrid construction. In this type of construction two or more steels of different strengths are used, the weaker steels being used where stresses are smaller and the stronger steels where stresses are higher.

Among the other factors that might lead to the use of high-strength steels are the following:

1. Superior corrosion resistance.
2. Possible savings in shipping, erection, and foundation costs caused by weight saving.
3. Use of shallower beams, permitting smaller floor depths.
4. Possible savings in fireproofing because smaller members can be used.

The first thought of most engineers in choosing a type of steel is the direct cost of the members. Such a comparison can be made quite easily, but determining which strength grade is most economical requires consideration of weights, sizes, deflections, maintenance, and fabrication. To make an accurate general comparison of the steels is probably impossible—rather, it is necessary to consider the specific job.

1.10   MEASUREMENT OF TOUGHNESS

The fracture toughness of steel is used as a general measure of its impact resistance, or its ability to absorb sudden increases in stress at a notch. The more ductile steel is, the greater will be its toughness. On the other hand, the lower its temperature, the higher will be its brittleness.

There are several procedures available for estimating notch toughness, but the Charpy V-notch test is the most commonly used. Although this test (which is described in ASTM Specification A6) is somewhat inaccurate, it does help identify brittle steels. With this test, the energy required to fracture a small bar of rectangular cross section with a specified notch (see Fig. 1.8) is measured.

The bar is fractured with a pendulum swung from a certain height. The amount of energy needed to fracture the bar is determined from the height to which the pendulum rises after the blow. The test may be repeated for different temperatures and the fracture energy plotted as a graph, as shown in Fig. 1.9. Such a graph clearly shows the relationship among temperature, ductility, and brittleness. The temperature at the point of steepest slope is referred to as the *transition temperature*.

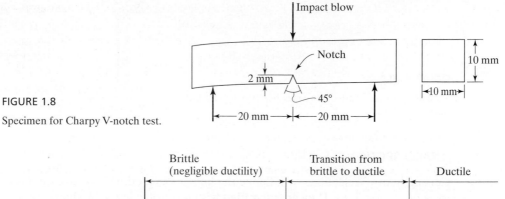

FIGURE 1.8

Specimen for Charpy V-notch test.

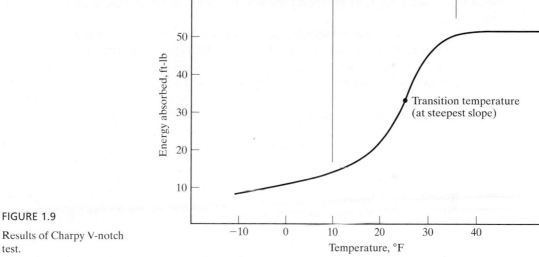

FIGURE 1.9

Results of Charpy V-notch test.

Although the Charpy test is well known, it actually provides a very poor measurement. Other methods for measuring the toughness of steel are considered in articles by Barsom and Rolfe.[8,9]

Different structural steels have different specifications for required absorbed energy levels (say, 20 ft-lb at 20°F), depending on the temperature, stress, and loading conditions under which they are to be used. The topic of brittleness is continued in the next section.

---

[8]J. M. Barsom, "Material Considerations in Structural Steel Design," *Engineering Journal*, AISC, 24, 3 (3rd Quarter 1987), pp. 127–139.

[9]S. T. Rolfe, "Fracture and Fatigue Control in Steel Structures," *Engineering Journal*, AISC, 14, 1 (1st Quarter 1977), pp. 2–15.

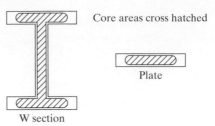

FIGURE 1.10

Core areas where brittle failure may be a problem in thick heavy members.

1.11   JUMBO SECTIONS

Certain heavy W sections with flange thicknesses exceeding 2 inches are often referred to as *jumbo sections*. They are identified with footnotes in the W shape, Table 1.1 of the Steel Manual.

Jumbo sections were originally developed for use as compression members and are quite satisfactory for that purpose. However, designers have frequently used them for tension or flexural members. In these applications, flange and web areas have had some serious cracking problems where welding or thermal cutting has been used. These cracks may result in smaller load-carrying capacities and problems related to fatigue.[10]

Thick pieces of steel tend to be more brittle than thin ones. Some of the reasons for this are that the core areas of thicker shapes (shown in Fig. 1.10) are subject to less rolling, have higher carbon contents (needed to produce required yield stresses), and have higher tensile stresses from cooling. These topics are discussed in later chapters.

Jumbo sections spliced with welds can be satisfactorily used for axial tension or flexural situations if the procedures listed in Specification A3.1c of the AISC Specification are carefully followed. Included among the requirements are the following:

1. The steel used must have certain absorbed energy levels, as determined by the Charpy V-notch test (20 ft-lb at a maximum temperature of 70°F). It is absolutely necessary that the tests be made on specimens taken from the core areas (shown in Fig. 1.10), where brittle fracture has proved to be a problem.
2. Temperature must be controlled during welding, and the work must follow a certain sequence.
3. Special splice details are required.

1.12   LAMELLAR TEARING

The steel specimens used for testing and developing stress–strain curves usually have their longitudinal axes in the direction that the steel was rolled. Should specimens be taken with their longitudinal axes transverse to the rolling direction "through the thickness" of the steel, the results will be lower ductility and toughness. Fortunately, this is a matter of little significance for almost all situations. It can, however, be quite important when thick plates

---

[10]R. Bjorhovde, "Solutions for the Use of Jumbo Shapes," *Proceedings 1988 National Steel Construction Conference*, AISC, Chicago, June 8–11, pp. 2–1 to 2–20.

and heavy structural shapes are used in highly restrained welded joints. (It can be a problem for thin members, too, but it is much more likely to give trouble in thick members.)

If a joint is highly restrained, the shrinkage of the welds in the through-the-thickness direction cannot be adequately redistributed, and the result can be a tearing of the steel called *lamellar tearing*. (*Lamellar* means "consisting of thin layers.") The situation is aggravated by the application of external tension. Lamellar tearing may show up as fatigue cracking after a number of cycles of load applications.

The lamellar tearing problem can be eliminated or greatly minimized with appropriate weld details and weld procedures. For example, the welds should be detailed so that shrinkage occurs as much as possible in the direction the steel was rolled. Several steel companies produce steels with enhanced through-the-thickness properties that provide much greater resistance to lamellar tearing. Even if such steels are used for heavy restrained joints, the special joint details mentioned before still are necessary.[11]

Figures 8-16 and 8-17 in the Steel Manual show preferred welded joint arrangements that reduce the possibility of lamellar tearing. Further information on the subject is provided in the ASTM A770 specification.

1.13   FURNISHING OF STRUCTURAL STEEL

The furnishing of structural steel consists of the rolling of the steel shapes, the fabrication of the shapes for the particular job (including cutting to the proper dimensions and punching the holes necessary for field connections), and their erection. Very rarely will a single company perform all three of these functions, and the average company performs only one or two of them. For instance, many companies fabricate structural steel and erect it, while others may be only steel fabricators or steel erectors. There are approximately 400 to 500 companies in the United States that make up the fabricating industry for structural steel. Most of them do both fabrication and erection.

Steel fabricators normally carry very little steel in stock because of the high interest and storage charges. When they get a job, they may order the shapes to certain lengths directly from the rolling mill, or they may obtain them from service centers. Service centers, which are an increasingly important factor in the supply of structural steel, buy and stock large quantities of structural steel, which they buy at the best prices they can find anywhere in the world.

Structural steel is usually designed by an engineer in collaboration with an architectural firm. The designer makes design drawings that show member sizes, controlling dimensions, and any unusual connections. The company that is to fabricate the steel makes the detailed drawings subject to the engineer's approval. These drawings provide all the information necessary to fabricate the members correctly. They show the dimensions for each member, the locations of holes, the positions and sizes of connections, and the like. A part of a typical detail drawing for a bolted steel beam is shown in

---

[11]"Commentary on Highly Restrained Welded Connections," *Engineering Journal*, AISC, vol. 10, no. 3 (3d quarter, 1973), pp. 61–73.

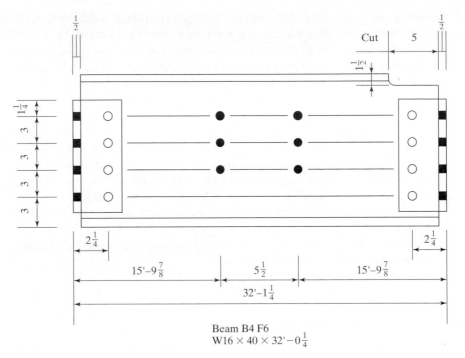

Beam B4 F6
W16 × 40 × 32'−0¼

FIGURE 1.11

Part of a detail drawing.

Fig. 1.11. There may be a few items included on this drawing that are puzzling to you since you have read only a few pages of this book. However, these items should become clear as you study the chapters to follow.

On actual detail drawings, details probably will be shown for several members. Here, the author has shown only one member, just to indicate the information needed so the shop can correctly fabricate the member. The darkened circles and rectangles indicate that the bolts are to be installed in the field, while the nondarkened ones show the connections which are to be made in the shop.

The erection of steel buildings is more a matter of assembly than nearly any other part of construction work. Each of the members is marked in the shop with letters and numbers to distinguish it from the other members to be used. The erection is performed in accordance with a set of erection plans. These plans are not detailed drawings, but are simple line diagrams showing the position of the various members in the building. The drawings show each separate piece or subassembly of pieces together with assigned shipping or erection marks, so that the steelworkers can quickly identify and locate members in their correct positions in the structure. (Persons performing steel erection often are called *ironworkers*, which is a name held over from the days before structural steel.) Directions (north, south, east, or west) usually are painted on column faces.

Sometimes, the erection drawing gives the sizes of the members, but this is not necessary. They may or may not be shown, depending on the practice of the particular fabricator.

Round Arch Hall at exhibition center, Leipzig, Germany. (© Klaws Hackenberg/Zefa/Corbis. Used by permission.)

Beams, girders, and columns will be indicated on the drawings by the letters B, G, or C, respectively, followed by the number of the particular member as B5, G12, and so on. Often, there will be several members of these same designations where members are repeated in the building.

Multistory steel frames often will have several levels of identical or nearly identical framing systems. Thus, one erection plan may be used to serve several floors. For such situations, the member designations for the columns, beams, and girders will have the level numbers incorporated in them. For instance, column C15(3–5) is column 15, third to fifth floors; while B4F6, or just B4(6), represents beam B4 for the sixth floor. A portion of a building erection drawing is shown in Fig. 1.12.

Next, we describe briefly the erection of the structural steel members for a building. Initially, a group of ironworkers, sometimes called the "raising gang," erects the steel members, installing only a sufficient number of bolts to hold the members in

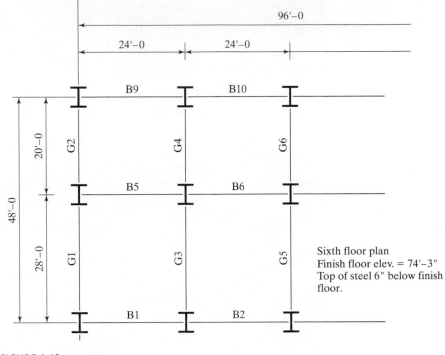

FIGURE 1.12

Part of an erection drawing, showing where each member is to be located.

position. In addition, they place any guy cables where needed for stability and plumbing of the steel frame.

Another group of ironworkers, who are sometimes referred to as the "detail gang," install the remaining bolts, carry out any needed field welding, and complete the plumbing of the structure. After the last two steps are completed, another crew installs the metal decking for the floor and roof slabs. They in turn are followed by the crews who place the necessary concrete reinforcing and the concrete for those slabs.[12]

1.14   THE WORK OF THE STRUCTURAL DESIGNER

The structural designer arranges and proportions structures and their parts so that they will satisfactorily support the loads to which they may feasibly be subjected. It might be said that he or she is involved with the general layout of structures; studies of the possible structural forms that can be used; consideration of loading conditions; analysis of stresses, deflections, and so on; design of parts; and the preparation of design drawings. More precisely, the word *design* pertains to the proportioning of the various parts of a structure after the forces have been calculated, and it is this process which will be emphasized throughout the text, using structural steel as the material.

1.15   RESPONSIBILITIES OF THE STRUCTURAL DESIGNER

The structural designer must learn to arrange and proportion the parts of structures so that they can be practically erected and will have sufficient strength and reasonable economy. These items are discussed briefly next.

1.15.1   Safety

Not only must the frame of a structure safely support the loads to which it is subjected, but it must support them in such a manner that deflections and vibrations are not so great as to frighten the occupants or to cause unsightly cracks.

1.15.2   Cost

The designer needs to keep in mind the factors that can lower cost without sacrifice of strength. These items, which are discussed in more detail throughout the text, include the use of standard-size members, simple connections and details, and members and materials that will not require an unreasonable amount of maintenance through the years.

1.15.3   Constructability

Another objective is the design of structures that can be fabricated and erected without great problems arising. Designers need to understand fabrication methods and should try to fit their work to the fabrication facilities available.

Designers should learn everything possible about the detailing, fabrication, and field erection of steel. The more the designer knows about the problems, tolerances, and clearances in shop and field, the more probable it is that reasonable, practical, and economical designs will be produced. This knowledge should include information concerning the transportation of the materials to the job site (such as the largest pieces that can be transported practically by rail or truck), labor conditions, and the equipment available for erection. Perhaps the designer should ask, "Could I get this thing together if I were sent out to do it?"

Finally, he or she needs to proportion the parts of the structure so that they will not unduly interfere with the mechanical features of the structure (pipes, ducts, etc.) or the architectural effects.

1.16   ECONOMICAL DESIGN OF STEEL MEMBERS

The design of a steel member involves much more than a calculation of the properties required to support the loads and the selection of the lightest section providing these properties. Although at first glance this procedure would seem to give the most economical designs, many other factors need to be considered.

Steel erection for Transamerica
Pyramid, San Francisco, CA.
(Courtesy of Kaiser Steel
Corporation.)

Today, the labor costs involved in the fabrication and erection of structural steel
are thought to run close to 60 percent of the total costs of steel structures. On the other
hand, material costs represent only about 25 percent of total costs. Thus, we can see
that any efforts we make to improve the economy of our work in structural steel
should be primarily concentrated in the labor area.

When designers are considering costs, they have a tendency to think only of
quantities of materials. As a result, they will sometimes carefully design a structure with
the lightest possible members and end up with some very expensive labor situations
with only minor material savings. Among the many factors that need to be considered
in providing economical steel structures are the following:

1. One of the best ways to achieve economy is to have open communications between designers, fabricators, erectors, and others involved in a particular project.
If this is done during the design process, the abilities and experience of each of

the parties may be utilized at a time when it is still possible to implement good economical ideas.

2. The designer needs to select steel sections of sizes that are usually rolled. Steel beams and bars and plates of unusual sizes will be difficult to obtain during boom periods and will be expensive during any period. A little study on the designer's part will enable him or her to avoid these expensive shapes. Steel fabricators are constantly supplied with information from the steel companies and the steel warehousers as to the sizes and lengths of sections available. (Most structural shapes can be procured in lengths from 60 to 75 ft, depending on the producer, while it is possible under certain conditions to obtain some shapes up to 120 ft in length.)

3. A blind assumption that the lightest section is the cheapest one may be in considerable error. A building frame designed by the "lightest-section" procedure will consist of a large number of different shapes and sizes of members. Trying to connect these many-sized members and fit them in the building will be quite complicated, and the pound price of the steel will, in all probability, be rather high. A more reasonable approach would be to smooth out the sizes by selecting many members of the same sizes, although some of them may be slightly overdesigned.

4. The beams usually selected for the floors in buildings will normally be the deeper sections, because these sections for the same weights have the largest moments of inertia and the greatest resisting moments. As building heights increase, however, it may be economical to modify this practice. As an illustration, consider the erection of a 20-story building, for which each floor has a minimum clearance. It is assumed that the depths of the floor beams may be reduced by 6 in without an unreasonable increase in beam weights. The beams will cost more, but the building height will be reduced by $20 \times 6 \text{ in} = 120 \text{ in}$, or 10 ft, with resulting savings in walls, elevator shafts, column heights, plumbing, wiring, and footings.[13]

5. The costs of erection and fabrication for structural steel beams are approximately the same for light and heavy members. Thus, beams should be spaced as far apart as possible to reduce the number of members that have to be fabricated and erected.

6. Structural steel members should be painted only if so required by the applicable specification. You should realize that steel should not be painted if it is to be in contact with concrete. Furthermore, the various fireproofing materials used for protecting steel members adhere better if the surfaces are unpainted.[14]

7. It is very desirable to keep repeating the same section over and over again. Such a practice will reduce the detailing, fabrication, and erection costs.

8. For larger sections, particularly the built-up ones, the designer needs to have information pertaining to transportation problems. The desired information includes the greatest lengths and depths that can be shipped by truck or rail

---

[13]H. Allison, "Low- and Medium-Rise Steel Buildings" (Chicago: AISC, 1991), pp. 1–5.
[14]Ibid, pp. 1–5.

(see Section 1.18), clearances available under bridges and power lines leading to the project, and allowable loads on bridges. It may be possible to fabricate a steel roof truss in one piece, but is it possible to transport it to the job site and erect it in one piece?

9. Sections should be selected that are reasonably easy to erect and which have no conditions that will make them difficult to maintain. As an example, it is necessary to have access to all exposed surfaces of steel bridge members so that they may be periodically painted (unless one of the special corrosion-resistant steels is used).

10. Buildings are often filled with an amazing conglomeration of pipes, ducts, conduits, and other items. Every effort should be made to select steel members that will fit in with the requirements made by these items.

11. The members of a steel structure are often exposed to the public, particularly in the case of steel bridges and auditoriums. Appearance may often be the major factor in selecting the type of structure, such as where a bridge is desired that will fit in and actually contribute to the appearance of an area. Exposed members may be surprisingly graceful when a simple arrangement, perhaps with curved members, is used; but other arrangements may create a terrible eyesore. The student has certainly seen illustrations of each case. It is very interesting to know that beautiful structures in steel are usually quite reasonable in cost.

The question is often asked, *How do we achieve economy in structural steel design?* The answer is simple: *It lies in what the steel fabricator does not have to do.* (In other words, economy can be realized when fabrication is minimized.)

The April 2000 issue of *Modern Steel Construction* provides several articles which present excellent material on the topic of economy in steel construction.[15] The student can very quickly learn a great deal of valuable information concerning the topic of economy in steel by reading these articles. The author thinks they are a "must read" for anyone practicing steel design.[16–19]

1.17   FAILURE OF STRUCTURES

Many people who are superstitious do not discuss flat tires or make their wills, because they fear they will be tempting fate. These same people would probably not care to discuss the subject of engineering failures. Despite the prevalence of this superstition, an

---

[15]*Modern Steel Construction,* April 2000, vol. 40, no. 4 (Chicago: American Institute of Steel Construction), pp. 6, 25–48, 60.

[16]C. J. Carter, T. M. Murray and W. A. Thornton, "Economy in Steel," in *Modern Steel Construction*, April 2000, vol. 40, no. 4 (Chicago: American Institute of Steel Construction).

[17]D. T. Ricker, "Value Engineering for Steel Construction," in *Modern Steel Construction*, April 2000, vol. 40, no. 4 (Chicago: American Institute of Steel Construction).

[18]J. E. Quinn, "Reducing Fabrication Costs," in *Modern Steel Construction*, April 2000, vol. 40, no. 4 (Chicago: American Institute of Steel Construction).

[19]Steel Joist Institute, "Reducing Joist Cost," in *Modern Steel Construction*, April 2000, vol. 40, no. 4 (Chicago: American Institute of Steel Construction).

Steel erection for Transamerica Pyramid, San Francisco, CA. (Courtesy of Kaiser Steel Corporation.)

awareness of the items that have most frequently caused failures in the past is invaluable to experienced and inexperienced designers alike. Perhaps a study of past failures is more important than a study of past successes. Benjamin Franklin supposedly made the observation that "a wise man learns more from failures than from success."

The designer with little experience particularly needs to know where the most attention should be given and where outside advice is needed. The vast majority of designers, experienced and inexperienced, select members of sufficient size and strength. The collapse of structures is usually due to insufficient attention to the details of connections, deflections, erection problems, and foundation settlement. Rarely, if ever, do steel structures fail due to faults in the material, but rather due to its improper use.

A frequent fault of designers is that after carefully designing the members of a structure, they carelessly select connections which may or may not be of sufficient size. They may even turn the job of selecting the connections over to drafters, who may not have sufficient understanding of the difficulties that can arise in connection design. Perhaps the most common mistake made in connection design is to neglect some of the forces acting on the connections, such as twisting moments. In a truss for which the members have been designed for axial forces only, the connections may be eccentrically loaded, resulting in moments that cause increasing stresses. These secondary stresses are occasionally so large that they need to be considered in design.

Another source of failure occurs when beams supported on walls have insufficient bearing or anchorage. Imagine a beam of this type supporting a flat roof on a rainy night when the roof drains are not functioning properly. As the water begins to form puddles on the roof, the beam tends to sag in the middle, causing a pocket to catch more rain, which creates more beam sag, and so on. As the beam deflects, it pushes out against the walls, possibly causing collapse of walls or slippage of beam ends off the wall. Picture a 60-ft steel beam, supported on a wall with only an inch or two of bearing, that contracts when the temperature drops 50 or 60 degrees overnight. A collapse due to a combination of beam contraction, outward deflection of walls, and vertical deflection caused by precipitation loads is not difficult to visualize; furthermore, actual cases in engineering literature are not difficult to find.

Foundation settlements cause a large number of structural failures, probably more than any other factor. Most foundation settlements do not result in collapse, but they very often cause unsightly cracks and depreciation of the structure. If all parts of the foundation of a structure settle equally, the stresses in the structure theoretically will not change. The designer, usually not able to prevent settlement, has the goal of designing foundations in such a manner that equal settlements occur. Equal settlements may be an impossible goal, and consideration should be given to the stresses that would be produced if settlement variations occurred. The student's background in structural analysis will tell him or her that uneven settlements in statically indeterminate structures may cause extreme stress variations. Where foundation conditions are poor, it is desirable, if feasible, to use statically determinate structures whose stresses are not appreciably changed by support settlements. (The student will learn in subsequent discussions that the ultimate strength of steel structures is usually affected only slightly by uneven support settlements.)

Some structural failures occur because inadequate attention is given to deflections, fatigue of members, bracing against swaying, vibrations, and the possibility of buckling of compression members or the compression flanges of beams. The usual structure, when completed, is sufficiently braced with floors, walls, connections, and special bracing, but there are times during construction when many of these items are not present. As previously indicated, the worst conditions may well occur during erection, and special temporary bracing may be required.

1.18   HANDLING AND SHIPPING STRUCTURAL STEEL

The following general rules apply to the sizes and weights of structural steel pieces that can be fabricated in the shop, shipped to the job, and erected:

1. The maximum weights and lengths that can be handled in the shop and at a construction site are roughly 90 tons and 120 ft, respectively.
2. Pieces as large as 8 ft high, 8 ft wide, and 60 ft long can be shipped on trucks with no difficulty (provided the axle or gross weights do not exceed the permissible values given by public agencies along the designated routes).
3. There are few problems in railroad shipment if pieces are no larger than 10 ft high, 8 ft wide, and 60 ft long, and weigh no more than 20 tons.
4. Routes should be carefully studied, and carriers consulted for weights and sizes exceeding the values given in (2) and (3).

1.19   CALCULATION ACCURACY

A most important point many students with their superb pocket calculators and personal computers have difficulty understanding is that structural design is not an exact science for which answers can confidently be calculated to eight significant figures. Among the reasons for this fact are that the methods of analysis are based on partly true assumptions, the strengths of materials used vary appreciably, and maximum loadings can be only approximated. With respect to this last reason, how many of the users of this book could estimate within 10 percent the maximum load in pounds per square foot that will ever occur on the building floor which they are now occupying? Calculations to more than two or three significant figures are obviously of little value and may actually be harmful in that they mislead the student by giving him or her a fictitious sense of precision. From a practical standpoint, it seems wise to carry all the digits on the calculator for intermediate steps and then round off the final answers.

1.20   COMPUTERS AND STRUCTURAL STEEL DESIGN

The availability of personal computers has drastically changed the way steel structures are analyzed and designed. In nearly every engineering school and design office, computers are used to perform structural analysis problems. Many of the structural analysis programs commercially available also can perform structural design.

Many calculations are involved in structural steel design, and many of these calculations are quite time-consuming. With the use of a computer, the design engineer can greatly reduce the time required to perform these calculations, and likely increase the accuracy of the calculations. In turn, this will then provide the engineer with more time to consider the implications of the design and the resulting performance of the structure, and more time to try changes that may improve economy or behavior.

Although computers do increase design productivity, they also tend to reduce the engineer's "feel" for the structure. This can be a particular problem for young engineers with very little design experience. Unless design engineers have this feel for system behavior, the use of computers can result in large, costly mistakes. Such situations may arise where anomalies and inconsistencies are not immediately apparent to the inexperienced engineer. Theoretically, the computer design of alternative systems for a few projects should substantially improve the engineer's judgment in a short span of time. Without computers, the development of this same judgment would likely require the engineer to work his or her way through numerous projects.

1.21   PROBLEMS FOR SOLUTION

1-1. List the three regions of a stress–strain diagram for mild or low-carbon structural steel.

1-2. List the specifying organization for the following types of steel:

    a. Cold-formed steel

    b. Hot-rolled steel

1-3. Define the following:

    a. Proportional limit

    b. Elastic limit

    c. Yield stress

1-4. List the preferred steel type (ASTM spec) for the following shapes:

    a. Plates

    b. W shapes

    c. C sections

1-5. List the two methods used to produce steel shapes.

1-6. List four advantages of steel as a structural material.

1-7. What type of steel (ASTM grade) has made the cost of 50 ksi the same as 36 ksi steel because of the use of scrap or recycled steel in the manufacturing process?

1-8. What are the differences between wrought iron, steel, and cast iron?

1-9. What is the range of carbon percentage for *mild* carbon steel?

1-10. List four disadvantages of steel as a structural material.

1-11. List four types of failures for structural steel structures.

C H A P T E R 2

Specifications, Loads, and Methods of Design

2.1 SPECIFICATIONS AND BUILDING CODES

The design of most structures is controlled by building codes and design specifications. Even if they are not so controlled, the designer will probably refer to them as a guide. No matter how many structures a person has designed, it is impossible for him or her to have encountered every situation. By referring to specifications, he or she is making use of the best available material on the subject. Engineering specifications that are developed by various organizations present the best opinion of those organizations as to what represents good practice.

Municipal and state governments concerned with the safety of the public have established building codes with which they control the construction of various structures within their jurisdiction. These codes, which are actually laws or ordinances, specify minimum design loads, design stresses, construction types, material quality, and other factors. They vary considerably from city to city, a fact that causes some confusion among architects and engineers.

Several organizations publish recommended practices for regional or national use. Their specifications are not legally enforceable, unless they are embodied in the local building code or made a part of a particular contract. Among these organizations are the AISC and AASHTO (American Association of State Highway and Transportation Officials). Nearly all municipal and state building codes have adopted the AISC Specification, and nearly all state highway and transportation departments have adopted the AASHTO Specifications.

Readers should note that logical and clearly written codes are quite helpful to design engineers. Furthermore, there are far fewer structural failures in areas that have good building codes that are strictly enforced.

Some people feel that specifications prevent engineers from thinking for themselves—and there may be some basis for the criticism. They say that the ancient engineers who built the great pyramids, the Parthenon, and the great Roman bridges were

South Fork Feather River Bridge in northern California, being erected by use of a 1626-ft-long cableway strung from 210-ft-high masts anchored on each side of the canyon. (Courtesy of Bethlehem Steel Corporation.)

controlled by few specifications, which is certainly true. On the other hand, it should be said that only a few score of these great projects have endured over many centuries, and they were, apparently, built without regard to cost of material, labor, or human life. They were probably built by intuition and by certain rules of thumb that the builders developed by observing the minimum size or strength of members, which would fail only under given conditions. Their likely numerous failures are not recorded in history; only their successes endured.

Today, however, there are hundreds of projects being constructed at any one time in the United States that rival in importance and magnitude the famous structures of the past. It appears that if all engineers in our country were allowed to design projects such as these, without restrictions, there would be many disastrous failures. *The important thing to remember about specifications, therefore, is that they are written, not for the purpose of restricting engineers, but for the purpose of protecting the public.*

No matter how many specifications are written, it is impossible for them to cover every possible design situation. As a result, no matter which building code or specification is or is not being used, the ultimate responsibility for the design of a safe structure lies with the structural designer. Obviously, the intent of these specifications is that the loading used for design be the one that causes the largest stresses.

Another very important code, the *International Building Code*[1] (IBC), was developed because of the need for a modern building code that emphasizes performance.

---

[1]International Code Council, Inc., *International Building Code* (Washington, DC, 2009).

It is intended to provide a model set of regulations to safeguard the public in all communities.

2.2 LOADS

Perhaps the most important and most difficult task faced by the structural engineer is the accurate estimation of the loads that may be applied to a structure during its life. No loads that may reasonably be expected to occur may be overlooked. After loads are estimated, the next problem is to determine the worst possible combinations of these loads that might occur at one time. For instance, would a highway bridge completely covered with ice and snow be simultaneously subjected to fast-moving lines of heavily loaded trailer trucks in every lane and to a 90-mile lateral wind, or is some lesser combination of these loads more likely?

Section B2 of the AISC Specification states that the nominal loads to be used for structural design shall be the ones stipulated by the applicable code under which the structure is being designed or as dictated by the conditions involved. If there is an absence of a code, the design loads shall be those provided in a publication of the American Society of Civil Engineers entitled *Minimum Design Loads for Buildings and Other Structures*.[2] This publication is commonly referred to as ASCE 7. It was originally published by the American National Standards Institute (ANSI) and referred to as the *ANSI 58.1 Standard*. The ASCE took over its publication in 1988.

In general, loads are classified according to their character and duration of application. As such, they are said to be *dead loads, live loads,* and *environmental loads*. Each of these types of loads are discussed in the next few sections.

2.3 DEAD LOADS

Dead loads are loads of constant magnitude that remain in one position. They consist of the structural frame's own weight and other loads that are permanently attached to the frame. For a steel-frame building, the frame, walls, floors, roof, plumbing, and fixtures are dead loads.

To design a structure, it is necessary for the weights, or dead loads, of the various parts to be estimated for use in the analysis. The exact sizes and weights of the parts are not known until the structural analysis is made and the members of the structure selected. The weights, as determined from the actual design, must be compared with the estimated weights. If large discrepancies are present, it will be necessary to repeat the analysis and design with better estimated weights.

Reasonable estimates of structure weights may be obtained by referring to similar types of structures or to various formulas and tables available in several publications. The weights of many materials are given in Part 17 of the Steel Manual. Even more detailed information on dead loads is provided in Tables C3-1 and C3-2 of ASCE 7-10. An experienced engineer can estimate very closely the weights of most materials and will spend little time repeating designs because of poor estimates.

---

[2]American Society of Civil Engineers, *Minimum Design Loads for Buildings and Other Structures*. ASCE 7-10. Formerly ANSI A58.1 (Reston, Va.:ASCE, 2010).

TABLE 2.1   Typical Dead Loads for Some Common Building Materials

Reinforced concrete	150 lb/cu ft
Structural steel	490 lb/cu ft
Plain concrete	145 lb/cu ft
Movable steel partitions	4 psf
Plaster on concrete	5 psf
Suspended ceilings	2 psf
5-Ply felt and gravel	6 psf
Hardwood flooring (7/8 in)	4 psf
$2 \times 12 \times 16$ in double wood floors	7 psf
Wood studs with 1/2 in gypsum each side	8 psf
Clay brick wythes (4 in)	39 psf

The approximate weights of some common building materials for roofs, walls, floors, and so on are presented in Table 2.1.

2.4   LIVE LOADS

Live loads are loads that may change in position and magnitude. They are caused when a structure is occupied, used, and maintained. Live loads that move under their own power, such as trucks, people, and cranes, are said to be *moving loads*. Those loads that may be moved are *movable loads*, such as furniture and warehouse materials. A great deal of information on the magnitudes of these various loads, along with specified minimum values, are presented in ASCE 7-10.

1. *Floor loads.* The minimum gravity live loads to be used for building floors are clearly specified by the applicable building code. Unfortunately, however, the values given in these various codes vary from city to city, and the designer must be sure that his or her designs meet the requirements of the locality in question. A few of the typical values for floor loadings are listed in Table 2.2. These values were adopted from ASCE 7-10. In the absence of a governing code, this is an excellent one to follow.

Quite a few building codes specify concentrated loads that must be considered in design. Section 4.4 of ASCE 7-10 and Section 1607.4 of IBC-2009 are two such examples. The loads specified are considered as alternatives to the uniform loads previously considered herein.

Some typical concentrated loads taken from Table 4-1 of ASCE 7-10 and Table 1607.1 of IBC-2009 are listed in Table 2.3. These loads are to be placed on floors or roofs at the positions where they will cause the most severe conditions. Unless otherwise specified, each of these concentrated loads is spread over an area 2.5×2.5 ft square ($6.25\ \text{ft}^2$).

2. *Traffic loads for bridges.* Bridges are subjected to series of concentrated loads of varying magnitude caused by groups of truck or train wheels.

3. *Impact loads.* Impact loads are caused by the vibration of moving or movable loads. It is obvious that a crate dropped on the floor of a warehouse or a truck

TABLE 2.2 Typical Minimum Uniform Live Loads
for Design of Buildings

Type of building	LL (psf)
Apartment houses	
Apartments	40
Public rooms	100
Dining rooms and restaurants	100
Garages (passenger cars only)	40
Gymnasiums, main floors, and balconies	100
Office buildings	
Lobbies	100
Offices	50
Schools	
Classrooms	40
Corridors, first floor	100
Corridors above first floor	80
Storage warehouses	
Light	125
Heavy	250
Stores (retail)	
First floor	100
Other floors	75

TABLE 2.3 Typical Concentrated Live Loads for Buildings

Hospitals—operating rooms, private rooms, and wards	1000 lb
Manufacturing building (light)	2000 lb
Manufacturing building (heavy)	3000 lb
Office floors	2000 lb
Retail stores (first floors)	1000 lb
Retail stores (upper floors)	1000 lb
School classrooms	1000 lb
School corridors	1000 lb

bouncing on uneven pavement of a bridge causes greater forces than would occur if the loads were applied gently and gradually. Cranes picking up loads and elevators starting and stopping are other examples of impact loads. Impact loads are equal to the difference between the magnitude of the loads actually caused and the magnitude of the loads had they been dead loads.

Section 4.6 of ASCE 7-10 Specification requires that when structures are supporting live loads that tend to cause impact, it is necessary for those loads to be increased by the percentages given in Table 2.4.

4. *Longitudinal loads.* Longitudinal loads are another type of load that needs to be considered in designing some structures. Stopping a train on a railroad bridge or a truck on a highway bridge causes longitudinal forces to be applied. It is not difficult to imagine the tremendous longitudinal force developed when the driver of a 40-ton-trailer truck traveling 60 mph suddenly has to apply the brakes while

TABLE 2.4   Live Load Impact Factors

Elevator machinery*	100%
Motor-driven machinery	20%
Reciprocating machinery	50%

*See Section C4.6, ASCE 7-10 Commentary.

crossing a highway bridge. There are other longitudinal load situations, such as ships bumping a dock during berthing and the movement of traveling cranes that are supported by building frames.

5. *Other live loads.* Among the other types of live loads with which the structural engineer will have to contend are *soil pressures* (such as the exertion of lateral earth pressures on walls or upward pressures on foundations); *hydrostatic pressures* (water pressure on dams, inertia forces of large bodies of water during earthquakes, and uplift pressures on tanks and basement structures); *blast loads* (caused by explosions, sonic booms, and military weapons); *thermal forces* (due to changes in temperature, causing structural deformations and resulting structural forces); and *centrifugal forces* (such as those on curved bridges and caused by trucks and trains, or similar effects on roller coasters, etc.).

Roof/bridge crane framing, Savannah, GA. (Courtesy of CMC South Carolina Steel.)

Hungry Horse Dam and Reservoir, Rocky Mountains, in northwestern Montana. (Courtesy of the Montana Travel Promotion Division.)

2.5 ENVIRONMENTAL LOADS

Environmental loads are caused by the environment in which a particular structure is located. For buildings, environmental loads are caused by rain, snow, wind, temperature change, and earthquakes. Strictly speaking, environmental loads are live loads, but they are the result of the environment in which the structure is located. Even though they do vary with time, they are not all caused by gravity or operating conditions, as is typical with other live loads. A few comments are presented in the paragraphs that follow concerning the different types of environmental loads:

1. *Snow*. In the colder states, snow loads are often quite important. One inch of snow is equivalent to a load of approximately 0.5 psf, but it may be higher at lower elevations where snow is denser. For roof designs, snow loads varying from 10 to 40 psf are commonly used, the magnitude depending primarily on the slope of the roof and, to a lesser degree, on the character of the roof surface. The larger values are used for flat roofs, the smaller ones for sloped roofs. Snow tends to slide off sloped roofs, particularly those with metal or slate surfaces. A load of approximately 10 psf might be used for 45° (degree) slopes and a 40-psf load for flat roofs. Studies of snowfall records in areas with severe winters may indicate the occurrence of snow loads much greater than 40 psf, with values as high as 200 psf in some western states.

   Snow is a variable load that may cover an entire roof or only part of it. The snow loads that are applied to a structure are dependent upon many factors, including geographic location, roof pitch, sheltering, and the shape of the roof. Chapter 7 of ASCE 7-10 provides a great deal of information concerning snow loads, including charts and formulas for estimating their magnitudes. There may be drifts against walls or buildup in valleys or between parapets. Snow may slide off one roof onto a lower one. The wind may blow it off one side of a sloping roof, or the snow may crust over and remain in position even during very heavy winds.

Bridges are generally not designed for snow loads, since the weight of the snow is usually not significant in comparison with truck and train loads. In any case, it is doubtful that a full load of snow and maximum traffic would be present at the same time. Bridges and towers are sometimes covered with layers of ice from 1 to 2 in thick. The weight of the ice runs up to about 10 psf. Another factor to be considered is the increased surface area of the ice-coated members, as it pertains to wind loads.

2. *Rain*. Though snow loads are a more severe problem than rain loads for the usual roof, the situation may be reversed for flat roofs, particularly those in warmer climates. If water on a flat roof accumulates faster than it runs off, the result is called *ponding*, because the increased load causes the roof to deflect into a dish shape that can hold more water, which causes greater deflections, and so on. This process continues until equilibrium is reached or until collapse occurs. Ponding is a serious matter, as illustrated by the large number of flat-roof failures that occur during rainstorms every year in the United States. It has been claimed that almost 50 percent of the lawsuits faced by building designers are concerned with roofing systems.[3] Ponding is one of the most common subjects of such litigation.

Ponding will occur on almost any flat roof to a certain degree, even though roof drains are present. Roof drains may very well be used, but they may be inadequate during severe storms, or they may become partially or completely clogged. The best method of preventing ponding is to have an appreciable slope of the roof (1/4 in/ft or more), together with good drainage facilities. In addition to the usual ponding, another problem may occur for very large flat roofs (with perhaps an acre or more of surface area). During heavy rainstorms, strong winds frequently occur. If there is a great deal of water on the roof, a strong wind may very well push a large quantity of it toward one end. The result can be a dangerous water depth as regards the load in psf on that end. For such situations, *scuppers* are sometimes used. These are large holes or tubes in walls or parapets that enable water above a certain depth to quickly drain off the roof.

Chapter 8 of ASCE 7-10 provides information for estimating the magnitude of rain loads that may accumulate on flat roofs.

3. *Wind loads*. A survey of engineering literature for the past 150 years reveals many references to structural failures caused by wind. Perhaps the most infamous of these have been bridge failures, such as those of the Tay Bridge in Scotland in 1879 (which caused the deaths of 75 persons) and the Tacoma Narrows Bridge in Tacoma, Washington, in 1940. But there have also been some disastrous building failures due to wind during the same period, such as that of the Union Carbide Building in Toronto in 1958. It is important to realize that a large percentage of building failures due to wind have occurred during the erection of the building.[4]

---

[3]Gary Van Ryzin, 1980, "Roof Design: Avoid Ponding by Sloping to Drain," *Civil Engineering* (New York, ASCE, January), pp. 77–81.

[4]"Wind Forces on Structures, Task Committee on Wind Forces. Committee on Loads and Stresses, Structural Division, ASCE, Final Report," *Transactions ASCE* 126, Part II (1961): 1124–1125.

A great deal of research has been conducted in recent years on the subject of wind loads. Nevertheless, a great deal more work needs to be done, as the estimation of these forces can by no means be classified as an exact science. The magnitudes of wind loads vary with geographical locations, heights above ground, types of terrain surrounding the buildings, the proximity and nature of other nearby structures, and other factors.

Wind pressures are frequently assumed to be uniformly applied to the windward surfaces of buildings and are assumed to be capable of coming from any direction. These assumptions are not very accurate, because wind pressures are not uniform over large areas, the pressures near the corners of buildings being probably greater than elsewhere due to wind rushing around the corners, and so on. From a practical standpoint, therefore, all of the possible variations cannot be considered in design, although today's specifications are becoming more and more detailed in their requirements.

When the designer working with large low-rise buildings makes poor wind estimates, the results are probably not too serious, but this is not the case when tall slender buildings (or long flexible bridges) are being designed. For many years, the average designer ignored wind forces for buildings whose heights were not at least twice their least lateral dimensions. For such cases as these, it was felt that the floors and walls provided sufficient lateral stiffnesses to eliminate the need for definite wind bracing systems. A better practice for designers to follow, however, is to consider all the possible loading conditions that a particular structure may have to resist. If one or more of these conditions (perhaps, wind loading) seem of little significance, they may then be neglected. Should buildings have their walls and floors constructed with modern lightweight materials and/or should they be subjected to unusually high wind loads (as in coastal or mountainous areas), they will probably have to be designed for wind loads even if the height/least lateral dimension ratios are less than two.

Building codes do not usually provide for the estimated forces during tornadoes. The average designer considers the forces created directly in the paths of tornadoes to be so violent that it is not economically feasible to design buildings to resist them. This opinion is undergoing some change, however, as it has been found that the wind resistance of structures (even of small buildings, including houses) can be greatly increased at very reasonable costs by better practices—tying the parts of structures together from the roofs through the walls to the footings—and by making appropriate connections of window frames to walls and, perhaps, other parts of the structure.[5,6]

Wind forces act as pressures on vertical windward surfaces, pressures or suction on sloping windward surfaces (depending on the slope), and suction on flat surfaces and on leeward vertical and sloping surfaces (due to the creation of negative pressures or vacuums). The student may have noticed this definite suction effect where shingles or other roof coverings have been lifted from the leeward roof surfaces of buildings during windstorms. Suction or uplift can easily be demonstrated by holding a piece of paper at two of its corners and blowing above it. For some common structures, uplift loads may be as large as 20 to 30 psf or even more.

[5] P. R. Sparks, "Wind Induced Instability in Low-Rise Buildings," *Proceedings of the 5th U.S. National Conference on Wind Engineering*, Lubbock, TX, November 6–18, 1985.

[6] P. R. Sparks, "The Risk of Progressive Collapse of Single-Story Buildings in Severe Storms," *Proceedings of the ASCE Structures Congress*, Orlando, FL, August 17–20, 1987.

Access bridge, Renton, WA. (Courtesy of the Bethlehem Steel Corporation.)

During the passing of a tornado or hurricane, a sharp reduction in atmospheric pressure occurs. This decrease in pressure does not penetrate airtight buildings, and the inside pressures, being greater than the external pressures, cause outward forces against the roofs and walls. Nearly everyone has heard stories of the walls of a building "exploding" outward during a storm.

As you can see, the accurate calculation of the most severe wind pressures that need to be considered for the design of buildings and bridges is quite an involved problem. Despite this fact, sufficient information is available today to permit the satisfactory estimation of these pressures in a reasonably efficient manner.

A procedure for estimating the wind pressures applied to buildings is presented in Chapters 26–31 of ASCE 7-10. Several factors are involved when we attempt to account for the effects of wind speed, shape and orientation of the building in question, terrain characteristics around the structure, importance of the building as to human life and welfare, and so on. Though the procedure seems rather complex, it is greatly simplified with the tables presented in the aforementioned specification.

4. *Earthquake loads*. Many areas of the world fall in "earthquake territory," and in those areas it is necessary to consider seismic forces in design for all types of structures. Through the centuries, there have been catastrophic failures of buildings, bridges, and other structures during earthquakes. It has been estimated that as many as 50,000 people lost their lives in the 1988 earthquake in Armenia.[7] The 1989 Loma Prieta and 1994 Northridge earthquakes in California caused many billions of dollars of property damage, as well as considerable loss of life.

---

[7]V. Fairweather, "The Next Earthquake," *Civil Engineering* (New York: ASCE, March 1990), pp. 54–57.

Risk of Major Earthquakes by 2050

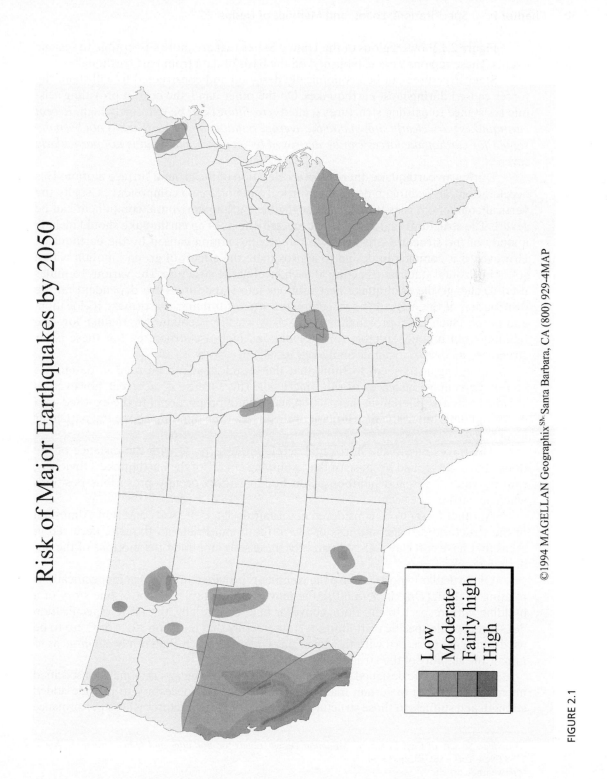

Low
Moderate
Fairly high
High

FIGURE 2.1

Figure 2.1 shows regions of the United States that are more susceptible to seismic events. These regions were established on the basis of data from past earthquakes.[8]

Steel structures can be economically designed and constructed to withstand the forces caused during most earthquakes. On the other hand, the cost of providing seismic resistance to existing structures (called *retrofitting*) can be extremely high. *Recent earthquakes have clearly shown that the average building or bridge that has not been designed for earthquake forces can be destroyed by an earthquake that is not particularly severe.*

During an earthquake, there is an acceleration of the ground surface motion. This acceleration can be broken down into vertical and horizontal components. Usually, the vertical component is assumed to be negligible, but the horizontal component can be severe. The structural analysis for the expected effects of an earthquake should include a study of the structure's response to the ground motion caused by the earthquake. However, it is common in design to approximate the effects of ground motion with a set of horizontal static loads acting at each level of the structure. The various formulas used to change the earthquake accelerations into static forces are dependent on the distribution of the mass of the structure, the type of framing, its stiffness, its location, and so on. Such an approximate approach is usually adequate for regular low-rise buildings, but it is not suitable for irregular and high-rise structures. For these latter structures, an overall dynamic analysis is usually necessary.

Some engineers seem to think that the seismic loads to be used in design are merely percentage increases of the wind loads. This surmise is incorrect, however, as seismic loads are different in their action and are not proportional to the exposed area of the building, but to the distribution of the mass of the building above the particular level being considered.

The forces due to the horizontal acceleration increase with the distance of the floor above the ground because of the "whipping effect" of the earthquake. Obviously, towers, water tanks, and penthouses on building roofs occupy precarious positions when an earthquake occurs.

Another factor to be considered in seismic design is the soil condition. Almost all of the structural damage and loss of life in the Loma Prieta earthquake occurred in areas that have soft clay soils. Apparently, these soils amplified the motions of the underlying rock.[9]

Of particular importance are the comments provided in the seismic specifications relating to drift. (*Drift* is defined as the movement or displacement of one story of a building with respect to the floor above or below.) Actually, the AISC Specification does not provide specific drift limits. It states merely that limits on story drift are to be used which are in accord with the governing code and which shall not be so large as to impair the stability of the structure.

If structures are designed so that computed drifts during an earthquake of specified intensity are limited to certain maximum values, it will be necessary to provide added strength and stiffness to those structures. The result will be structures whose performance

---

[8]*American Society of Civil Engineers Minimum Design Loads for Buildings and Other Structures*, ASCE 7–88 (New York: ASCE), pp. 33, 34.

[9]Fairweather, op. cit.

is substantially improved during earthquakes. The AISC Manual does not provide detailed specifications for designing structures subject to seismic loads, but such information is presented in the companion *AISC Seismic Design Manual*,[10] as well as in ASCE 7-10.

Sample calculations of snow, rain, wind, and seismic loads as required by the ASCE 7 Specification are presented in a textbook entitled *Structural Analysis Using Classical and Matrix Methods*.[11]

2.6   LOAD AND RESISTANCE FACTOR DESIGN (LRFD) AND ALLOWABLE STRENGTH DESIGN (ASD)

The AISC Specification provides two acceptable methods for designing structural steel members and their connections. These are **Load and Resistance Factor Design** (LRFD) and **Allowable Strength Design** (ASD). As we will learn in this textbook, both procedures are based on limit states design principles, which provide the boundaries of structural usefulness.

The term **limit state** is used to describe a condition at which a structure or part of a structure ceases to perform its intended function. There are two categories of limit states: strength and serviceability.

Strength limit states define load-carrying capacity, including excessive yielding, fracture, buckling, fatigue, and gross rigid body motion. Serviceability limit states define performance, including deflection, cracking, slipping, vibration, and deterioration. All limit states must be prevented.

Structural engineers have long recognized the inherent uncertainty of both the magnitude of the loads acting on a structure and the ability of the structure to carry those loads. Usually, the effects of multiple loads are additive, but in some cases—for example, a beam column—one load can magnify the effect of another load.

In the best of cases, the combined effect of multiple loads, related to a particular limit state or failure mode, can be described with a mathematical probability density function. In addition, the structural limit state can be described by another mathematical probability density function. For this ideal case, the two probability density functions yield a mathematical relation for either the difference between or the ratio of the two means, and the possibility that the load will exceed the resistance.

The margin established between resistance and load in real cases is intended to reduce the probability of failure, depending on the consequences of failure or unserviceability. The question we have is how to achieve this goal when there usually is insufficient information available for a completely mathematical description of either load or resistance. LRFD is one approach; ASD is another. Both methods have as their goal the obtaining of a numerical margin between resistance and load that will result in an acceptably small probability of unacceptable structural response.

There are two major differences between LRFD and ASD. The first pertains to the method used for calculating the design loads. This difference is explained in Sections 2.9, 2.10, and 2.11. The second difference pertains to the use of resistance factors (ϕ in LRFD) and safety factors (Ω in ASD). This difference is explained in

---

[10]American Institute of Steel Construction, 2006 (Chicago: AISC).

[11]J. C. McCormac, 2007 (Hoboken, NJ: John Wiley & Sons, Inc.), pp. 24–40.

Sections 2.12 and 2.13. These five sections should clearly fix in the reader's mind an understanding of the differences between LRFD and ASD. It is also important to realize that both LRFD and ASD employ the same methods of structural analysis. Obviously, the behavior of a given structure is independent of the method by which it is designed.

With both the LRFD procedure and the ASD procedure, expected values of the individual loads (dead, live, wind, snow, etc.) are estimated in exactly the same manner as required by the applicable specification. These loads are referred to as **service** or **working loads** throughout the text. Various combinations of these loads, which may feasibly occur at the same time, are grouped together and the largest values so obtained used for analysis and design of structures. The largest load group (in ASD) or the largest linear combination of loads in a group (in LRFD) is then used for analysis and design.

2.7   NOMINAL STRENGTHS

With both LRFD and ASD, the term **nominal strength** is constantly used. The nominal strength of a member is its calculated theoretical strength, with no safety factors (Ω_s) or resistance factors (ϕ_s) applied. In LRFD, a resistance factor, usually less than 1.0, is multiplied by the nominal strength of a member, or in ASD, the nominal strength is divided by a safety factor, usually greater than 1.0, to account for variations in material strength, member dimensions, and workmanship as well as the manner and consequences of failure. The calculation of nominal strengths for tension members is illustrated in Chapter 3, and in subsequent chapters for other types of members.

2.8   SHADING

Though the author does not think the reader will have any trouble whatsoever distinguishing between and making the calculations for the LRFD and ASD methods, in much of the book he has kept them somewhat separate by shading the ASD materials. He selected the ASD method to be shaded because the numbers in the Steel Manual pertaining to that method are shaded (actually, with a green color there).

2.9   COMPUTATION OF LOADS FOR LRFD AND ASD

With both the LRFD and the ASD procedures, expected values of the individual loads (dead, live, wind, snow, etc.) are first estimated in exactly the same manner as required by the applicable specification. These loads are referred to as **service** or **working loads** throughout the text. Various combinations of these loads that feasibly may occur at the same time are grouped together. The largest load group (in ASD) or the largest linear combination of loads in a group (in LRFD) is then used for analysis and design.

In this section and the next two sections, the loading conditions used for LRFD and ASD are presented. In both methods the individual loads (dead, live, and environmental) are estimated in exactly the same manner. After the individual loads are estimated, the next problem is to decide the worst possible combinations of those loads which might occur at the same time and which should be used in analysis and design.

LRFD Load Combinations

With the LRFD method, possible service load groups are formed, and each service load is multiplied by a load factor, normally larger than 1.0. The magnitude of the load factor reflects the uncertainty of that particular load. The resulting linear combination of service loads in a group, each multiplied by its respective load factor, is called a **factored load**. The largest values determined in this manner are used to compute the moments, shears, and other forces in the structure. These controlling values may not be larger than the nominal strengths of the members multiplied by their reduction or ϕ factors. Thus, the factors of safety have been incorporated in the load factors, and we can say

$$(\text{Reduction factor } \phi)(\text{Nominal strength of a member}) \geq$$
$$\text{computed factored force in member, } R_u$$
$$\phi \, R_n \geq R_u$$

ASD Load Combinations

With ASD, the service loads are generally not multiplied by load factors or safety factors. Rather, they are summed up, as is, for various feasible combinations, and the largest values so obtained are used to compute the forces in the members. These total forces may not be greater than the nominal strengths of the members, divided by appropriate safety factors. In equation form, the statement may be written as

$$\frac{\text{Nominal strength of member}}{\text{Safety factor } \Omega} \geq \text{largest computed force in member, } R_a.$$

$$\frac{R_n}{\Omega} \geq R_a$$

2.10 COMPUTING COMBINED LOADS WITH LRFD EXPRESSIONS

In Part 2 of the Steel Manual, entitled "General Design Considerations," load factors are calculated to increase the magnitudes of service loads to use with the LRFD procedure. The purpose of these factors is to account for the uncertainties involved in estimating the magnitudes of dead and live loads. To give the reader an idea of what we are talking about, the author poses the following question: "How close, in percent, can you estimate the worst wind or snow load that will ever be applied to the building you are now occupying?" As you think about that for a while, you will probably begin to run up your values considerably.

The required strength of a member for LRFD is determined from the load combinations given in the applicable building code. In the absence of such a code, the values given in ASCE 7 seem to be good ones to use. Part 2 of the AISC Manual provides the following load factors for buildings, which are based on ASCE 7 and are the values used in this text:

1. $U = 1.4D$
2. $U = 1.2D + 1.6L + 0.5\,(L_r \text{ or } S \text{ or } R)$
3. $U = 1.2D + 1.6(L_r \text{ or } S \text{ or } R) + (L^* \text{ or } 0.5W)$
4. $U = 1.2D + 1.0W + L^* + 0.5(L_r \text{ or } S \text{ or } R)$

5. $U = 1.2D + 1.0E + L^* + 0.2S$

6. $U = 0.9D + 1.0W$

7. $U = 0.9D + 1.0E$

*The load factor on L in combinations (3.), (4.), and (5.) is to be taken as 1.0 for floors in places of public assembly, for live loads in excess of 100 psf and for parking garage live load. The load factor is permitted to equal 0.5 for other live loads.

In these load combinations, the following abbreviations are used:

U = the design or factored load

D = dead load

L = live load due to occupancy

L_r = roof live load

S = snow load

R = nominal load due to initial rainwater or ice, exclusive of the ponding contribution

W = wind load

E = earthquake load

The load factors for dead loads are smaller than the ones for live loads, because designers can estimate so much more accurately the magnitudes of dead loads than of live loads. In this regard, the student will notice that loads which remain in place for long periods will be less variable in magnitude, while those that are applied for brief periods, such as wind loads, will have larger variations.

It is hoped that the discussion of these load factors will make the designer more conscious of load variations.

The service load values D, L, L_r, S, R, W, and E are all mean values. The different load combinations reflect 50-year recurrence values for different transient loads. In each of these equations, one of the loads is given its maximum estimated value in a 50-year period, and that maximum is combined with several other loads whose magnitudes are estimated at the time of that particular maximum load. You should notice in Equations 4, 5, 6, and 7 that the wind and seismic load factors are given as 1.0. Usually, building codes convert wind and seismic loads to ultimate or factored values. Thus, they have already been multiplied by a load factor. If that is not the case, a load factor larger than 1.0 must be used.

The preceding load factors do not vary in relation to the seriousness of failure. The reader may feel that a higher load factor should be used for a hospital than for a cattle barn, but this is not required. It is assumed, however, that the designer will consider the seriousness of failure when he or she specifies the magnitudes of the service loads. It also should be realized that the ASCE 7 load factors are minimum values, and the designer is perfectly free to use larger ones if it is deemed prudent.

The following are several additional comments regarding the application of the LRFD load combination expressions:

1. It is to be noted that in selecting design loads, adequate allowance must be made for impact conditions before the loads are substituted into the combination expressions.

2. Load combinations 6 and 7 are used to account for the possibilities of uplift. Such a condition is included to cover cases in which tension forces develop, owing to overturning moments. It will govern only for tall buildings where high lateral loads are present. In these combinations, the dead loads are reduced by 10 percent to take into account situations where they may have been overestimated.

3. It is to be clearly noted that the wind and earthquake forces have signs—that is, they may be compressive or they may be tensile (that is, tending to cause uplift). Thus, the signs must be accounted for in substituting into the load combinations. The ± signs aren't so much a matter of tension or compression as they are of saying that wind and earthquake loads can be in any horizontal and sometimes vertical direction. Load combinations 6 and 7 apply specifically to the case in which loads in a member due to wind or earthquake and gravity dead load counteract each other. For a particular column, the maximum tensile W or E force will, in all probability, be different from its maximum compressive force.

4. The magnitudes of the loads (D, L, L_r, etc.) should be obtained from the governing building code or from ASCE 7-10. Wherever applicable, the live loads used for design should be the reduced values specified for large floor areas, multistory buildings, and so on.

Examples 2-1 to 2-3 show the calculation of the factored loads, using the applicable LRFD load combinations. The largest value obtained is referred to as the *critical* or *governing load combination* and is to be used in design.

Example 2-1

The interior floor system shown in Figure 2.2 has W24 × 55 sections spaced 8 ft on center and is supporting a floor dead load of 50 psf and a live floor load of 80 psf. Determine the governing load in lb/ft that each beam must support.

Solution. Note that each foot of the beam must support itself (a dead load) plus $8 \times 1 = 8$ ft^2 of the building floor.

$$D = 55 \text{ lb/ft} + (8 \text{ ft})(50 \text{ psf}) = 455 \text{ lb/ft}$$
$$L = (8 \text{ ft})(80 \text{ psf}) = 640 \text{ lb/ft}$$

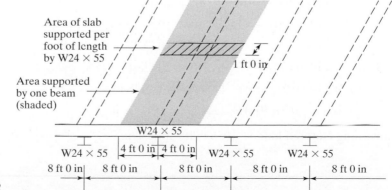

FIGURE 2.2

Computing factored loads, using the LRFD load combinations. In this substitution, the terms having no values are omitted. Note that with a floor live load of 80 psf a load factor of 0.5 has been added to load combinations (3.), (4.), and (5.) per the exception stated in ASCE 7-10 and this text for floor live loads.

1. $W_u = (1.4)(455) = 637$ lb/ft
2. $W_u = (1.2)(455) + (1.6)(640) = 1570$ lb/ft
3. $W_u = (1.2)(455) + (0.5)(640) = 866$ lb/ft
4. $W_u = (1.2)(455) + (0.5)(640) = 866$ lb/ft
5. $W_u = (1.2)(455) + (0.5)(640) = 866$ lb/ft
6. $W_u = (0.9)(455) = 409.5$ lb/ft
7. $W_u = (0.9)(455) = 409.5$ lb/ft

Governing factored load = 1570 lb/ft to be used for design.

Ans. 1570 lb/ft

Example 2-2

A roof system with W16 × 40 sections spaced 9 ft on center is to be used to support a dead load of 40 psf; a roof live, snow, or rain load of 30 psf; and a wind load of ±32 psf. Compute the governing factored load per linear foot.

Solution.

$$D = 40 \text{ lb/ft} + (9 \text{ ft})(40 \text{ psf}) = 400 \text{ lb/ft}$$
$$L = 0$$
$$L_r \text{ or } S \text{ or } R = (9 \text{ ft})(30 \text{ psf}) = 270 \text{ lb/ft}$$
$$W = (9 \text{ ft})(32 \text{ psf}) = 288 \text{ lb/ft}$$

Substituting into the load combination expressions and noting that the wind can be downward, − or uplift, + in Equation 6, we derive the following loads:

1. $W_u = (1.4)(400) = 560$ lb/ft
2. $W_u = (1.2)(400) + (0.5)(270) = 615$ lb/ft
3. $W_u = (1.2)(400) + (1.6)(270) + (0.5)(288) = 1056$ lb/ft
4. $W_u = (1.2)(400) + (1.0)(288) + (0.5)(270) = 903$ lb/ft
5. $W_u = (1.2)(400) + (0.2)(270) = 534$ lb/ft
6. (a) $W_u = (0.9)(400) + (1.0)(288) = 648$ lb/ft
   (b) $W_u = (0.9)(400) + (1.0)(-288) = 72$ lb/ft

Governing factored load = 1056 lb/ft for design.

Ans. 1056 lb/ft

Example 2-3

The various axial loads for a building column have been computed according to the applicable building code, with the following results: dead load = 200 k; load from roof = 50 k (roof live load); live load from floors (reduced as applicable for large floor area and multistory columns) = 250 k; compression wind = 128 k; tensile wind = 104 k; compression earthquake = 60 k; and tensile earthquake = 70 k.

Determine the critical design column load, P_u, using the LRFD load combinations.

Solution.

This problem solution assumes the column floor live load meets the exception for the use of the load factor of 0.5 in load combinations (3.), (4.), and (5.)

1. $P_u = (1.4)(200) = 280$ k
2. $P_u = (1.2)(200) + (1.6)(250) + (0.5)(50) = 665$ k
3. (a) $P_u = (1.2)(200) + (1.6)(50) + (0.5)(250) = 445$ k
   (b) $P_u = (1.2)(200) + (1.6)(50) + (0.5)(128) = 384$ k
4. (a) $P_u = (1.2)(200) + (1.0)(128) + (0.5)(250) + (0.5)(50) = 518$ k
   (b) $P_u = (1.2)(200) - (1.0)(104) + (0.5)(250) + (0.5)(50) = 286$ k
5. (a) $P_u = (1.2)(200) + (1.0)(60) + (0.5)(250) = 425$ k
   (b) $P_u = (1.2)(200) - (1.0)(70) + (0.5)(250) = 295$ k
6. (a) $P_u = (0.9)(200) + (1.0)(128) = 308$ k
   (b) $P_u = (0.9)(200) - (1.0)(104) = 76$ k
7. (a) $P_u = (0.9)(200) + (1.0)(60) = 240$ k
   (b) $P_u = (0.9)(200) - (1.0)(70) = 110$ k

The critical factored load combination, or design strength, required for this column is 665 k, as determined by load combination (2). It will be noted that the results of combination (6a) and (6b) do not indicate an uplift problem.

Ans. 665 k

2.11   COMPUTING COMBINED LOADS WITH ASD EXPRESSIONS

In Part 2 of the 2011 edition of the Steel Manual, the load combinations shown next are presented for ASD analysis and design. The resulting values are not interchangeable with the LRFD values.

1. D
2. D + L
3. D + (L_r or S or R)
4. D + 0.75L + 0.75(L_r or S or R)

5. $D + (0.6W \text{ or } 0.7E)$

6. (a) $D + 0.75L + 0.75(0.6W) + 0.75(L_r \text{ or } S \text{ or } R)$

   (b) $D + 0.75L + 0.75(0.7E) + 0.75(S)$

7. $0.6D + 0.6W$

8. $0.6D + 0.7E$

In the seventh and eighth expressions, the reader should note that the full dead load is not used. The variable loads W and E have lateral components and tend to cause the structure to overturn. On the other hand, the dead load is a gravity load, which tends to prevent overturning. Therefore, it can be seen that a more severe condition occurs if for some reason the full dead load is not present.

The student must realize that the AISC Specification provides what the AISC deems to be the maximum loads to be considered for a particular structure. If in the judgment of the designer the loads will be worse than the recommended values, then the values may certainly be increased. As an illustration, if the designer feels that the maximum values of the wind and rain may occur at the same time in his or her area, the 0.75 factor may be neglected. The designer should carefully consider whether the load combinations specified adequately cover all the possible combinations for a particular structure. If it is thought that they do not, he or she is free to consider additional loads and combinations as may seem appropriate. This is true for LRFD and ASD.

Example 2-4, which follows, presents the calculation of the governing ASD load to be used for the roof system of Example 2-2.

Example 2-4

Applying the ASD load combinations recommended by the AISC, determine the load to be used for the roof system of Example 2-2, where $D = 400$ lb/ft, L_r or S or $R = 270$ lb/ft, and $W = 300$ lb/ft. Assume that wind can be plus or minus.

Solution.

1. $W_a = 400$ lb/ft

2. $W_a = 400$ lb/ft

3. $W_a = 400 + 270 = 670$ lb/ft

4. $W_a = 400 + (0.75)(270) = 602.5$ lb/ft

5. $W_a = 400 + (0.6)(300) = 580$ lb/ft

6. (a) $W_a = 400 + 0.75[(0.6)(300)] + 0.75(270) = 737.5$ lb/ft

   (b) $W_a = 400 + 0.75(270) = 602.5$ lb/ft

7. $W_a = (0.6)(400) + (0.6)(-300) = 60$ lb/ft

8. $W_a = (0.6)(400) = 240$ lb/ft

Governing load = 737.5 lb/ft.

2.12   TWO METHODS OF OBTAINING AN ACCEPTABLE LEVEL OF SAFETY

The margin established between resistance and load in real cases is intended to reduce the probability of failure or unserviceability to an acceptably small value, depending on the consequences of failure or unserviceability. The question we have is how to achieve this goal when there usually is insufficient information for a complete mathematical description of either load or resistance. LRFD is one approach; ASD is another. Both methods have as their goal the obtaining of a numerical margin between resistance and load that will result in an acceptably small chance of unacceptable structural response.

A safety factor, Ω, is a number usually greater than 1.0 used in the ASD method. The nominal strength for a given limit state is divided by Ω and the result compared with the applicable service load condition.

A resistance factor, ϕ, is a number usually less than 1.0, used in the LRFD method. The nominal strength for a given limit state is multiplied by ϕ and the result compared with the applicable factored load condition.

The relationship between the safety factor Ω and the resistance factor ϕ is one we should remember. In general $\Omega = \dfrac{1.5}{\phi}$. (For instance if $\phi = 0.9$, Ω will equal $\dfrac{1.5}{0.9} = 1.67$. If $\phi = 0.75$, Ω will equal $\dfrac{1.50}{0.75} = 2.00$.)

The load factors in the linear combination of loads in a service load group do not have a standard symbol in the AISC Manual, but the symbol λ will be used here.

Thus if we set

$$Q_i = \text{one of N service loads in a group}$$
$$\lambda_i = \text{load factor associated with loads in LRFD}$$
$$R_n = \text{nominal structural strength}$$

Then for LRFD

$$\phi R_n \geq \sum_{i=1}^{N} \lambda_i Q_i$$

And for ASD

$$\frac{R_n}{\Omega} \geq \sum_{i=1}^{N} Q_i$$

2.13   DISCUSSION OF SIZES OF LOAD FACTORS AND SAFETY FACTORS

Students may sometimes feel that it is foolish to design structures with such large load factors in LRFD design and such large safety factors in ASD design. As the years go by, however, they will learn that these values are subject to so many uncertainties that they may very well spend many sleepless nights wondering if those they have used are sufficient (and they may join other designers in calling them "factors of ignorance"). Some of the uncertainties affecting these factors are as follows:

1. Material strengths may initially vary appreciably from their assumed values, and they will vary more with time due to creep, corrosion, and fatigue.
2. The methods of analysis are often subject to considerable errors.

3. The so-called vagaries of nature, or acts of God (hurricanes, earthquakes, etc.), cause conditions difficult to predict.

4. The stresses produced during fabrication and erection are often severe. Laborers in shop and field seem to treat steel shapes with reckless abandon. They drop them. They ram them. They force the members into position to line up the bolt holes. In fact, the stresses during fabrication and erection may exceed those that occur after the structure is completed. The floors for the rooms of apartment houses and office buildings are probably designed for service live loads varying from 40 to 80 psf (pounds per square foot). During the erection of such buildings, the contractor may have 10 ft of bricks or concrete blocks or other construction materials or equipment piled up on some of the floors (without the knowledge of the structural engineer), causing loads of several hundred pounds per square foot. This discussion is not intended to criticize the practice (not that it is a good one), but rather to make the student aware of the things that happen during construction. (It is probable that the majority of steel structures are overloaded somewhere during construction, but hardly any of them fail. On many of these occasions, the ductility of the steel has surely saved the day.)

5. There are technological changes that affect the magnitude of live loads. The constantly increasing traffic loads applied to bridges through the years is one illustration. The wind also seems to blow harder as the years go by, or at least, building codes keep raising the minimum design wind pressures as more is learned about the subject.

6. Although the dead loads of a structure can usually be estimated quite closely, the estimate of the live loads is more inaccurate. This is particularly true in estimating the worst possible combination of live loads occurring at any one time.

7. Other uncertainties are the presence of residual stresses and stress concentrations, variations in dimensions of member cross sections, and so on.

2.14 AUTHOR'S COMMENT

Should designs be made by both LRFD and ASD, the results will be quite close to each other. On some occasions, the LRFD designs will be slightly more economical. In effect, the smaller load factor used for dead loads in LRFD designs, compared with the load factors used for live loads, gives LRFD a little advantage. With ASD design, the safety factor used for both dead and live loads is constant for a particular problem.

2.15 PROBLEMS FOR SOLUTION

For Probs. 2-1 through 2-4 determine the maximum combined loads using the recommended AISC expressions for LRFD.

2-1 $D = 100$ psf, $L = 70$ psf, $R = 12$ psf, $L_r = 20$ psf and $S = 30$ psf (*Ans.* 247 psf)

2-2 $D = 12,000$ lb, $W = \pm 52,000$ lb

2-3 $D = 9000$ lb, $L = 5000$ lb, $L_r = 2500$ lb, $E = \pm 6500$ lb (*Ans.* 20,050 lb)

2-4 $D = 24$ psf, $L_r = 16$ psf and $W = \pm 42$ psf

2-5 Structural steel beams are to be placed at 7 ft 6 in on center under a reinforced concrete floor slab. If they are to support a service dead load D = 64 psf of floor area and a service live load L = 100 psf of floor area, determine the factored uniform load per foot which each beam must support. (*Ans.* 1776 plf)

2-6 A structural steel beam supports a roof that weighs 20 psf. An analysis of the loads has the following: S = 12 psf, L_r = 18 psf and W = 38 psf (upwards) or 16 psf (downwards). If the beams are spaced 6 ft 0 in apart, determine the factored uniformly distributed loads per foot (upward and downward, if appropriate) by which each beam should be designed.

For Probs. 2-7 through 2-10 compute the maximum combined loads using the recommended ASD expressions from the AISC.

2-7 Rework Problem 2-1. (*Ans.* 175 psf)

2-8 Rework Problem 2-2.

2-9 Rework Problem 2-3. (*Ans.* 18,037.5 lb)

2-10 Rework Problem 2-4.

2-11 Structural steel beams are to be placed at 7 ft 6 in on center under a reinforced concrete floor slab. If they are to support a service dead load D = 64 psf of floor area and a service live load L = 100 psf of floor area, determine the uniform load per foot which each beam must support using the ASD expressions. (*Ans.* 1230 plf)

2-12 A structural steel beam supports a roof that weighs 20 psf. An analysis of the loads has the following: S = 12 psf, L_r = 18 psf and W = 38 psf (upwards) or 16 psf (downwards). If the beams are spaced 6 ft 0 in apart, determine the uniformly distributed loads per foot (upward and downward, if appropriate) by which each beam should be designed using the ASD expressions.

CHAPTER 3

Analysis of Tension Members

3.1 INTRODUCTION

Tension members are found in bridge and roof trusses, towers, and bracing systems, and in situations where they are used as tie rods. The selection of a section to be used as a tension member is one of the simplest problems encountered in design. As there is no danger of the member buckling, the designer needs to determine only the load to be supported, as previously described in Chapter 2. Then the area required to support that load is calculated as described in Chapter 4, and finally a steel section is selected that provides the required area. Though these introductory calculations for tension members are quite simple, they serve the important tasks of starting students off with design ideas and getting their "feet wet" with the massive Steel Manual.

One of the simplest forms of tension members is the circular rod, but there is some difficulty in connecting it to many structures. The rod has been used frequently in the past, but finds only occasional uses today in bracing systems, light trusses, and timber construction. One important reason rods are not popular with designers is that they have been used improperly so often in the past that they have a bad reputation; however, if designed and installed correctly, they are satisfactory for many situations.

The average-size rod has very little bending stiffness and may quite easily sag under its own weight, injuring the appearance of the structure. The threaded rods formerly used in bridges often worked loose and rattled. Another disadvantage of rods is the difficulty of fabricating them with the exact lengths required and the consequent difficulties of installation.

When rods are used in wind bracing, it is a good practice to produce initial tension in them, as this will tighten up the structure and reduce rattling and swaying. Prestressing the rods limits the amount of compression they will experience during load reversal. (In a similar manner, bicycle wheel spokes are prestressed in tension to prevent the development of compression in them.) To obtain initial tension, the members may be detailed shorter than their required lengths, a method that gives the steel fabricator very little trouble. A common rule of thumb is to detail the rods about 1/16 in short for each 20 ft of length. (Approximate stress

$$f = \epsilon E = \frac{\left(\dfrac{1}{16}\,\text{in}\right)}{\left(12\,\dfrac{\text{in}}{\text{ft}}\right)(20\,\text{ft})}(29 \times 10^6\,\text{psi}) = 7550\,\text{psi.})$$ Another very satisfactory method

involves tightening the rods with some sort of sleeve nut or turnbuckle. Table 15-5 of the Steel Manual provides detailed information for such devices.

The preceding discussion on rods should illustrate why rolled shapes such as angles have supplanted rods for most applications. In the early days of steel structures, tension members consisted of rods, bars, and perhaps cables. Today, although the use of cables is increasing for suspended-roof structures, tension members usually consist of single angles, double angles, tees, channels, W sections, or sections built up from plates or rolled shapes. These members look better than the old ones, are stiffer, and are easier to connect. Another type of tension section often used is the welded tension plate, or flat bar, which is very satisfactory for use in transmission towers, signs, foot bridges, and similar structures.

A few of the various types of tension members in general use are illustrated in Fig. 3.1. In this figure, the dotted lines represent the intermittent tie plates or bars used to connect the shapes.

The tension members of steel roof trusses may consist of single angles as small as 2 1/2 × 2 × 1/4 for minor members. A more satisfactory member is made from two angles placed back to back, with sufficient space between them to permit the insertion of plates (called gusset plates) for connection purposes. Where steel sections are used back-to-back in this manner, they should be connected to each other every 4 or 5 ft to prevent rattling, particularly in bridge trusses. Single angles and double angles are probably the most common types of tension members in use. Structural tees make very satisfactory chord members for welded trusses, because the angles just mentioned can conveniently be connected to the webs of the tees.

For bridges and large roof trusses, tension members may consist of channels, W or S shapes, or even sections built up from some combination of angles, channels, and plates. Single channels are frequently used, as they have little eccentricity and are conveniently connected. Although, for the same weight, W sections are stiffer than S sections, they may have a connection disadvantage in their varying depths. For instance,

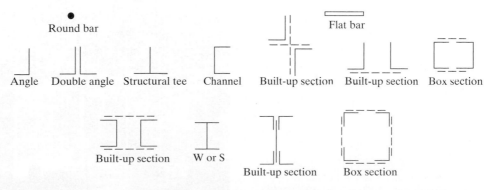

FIGURE 3.1

Types of tension members.

the W12 × 79, W12 × 72, and W12 × 65 all have slightly different depths (12.4 in, 12.3 in, and 12.1 in, respectively), while the S sections of a certain nominal size all have the same depths. For instance, the S12 × 50, the S12 × 40.8, and the S12 × 35 all have 12.00-in depths.

Although single structural shapes are a little more economical than built-up sections, the latter are occasionally used when the designer is unable to obtain sufficient area or rigidity from single shapes. It is important to remember that where built-up sections are used, field connections will have to be made and paint applied; therefore, sufficient space must be available to accomplish these things.

Members consisting of more than one section need to be tied together. Tie plates (also called *tie bars*), located at various intervals, or perforated cover plates serve to hold the various pieces in their correct positions. These plates correct any unequal distribution of loads between the various parts. They also keep the slenderness ratios of the individual parts within limitations, and they may permit easier handling of the built-up members. Long individual members such as angles may be inconvenient to handle due to flexibility, but when four angles are laced together into one member, as shown in Fig. 3.1, the member has considerable stiffness. None of the intermittent tie plates may be considered to increase the effective cross-sectional areas of the sections. As they do not theoretically carry portions of the force in the main sections, their sizes are usually governed by specifications and perhaps by some judgment on the designer's part. Perforated cover plates (see Fig. 6.9) are an exception to this rule, as part of their areas can be considered effective in resisting axial load.

A building framework under construction.
(Courtesy of Bethlehem Steel Corporation.)

Steel cables are made with special steel alloy wire ropes that are cold-drawn to the desired diameter. The resulting wires with strengths of about 200,000 to 250,000 psi can be economically used for suspension bridges, cable supported roofs, ski lifts, and other similar applications.

Normally, to select a cable tension member, the designer uses a manufacturer's catalog. The yield stress of the steel and the cable size required for the design force are determined from the catalog. It is also possible to select clevises or other devices to use for connectors at the cable ends. (See AISC Manual Table 15-3.)

3.2   NOMINAL STRENGTHS OF TENSION MEMBERS

A ductile steel member without holes and subject to a tensile load can resist without fracture a load larger than its gross cross-sectional area times its yield stress because of strain hardening. However, a tension member loaded until strain hardening is reached will lengthen a great deal before fracture—a fact that will, in all probability, end its usefulness and may even cause failure of the structural system of which the member is a part.

The skeleton of the roof of a Ford building under construction. (Courtesy of Bethlehem Steel Corporation.)

If, on the other hand, we have a tension member with bolt holes, it can possibly fail by fracture at the net section through the holes. This failure load may very well be smaller than the load required to yield the gross section, apart from the holes. It is to be realized that the portion of the member where we have a reduced cross-sectional area due to the presence of holes normally is very short compared with the total length of the member. Though the strain-hardening situation is quickly reached at the net section portion of the member, yielding there may not really be a limit state of significance, because the overall change in length of the member due to yielding in this small part of the member length may be negligible.

As a result of the preceding information, the AISC Specification (D2) states that the nominal strength of a tension member, P_n, is to be the smaller of the values obtained by substituting into the following two expressions:

For the limit state of yielding in the gross section (which is intended to prevent excessive elongation of the member),

$$P_n = F_y A_g \qquad \text{(AISC Equation D2-1)}$$

$$\phi_t P_n = \phi_t F_y A_g = \text{design tensile strength by LRFD } (\phi_t = 0.9)$$

$$\boxed{\frac{P_n}{\Omega_t} = \frac{F_y A_g}{\Omega_t} = \text{allowable tensile strength for ASD } (\Omega_t = 1.67)}$$

For tensile rupture in the net section, as where bolt or rivet holes are present,

$$P_n = F_u A_e \qquad \text{(AISC Equation D2-2)}$$

$$\phi_t P_n = \phi_t F_u A_e = \text{design tensile rupture strength for LRFD } (\phi_t = 0.75)$$

$$\boxed{\frac{P_n}{\Omega_t} = \frac{F_u A_e}{\Omega_t} \text{ allowable tensile rupture strength for ASD } (\Omega_t = 2.00)}$$

In the preceding expressions, F_y and F_u are the specified minimum yield and tensile stresses, respectively, A_g is the gross area of the member, and A_e is the effective net area that can be assumed to resist tension at the section through the holes. This area may be somewhat smaller than the actual net area, A_n, because of stress concentrations and other factors that are discussed in Section 3.5. Values of F_y and F_u are provided in Table 1.1 of this text (Table 2-4 in AISC Manual) for the ASTM structural steels on the market today.

For tension members consisting of rolled steel shapes, there actually is a third limit state, block shear, a topic presented in Section 3.7.

The design and allowable strengths presented here are not applicable to threaded steel rods or to members with pin holes (as in eyebars). These situations are discussed in Sections 4.3 and 4.4.

It is not likely that stress fluctuations will be a problem in the average building frame, because the changes in load in such structures usually occur only occasionally and produce relatively minor stress variations. Full design wind or earthquake loads occur so infrequently that they are not considered in fatigue design. Should there, however, be frequent variations or even reversals in stress, the matter of fatigue must be considered. This subject is presented in Section 4.5.

3.3   NET AREAS

The presence of a hole obviously increases the unit stress in a tension member, even if the hole is occupied by a bolt. (When fully tightened high-strength bolts are used, there may be some disagreement with this statement under certain conditions.) There is still less area of steel to which the load can be distributed, and there will be some concentration of stress along the edges of the hole.

Tension is assumed to be uniformly distributed over the net section of a tension member, although photoelastic studies show there is a decided increase in stress intensity around the edges of holes, sometimes equaling several times what the stresses would be if the holes were not present. For ductile materials, however, a uniform stress distribution assumption is reasonable when the material is loaded beyond its yield stress. Should the fibers around the holes be stressed to their yield stress, they will yield without further stress increase, with the result that there is a redistribution, or balancing, of stresses. At ultimate load, it is reasonable to assume a uniform stress distribution. The influence of ductility on the strength of bolted tension members has been clearly demonstrated in tests. Tension members (with bolt holes) made from ductile steels have proved to be as much as one-fifth to one-sixth stronger than similar members made from brittle steels with the same strengths. We have already shown in Chapter 1 that it is possible for steel to lose its ductility and become subject to brittle fracture. Such a condition can be created by fatigue-type loads and by very low temperatures.

This initial discussion is applicable only for tension members subjected to relatively static loading. Should tension members be designed for structures subjected to fatigue-type loadings, considerable effort should be made to minimize the items causing stress concentrations, such as points of sudden change of cross section, and sharp corners. In addition, as described in Section 4.5, the members may have to be enlarged.

The term "net cross-sectional area," or simply, "net area," refers to the gross cross-sectional area of a member, minus any holes, notches, or other indentations. In considering the area of such items, it is important to realize that it is usually necessary to subtract an area a little larger than the actual hole. For instance, in fabricating structural steel that is to be connected with bolts, the long-used practice was to punch holes with a diameter 1/16 in larger than that of the bolts. When this practice was followed, the punching of a hole was assumed to damage or even destroy 1/16 in more of the surrounding metal. As a result, the diameter of the hole subtracted was 1/8 in larger than the diameter of the bolt. The area of the hole was rectangular and equalled the diameter of the bolt plus 1/8 in times the thickness of the metal.

Today, drills made from very much improved steels enable fabricators to drill very large numbers of holes without resharpening. As a result, a large proportion of bolt holes are now prepared with numerically controlled drills. Though it seems reasonable to add only 1/16 in to the bolt diameters for such holes, to be consistent, the author adds 1/8 in for all standard bolt holes mentioned in this text. (Should the holes be slotted as described in Chapter 12, the usual practice is to add 1/16 in to the actual width of the holes.)

For steel much thicker than bolt diameters, it is difficult to punch out the holes to the full sizes required without excessive deformation of the surrounding material. These holes may be subpunched (with diameters 3/16 in undersized) and then reamed out to full size after the pieces are assembled. Very little material is damaged by this quite expensive process, as the holes are even and smooth, and it is considered unnec-essary to subtract the 1/16 in for damage to the sides.

It may be necessary to have an even greater latitude in meeting dimensional tol-erances during erection and for high-strength bolts larger than 5/8 in in diameter. For such a situation, holes larger than the standard ones may be used without reducing the performance of the connections. These oversized holes can be short-slotted or long-slotted, as described in Section 12.9.

Example 3-1 illustrates the calculations necessary for determining the net area of a plate type of tension member.

Example 3-1

Determine the net area of the 3/8 × 8-in plate shown in Fig. 3.2. The plate is connected at its end with two lines of 3/4-in bolts.

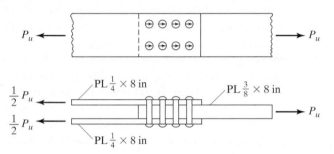

FIGURE 3.2

Solution

$$A_n = \left(\frac{3}{8}\,\text{in}\right)(8\,\text{in}) - 2\left(\frac{3}{4}\,\text{in} + \frac{1}{8}\,\text{in}\right)\left(\frac{3}{8}\,\text{in}\right) = 2.34\,\text{in}^2\ (1510\,\text{mm}^2)$$

Ans. 2.34 in^2

The connections of tension members should be arranged so that no eccentricity is present. (An exception to this rule is permitted by the AISC Specification for certain bolted and welded connections, as described in Chapters 13 and 14.) If this arrange-ment is possible, the stress is assumed to be spread uniformly across the net section of a member. Should the connections have eccentricities, moments will be produced that will cause additional stresses in the vicinity of the connection. Unfortunately, it is often quite difficult to arrange connections without eccentricity. Although specifications cover some situations, the designer may have to give consideration to eccentricities in some cases by making special estimates.

The centroidal axes of truss members meeting at a joint are assumed to coincide. Should they not coincide, eccentricity is present and secondary stresses are the result. The centroidal axes of truss members are assumed to coincide with the lines of action of their respective forces. No problem is present in a symmetrical member, as its cen-troidal axis is at its center line; but for unsymmetrical members, the problem is a little more difficult. For these members, the center line is not the centroidal axis, but the usual practice is to arrange the members at a joint so that their gage lines coincide. If a

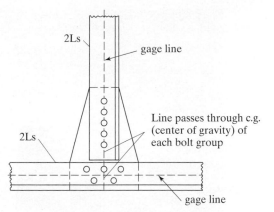

FIGURE 3.3

Lining up centroidal axes of members.

member has more than one gage line, the one closest to the actual centroidal axis of the member is used in detailing. Figure 3.3 shows a truss joint in which the c.g.s coincide.

3.4 EFFECT OF STAGGERED HOLES

Should there be more than one row of bolt holes in a member, it is often desirable to stagger them in order to provide as large a net area as possible at any one section to resist the load. In the preceding paragraphs, tensile members have been assumed to fail transversely, as along line AB in either Fig. 3.4(a) or 3.4(b). Figure 3.4(c) shows a member in which a failure other than a transverse one is possible. The holes are staggered, and failure along section $ABCD$ is possible unless the holes are a large distance apart.

To determine the critical net area in Fig. 3.4(c), it might seem logical to compute the area of a section transverse to the member (as ABE), less the area of one hole, and then the area along section $ABCD$, less two holes. The smallest value obtained along these sections would be the critical value. This method is faulty, however. Along the diagonal line from B to C, there is a combination of direct stress and shear, and a somewhat smaller area should be used. The strength of the member along section $ABCD$ is obviously somewhere between the strength obtained by using a net area computed by subtracting one hole from the transverse cross-sectional area and the value obtained by subtracting two holes from section $ABCD$.

Tests on joints show that little is gained by using complicated theoretical formulas to consider the staggered-hole situation, and the problem is usually handled with an empirical equation. The AISC Specification (B4.3b) and other specifications offer a very simple method for computing the net width of a tension member along a zigzag

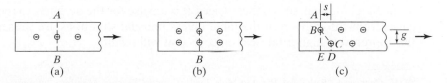

FIGURE 3.4

Possible failure sections in plates.

Trans-World Dome, St. Louis, MO. (Courtesy of Trade ARBED.)

section.[1] The method is to take the gross width of the member, regardless of the line along which failure might occur, subtract the diameter of the holes along the zigzag section being considered, and add for each inclined line the quantity given by the expression $s^2/4g$. (Since this simple expression was introduced in 1922, many investigators have proposed other, often quite complicated rules. However, none of them seems to provide significantly better results.)

In this expression, s is the longitudinal spacing (or pitch) of any two holes and g is the transverse spacing (or gage) of the same holes. The values of s and g are shown in Fig. 3.4(c). There may be several paths, any one of which may be critical at a particular joint. Each possibility should be considered, and the one giving the least value should be used. The smallest net width obtained is multiplied by the plate thickness to give the net area, A_n. Example 3-2 illustrates the method of computing the critical net area of a section that has three lines of bolts. (For angles, the gage for holes in opposite legs is considered to be the sum of the gages from the back of the angle minus the thickness of the angle.)

Holes for bolts and rivets are normally drilled or punched in steel angles at certain standard locations. These locations or gages are dependent on the angle-leg widths and on the number of lines of holes. Table 3.1, which is taken from Table 1-7A, p. 1–48 of the Steel Manual, shows these gages. It is unwise for the designer to require different gages from those given in the table unless unusual situations are present, because of the appreciably higher fabrication costs that will result.

---

[1] V. H. Cochrane, "Rules for Riveted Hole Deductions in Tension Members," *Engineering News-Record* (New York, November 16, 1922), pp. 847–848.

TABLE 3.1   Workable Gages for Angles, in Inches

	Leg	8	7	6	5	4	$3\frac{1}{2}$	3	$2\frac{1}{2}$	2	$1\frac{3}{4}$	$1\frac{1}{2}$	$1\frac{3}{8}$	$1\frac{1}{4}$	1
	g	$4\frac{1}{2}$	4	$3\frac{1}{2}$	3	$2\frac{1}{2}$	2	$1\frac{3}{4}$	$1\frac{3}{8}$	$1\frac{1}{8}$	1	$\frac{7}{8}$	$\frac{7}{8}$	$\frac{3}{4}$	$\frac{5}{8}$
	g_1	3	$2\frac{1}{2}$	$2\frac{1}{4}$	2										
	g_2	3	3	$2\frac{1}{2}$	$1\frac{3}{4}$										

Example 3-2

Determine the critical net area of the 1/2-in-thick plate shown in Fig. 3.5, using the AISC Specification (Section B4.3b). The holes are punched for 3/4-in bolts.

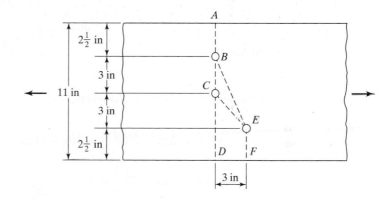

FIGURE 3.5

Solution. The critical section could possibly be *ABCD*, *ABCEF*, or *ABEF*. Hole diameters to be subtracted are $3/4 + 1/8 = 7/8$ in. The net areas for each case are as follows:

$$ABCD = (11\text{ in})\left(\frac{1}{2}\text{ in}\right) - 2\left(\frac{7}{8}\text{ in}\right)\left(\frac{1}{2}\text{ in}\right) = 4.63\text{ in}^2$$

$$ABCEF = (11\text{ in})\left(\frac{1}{2}\text{ in}\right) - 3\left(\frac{7}{8}\text{ in}\right)\left(\frac{1}{2}\text{ in}\right) + \frac{(3\text{ in})^2}{4(3\text{ in})}\left(\frac{1}{2}\text{ in}\right) = 4.56\text{ in}^2 \longleftarrow$$

$$ABEF = (11\text{ in})\left(\frac{1}{2}\text{ in}\right) - 2\left(\frac{7}{8}\text{ in}\right)\left(\frac{1}{2}\text{ in}\right) + \frac{(3\text{ in})^2}{4(6\text{ in})}\left(\frac{1}{2}\text{ in}\right) = 4.81\text{ in}^2$$

The reader should note that it is a waste of time to check path *ABEF* for this plate. Two holes need to be subtracted for routes *ABCD* and *ABEF*. As *ABCD* is a shorter route, it obviously controls over *ABEF*.

Ans. 4.56 in^2

The problem of determining the minimum pitch of staggered bolts such that no more than a certain number of holes need be subtracted to determine the net section is handled in Example 3-3.

Example 3-3

For the two lines of bolt holes shown in Fig. 3.6, determine the pitch that will give a net area $DEFG$ equal to the one along ABC. The problem may also be stated as follows: Determine the pitch that will give a net area equal to the gross area less one bolt hole. The holes are punched for 3/4-in bolts.

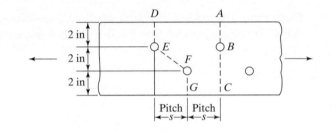

FIGURE 3.6

Solution. The hole diameters to be subtracted are 3/4 in + 1/8 in = 7/8 in.

$$ABC = 6 \text{ in} - (1)\left(\frac{7}{8}\text{ in}\right) = 5.13 \text{ in}$$

$$DEFG = 6 \text{ in} - 2\left(\frac{7}{8}\text{ in}\right) + \frac{s^2}{4(2 \text{ in})} = 4.25 \text{ in} + \frac{s^2}{8 \text{ in}}$$

$$ABC = DEFG$$

$$5.13 = 4.25 + \frac{s^2}{8}$$

$$s = 2.65 \text{ in}$$

The $s^2/4g$ rule is merely an approximation or simplification of the complex stress variations that occur in members with staggered arrangements of bolts. Steel specifications can provide only minimum standards, and designers will have to logically apply such information to complicated situations, which the specifications could not cover, in their attempts at brevity and simplicity. The next few paragraphs present a discussion and numerical examples of the $s^2/4g$ rule applied to situations not specifically addressed in the AISC Specification.

The AISC Specification does not include a method for determining the net widths of sections other than plates and angles. For channels, W sections, S sections, and others, the web and flange thicknesses are not the same. As a result, it is necessary to work with net areas rather than net widths. If the holes are placed in straight lines across such a member, the net area can be obtained by simply subtracting the cross-sectional areas of the holes from the gross area of the member. If the holes are staggered, the $\dfrac{s^2}{4g}$ values must be multiplied by the applicable thickness to change it to an area. Such a procedure is illustrated for a W section in Example 3-4, where bolts pass through the web only.

Example 3-4

Determine the net area of the W12 × 16 ($A_g = 4.71$ in^2) shown in Fig. 3.7, assuming that the holes are for 1-in bolts.

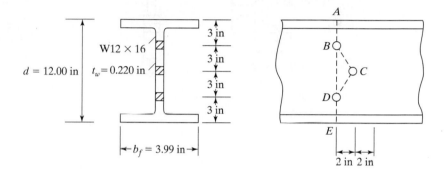

FIGURE 3.7

Solution. Net areas: hole ϕ is 1 in $+\frac{1}{8}$ in $= 1\frac{1}{8}$ in

$$ABDE = 4.71 \text{ in}^2 - 2\left(1\frac{1}{8}\text{ in}\right)(0.220 \text{ in}) = 4.21 \text{ in}^2$$

$$ABCDE = 4.72 \text{ in}^2 - 3\left(1\frac{1}{8}\text{ in}\right)(0.220 \text{ in}) + (2)\frac{(2\text{ in})^2}{4(3\text{ in})}(0.220 \text{ in}) = 4.11 \text{ in}^2 \leftarrow$$

If the zigzag line goes from a web hole to a flange hole, the thickness changes at the junction of the flange and web. In Example 3-5, the author has computed the net area of a channel that has bolt holes staggered in its flanges and web. The channel is assumed to be flattened out into a single plate, as shown in parts (b) and (c) of Fig. 3.8. The net area along route $ABCDEF$ is determined by taking the area of the channel minus the area of the holes along the route in the flanges and web plus the $s^2/4g$ values for each zigzag line times the appropriate thickness. For line CD, $s^2/4g$ has been multiplied by the thickness of the web. *For lines BC and DE (which run from holes in the web to holes in the flange), an approximate procedure has been used in which the s²/4g values have been multiplied by the average of the web and flange thicknesses.*

Example 3-5

Determine the net area along route $ABCDEF$ for the C15 × 33.9 ($A_g = 10.00$ in^2) shown in Fig. 3.8. Holes are for $\frac{3}{4}$-in bolts.

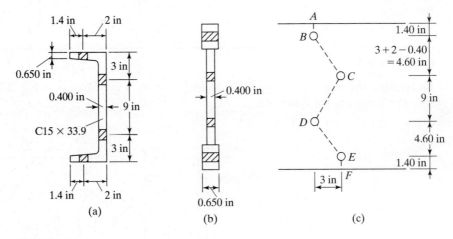

FIGURE 3.8

Solution

Approximate net A along

$$ABCDEF = 10.00 \text{ in}^2 - 2\left(\frac{7}{8}\text{ in}\right)(0.650 \text{ in})$$
$$- 2\left(\frac{7}{8}\text{ in}\right)(0.400 \text{ in})$$
$$+ \frac{(3 \text{ in})^2}{4(9 \text{ in})}(0.400 \text{ in})$$
$$+ (2)\frac{(3 \text{ in})^2}{(4)(4.60 \text{ in})}\left(\frac{0.650 \text{ in} + 0.400 \text{ in}}{2}\right)$$
$$= 8.78 \text{ in}^2$$

Ans. 8.78 in^2

3.5 EFFECTIVE NET AREAS

When a member other than a flat plate or bar is loaded in axial tension until failure occurs across its net section, its actual tensile failure stress will probably be less than the coupon tensile strength of the steel, *unless all of the various elements which make up the section are connected so that stress is transferred uniformly across the section.*

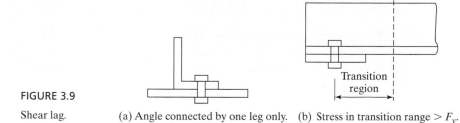

FIGURE 3.9

Shear lag.   (a) Angle connected by one leg only.   (b) Stress in transition range $> F_y$.

If the forces are not transferred uniformly across a member cross section, there will be a transition region of uneven stress running from the connection out along the member for some distance. This is the situation shown in Fig. 3.9(a), where a single angle tension member is connected by one leg only. At the connection more of the load is carried by the connected leg, and it takes the transition distance shown in part (b) of the figure for the stress to spread uniformly across the whole angle.

In the transition region the stress in the connected part of the member may very well exceed F_y and go into the strain-hardening range. Unless the load is reduced, the member may fracture prematurely. The farther we move out from the connection, the more uniform the stress becomes. In the transition region, the shear transfer has "lagged" and the phenomenon is referred to as *shear lag*.

In such a situation, the flow of tensile stress between the full member cross section and the smaller connected cross section is not 100 percent effective. As a result, the AISC Specification (D.3) states that the effective net area, A_e, of such a member is to be determined by multiplying an area A (which is the net area or the gross area or the directly connected area, as described in the next few pages) by a reduction factor U. The use of a factor such as U accounts for the nonuniform stress distribution, in a simple manner.

$$A_e = A_n U \qquad \text{(AISC Equation D3-1)}$$

The value of the reduction coefficient, U, is affected by the cross section of the member and by the length of its connection. An explanation of the way in which U factors are determined follows.

The angle shown in Fig. 3.10(a) is connected at its ends to only one leg. You can easily see that its area effective in resisting tension can be appreciably increased by shortening the width of the unconnected leg and lengthening the width of the connected leg, as shown in Fig. 3.10(b).

Investigators have found that one measure of the effectiveness of a member such as an angle connected by one leg is the distance $\bar{x}$ measured from the plane of the connection to the centroid of the area of the whole section.[2,3] The smaller the value of $\bar{x}$, the larger is the effective area of the member, and thus the larger is the member's design strength.

---

[2] E. H. Gaylord, Jr., and C. N. Gaylord, *Design of Steel Structures*, 2d ed. (New York: McGraw-Hill Book Company, 1972), pp. 119–123.

[3] W. H. Munse and E. Chesson, Jr., "Riveted and Bolted Joints: Net Section Design," *Journal of the Structural Division,* ASCE, *89*, STI (February 1963).

FIGURE 3.10

Reducing shear lag by reducing length of unconnected leg and and thus $\bar{x}$.

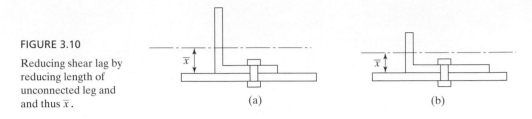

(a)   (b)

Another measure of the effectiveness of a member is the length of its connection, L. The greater this length, the smoother will be the transfer of stress to the member's unconnected parts. In other words, if 3 bolts at 3 inches on center are used, the effective area of the member will be less than if 3 bolts at 4 inches on center are used.

The effect of these two parameters, $\bar{x}$ and L, is expressed empirically with the reduction factor

$$U = 1 - \frac{\bar{x}}{L}$$

From this expression, we can see that the smaller the value of $\bar{x}$ and the larger the value of L, the larger will be the value of U, and thus the larger will be the effective area of the member. Section D3 of the AISC Commentary for Section D of the specification has additional explanation of the shear lag effect. Figures C-D3.1 through C-D3.4 show how $\bar{x}$ and L are determined for various bolted and welded tension members.

3.5.1 Bolted Members

Should a tension load be transmitted by bolts, the gross area is reduced to the net area A_n of the member, and U is computed as follows:

$$U = 1 - \frac{\bar{x}}{L}$$

The length L used in this expression is equal to the distance between the first and last bolts in the line. When there are two or more lines of bolts, L is the length of the line with the maximum number of bolts. Should the bolts be staggered, L is the out-to-out dimension between the extreme bolts in a line. Note that the longer the connection (L) becomes, the larger U will become, as will the effective area of the member. (On the other hand, we will learn in the connection chapters of this text that the effectiveness of connectors is somewhat reduced if very long connections are used.) Insufficient data are available for the case in which only one bolt is used in each line. It is thought that a reasonable approach for this case is to let $A_e = A_n$ of the connected element. Table 3.2 provides a detailed list of shear lag or U factors for different situations. This table is a copy of Table D3.1 of the AISC Specification.

For some problems herein, the authors calculate U with the $1 - \dfrac{\bar{x}}{L}$ expression, Case 2 from Table 3.2, and then compares it with the value from Case 7 for W, M, S, HP, or tees cut from these shapes and from Case 8 for single angles. He then uses the larger of the two values in his calculations, as permitted by the AISC Specification.

TABLE 3.2 Shear Lag Factors for Connections to Tension Members

Case	Description of Element		Shear Lag Factor, U	Example
1	All tension members where the tension load is transmitted directly to each of the cross-sectional elements by fasteners or welds (except as in Cases 4, 5 and 6).		$U = 1.0$	——
2	All tension members, except plates and HSS, where the tension load is transmitted to some but not all of the cross-sectional elements by fasteners or longitudinal welds or by longitudinal welds in combination with transverse welds. (Alternatively, for W, M, S and HP, Case 7 may be used. For angles, Case 8 may be used.)		$U = 1 - \bar{x}/l$	
3	All tension members where the tension load is transmitted only by transverse welds to some but not all of the cross-sectional elements.		$U = 1.0$ and A_n = area of the directly connected elements	——
4	Plates where the tension load is transmitted by longitudinal welds only.		$l \geq 2w \ldots U = 1.0$ $2w > l \geq 1.5w \ldots U = 0.87$ $1.5w > l \geq w \ldots U = 0.75$	
5	Round HSS with a single concentric gusset plate		$l \geq 1.3D \ldots U = 1.0$ $D \leq l < 1.3D \ldots U = 1 - \bar{x}/l$ $\bar{x} = D/\pi$	
6	Rectangular HSS	with a single concentric gusset plate	$l \geq H \ldots U = 1 - \bar{x}/l$ $\bar{x} = \dfrac{B^2 + 2BH}{4(B + H)}$	
		with two side gusset plates	$l \geq H \ldots U = 1 - \bar{x}/l$ $\bar{x} = \dfrac{B^2}{4(B + H)}$	
7	W, M, S or HP Shapes or Tees cut from these shapes. (If U is calculated per Case 2, the larger value is permitted to be used.)	with flange connected with 3 or more fasteners per line in the direction of loading	$b_f \geq 2/3d \ldots U = 0.90$ $b_f < 2/3d \ldots U = 0.85$	——
		with web connected with 4 or more fasteners per line in the direction of loading	$U = 0.70$	——
8	Single and double angles (If U is calculated per Case 2, the larger value is permitted to be used.)	with 4 or more fasteners per line in the direction of loading	$U = 0.80$	——
		with 3 fasteners per line in the direction of loading (With fewer than 3 fasteners per line in the direction of loading, use Case 2.)	$U = 0.60$	——

l = length of connection, in. (mm); w = plate width, in. (mm); $\bar{x}$ = eccentricity of connection, in. (mm); B = overall width of rectangular HSS member, measured 90° to the plane of the connection, in. (mm); H = overall height of rectangular HSS member, measured in the plane of the connection, in. (mm)

Source: AISC Specification, Table D3.1, p. 16.1-28, June 22, 2010. Copyright © American Institute of Steel Construction. Reprinted with permission. All rights reserved.

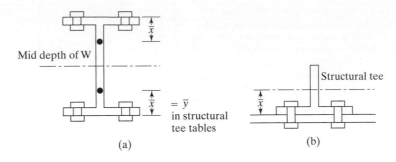

FIGURE 3.11

Values of $\bar{x}$ for different shapes.

In order to calculate U for a W section connected by its flanges only, we will assume that the section is split into two structural tees. Then the value of $\bar{x}$ used will be the distance from the outside edge of the flange to the c.g. of the structural tee, as shown in parts (a) and (b) of Fig. 3.11.

The AISC Specification permits the designer to use larger values of U than obtained from the equation if such values can be justified by tests or other rational criteria.

Section D3 of the AISC Commentary provides suggested $\bar{x}$ values for use in the equation for U for several situations not addressed in the Specification. Included are values for W and C sections bolted only through their webs. Also considered are single angles with two lines of staggered bolts in one of their legs. The basic idea for computing $\bar{x}$ for these cases is presented in the next paragraph.[4]

The channel of Fig. 3.12(a) is connected with two lines of bolts through its web. The "angle" part of this channel above the center of the top bolt is shown darkened in part (b) of the figure. This part of the channel is unconnected. For shear lag purpose, we can determine the horizontal distance from the outside face of the web to the channel centroid. This distance, which is given in the Manual shape tables, will be the $\bar{x}$ used in the equation. It is felt that with this idea in mind, the reader will be able to understand the values shown in the Commentary for other sections.

Example 3-6 illustrates the calculations necessary for determining the effective net area of a W section bolted through its flanges at each end.

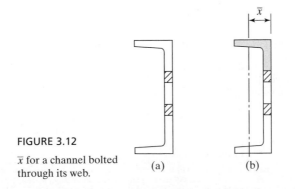

FIGURE 3.12

$\bar{x}$ for a channel bolted through its web.

(a)   (b)

---

[4]W. S. Easterling and L. G. Giroux, "Shear Lag Effects in Steel Tension Members," *Engineering Journal*, AISC, no. 3 (3rd Quarter, 1993), pp. 77–89.

Example 3-6

Determine the LRFD design tensile strength and the ASD allowable design tensile strength for a W10 × 45 with two lines of $\frac{3}{4}$-in diameter bolts in each flange using A572 Grade 50 steel, with $F_y = 50$ ksi and $F_u = 65$ ksi, and the AISC Specification. There are assumed to be at least three bolts in each line 4 in on center, and the bolts are not staggered with respect to each other.

Solution. Using a W10 × 45 ($A_g = 13.3$ in², $d = 10.10$ in, $b_f = 8.02$ in, $t_f = 0.620$ in)

Nominal or available tensile strength of section $P_n = F_y A_g = (50 \text{ ksi})(13.3 \text{ in}^2) = 665$ k

(a) Gross section yielding

LRFD with $\phi_t = 0.9$	ASD with $\Omega_t = 1.67$
$\phi_t P_n = (0.9)(665 \text{ k}) = 598.5 \text{ k}$	$\dfrac{P_n}{\Omega_t} = \dfrac{665 \text{ k}}{1.67} = 398.2 \text{ k}$

(b) Tensile rupture strength

$$A_n = 13.3 \text{ in}^2 - (4)\left(\frac{3}{4} \text{ in} + \frac{1}{8} \text{ in}\right)(0.620 \text{ in}) = 11.13 \text{ in}^2$$

Referring to tables in Manual for one-half of a W10 × 45 (or, that is, a WT5 × 22.5), we find that

$$\bar{x} = 0.907 \text{ in} \ (\bar{y} \text{ from AISC Manual Table 1-8})$$

Length of connection, $L = 2 (4 \text{ in}) = 8 \text{ in}$

From Table 3.2 (Case 2), $U = 1 - \dfrac{\bar{x}}{L} = 1 - \dfrac{0.907 \text{ in}}{8 \text{ in}} = 0.89$

But $b_f = 8.02 \text{ in} > \dfrac{2}{3}d = \left(\dfrac{2}{3}\right)(10.1) = 6.73 \text{ in}$

$\therefore$ U from Table 3.2 (Case 7) is 0.90 $\leftarrow$

$$A_e = U A_n = (0.90)(11.13 \text{ in}^2) = 10.02 \text{ in}^2$$
$$P_n = F_u A_e = (65 \text{ ksi})(10.02 \text{ in}^2) = 651.3 \text{ k}$$

LRFD with $\phi_t = 0.75$	ASD with $\Omega_t = 2.00$
$\phi_t P_n = (0.75)(651.3 \text{ k}) = 488.5 \text{ k}$ $\leftarrow$	$\dfrac{P_n}{\Omega_t} = \dfrac{651.3 \text{ k}}{2.00} = 325.6 \text{ k}$ $\leftarrow$

Ans.   LRFD $= 488.5$ k (Rupture controls)   ASD $= 325.6$ k (Rupture controls)

Notes:

1. In Table 3.2 (Case 7), it is stated that U may be used equal to 0.90 for W sections if $b_f \geq 2/3d$.
2. Answers to tensile strength problems like this one may be derived from Table 5-1 of the Manual. However, the values in this table are based on the assumptions that $U = 0.9$ and $A_e = 0.75A_g$. As a result, values will vary a little from those determined with calculated values of U and A_e. For this problem, the LRFD values from AISC Table 5-9 are 599 k for tensile yielding and 487 k for tensile rupture. For ASD, the allowable values are 398 k and 324 k, respectively.

Example 3-7

Determine the LRFD design tensile strength and the ASD allowable tensile strength for an A36 ($F_y = 36$ ksi and $F_u = 58$ ksi) L6 × 6 × 3/8 in that is connected at its ends with one line of four 7/8-in-diameter bolts in standard holes 3 in on center in one leg of the angle.

Solution. Using an L6 × 6 × $\dfrac{3}{8}$ ($A_g = 4.38$ in^2, $\bar{y} = \bar{x} = 1.62$ in) nominal or available tensile strength of the angle

$$P_n = F_y A_g = (36 \text{ ksi})(4.38 \text{ in}^2) = 157.7 \text{ k}$$

(a) Gross section yielding

LRFD with $\phi_t = 0.9$	ASD with $\Omega_t = 1.67$
$\phi_t P_n = (0.9)(157.7 \text{ k}) = 141.9 \text{ k} \leftarrow$	$\dfrac{P_n}{\Omega_t} = \dfrac{157.7 \text{ k}}{1.67} = 94.4 \text{ k} \leftarrow$

(b) Tensile rupture strength

$$A_n = 4.38 \text{ in}^2 - (1)\left(\frac{7}{8}\text{in} + \frac{1}{8}\text{in}\right)\left(\frac{3}{8}\text{in}\right) = 4.00 \text{ in}^2$$

Length of connection, $L = (3)(3 \text{ in}) = 9 \text{ in}$

$$U = 1 - \frac{\bar{x}}{L} = 1 - \frac{1.62 \text{ in}}{9 \text{ in}} = 0.82$$

From Table 3.2, Case 8, for 4 or more fasteners in the direction of loading, $U = 0.80$. Use calculated $U = 0.82$.

$$A_e = A_n U = (4.00 \text{ in}^2)(0.82) = 3.28 \text{ in}^2$$
$$P_n = F_u A_e = (58 \text{ ksi})(3.28 \text{ in}^2) = 190.2 \text{ k}$$

LRFD with $\phi_t = 0.75$	ASD with $\Omega_t = 2.00$
$\phi_t P_n = (0.75)(190.2\text{ k}) = 142.6\text{ k}$	$\dfrac{P_n}{\Omega_t} = \dfrac{190.2\text{ k}}{2.00} = 95.1\text{ k}$

Ans. LRFD = 141.9 k (Yielding controls)   ASD = 94.4 k (Yielding controls)

3.5.2 Welded Members

When tension loads are transferred by welds, the rules from AISC Table D-3.1, Table 3.2 in this text, that are to be used to determine values for A and U (A_e as for bolted connections $= AU$) are as follows:

1. Should the load be transmitted only by longitudinal welds to other than a plate member, or by longitudinal welds in combination with transverse welds, A is to equal the gross area of the member A_g (Table 3.2, Case 2).
2. Should a tension load be transmitted only by transverse welds, A is to equal the area of the directly connected elements and U is to equal 1.0 (Table 3.2, Case 3).
3. Tests have shown that when flat plates or bars connected by longitudinal fillet welds (a term described in Chapter 14) are used as tension members, they may fail prematurely by shear lag at the corners if the welds are too far apart. Therefore, the AISC Specification states that when such situations are encountered, the length of the welds may not be less than the width of the plates or bars. The letter A represents the area of the plate, and UA is the effective net area. For such situations, the values of U to be used (Table 3.2, Case 4) are as follows:

When $l \geq 2w$	$U = 1.0$
When $2w > l \geq 1.5w$	$U = 0.87$
When $1.5w > l \geq w$	$U = 0.75$

   Here, l = weld length, in
   $\qquad w$ = plate width (distance between welds), in

For combinations of longitudinal and transverse welds, l is to be used equal to the length of the longitudinal weld, because the transverse weld has little or no effect on the shear lag. (That is, it does little to get the load into the unattached parts of the member.)

Examples 3-8 and 3-9 illustrate the calculations of the effective areas, the LRFD tensile design strengths, and the ASD allowable design strengths of two welded members.

Example 3-8

The 1 × 6 in plate shown in Fig. 3.13 is connected to a 1 × 10 in plate with longitudinal fillet welds to transfer a tensile load. Determine the LRFD design tensile strength and the ASD allowable tensile strength of the member if $F_y = 50$ ksi and $F_u = 65$ ksi.

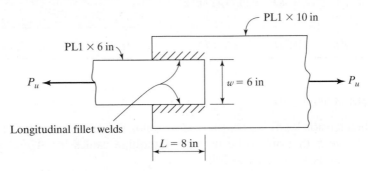

FIGURE 3.13

Solution. Considering the nominal or available tensile strength of the smaller PL 1 in $\times$ 6 in

$$P_n = F_y A_g = (50 \text{ ksi})(1 \text{ in} \times 6 \text{ in}) = 300 \text{ k}$$

(a) Gross section yielding

LRFD with $\phi_t = 0.9$	ASD with $\Omega_t = 1.67$
$\phi_t P_n = (0.9)(300 \text{ k}) = 270 \text{ k}$	$\dfrac{P_n}{\Omega_t} = \dfrac{300 \text{ k}}{1.67} = 179.6 \text{ k}$

(b) Tensile rupture strength

$$1.5w = 1.5 \times 6 \text{ in} = 9 \text{ in} > L = 8 \text{ in} > w = 6 \text{ in}$$
$$\therefore U = 0.75 \text{ from Table 3.2, Case 4}$$
$$A_e = A_n U = (6.0 \text{ in}^2)(0.75) = 4.50 \text{ in}^2$$
$$P_n = F_u A_e = (65 \text{ ksi})(4.50 \text{ in}^2) = 292.5 \text{ k}$$

LRFD with $\phi_t = 0.75$	ASD with $\Omega_t = 2.00$
$\phi_t P_n = (0.75)(292.5 \text{ k}) = 219.4 \text{ k} \leftarrow$	$\dfrac{P_n}{\Omega_t} = \dfrac{292.5 \text{ k}}{2.00} = 146.2 \text{ k} \leftarrow$

Ans. LRFD = 219.4 k (Rupture controls)   ASD = 146.2 k (Rupture controls)

Sometimes an angle has one of its legs connected with both longitudinal and transverse welds, but no connections are made to the other leg. To determine U from Table 3.2 for such a case is rather puzzling. The author feels that Case 2 of Table 3.2 (that is, $U = 1 - \dfrac{\bar{x}}{L}$) should be used for this situation. This is done in Example 3-9.

Example 3-9

Compute the LRFD design tensile strength and the ASD allowable tensile strength of the angle shown in Fig. 3.14. It is welded on the end (transverse) and sides (longitudinal) of the 8-in leg only. $F_y = 50$ ksi and $F_u = 70$ ksi.

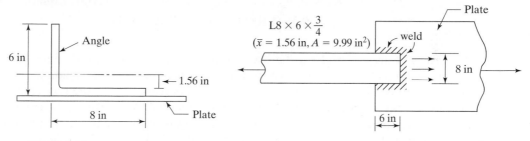

L8 × 6 × $\dfrac{3}{4}$
($\bar{x} = 1.56$ in, $A = 9.99$ in²)

FIGURE 3.14

Solution. Nominal or available tensile strength of the angle

$$= P_n = F_y A_g = (50 \text{ ksi})(9.99 \text{ in}^2) = 499.5 \text{ k}$$

(a) Gross section yielding

LRFD with $\phi_t = 0.9$	ASD with $\Omega_t = 1.67$
$\phi_t P_n = (0.9)(499.5 \text{ k}) = 449.5 \text{ k}$	$\dfrac{P_n}{\Omega_t} = \dfrac{499.5 \text{ k}}{1.67} = 299.1 \text{ k}$

(b) Tensile rupture strength (As only one leg of L is connected, a reduced effective area needs to be computed.) Use Table 3.2 (Case 2)

$$U = 1 - \frac{\bar{x}}{L} = 1 - \frac{1.56 \text{ in}}{6 \text{ in}} = 0.74$$
$$A_e = A_g U = (9.99 \text{ in}^2)(0.74) = 7.39 \text{ in}^2$$
$$P_n = F_u A_e = (70 \text{ ksi})(7.39 \text{ in}^2) = 517.3 \text{ k}$$

LRFD with $\phi_t = 0.75$	ASD with $\Omega_t = 2.00$
$\phi_t P_n = (0.75)(517.3 \text{ k}) = 388.0 \text{ k} \leftarrow$	$P_n = \dfrac{517.3 \text{ k}}{2.00} = 258.6 \text{ k} \leftarrow$

Ans.   LRFD = 388.0 k (Rupture controls)   ASD = 258.6 k (Rupture controls)

3.6   CONNECTING ELEMENTS FOR TENSION MEMBERS

When splice or gusset plates are used as statically loaded tensile connecting elements, their strength shall be determined as follows:

(a) For tensile yielding of connecting elements

$$R_n = F_y A_g \qquad \text{(AISC Equation J4-1)}$$
$$\phi = 0.90 \text{ (LRFD)} \qquad \Omega = 1.67 \text{ (ASD)}$$

(b) For tensile rupture of connecting elements

$$R_n = F_u A_e \qquad \text{(AISC Equation J4-2)}$$
$$\phi = 0.75 \text{ (LRFD)} \qquad \Omega = 2.00 \text{ (ASD)}.$$

The net area $A = A_n$, to be used in the second of these expressions may not exceed 85 percent of A_g. Tests have shown for decades that bolted tension connection elements rarely have an efficiency greater than 85 percent, even if the holes represent a very small percentage of the gross area of the elements. The lengths of the connecting elements are rather small, compared with the lengths of the members; as a result, inelastic deformations of the gross sections are limited. In Example 3-10, the strength of a pair of tensile connecting plates is determined.

Example 3-10

The tension member ($F_y = 50$ ksi and $F_u = 65$ ksi) of Example 3-6 is assumed to be connected at its ends with two 3/8 × 12-in plates, as shown in Fig. 3.15. If two lines of 3/4-in bolts are used in each plate, determine the LRFD design tensile force and the ASD allowable tensile force that the two plates can transfer.

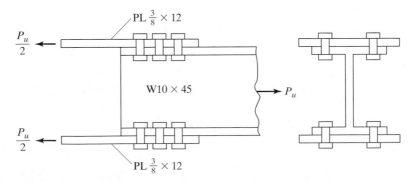

FIGURE 3.15

Solution. Nominal strength of plates

$$R_n = F_y A_g = (50 \text{ ksi})\left(2 \times \frac{3}{8} \text{ in} \times 12 \text{ in}\right) = 450 \text{ k}$$

(a) Tensile yielding of connecting elements

LRFD with $\phi = 0.90$	ASD with $\Omega = 1.67$
$\phi R_n = (0.90)(450 \text{ k}) = 405 \text{ k}$	$\dfrac{R_n}{\Omega} = \dfrac{450 \text{ k}}{1.67} = 269.5 \text{ k}$

(b) Tensile rupture of connecting elements

$$A_n \text{ of 2 plates} = 2\left[\left(\frac{3}{8} \text{ in} \times 12 \text{ in}\right) - 2\left(\frac{3}{4} \text{ in} + \frac{1}{8} \text{ in}\right)\left(\frac{3}{8} \text{ in}\right)\right] = 7.69 \text{ in}^2$$

$$0.85 A_g = (0.85)\left(2 \times \frac{3}{8} \text{ in} \times 12 \text{ in}\right) = 7.65 \text{ in}^2 \leftarrow$$

$$R_n = F_u A_e = (65 \text{ ksi})(7.65 \text{ in}^2) = 497.2 \text{ k}$$

LRFD with $\phi = 0.75$	ASD with $\Omega = 2.00$
$\phi R_n = (0.75)(497.2 \text{ k}) = 372.9 \text{ k} \leftarrow$	$\dfrac{R_n}{\Omega} = \dfrac{497.2 \text{ k}}{2.00} = 248.6 \text{ k} \leftarrow$

Ans. LRFD $= 372.9$ k (Rupture controls)   ASD $= 248.6$ k (Rupture controls)

3.7 BLOCK SHEAR

The LRFD design strength and the ASD allowable strengths of tension members are not always controlled by tension yielding, tension rupture, or by the strength of the bolts or welds with which they are connected. They may instead be controlled by *block shear* strength, as described in this section.

The failure of a member may occur along a path involving tension on one plane and shear on a perpendicular plane, as shown in Fig. 3.16, where several possible block shear failures are illustrated. For these situations, it is possible for a "block" of steel to tear out.

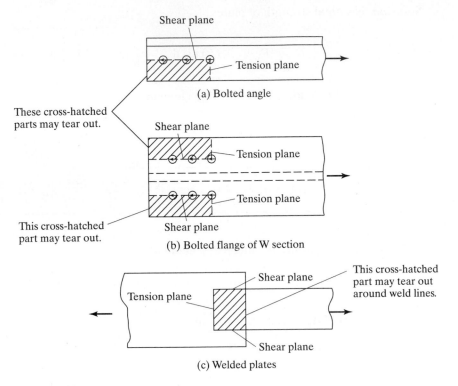

These cross-hatched parts may tear out.

This cross-hatched part may tear out.

FIGURE 3.16
Block shear.

When a tensile load applied to a particular connection is increased, the fracture strength of the weaker plane will be approached. That plane will not fail then, because it is restrained by the stronger plane. The load can be increased until the fracture strength of the stronger plane is reached. During this time, the weaker plane is yielding. The total strength of the connection equals the fracture strength of the stronger plane plus the yield strength of the weaker plane.[5] Thus, it is not realistic to add the fracture strength of one plane to the fracture strength of the other plane to determine the block shear resistance of a particular member. *You can see that block shear is a tearing, or rupture, situation and not a yielding situation.*

The member shown in Fig. 3.17(a) has a large shear area and a small tensile area; thus, the primary resistance to a block shear failure is shearing and not tensile. The AISC Specification states that it is logical to assume that when shear fracture occurs on this large shear-resisting area, the small tensile area has yielded.

Part (b) of Fig. 3.17 shows, considerably enlarged, a free body of the block that tends to tear out of the angle of part (a). You can see in this sketch that the block shear is caused by the bolts bearing on the back of the bolt holes.

[5]L. B. Burgett, "Fast Check for Block Shear," *Engineering Journal*, AISC, vol. 29, no. 4 (4th Quarter, 1992), pp. 125–127.

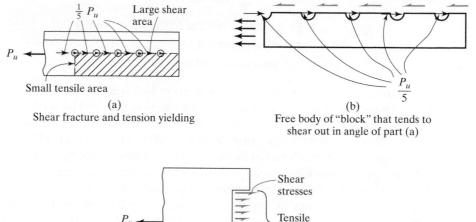

(a)

Shear fracture and tension yielding

(b)

Free body of "block" that tends to shear out in angle of part (a)

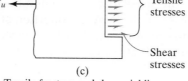

(c)

Tensile fracture and shear yielding

FIGURE 3.17

Block shear.

In part (c) of Fig. 3.17, a member is represented that, so far as block shear goes, has a large tensile area and a small shear area. The AISC feels that for this case the primary resisting force against a block shear failure will be tensile and not shearing. Thus, a block shear failure cannot occur until the tensile area fractures. At that time, it seems logical to assume that the shear area has yielded.

Based on the preceding discussion, the AISC Specification (J4.3) states that the block shear design strength of a particular member is to be determined by (1) computing the tensile fracture strength on the net section in one direction and adding to that value the shear yield strength on the gross area on the perpendicular segment, and (2) computing the shear fracture strength on the gross area subject to tension and adding it to the tensile yield strength on the net area subject to shear on the perpendicular segment. The expression to apply is the one with the larger rupture term.

Test results show that this procedure gives good results. Furthermore, it is consistent with the calculations previously used for tension members, where gross areas are used for one limit state of yielding ($F_y A_g$), and net area for the fracture limit state ($F_u A_e$).

The AISC Specification (J4.3) states that the available strength R_n for the block shear rupture design strength is as follows:

$$R_n = 0.6F_u A_{nv} + U_{bs}F_u A_{nt} \le 0.6F_y A_{gv} + U_{bs}F_u A_{nt}$$

(AISC Equation J4-5)

$$\phi = 0.75 \,(\text{LRFD}) \qquad \Omega = 2.00 \,(\text{ASD})$$

in which

$$A_{gv} = \text{gross area subjected to shear, in}^2 \text{ (mm}^2)$$
$$A_{nv} = \text{net area subjected to shear, in}^2 \text{ (mm}^2)$$
$$A_{nt} = \text{net area subjected to tension, in}^2 \text{ (mm}^2).$$

Another value included in AISC Equation J4-5 is a reduction factor U_{bs}. Its purpose is to account for the fact that stress distribution may not be uniform on the tensile plane for some connections. Should the tensile stress distribution be uniform, U_{bs} will be taken equal to 1.0, according to the AISC Specification (J4.3). The tensile stress is generally considered to be uniform for angles, gusset (or connection) plates, and for coped beams with one line of bolts. The connections of part (a) of Fig. 3.18 fall into this class. Should the tensile stress be nonuniform, U_{bs} is to be set equal to 0.5. Such a situation occurs in coped beams with two lines of bolts as illustrated in part (b) of the figure. The stress there is nonuniform because the row of bolts nearer the end of the beam picks up the largest proportion of the shear load. Should the bolts for coped beams be placed at nonstandard distances from beam ends, the same situation of nonuniform tensile stress can occur, and a U_{bs} value of 0.5 should be used.

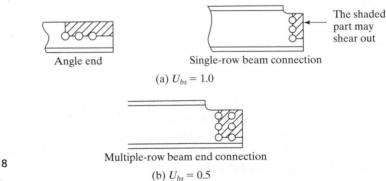

Angle end   Single-row beam connection

The shaded part may shear out

(a) $U_{bs} = 1.0$

Multiple-row beam end connection

(b) $U_{bs} = 0.5$

FIGURE 3.18

Block shear.

Examples 3-11 to 3-13 illustrate the determination of the block shear strengths for three members. The topic of block shear is continued in the connection chapters of this text, where we will find that it is absolutely necessary to check beam connections where the top flange of the beams are coped, or cut back, as illustrated in Figs. 10.2(c), 10.6, and 15.6(b). **Should the block shear strength of a connection be insufficient, it may be increased by increasing the edge distance and/or the bolt spacing.**

Example 3-11

The A572 Grade 50 ($F_u = 65$ ksi) tension member shown in Fig. 3.19 is connected with three 3/4-in bolts. Determine the LRFD block shear rupture strength and the ASD allowable block-shear rupture strength of the member. Also calculate the LRFD design tensile strength and the ASD allowable tensile strength of the member.

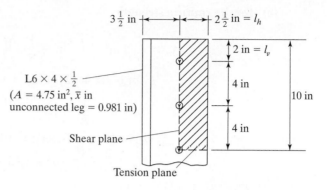

FIGURE 3.19

Solution

$$A_{gv} = (10 \text{ in})\left(\frac{1}{2}\text{in}\right) = 5.0 \text{ in}^2$$

$$A_{nv} = \left[10 \text{ in} - (2.5)\left(\frac{3}{4}\text{in} + \frac{1}{8}\text{in}\right)\right]\left(\frac{1}{2}\text{in}\right) = 3.91 \text{ in}^2$$

$$A_{nt} = \left[2.5 \text{ in} - \left(\frac{1}{2}\right)\left(\frac{3}{4}\text{in} + \frac{1}{8}\text{in}\right)\right]\left(\frac{1}{2}\text{in}\right) = 1.03 \text{ in}^2$$

$$U_{bs} = 1.0$$

$$R_n = (0.6)(65 \text{ ksi})(3.91 \text{ in}^2) + (1.0)(65 \text{ ksi})(1.03 \text{ in}^2) = 219.44 \text{ k}$$
$$\leq (0.6)(50 \text{ ksi})(5.0 \text{ in}^2) + (1.0)(65 \text{ ksi})(1.03 \text{ in}^2) = 216.95 \text{ k}$$
$$219.44 \text{ k} > 216.95 \text{ k}$$
$$\therefore R_n = 216.95 \text{ k}$$

(a) Block shear strength

LRFD with $\phi = 0.75$	ASD with $\Omega = 2.00$
$\phi R_n = (0.75)(216.95 \text{ k}) = 162.7 \text{ k} \leftarrow$	$\dfrac{R_n}{\Omega} = \dfrac{216.95 \text{ k}}{2.00} = 108.5 \text{ k} \leftarrow$

(b) Nominal or available tensile strength of angle

$$P_n = F_y A_g = (50 \text{ ksi})(4.75 \text{ in}^2) = 237.5 \text{ k}$$

Gross section yielding

LRFD with $\phi_t = 0.9$	ASD with $\Omega_t = 1.67$
$\phi_t P_n = (0.9)(237.5 \text{ k}) = 213.7 \text{ k}$	$\dfrac{P_n}{\Omega_t} = \dfrac{237.5 \text{ k}}{1.67} = 142.2 \text{ k}$

(c) Tensile rupture strength

$$A_n = 4.75 \text{ in}^2 - \left(\frac{3}{4}\text{in} + \frac{1}{8}\text{in}\right)\left(\frac{1}{2}\text{in}\right) = 4.31 \text{ in}^2$$

$$L \text{ for bolts} = (2)(4 \text{ in}) = 8 \text{ in}$$

$$U = 1 - \frac{\overline{x}}{L} = 1 - \frac{0.981 \text{ in}}{8 \text{ in}} = 0.88$$

$$A_e = UA_n = (0.88)(4.31 \text{ in}^2) = 3.79 \text{ in}^2$$

$$P_n = F_u A_e = (65 \text{ ksi})(3.79 \text{ in}^2) = 246.4 \text{ k}$$

LRFD with $\phi_t = 0.75$	ASD with $\Omega_t = 2.00$
$\phi_t P_n = (0.75)(246.4 \text{ k}) = 184.8 \text{ k}$	$\dfrac{P_n}{\Omega_t} = \dfrac{246.4 \text{ k}}{2.00} = 123.2 \text{ k}$

Ans. LRFD = 162.7 k (Block shear controls)   ASD = 108.5 k (Block shear controls)

Example 3-12

Determine the LRFD design strength and the ASD allowable strength of the A36 ($F_y = 36$ ksi, $F_u = 58$ ksi) plates shown in Fig. 3.20. Include block shear strength in the calculations.

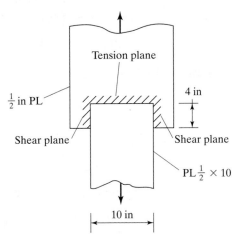

$\frac{1}{2}$ in PL

Tension plane

4 in

Shear plane

Shear plane

PL $\frac{1}{2} \times 10$

10 in

FIGURE 3.20

Solution

(a) Gross section yielding

$$P_n = F_y A_g = (36 \text{ ksi})\left(\frac{1}{2}\text{in} \times 10 \text{ in}\right) = 180 \text{ k}$$

LRFD with $\phi_t = 0.9$	ASD with $\Omega_t = 1.67$
$\phi_t P_n = (0.9)(180\text{ k}) = 162\text{ k} \leftarrow$	$\dfrac{P_n}{\Omega_t} = \dfrac{180\text{ k}}{1.67} = 107.8\text{ k} \leftarrow$

(b) Tensile rupture strength

$$U = 1.0 \text{ (Table 3.2, Case 1)}$$
$$A_e = (1.0)\left(\frac{1}{2}\text{ in} \times 10\text{ in}\right) = 5.0\text{ in}^2$$
$$P_n = F_u A_e = (58\text{ ksi})(5.0\text{ in}^2) = 290\text{ k}$$

LRFD with $\phi_t = 0.75$	ASD with $\Omega_t = 2.00$
$\phi_t P_n = (0.75)(290\text{ k}) = 217.5\text{ k}$	$\dfrac{P_n}{\Omega_t} = \dfrac{290\text{ k}}{2.00} = 145\text{ k}$

(c) Block shear strength

$$A_{gv} = \left(\frac{1}{2}\text{ in}\right)(2 \times 4\text{ in}) = 4.00\text{ in}^2$$
$$A_{nv} = 4.00\text{ in}^2$$
$$A_{nt} = \left(\frac{1}{2}\text{ in}\right)(10\text{ in}) = 5.0\text{ in}^2$$
$$U_{bs} = 1.0$$
$$R_n = (0.6)(58\text{ ksi})(4.0\text{ in}^2) + (1.00)(58\text{ ksi})(5.0\text{ in}^2) = 429.2\text{ k}$$
$$\leq (0.6)(36\text{ ksi})(4.0\text{ in}^2) + (1.00)(58\text{ ksi})(5.0\text{ in}^2) = 376.4\text{ k}$$
$$429.2\text{ k} > 376.4\text{ k}$$
$$\therefore R_n = 376.4\text{ k}$$

LRFD with $\phi = 0.75$	ASD with $\Omega = 2.00$
$\phi R_n = (0.75)(376.4\text{ k}) = 282.3\text{ k}$	$\dfrac{R_n}{\Omega} = \dfrac{376.4\text{ k}}{2.00} = 188.2\text{ k}$

Ans.   LRFD = 162 k (Yielding controls)   ASD = 107.8 k (Yielding controls)

Example 3-13

Determine the LRFD tensile design strength and the ASD tensile strength of the W12 × 30 (F_y = 50 ksi, F_u = 65 ksi) shown in Fig. 3.21 if $\frac{7}{8}$-in bolts are used in the connection. Include block shear calculations for the flanges.

Solution

(a) Gross section yielding

$$P_n = F_y A_g = (50)(8.79) = 439.5 \text{ k}$$

LRFD with ϕ_t = 0.9	ASD with Ω_t = 1.67
$\phi_t P_n = (0.9)(439.5) = 395.5 \text{ k}$	$\dfrac{P_n}{\Omega_t} = \dfrac{439.5}{1.67} = 263.2 \text{ k}$

(b) Tensile rupture strength

$$A_n = 8.79 \text{ in}^2 - (4)\left(\frac{7}{8}\text{ in} + \frac{1}{8}\text{ in}\right)(0.440 \text{ in}) = 7.03 \text{ in}^2$$

$$\bar{x} = \bar{y} \text{ in table} = 1.27 \text{ in for WT6} \times 15$$

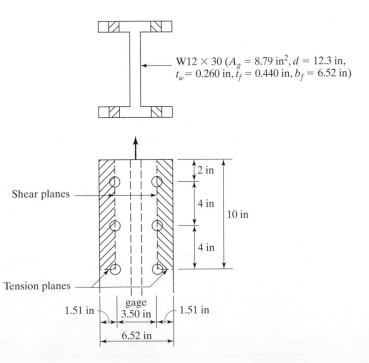

W12 × 30 (A_g = 8.79 in², d = 12.3 in, t_w = 0.260 in, t_f = 0.440 in, b_f = 6.52 in)

Shear planes

2 in

4 in

10 in

4 in

Tension planes

1.51 in   gage 3.50 in   1.51 in

6.52 in

FIGURE 3.21

$$U = 1 - \frac{\overline{x}}{L} = 1 - \frac{1.27 \text{ in}}{2 \times 4 \text{ in}} = 0.84$$

$$b_f = 6.52 \text{ in} < \frac{2}{3} \times 12.3 = 8.20 \text{ in}$$

$$\therefore \ \text{Use } U = 0.85 \text{ for Case 7 in Table 3.2}$$

$$A_e = UA_n = (0.85)(7.03 \text{ in}^2) = 5.98 \text{ in}^2$$

$$P_n = F_u A_e = (65 \text{ ksi})(5.98 \text{ in}^2) = 388.7 \text{ k}$$

LRFD with $\phi_t = 0.75$	ASD with $\Omega_t = 2.00$
$\phi_t P_n = (0.75)(388.7 \text{ k}) = 291.5 \text{ k} \leftarrow$	$\dfrac{P_n}{\Omega_t} = \dfrac{388.7 \text{ k}}{2.00} = 194.3 \text{ k} \leftarrow$

(c) Block shear strength considering both flanges

$$A_{gv} = (4)(10 \text{ in})(0.440 \text{ in}) = 17.60 \text{ in}^2$$

$$A_{nv} = (4)\left[10 \text{ in} - (2.5)\left(\frac{7}{8} \text{ in} + \frac{1}{8} \text{ in}\right)\right]0.440 \text{ in} = 13.20 \text{ in}^2$$

$$A_{nt} = (4)\left[1.51 \text{ in} - \left(\frac{1}{2}\right)\left(\frac{7}{8} \text{ in} + \frac{1}{8} \text{ in}\right)\right]0.440 \text{ in} = 1.78 \text{ in}^2$$

$$R_n = (0.6)(65 \text{ ksi})(13.20 \text{ in}^2) + (1.00)(65 \text{ ksi})(1.78 \text{ in}^2) = 630.5 \text{ k}$$

$$\leq (0.6)(50 \text{ ksi})(17.60 \text{ in}^2) + (1.00)(65 \text{ ksi})(1.78 \text{ in}^2) = 643.7 \text{ k}$$

$$630.5 \text{ k} < 643.7 \text{ k}$$

$$\therefore \ R_n = 630.5 \text{ k}$$

LRFD with $\phi = 0.75$	ASD with $\Omega = 2.00$
$\phi R_n = (0.75)(630.5 \text{ k}) = 472.9 \text{ k}$	$\dfrac{R_n}{\Omega} = \dfrac{630.5 \text{ k}}{2.00} = 315.2 \text{ k}$

Ans.   LRFD $= 291.5$ k (Rupture controls)   ASD $= 194.3$ k (Rupture controls)

3.8 PROBLEMS FOR SOLUTION (USE STANDARD-SIZE BOLT HOLES FOR ALL PROBLEMS.)

3-1 to 3-12. *Compute the net area of each of the given members.*

3-1. (*Ans.* 5.34 in^2)

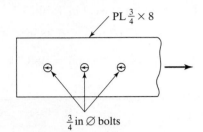

FIGURE P3-1

3-2.

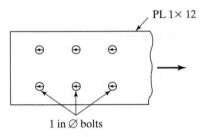

FIGURE P3-2

3-3. (*Ans.* 9.38 in^2)

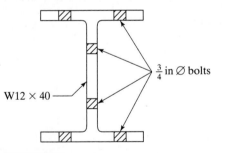

FIGURE P3-3

3-4.

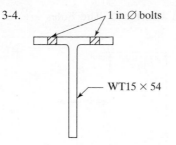

1 in Ø bolts

WT15 × 54

FIGURE P3-4

3-5.   An L8 × 4 × 3/4 with two lines of $\frac{3}{4}$-in Ø bolts in the long leg and one line of $\frac{3}{4}$-in Ø bolts in the short leg. (*Ans.* 6.52 in²)

3-6.   A pair of Ls 4 × 4 × $\frac{1}{4}$, with one line of $\frac{7}{8}$-in Ø bolts in each leg.

3-7.   A W18 × 35 with two holes in each flange and one in the web, all for $\frac{7}{8}$-in Ø bolts. (*Ans.* 8.30 in²)

3-8.   The built-up section shown in Fig. P3-8 for which $\frac{3}{4}$-in Ø bolts are used.

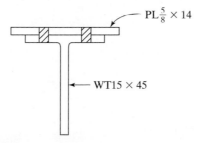

PL$\frac{5}{8}$ × 14

WT15 × 45

FIGURE P3-8

3-9.   The 1 × 8 plate shown in Fig. P3-9. The holes are for $\frac{3}{4}$-in Ø bolts. (*Ans.* 6.44 in²)

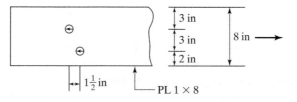

3 in

3 in

2 in

8 in ⟶

$1\frac{1}{2}$ in

PL 1 × 8

FIGURE P3-9

3-10.   The 3/4 × 10 plate shown in Fig. P3-10. The holes are for 7/8-in Ø bolts.

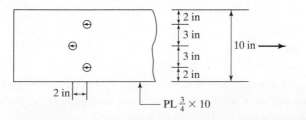

2 in

3 in

3 in

2 in

10 in ⟶

2 in

PL $\frac{3}{4}$ × 10

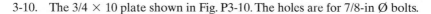

FIGURE P3-10

3-11.   The $7/8 \times 14$ plate shown in Fig. P3-11. The holes are for 7/8-in Ø bolts. (*Ans.* 10.54 in^2)

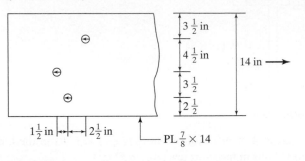

FIGURE P3-11

3-12.   The $6 \times 4 \times 1/2$ angle shown has one line of 3/4-in Ø bolts in each leg. The bolts are 4-in on center in each line and are staggered 2 in with respect to each other.

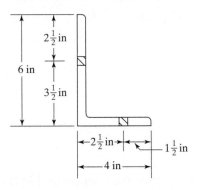

FIGURE P3-12

3-13.   The tension member shown in Fig. P3-13 contains holes for 3/4-in Ø bolts. At what spacing, *s*, will the net area for the section through one hole be the same as a rupture line passing though two holes? (*Ans.* 3.24 in)

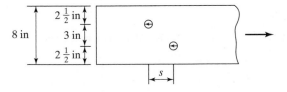

FIGURE P3-13

3-14.   The tension member shown in Figure P3-14 contains holes for 7/8-in Ø bolts. At what spacing, *s*, will the net area for the section through two holes be the same as a rupture line passing through all three holes?

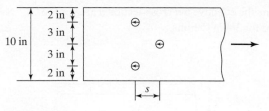

FIGURE P3-14

3-15. An L6 × 6 × 1/2 is used as a tension member, with one gage line of 3/4-in Ø bolts in each leg at the usual gage location (see Table 3.1). What is the minimum amount of stagger, s, necessary so that only one bolt need be subtracted from the gross area of the angle? Compute the net area of this member if the lines of holes are staggered at 3 in. (*Ans.* $s = 4.77$ in, $A_n = 5.05$ in^2)

3-16. An L8 × 4 × 3/4 is used as a tension member, with 7/8-in Ø bolts in each leg at the usual gage location (see Table 3.1). Two rows of bolts are used in the long leg, and one in the short leg. Determine the minimum stagger, s, necessary so that only two holes need be subtracted in determining the net area. What is the net area?

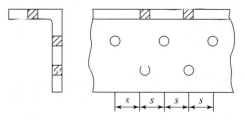

FIGURE P3-16

3-17. Determine the smallest net area of the tension member shown in Fig. P3-17. The holes are for 3/4-in Ø bolts at the usual gage locations. The stagger is 1 1/2 in. (*Ans.* 2.98 in^2)

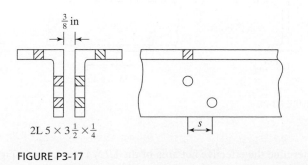

$\frac{3}{8}$ in

2L 5 × 3$\frac{1}{2}$ × $\frac{1}{4}$

FIGURE P3-17

3-18. Determine the effective net cross-sectional area of the C12 $\times$ 25 shown in Fig. P3-18. Holes are for 3/4 in Ø bolts.

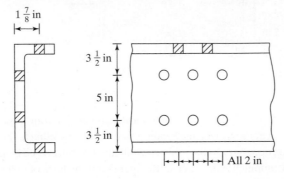

FIGURE P3-18

3-19. Compute the effective net area of the built-up section shown in Fig. P3-19 if the holes are punched for 3/4-in Ø bolts. Assume U = 0.90. (*Ans.* 20.18 in²)

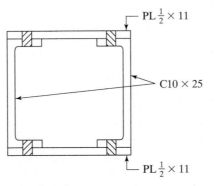

FIGURE P3-19

3-20 to 3-22. *Determine the effective net areas of the sections shown by using the U values given in Table 3.2 of this chapter.*

3-20.

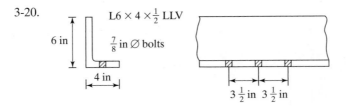

FIGURE P3-20

3-21. Determine the effective net area of the L7 $\times$ 4 $\times\frac{1}{2}$ shown in Fig. P3-21. Assume the holes are for 1-in Ø bolts. (*Ans.* 3.97 in²)

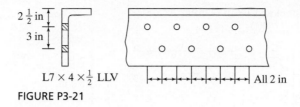

$2\frac{1}{2}$ in

3 in

L7 × 4 × $\frac{1}{2}$ LLV
All 2 in

FIGURE P3-21

3-22. An MC12 × 45 is connected through its web with 3 gage lines of 7/8-in Ø bolts. The gage lines are 3 in on center and the bolts are spaced 3 in on center along the gage line. If the center row of bolts is staggered with respect to the outer row, determine the effective net cross-sectional area of the channel. Assume there are four bolts in each line.

3-23. Determine the effective net area of the W16 × 40 shown in Fig. P3-23. Assume the holes are for 3/4-in Ø bolts. (*Ans.* 8.53 in^2)

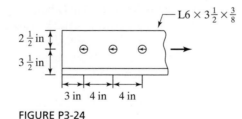

W16 × 40 $\frac{3}{4}$ in Ø bolts

All 3 $\frac{1}{2}$ in

FIGURE P3-23

3 24 to 3-34. *Determine the LRFD design strength and the ASD allowable strength of sections given. Neglect block shear.*

3-24. A36 steel and 7/8-in Ø bolts

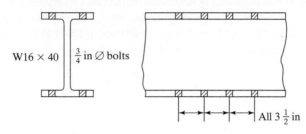

L6 × 3$\frac{1}{2}$ × $\frac{3}{8}$

$2\frac{1}{2}$ in

$3\frac{1}{2}$ in

3 in   4 in   4 in

FIGURE P3-24

3-25. A36 steel and 3/4-in Ø bolts (*Ans.* LRFD 170.42 k, ASD 113.39 k)

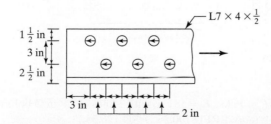

L7 × 4 × $\frac{1}{2}$

$1\frac{1}{2}$ in

3 in

$2\frac{1}{2}$ in

3 in

2 in

FIGURE P3-25

3-26. A36 steel and 7/8-in Ø bolts

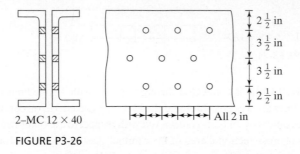

2 $\frac{1}{2}$ in

3 $\frac{1}{2}$ in

3 $\frac{1}{2}$ in

2 $\frac{1}{2}$ in

2–MC 12 × 40   All 2 in

FIGURE P3-26

3-27. A W18×40 consisting of A992 steel and having two lines of 1-in Ø bolts in each flange. There are 4 bolts in each line, 3 in on center. (*Ans.* LRFD 391.1 k, ASD 260.7 k)

3-28. A WT8 × 50 of A992 steel having two lines of 7/8-in Ø bolts as shown in Fig. P3-28. There are 4 bolts in each line, 3 in on center.

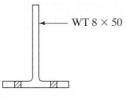

WT 8 × 50

FIGURE P3-28

3-29. A W8 × 40 of A992 steel having two lines of 3/4-in Ø bolts in each flange. There are 3 bolts in each line, 4 in on center. (*Ans.* LRFD 431.2 k, ASD 287.4 k)

3-30. A double angle, ⌐⌐ 7 × 4 × 3/4 in with two gage lines in its long leg and one in its short leg, for 7/8-in Ø bolts as shown in Fig. P3-30. Standard gages are to be used as determined from Table 3.1 in this chapter. A36 steel is used.

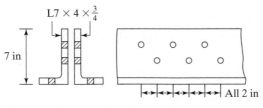

L7 × 4 × $\frac{3}{4}$

7 in

All 2 in

FIGURE P3-30

3-31. A C9×20 ($F_y = 36$ ksi, $F_u = 58$ ksi) with 2 lines of 7/8-in Ø bolts in the web as shown in Fig. P3-31. (*Ans.* LRFD 190.2 k, ASD 126.5 k)

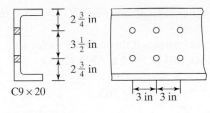

FIGURE P3-31

3-32. A WT5 × 15 consisting of A992 steel with a transverse weld to its flange only as shown in Fig. P3-32.

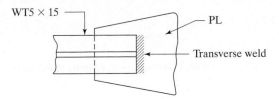

FIGURE P3-32

3-33. A C6 × 10.5 consisting of A36 steel with two longitudinal welds shown in Fig. P3-33 (*Ans*. LRFD 99.5 k, ASD 66.2 k)

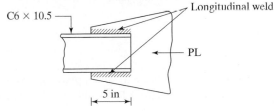

FIGURE P3-33

3-34. A $\frac{3}{8}$ × 5 plate consisting of A36 steel with two longitudinal welds as shown in Fig. P3-34.

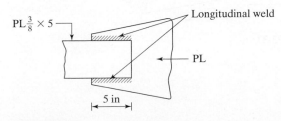

FIGURE P3-34

3-35 to 3-37.   *Determine the LRFD design strength and the ASD allowable strength of the sections given, including block shear.*

3-35.   A WT6 × 26.5, A992 steel, attached through the flange with six – 1-in Ø bolts as shown in Fig. P3-35. (*Ans.* LRFD 269.2 k, ASD 179.5 k)

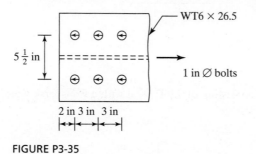

FIGURE P3-35

3-36.   A C9 × 15 (A36 steel) with 2 lines of 3/4-in Ø bolts in the web as shown in Fig. P3-36.

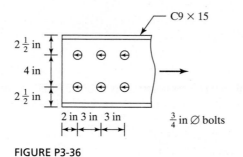

FIGURE P3-36

3-37.   An 6 × 6 × 3/8 angle welded to a 3/8 in gusset plate as shown in Fig. P3-37. All steel is F_y = 36 ksi and F_u = 58 ksi. (*Ans.* LRFD 139.1 k, ASD 92.7 k)

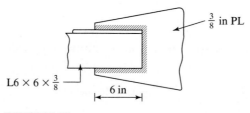

FIGURE P3-37

CHAPTER 4

Design of Tension Members

4.1 SELECTION OF SECTIONS

The determination of the design strengths of various tension members was presented in Chapter 3. In this chapter, the selection of members to support given tension loads is described. Although the designer has considerable freedom in the selection, the resulting members should have the following properties: (a) compactness, (b) dimensions that fit into the structure with reasonable relation to the dimensions of the other members of the structure, and (c) connections to as many parts of the sections as possible to minimize shear lag.

The choice of member type is often affected by the type of connections used for the structure. Some steel sections are not very convenient to bolt together with the required gusset or connection plates, while the same sections may be welded together with little difficulty. Tension members consisting of angles, channels, and W or S sections will probably be used when the connections are made with bolts, while plates, channels, and structural tees might be used for welded structures.

Various types of sections are selected for tension members in the examples to follow, and in each case where bolts are used, some allowance is made for holes. Should the connections be made *entirely* by welding, no holes have to be added to the net areas to give the required gross area. *The student should realize, however, that, very often, welded members may have holes punched in them for temporary bolting during field erection before the permanent field welds are made. These holes need to be considered in design.* It must also be remembered that in AISC Equation D2-2 ($P_n = F_u A_e$) the value of A_e may be less than A_g, even though there are no holes, depending on the arrangement of welds and on whether all of the parts of the members are connected.

The slenderness ratio of a member is the ratio of its unsupported length to its least radius of gyration. Steel specifications give preferable maximum values of slenderness ratios for both tension and compression members. The purpose of such limitations for tension members is to ensure the use of sections with stiffness sufficient to prevent undesirable lateral deflections or vibrations. Although tension members are not subject to buckling under normal loads, stress reversal may occur during shipping and erection and perhaps due to wind or earthquake loads. Specifications usually recommend that slenderness ratios be kept below certain maximum values in order that some minimum compressive strengths be provided in the members. For tension members other than rods, the AISC Specification does not provide a maximum slenderness ratio for tension members, but Section D.1 of the specification suggests that a maximum value of 300 be used.

It should be noted that out-of-straightness does not affect the strength of tension members very much, because the tension loads tend to straighten the members. (The same statement cannot be made for compression members.) For this reason, the AISC Specification is a little more liberal in its consideration of tension members, including those subject to some compressive forces due to transient loads such as wind or earthquake.

The *recommended* maximum slenderness ratio of 300 is not applicable to tension rods. Maximum L/r values for rods are left to the designer's judgment. If a maximum value of 300 were specified for them, they would seldom be used, because of their extremely small radii of gyration, and thus very high slenderness ratios.

The AASHTO Specifications provide mandatory maximum slenderness ratios of 200 for main tension members and 240 for secondary members. (A *main member* is defined by the AASHTO as one in which stresses result from dead and/or live loads, while *secondary members* are those used to brace structures or to reduce the unbraced length of other members—main or secondary.) *No such distinction is made in the AISC Specification between main and secondary members.* The AASHTO also requires that the maximum slenderness ratio permitted for members subjected to stress reversal be 140.

The design of steel members is, in effect, a trial-and-error process, although tables such as those given in the Steel Manual often enable us to directly select a desirable section. For a tension member, we can estimate the area required, select a section from the Manual providing the corresponding area, and check the section's strength, as described in the previous chapter. After this is done, it may be necessary to try a slightly larger or perhaps smaller section and repeat the checking process. The goal of the design process is to size members such that they are safe by satisfying the failure conditions listed in the AISC Specification. The student must realize that this process is iterative and that there will be some rounding up or down in the process of selecting the final section. The area needed for a particular tension member can be estimated with the LRFD equations or the ASD equations, as described next.

If the LRFD equations are used, the design strength of a tension member is the least of $\phi_t F_y A_g$, $\phi_t F_u A_e$, or its block shear strength. In addition, the slenderness ratio should, preferably, not exceed 300.

    a. To satisfy the first of these expressions, the minimum gross area must be at least equal to

$$\min A_g = \frac{P_u}{\phi_t F_y}. \tag{4.1}$$

Transfer truss, 150 Federal Street, Boston, MA. (Courtesy Owen Steel Company, Inc.)

b. To satisfy the second expression, the minimum value of A_e must be at least

$$\text{min } A_e = \frac{P_u}{\phi_t F_u}.$$

And since $A_e = U A_n$ for a bolted member, the minimum value of A_n is

$$\text{min } A_n = \frac{\text{min } A_e}{U} = \frac{P_u}{\phi_t F_u U}.$$

Then the minimum A_g is

$$= \text{min } A_n + \text{estimated area of holes}$$

$$= \frac{P_u}{\phi_t F_u U} + \text{estimated area of holes} \tag{4.2}$$

c. The third expression can be evaluated, once a trial shape has been selected and the other parameters related to the block shear strength are known.

The designer can substitute into Equations 4.1 and 4.2, taking the larger value of A_g so obtained for an initial size estimate. It is, however, well to notice that the maximum preferable slenderness ratio L/r is 300. From this value, it is easy to compute the smallest preferable value of r with respect to each principal axis of the cross section for a particular design—that is, the value of r for which the slenderness ratio will be exactly 300. It is undesirable to consider a section whose least r is less than this value, because its slenderness ratio would exceed the preferable maximum value of 300:

$$\min r = \frac{L}{300} \tag{4.3}$$

If the ASD equations are used for tension member design, the allowable strength is the lesser of $\dfrac{F_y A_g}{\Omega_t}$ and $\dfrac{F_u U A_n}{\Omega_t}$. From these expressions, the minimum gross areas required are as follows:

$$\min A_g = \frac{\Omega_t P_a}{F_y} \tag{4.1a}$$

$$\min A_g = \frac{\Omega_t P_a}{F_u U} + \text{estimated area of holes} \tag{4.2a}$$

In the expressions for LRFD (4.1 and 4.2), P_u represents the factored load forces; in ASD (4.1a and 4.2a), P_a represents the result of our application of the load combinations for ASD design. The estimated areas required by these two methods will normally vary a little from each other.

Example 4-1 illustrates the design of a bolted tension member with a W section, while Example 4-2 illustrates the selection of a bolted single angle tension member. In both problems, the areas are estimated with the LRFD expressions. After sections are selected from the Manual, they are checked for their LRFD design strengths and for their allowable ASD strengths. Whichever of the two methods is being used, it may be necessary to try a larger or smaller section and go through the calculations again.

For many of the example design problems presented in this text, the author has used only the LRFD expressions for estimating preliminary member sizes. He could just as well have used only the ASD design expressions. The results by the two methods will be very close to each other. Whatever the estimated sizes, they are carefully checked with both the appropriate LRFD and ASD equations. If the equations are not satisfied, new member sizes will be estimated and checked. We will have the same final results whether we dreamed up an estimated first size out of the blue or used some equation for estimating.

You will find on some occasions that a slightly smaller section will satisfy the LRFD equations than will satisfy the ASD equations. One reason for this is the fact that the load factors required for dead loads are much smaller than those required for live loads. Such is not the case with ASD and its safety factors.

Usually, for the examples in this text only, D and L loads are specified so that we will not have to go through all of the load combination expressions. For such problems, then, we will need only to use the following load combinations:

For LRFD	For ASD
$P_u = 1.4D$	$P_a = D + L$
$P_u = 1.2D + 1.6L$	

As the first of the LRFD expressions will not control unless the dead load is more than eight times as large as the live load, the first expression is omitted for the remaining problems in this text (unless $D > 8L$).

In Example 4-1, a W section is selected for a given set of tensile loads. For this first application of the tension design formulas, the authors have narrowed the problem down to one series of W shapes so that the reader can concentrate on the application of the formulas and not become lost in considering W8s, W10s, W14s, and so on. Exactly the same procedure can be used for trying these other series, as is used here for the W12.

Example 4-1

Select a 30-ft-long W12 section of A992 steel to support a tensile service dead load $P_D = 130$ k and a tensile service live load $P_L = 110$ k. As shown in Fig. 4.1, the member is to have two lines of bolts in each flange for 7/8-in bolts (at least three in a line 4 in on center).

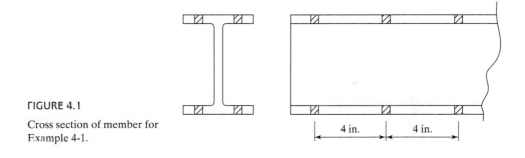

FIGURE 4.1

Cross section of member for Example 4-1.

4 in.   4 in.

Solution

(a) Considering the necessary load combinations

LRFD	ASD
$P_u = 1.4D = (1.4)(130\,\text{k}) = 182\,\text{k}$ $P_u = 1.2D + 1.6L = (1.2)(130\,\text{k}) + (1.6)(110\,\text{k}) = 332\,\text{k}$	$P_a = D + L = 130\,\text{k} + 110\,\text{k}$ $= 240\,\text{k}$

(b) Computing the minimum A_g required, using LRFD Equations 4.1 and 4.2

1. $\min A_g = \dfrac{P_u}{\phi_t F_y} = \dfrac{332\,\text{k}}{(0.90)(50\,\text{ksi})} = 7.38\,\text{in}^2$

2. $\min A_g = \dfrac{P_u}{\phi_t F_u U} + \text{estimated hole areas}$

Assume that $U = 0.85$ from Table 3.2, Case 7, and assume that flange thickness is about 0.380 in after looking at W12 sections in the LRFD Manual which have areas of 7.38 in^2 or more. $U = 0.85$ was assumed since b_f appears to be less than 2/3 d.

$$\text{min } A_g = \frac{332 \text{ k}}{(0.75)(65 \text{ ksi})(0.85)} + (4)\left(\frac{7}{8} \text{ in} + \frac{1}{8} \text{ in}\right)(0.380 \text{ in}) = 9.53 \text{ in}^2 \leftarrow$$

(c) Preferable minimum r

$$\text{min } r = \frac{L}{300} = \frac{(12 \text{ in/ft})(30 \text{ ft})}{300} = 1.2 \text{ in}$$

Try W12 × 35 ($A_g = 10.3 \text{ in}^2, d = 12.50 \text{ in}, b_f = 6.56 \text{ in},$
$t_f = 0.520 \text{ in}, r_{min} = r_y = 1.54 \text{ in}$)

Checking

(a) Gross section yielding

$$P_n = F_y A_g = (50 \text{ ksi})(10.3 \text{ in}^2) = 515 \text{ k}$$

LRFD with $\phi_t = 0.9$	ASD with $\Omega_t = 1.67$
$\phi_t P_n = (0.9)(515 \text{ k}) = 463.5 \text{ k} > 332 \text{ k}$ **OK**	$\dfrac{P_n}{\Omega_t} = \dfrac{515 \text{ k}}{1.67} = 308.4 > 240 \text{ k}$ **OK**

(b) Tensile rupture strength

From Table 3.2, Case 2
$\bar{x}$ for half of W12 × 35 or, that is, a WT6 × 17.5 = 1.30 in

$$L = (2)(4 \text{ in}) = 8 \text{ in}$$

$$U = \left(1 - \frac{\bar{x}}{L}\right) = \left(1 - \frac{1.30 \text{ in}}{8 \text{ in}}\right) = 0.84$$

From Table 3.2, Case 7
$U = 0.85$, since $b_f = 6.56 \text{ in} < \frac{2}{3}d = \left(\frac{2}{3}\right)(12.50 \text{ in}) = 8.33 \text{ in},$

$$A_n = 10.3 \text{ in}^2 - (4)\left(\frac{7}{8} \text{ in} + \frac{1}{8} \text{ in}\right)(0.520 \text{ in}) = 8.22 \text{ in}^2$$

$$A_e = (0.85)(8.22 \text{ in}^2) = 6.99 \text{ in}^2$$

$$P_n = F_u A_e = (65 \text{ ksi})(6.99 \text{ in}^2) = 454.2 \text{ k}$$

LRFD with $\phi_t = 0.75$	ASD with $\Omega_t = 2.00$
$\phi_t P_n = (0.75)(454.2 \text{ k}) = 340.7 \text{ k} > 332 \text{ k}$ **OK**	$\dfrac{P_n}{\Omega_t} = \dfrac{454.2 \text{ k}}{2.00} = 227.1 \text{ k} < 240 \text{ k}$ **N.G.**

(c) Slenderness ratio

$$\frac{L_y}{r_y} = \frac{12 \text{ in/ft} \times 30 \text{ ft}}{1.54 \text{ in}} = 234 < 300, \textbf{OK} \qquad \text{OK}$$

Ans. By LRFD, use W12 × 35.   By ASD, use next larger section W12 × 40.

Bridge over Allegheny River at Kittaning, PA. (Courtesy of the American Bridge Company.)

In Example 4-2, a broader situation is presented, in that the lightest satisfactory angle in the Steel Manual is selected for a given set of tensile loads.

Example 4-2

Design a 9-ft single-angle tension member to support a dead tensile working load of 30 k and a live tensile working load of 40 k. The member is to be connected to one leg only with 7/8-in bolts (at least four in a line 3 in on center). Assume that only one bolt is to be located at any one cross section. Use A36 steel with $F_y = 36$ ksi and $F_u = 58$ ksi.

Solution

LRFD	ASD
$P_u = (1.2)(30) + (1.6)(40) = 100$ k	$P_a = 30 + 40 = 70$ k

1. min A_g required $= \dfrac{P_u}{\phi_t F_y} = \dfrac{100}{(0.9)(36)} = 3.09$ in^2

2. Assume that $U = 0.80$, Table 3.2 (Case 8)

$$\text{min } A_n \text{ required} = \frac{P_u}{\phi_t F_u U} = \frac{100 \text{ k}}{(0.75)(58 \text{ ksi})(0.80)} = 2.87 \text{ in}^2$$

$$\text{min } A_g \text{ required} = 2.87 \text{ in}^2 + \text{bolt hole area} = 2.87 \text{ in}^2 + \left(\frac{7}{8}\text{ in} + \frac{1}{8}\text{ in}\right)(t)$$

3. Min r required $= \dfrac{(12 \text{ in/ft})(9 \text{ ft})}{300} = 0.36$ in

Angle $t_{(in)}$	Area of one 1-in bolt hole (in^2)	Gross area required = larger of $P_u/\phi_t F_y$ or $P_u/\phi_t F_u U$ + est. hole area (in^2)	Lightest angles available, their areas (in^2) and least radii of gyration (in)
5/16	0.312	3.18	$6 \times 6 \times \frac{5}{16} (A = 3.67, r_z = 1.19)$
3/8	0.375	3.25	$6 \times 3\frac{1}{2} \times \frac{3}{8} (A = 3.44, r_z = 0.763)$
7/16	0.438	3.30	$4 \times 4 \times \frac{7}{16} (A = 3.30, r_z = 0.777) \leftarrow$
			$5 \times 3 \times \frac{7}{16} (A = 3.31, r_z = 0.644)$
1/2	0.500	3.37	$4 \times 3\frac{1}{2} \times \frac{1}{2} (A = 3.50, r_z = 0.716)$
5/8	0.625	3.50	$4 \times 3 \times \frac{5}{8} (A = 3.99, r_z = 0.631)$
	Try L4 × 4 × $\dfrac{7}{16}$ ($\bar{x} = 1.15$ in)		

Checking

(a) Gross section yielding

$$P_n = F_y A_g = (36 \text{ ksi})(3.30 \text{ in}^2) = 118.8 \text{ k}$$

LRFD with $\phi_t = 0.9$	ASD with $\Omega_t = 1.67$
$\phi_t P_n = (0.9)(118.8 \text{ k}) = 106.9 \text{ k} > 100 \text{ k}$ OK	$\dfrac{P_n}{\Omega_t} = \dfrac{118.8 \text{ k}}{1.67} = 71.1 \text{ k} > 70 \text{ k}$ OK

(b) Tensile rupture strength

$$A_n = 3.30 \text{ in}^2 - (1)\left(\frac{7}{16} \text{ in}\right) = 2.86 \text{ in}^2$$

$$U = 1 - \frac{\overline{x}}{L} = 1 - \frac{1.15 \text{ in}}{(3)(3 \text{ in})} = 0.87 \leftarrow$$

U from Table 3.2 (Case 8) = 0.80
$A_e = U A_n = (0.87)(2.86 \text{ in}^2) = 2.49 \text{ in}^2$
$P_n = F_u A_e = (58 \text{ ksi})(2.49 \text{ in}^2) = 144.4 \text{ k}$

LRFD with $\phi_t = 0.75$	ASD with $\Omega_t = 2.00$
$\phi_t P_n = (0.75)(144.4 \text{ k}) = 108.3 \text{ k} > 100 \text{ k}$ OK	$\dfrac{P_n}{\Omega_t} = \dfrac{144.4 \text{ k}}{2.00} = 72.2 \text{ k} > 70 \text{ k}$ OK

Ans. By LRFD, use L4 × 4 × $\frac{7}{16}$. By ASD, select L4 × 4 × $\frac{7}{16}$.

In the CD enclosed with the Manual are tension member designs for other steel sections. Included are WT, rectangular and round HSS, and double angle sections.

4.2 BUILT-UP TENSION MEMBERS

Sections D4 and J3.5 of the AISC Specification provide a set of definite rules describing how the different parts of built-up tension members are to be connected together.

1. When a tension member is built up from elements in continuous contact with each other, such as a plate and a shape, or two plates, the longitudinal spacing of connectors between those elements must not exceed 24 times the thickness of the thinner plate—or 12 in if the member is to be painted, or if it is not to be painted and not to be subjected to corrosive conditions.

2. Should the member consist of unpainted weathering steel elements in continuous contact and be subject to atmospheric corrosion, the maximum permissible connector spacings are 14 times the thickness of the thinner plate, or 7 in.

3. Should a tension member be built up from two or more shapes separated by intermittent fillers, the shapes preferably should be connected to each other at intervals such that the slenderness ratio of the individual shapes between the fasteners does not exceed 300.

4. The distance from the center of any bolts to the nearest edge of the connected part under consideration may not be larger than 12 times the thickness of the connected part, or 6 in.

5. For elements in continuous contact with each other, the spacing of connectors are given in Sections J3.3 through J3.5 of the AISC Specification.

Example 4-3 illustrates the review of a tension member that is built up from two channels which are separated from each other. Included in the problem is the design of tie plates or tie bars to hold the channels together, as shown in Fig. 4.2(b). These plates, which are used to connect the parts of built-up members on their open sides, result in more uniform stress distribution among the various parts. Section D4 of the AISC Specification provides empirical rules for their design. (Perforated cover plates may also be used.) The rules are based on many decades of experience with built-up tension members.

In the "Dimensions and Properties" section of Part 1 of the Manual, the usual positions for placing bolts in the flanges of Ws, Cs, WTs, etc., are listed under the heading "Workable Gage." For the channels used in this example, the gage g is given as $1\frac{3}{4}$ in and is shown in Fig. 4.2.

In Fig. 4.2, the distance between the lines of bolts connecting the tie plates in the channels can be seen to equal 8.50 in. The AISC Specification (D4) states that the length of tie plates (lengths are always measured parallel to the long direction of the members, in this text) may not be less than two-thirds the distance between the lines of connectors. Furthermore, their thickness may not be less than one-fiftieth of this distance.

The minimum permissible width of tie plates (not mentioned in the specification) is the width between the lines of connectors plus the necessary edge distance on each side, to keep the bolts from splitting the plate. For this example, this minimum edge distance is taken as $1\frac{1}{2}$ in, from Table J3.4 of the AISC Specification. (Detailed information concerning edge distances for bolts is provided in Chapter 12 .) The plate dimensions are rounded off to agree with the plate sizes available from the steel mills, as given in the Bars and Plates section of Part 1 of the Steel Manual. It is much cheaper to select standard thicknesses and widths rather than to pick odd ones that will require cutting or other operations.

The AISC Specification (D4) provides a maximum spacing between tie plates by stating that the L/r of each individual component of a built-up member running along by itself between tie plates preferably should not exceed 300. If the designer substitutes into this expression $(L/r = 300)$, the least r of an individual component of the built-up member, the value of L may be computed. This will be the maximum spacing of the tie plates preferred by the AISC Specification for this member.

Example 4-3

The two C12 × 30s shown in Fig. 4.2 have been selected to support a dead tensile working load of 120 k and a 240-k live tensile working load. The member is 30 ft long, consists of A36 steel, and has one line of three 7/8-in bolts in each channel flange 3 in on center. Using the AISC Specification, determine whether the member is satisfactory and design the necessary tie plates. Assume centers of bolt holes are 1.75 in from the backs of the channels.

Solution.   Using C12 × 30s $(A_g = 8.81$ in^2 each, $t_f = 0.501$ in, $I_x = 162$ in^4 each, $I_y = 5.12$ in^4 each, y axis 0.674 in from back of C, $r_y = 0.762$ in)

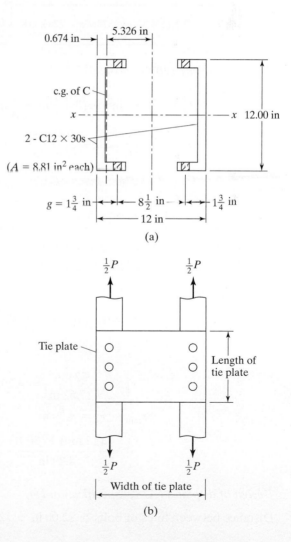

FIGURE 4.2

Built-up section for Example 4-3.

Solution

Loads to be resisted

LRFD	ASD
$P_u = (1.2)(120\text{ k}) + (1.6)(240\text{ k}) = 528\text{ k}$	$P_a = 120\text{ k} + 240\text{ k} = 360\text{ k}$

(a) Gross section yielding

$$P_n = F_y A_g = (36\text{ ksi})(2 \times 8.81\text{ in}^2) = 634.3\text{ k}$$

LRFD with $\phi_t = 0.9$	ASD with $\Omega_t = 1.67$
$\phi_t P_n = (0.9)(634.3\text{ k}) = 570.9\text{ k} > 528\text{ k }\textbf{OK}$	$\dfrac{P_n}{\Omega_t} = \dfrac{634.3\text{ k}}{1.67} = 379.8\text{ k} > 360\text{ k }\textbf{OK}$

(b) Tensile rupture strength

$$A_n = 2\left[8.81\text{ in}^2 - (2)\left(\frac{7}{8}\text{ in} + \frac{1}{8}\text{ in}\right)(0.501\text{ in})\right] = 15.62\text{ in}^2$$

$$U = 1 - \frac{\overline{x}}{L} = 1 - \frac{0.674\text{ in}}{(2)(3\text{ in})} = 0.89 \text{ from Table 3.2 (Case 2)}$$

$$P_n = F_u U A_n = (58\text{ ksi})(15.62\text{ in}^2)(0.89) = 806.3\text{ k}$$

LRFD with $\phi_t = 0.75$	ASD with $\Omega_t = 2.00$
$\phi_t P_n = (0.75)(806.3\text{ k}) = 604.7\text{ k} > 528\text{ k }\textbf{OK}$	$\dfrac{P_n}{\Omega_t} = \dfrac{806.3\text{ k}}{2.00} = 403.1\text{ k} > 360\text{ k }\textbf{OK}$

Slenderness ratio

$$I_x = (2)(162\text{ in}^4) = 324\text{ in}^4$$

$$I_y = (2)(5.12\text{ in}^4) + (2)(8.81\text{ in}^2)(5.326\text{ in})^2 = 510\text{ in}^4$$

$$r_x = \sqrt{\frac{324\text{ in}^4}{17.62\text{ in}^2}} = 4.29\text{ in} < r_y = \sqrt{\frac{510}{17.62}} = 5.38\text{ in}$$

$$\therefore r_{\min} = r_x = 4.29\text{ in}$$

$$\frac{L_x}{r_x} = \frac{(12\text{ in/ft} \times 30\text{ ft})}{4.29\text{ in}} = 83.9 < 300$$

Design of tie plates (AISC Specification D4)

Distance between lines of bolts $= 12.00\text{ in} - (2)\left(1\frac{3}{4}\text{ in}\right) = 8.50\text{ in}$

Minimum length of tie plates $= \left(\frac{2}{3}\right)(8.50 \text{ in}) = 5.67 \text{ in } (say\ 6\ in)$

Minimum thickness of tie plates $= \left(\frac{1}{50}\right)(8.50 \text{ in}) = 0.17 \text{ in} \left(say\ \frac{3}{16}\ in\right)$

Minimum width of tie plates $= 8.50 \text{ in} + (2)\left(1\frac{1}{2}\text{ in}\right) = 11.5 \text{ in } (say\ 12\ in)$

Maximum preferable spacing of tie plates

Least r of one $C = 0.762 \text{ in} = r_y$

Maximum preferable $\dfrac{L}{r} = 300$

$$\frac{(12 \text{ in/ft})(L)}{0.762 \text{ in}} = 300$$

$$L = 19.05 \text{ ft } (say\ 15\ ft)$$

Use $\frac{3}{16} \times 6 \times 1$ ft 0 in tie plate 15 ft 0 in on center.

4.3   RODS AND BARS

When rods and bars are used as tension members, they may be simply welded at their ends, or they may be threaded and held in place with nuts. The AISC nominal tensile design stress for threaded rods, F_{nt}, is given in AISC Table J3.2 and equals $0.75F_u$. This is to be applied to the gross area of the rod A_D, computed with the major thread diameter — that is, the diameter to the outer extremity of the thread. The area required for a particular tensile load can then be calculated as follows:

$$R_n = F_{nt}A_D = 0.75\ F_u A_D$$

$\phi = 0.75$ LRFD	$\Omega = 2.00$ ASD
$A_D \geq \dfrac{P_u}{\phi\, 0.75 F_u}$	$A_D \geq \dfrac{\Omega P_a}{0.75 F_u}$

In Table 7-18 of the Manual, entitled "Threading Dimensions for High-Strength and Non-High-Strength Bolts," properties of standard threaded rods are presented. Example 4-4 illustrates the selection of a rod by the use of this table.

Example 4-4

Using the AISC Specification, select a standard threaded rod of A36 steel to support a tensile working dead load of 10 k and a tensile working live load of 20 k.

Solution

LRFD	ASD
$P_u = (1.2)(10 \text{ k}) + (1.6)(20 \text{ k}) = 44 \text{ k}$	$P_a = 10 \text{ k} + 20 \text{ k} = 30 \text{ k}$

$$A_D \geq \frac{P_u}{\phi \, 0.75 F_u} = \frac{44 \text{ k}}{(0.75)(0.75)(58 \text{ ksi})} = 1.35 \text{ in}^2$$

Try $1\frac{3}{8}$ in diameter rod from AISC Table 7-17 using the gross area of the rod 1.49 in².

$$R_n = 0.75 F_u A_D = (0.75)(58 \text{ ksi})(1.49 \text{ in}^2) = 64.8 \text{ k}$$

LRFD $\phi = 0.75$	ASD $\Omega = 2.00$
$\phi R_n = (0.75)(64.8 \text{ k}) = 48.6 \text{ k} > 44 \text{ k}$ **OK**	$\dfrac{R_n}{\Omega} = \dfrac{64.8 \text{ k}}{2.00} = 32.4 > 30 \text{ k}$ **OK**

Use $1\frac{3}{8}$ in-diameter rod with 6 threads per in.

As shown in Fig. 4.3, upset rods sometimes are used, where the rod ends are made larger than the regular rod and the threads are placed in the upset ends. Threads, obviously, reduce the cross-sectional area of a rod. If a rod is upset and the threads are placed in that part of the rod, the result will be a larger cross-sectional area at the root of the thread than we would have if the threads were placed in the regular part of the rod.

Table J3.2 footnote (d) in the AISC Specification states that the nominal tensile strength of the threaded portion of the upset ends is equal to $0.75F_u A_D$, where A_D is the cross-sectional area of the rod at its major thread diameter. This value must be larger than the nominal body area of the rod (before upsetting) times F_y, so that the net section fracture strength exceeds the gross section yield strength.

Upsetting permits the designer to use the entire area of the regular part of the bar for strength calculations. Nevertheless, the use of upset rods probably is not economical and should be avoided unless a large order is being made.

One situation in which tension rods are sometimes used is in steel-frame industrial buildings with purlins running between their roof trusses to support the roof surface. These types of buildings will also frequently have girts running between the columns along the vertical walls. (Girts are horizontal beams used on the sides of buildings, usually industrial, to resist lateral bending due to wind. They also are often used to support corrugated or other types of siding.) Sag rods may be required to provide support for the purlins parallel to the roof surface and vertical support for the girts along the walls. For roofs with steeper slopes than one vertically to four horizontally, sag rods are often considered necessary to provide lateral support for the purlins, particularly where the purlins consist of steel channels. Steel channels are commonly used as purlins, but they have very little resistance to lateral bending. Although the resisting moment needed parallel to the roof surface is small, an extremely large channel is required to provide

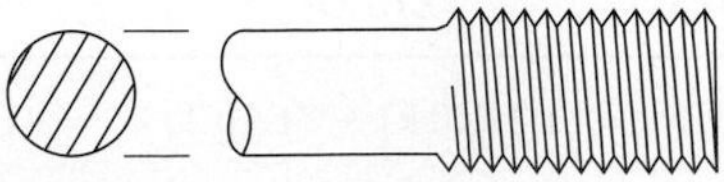

FIGURE 4.3

A round upset rod.

New Albany Bridge crossing the Ohio River between Louisville, KY, and New Albany, IN. (Courtesy of the Lincoln Electric Company.)

such a moment. The use of sag rods for providing lateral support to purlins made from channels usually is economical because of the bending weakness of channels about their y axes. For light roofs (e.g., where trusses support corrugated steel roofs), sag rods will probably be needed at the one-third points if the trusses are more than 20 ft on centers. Sag rods at the midpoints are usually sufficient if the trusses are less than 20 ft on centers. For heavier roofs, such as those made of slate, cement tile, or clay tile, sag rods will probably be needed at closer intervals. The one-third points will likely be necessary if the trusses are spaced at greater intervals than 14 ft, and the midpoints will be satisfactory if truss spacings are less than 14 ft. Some designers assume that the load components parallel to the roof surface can be taken by the roof, particularly if it consists of corrugated steel sheets, and that tie rods are unnecessary. This assumption, however, is open to some doubt and definitely should not be followed if the roof is very steep.

Designers have to use their own judgment in limiting the slenderness values for rods, as they will usually be several times the limiting values mentioned for other types of tension members. A common practice of many designers is to select rod diameters no less than 1/500th of their lengths, to obtain some rigidity, even though design calculations may permit smaller sizes.

It is typically desirable to limit the minimum size of sag rods to 5/8 in, because smaller rods than these are often damaged during construction. The threads on smaller rods are quite easily damaged by overtightening, which seems to be a frequent habit of construction workers. Sag rods are designed for the purlins of a roof truss in Example 4-5. The rods are assumed to support the simple beam reactions for the components of the gravity loads (roofing, purlins, snow, and ice) parallel to the roof surface. Wind forces are assumed to act perpendicular to the roof surfaces and, theoretically, will not affect the sag rod forces. The maximum force in a sag rod will occur in the top sag rod, because it must support the sum of the forces in the lower sag rods. It is theoretically possible to use smaller rods for the lower sag rods, but this reduction in size will probably be impractical.

Example 4-5

Design the sag rods for the purlins of the truss shown in Fig. 4.4. Purlins are to be supported at their one-third points between the trusses, which are spaced 21 ft on centers. Use A36 steel and assume that a minimum-size rod of 5/8 in is permitted. A clay tile roof weighing 16 psf (0.77 kN/m^2) of roof surface is used and supports a snow load of 20 psf (0.96 kN/m^2) of horizontal projection of roof surface. Details of the purlins and the sag rods and their connections are shown in Figs. 4.4 and 4.5. In these figures, the dotted lines represent ties and struts in the end panels in the plane of the roof, commonly used to give greater resistance to loads located on one side of the roof (a loading situation that might occur when snow is blown off one side of the roof during a severe windstorm).

Solution.   Gravity loads in psf of roof surface are as follows:

Average weight in psf of the seven purlins on each side of the roof

$$= \frac{(7)(11.5 \text{ lb/ft})}{37.9 \text{ ft}} = 2.1 \text{ psf}$$

$$\text{Snow} = 20 \text{ psf} \left(\frac{3}{\sqrt{10}} \right) = 19.0 \text{ psf of roof surface}$$

Tile roofing $= 16.0$ psf

$$w_u = (1.2)(2.1 + 16.0) + (1.6)(19.0) = 52.1 \text{ psf}$$

You can see in Figs. 4.4 and 4.5 that half of the load component parallel to the roof surface between the top two purlins on each side of the truss is carried directly to the horizontal sag rod between the purlins. In this example, there are seven purlins (with six spaces between them) on each side of the truss. Thus, 1/12th of the total inclined load goes directly to the horizontal sag rod.

LRFD	ASD
LRFD load on top inclined sag rod, using controlling load factor equation $w_u = (1.2)(2.1 \text{ psf} + 16.0 \text{ psf})$ $+ (1.6)(19.0 \text{ psf})$ $= 52.1 \text{ psf}$	ASD load on top inclined sag rod, using controlling ASD load equation $w = 2.1 \text{ psf} + 16.0 \text{ psf} + 19.0 \text{ psf}$ $= 37.1 \text{ psf}$
Component of loads parallel to roof surface $= \left(\frac{1}{\sqrt{10}} \right)(52.1 \text{ psf}) = 16.5 \text{ psf}$	Component of loads parallel to roof surface $= \left(\frac{1}{\sqrt{10}} \right)(37.1) = 11.7 \text{ psf}$
Load on top inclined sag rod $= \left(\frac{11}{12} \right)(37.9 \text{ ft})(7 \text{ ft})(16.5 \text{ psf})$ $= 4013 \text{ lbs} = 4.01 \text{ k} = P_u$	Load on top inclined sag rod $= \left(\frac{11}{12} \right)(37.9 \text{ ft})(7 \text{ ft})(11.7 \text{ psf})$ $= 2845 \text{ lbs} = 2.85 \text{ k} = P$

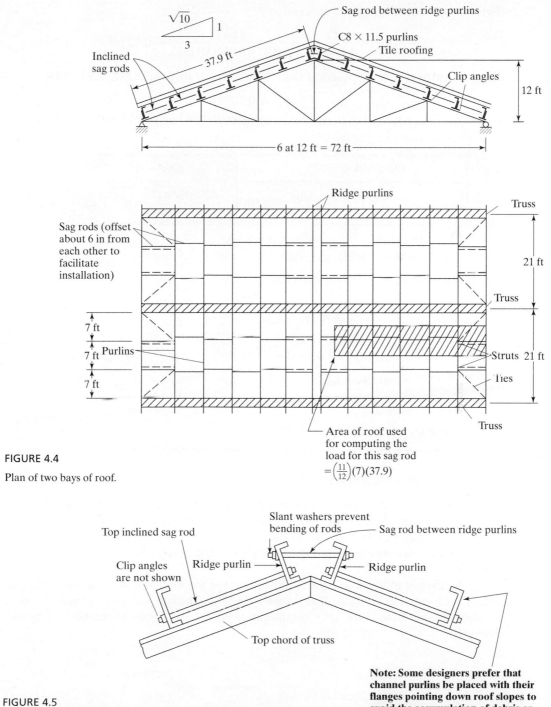

FIGURE 4.4

Plan of two bays of roof.

FIGURE 4.5

Details of sag rod connections.

Selecting section with LRFD expression

$$A_D = \frac{P_u}{\phi\, 0.75 F_u} = \frac{4.01\text{ k}}{(0.75)(0.75)(58\text{ ksi})} = 0.12\text{ in}^2$$

Try $\frac{5}{8}$-in rod as minimum practical size, 11 threads per in, from AISC Table 7-18.

$$A_D = 0.307\text{ in}^2$$
$$R_n = 0.75 F_u A_D = (0.75)(58\text{ ksi})(0.307\text{ in}^2) = 13.36\text{ k}$$

LRFD with $\phi = 0.75$	ASD with $\Omega = 2.00$
$\phi R_n = (0.75)(13.36\text{ k}) = 10.02\text{ k} > 4.01\text{ k }$ **OK**	$\dfrac{R_n}{\Omega} = \dfrac{13.36\text{ k}}{2.00} = 6.68\text{ k} > 2.85\text{ k }$ **OK**

Use $\frac{5}{8}$-in rod for both LRFD and ASD.

Checking force in tie rods between ridge purlins

LRFD	ASD
$P_u = (37.9\text{ ft})(7\text{ ft})(16.5\text{ psf})\left(\dfrac{\sqrt{10}}{3}\right)$	$P_a = (37.9\text{ ft})(7\text{ ft})(11.7\text{ psf})\left(\dfrac{\sqrt{10}}{3}\right)$
$= 4614\text{ lbs} = 4.61\text{ k} < 10.02\text{ k }$ **OK**	$= 3280\text{ lbs} = 3.28\text{ k} < 6.68\text{ k }$ **OK**

Use $\frac{5}{8}$-in rod for both LRFD and ASD.

4.4   PIN-CONNECTED MEMBERS

Until the early years of the twentieth century nearly all bridges in the United States were pin-connected, but today pin-connected bridges are used infrequently because of the advantages of bolted and welded connections. One trouble with the old pin-connected trusses was the wearing of the pins in the holes, which caused looseness of the joints.

An eyebar is a special type of pin-connected member whose ends where the pin holes are located are enlarged, as shown in Fig. 4.6. Though just about obsolete today, eyebars at one time were very commonly used for the tension members of bridge trusses.

FIGURE 4.6

End of an eyebar.

Pin-connected eyebars are still used occasionally as tension members for long-span bridges and as hangers for some types of bridges and other structures, where they are normally subjected to very large dead loads. As a result, the eyebars are usually prevented from rattling and wearing, as they would under live loads.

Eyebars are not generally made by forging but are thermally cut from plates. As stated in the AISC Commentary (D6), extensive testing has shown that thermally cut members result in more balanced designs. The heads of eyebars are specially shaped so as to provide optimum stress flow around the holes. These proportions are set on the basis of long experience and testing with forged eyebars, and the resulting standards are rather conservative for today's thermally cut members.

The AISC Specification (D5) provides detailed requirements for pin-connected members as to strength and proportions of the pins and plates. The design strength of such a member is the lowest value obtained from the following equations, where reference is made to Fig. 4.7:

1. **Tension rupture on the net effective area. See Fig. 4.7(a).**

$$P_n = 2tb_e F_u \qquad \text{(AISC Equation D5–1)}$$
$$\phi = 0.75 \,(\text{LRFD}) \qquad \Omega = 2.00 \,\text{ASD}$$

in which t = plate thickness and $b_e = 2t + 0.63$, but may not exceed the distance from the hole edge to the edge of the part measured perpendicular to the line of force.

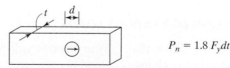

$$P_n = (2t)(2t + 0.63)(F_u)$$

(a) Tensile rupture strength on net effective area

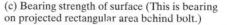

$$P_n = (0.6)(2t)\left(a + \tfrac{d}{2}\right)(F_u)$$

(b) Shear rupture strength on effective area

$$P_n = 1.8\,F_y dt$$

(c) Bearing strength of surface (This is bearing on projected rectangular area behind bolt.)

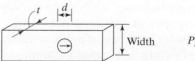

Width $\qquad P_n = (F_y)(\text{width})(t)$

FIGURE 4.7

Strength of pin-connected tension members.   (d) Tensile yielding strength in gross section

2. **For shear rupture on the effective area. See Fig. 4.7(b).**

$$P_n = 0.6F_u A_{sf} \qquad \text{(AISC Equation D5–2)}$$
$$\phi = 0.75 \,(\text{LRFD}) \qquad \Omega = 2.00 \,\text{ASD}$$

in which $A_{sf} = 2t(a + d/2)$, where a is the shortest distance from the edge of the pin hole to the member edge, measured parallel to the force.

3. **Strength of surfaces in bearing. See Fig. 4.7(c).**

$$R_n = 1.8F_y A_{pb} \qquad \text{(AISC Equation J7–1)}$$
$$\phi = 0.75 \,(\text{LRFD}) \qquad \Omega = 2.00 \,\text{ASD}$$

in which A_{pb} = projected bearing area = dt. Notice that LRFD Equation J7-1 applies to milled surfaces; pins in reamed, drilled, or bored holes; and ends of fitted bearing stiffeners. (AISC Specification J7(b) provides other equations for determining the bearing strength for expansion rollers and rockers.)

4. **Tensile yielding in the gross section. See Fig. 4.7(d).**

$$P_n = F_y A_g \qquad \text{(AISC Equation D2–1)}$$
$$\phi_t = 0.90 \,(\text{LRFD}) \qquad \Omega_t = 1.67 \,(\text{ASD})$$

The AISC Specification D6.2 says that thicknesses < 1/2 in for both eyebars and pin-connected plates are permissible only when external nuts are provided to tighten the pin plates and filler plates into snug contact. The bearing design strength of such plates is provided in AISC Specification J7.

In addition to the other requirements mentioned, AISC Specification D5 specifies certain proportions between the pins and the eyebars. These values are based on long experience in the steel industry, and on experimental work by B. G. Johnston.[1] It has been found that when eyebars and pin-connected members are made from steels with yield stresses greater than 70 kips per square inch (ksi), there may be a possibility of **dishing** (a complicated inelastic stability failure where the head of the eyebar tends to curl laterally into a dish-like shape). For this reason, the AISC Specification requires stockier member proportions for such situations—hole diameter not to exceed five times the plate thickness, and the width of the eyebar to be reduced accordingly.

4.5 DESIGN FOR FATIGUE LOADS

It is not likely that fatigue stresses will be a problem in the average building frame, because the changes in load in such structures usually occur only occasionally and produce relatively minor stress variations. Should there, however, be frequent variations in, or even reversals of, stress, the matter of fatigue must be considered. Fatigue can be a problem in buildings when crane runway girders or heavy vibrating or moving machinery or equipment are supported.

If steel members are subjected to loads that are applied, and then removed, or changed significantly many thousands of times, cracks may occur, and they may spread

---

[1]B. G. Johnston, "Pin-Connected Plate Links," *Transactions ASCE*, 104 (1939).

so much as to cause failure. The steel must be subjected either to stress reversals or to variations in tension stress, because fatigue problems occur only when tension is present. (Sometimes, however, fatigue cracks occur in members that are subjected only to calculated compression stresses if parts of those members have high residual tension stresses.) The results of fatigue loading is that steel members may fail at stresses well below the stresses at which they would fail if they were subject to static loads. The fatigue strength of a particular member is dependent on the number of cycles of stress change, the range of stress change, and the size of flaws.

In Appendix 3 of the AISC Specification, a simple design method is presented for considering fatigue stresses. For this discussion, the term *stress range* is defined as the magnitude of the change in stress in a member due to the application or removal of service live loads. Should there be stress reversal, the stress range equals the numerical sum of the maximum repeated tensile and compressive stresses.

The fatigue life of members increases as the stress range is decreased. Furthermore, at very low stress ranges, the fatigue life is very large. In fact, there is a range at which the member life appears to be infinite. This range is called the *threshold fatigue stress range*.

If it is anticipated that there will be fewer than 20,000 cycles of loading, no consideration needs to be given to fatigue. (Note that three cycles per day for 25 years equals 27,375 cycles.) If the number of cycles is greater than 20,000, an allowable *stress range*, is calculated as specified in Appendix 3.3 of the AISC Specification. Should a member be selected and found to have a design stress range below the actual stress range, it will be necessary to select a larger member.

The following two additional notes pertain to the AISC fatigue design procedure:

1. The design stress range determined in accordance with the AISC requirements is applicable only to the following situations:
   a. Structures for which the steel has adequate corrosion protection for the conditions expected in that locality.
   b. Structures for which temperatures do not exceed 300°F.
2. The provisions of the AISC Specification apply to stresses that are calculated with service loads, and the maximum permitted stress due to these loads is $0.66F_y$.

   Formulas are given, in Appendix 3 of the AISC Specification, for computing the allowable stress range. For the stress categories A, B, B′, C, D, E, and E′ listed in AISC Appendix Table A3.1,

$$F_{SR} = \left(\frac{C_f}{n_{SR}} \right)^{0.333} \geq F_{TH} \qquad \text{(AISC Equation A-3-1)}$$

in which

F_{SR} = allowable stress range, ksi
C_f = constant from Table A-3.1 in AISC Appendix A
n_{SR} = number of stress range fluctuations in design life
    = number of stress range fluctuations per day $\times$ 365 $\times$ years of design life
F_{TH} = threshold allowable stress range, maximum stress range for indefinite design life from AISC Appendix Table A-3.1, ksi

Example 4-6 presents the design of a tension member subjected to fluctuating loads, using Appendix 3 of the AISC Specification. Stress fluctuations and reversals are an everyday problem in the design of bridge structures. The AASHTO Specifications provide allowable stress ranges, determined in a manner similar to that of the AISC Specification.

Example 4-6

A tension member is to consist of a W12 section (F_y = 50 ksi) with fillet-welded end connections. The service dead load is 40 k, while it is estimated that the service live load will vary from a compression of 20 k to a tension of 90 k fifty times per day for an estimated design life of 25 years. Select the section, using the AISC procedure.

Solution

$$P_u = (1.2)(40 \text{ k}) + (1.6)(90 \text{ k}) = 192 \text{ k}$$

Estimated section size for gross section tension yield

$$A_g \geq \frac{P_u}{\phi_t F_y} = \frac{192 \text{ k}}{(0.9)(50 \text{ ksi})} = 4.27 \text{ in}^2$$

Try W12 × 16(A_g = 4.71 in²)

$$n_{SR} = (50)(365)(25) = 456{,}250$$

According to Table A-3.1 in Appendix 3 of the AISC Specification, the member falls into Section 1 of the table and into stress category A.

$$C_f = 250 \times 10^8 \text{ from table}$$

$$F_{TH} = 24 \text{ ksi from table}$$

$$F_{SR} = \left(\frac{C_f}{n_{SR}}\right)^{0.333} = \left(\frac{250 \times 10^8}{456{,}250}\right)^{0.333} = 37.84 \text{ ksi}$$

$$\text{Max service load tension} = \frac{40 \text{ k} + 90 \text{ k}}{4.71 \text{ in}^2} = 27.60 \text{ ksi}$$

$$\text{Min service load tension} = \frac{40 \text{ k} - 20 \text{ k}}{4.71 \text{ in}^2} = 4.25 \text{ ksi}$$

$$\text{Actual stress range} = 27.60 - 4.25 = 23.35 \text{ ksi}$$

$$< F_{SR} = 37.84 \text{ ksi} \qquad \text{(OK)}$$

Use W12 × 16.

4.6   PROBLEMS FOR SOLUTION

For all these problems, select sizes with LRFD expressions and check the selected sections with both the LRFD and the ASD expressions.

4-1 to 4-8.   *Select sections for the conditions described, using $F_y = 50$ ksi and $F_u = 65$ ksi, unless otherwise noted, and neglecting block shear.*

4-1.   Select the lightest W12 section available to support working tensile loads of $P_D = 120$ k and $P_W = 288$ k. The member is to be 20 ft long and is assumed to have two lines of holes for 3/4-in Ø bolts in each flange. There will be at least three holes in each line 3 in on center. (*Ans.* W12 × 45 LRFD and ASD)

4-2.   Repeat Prob. 4-1 selecting a W10 section.

4-3.   Select the lightest WT7 available to support a factored tensile load $P_u = 250$ k, $P_a = 160$ k. Assume there are two lines of 7/8-in Ø bolts in the flange (at least three bolts in each line 4 in on center). The member is to be 30 ft long. (*Ans.* WT7 × 26.5 LRFD, WT7 × 24 ASD)

4-4.   Select the lightest S section that will safely support the service tensile loads $P_D = 75$ k and $P_L = 40$ k. The member is to be 20 ft long and is assumed to have one line of holes for 3/4-in Ø bolts in each flange. Assume that there are at least three holes in each line 4 in on center. Use A36 steel.

4-5.   Select the lightest C section that will safely support the service tensile loads $P_D = 65$ k and $P_L = 50$ k. The member is to be 14 ft long and is assumed to have two lines of holes for 3/4-in Ø bolts in the web. Assume that there are at least three holes in each line 3 in on center. Use A36 steel. (*Ans.* C8 × 18.75 LRFD and ASD)

4-6.   Select the lightest W10 section that will resist a service tensile load, $P_D = 175$ k and $P_L = 210$ k. The member is to be 25 ft long and is assumed to have two lines of holes in each flange and two lines of holes in the web. Assume there are four bolts in each line 3 in on center. All holes are for 7/8-in Ø bolts. Use A992 - Grade 50 steel.

4-7.   Select the lightest C section that will safely support the service tensile loads $P_D = 20$ k and $P_L = 34$ k. The member is to be 12 ft long and is assumed to have only a transverse weld at the end of the channel. A36 steel is used. (*Ans.* C6 × 10.5 LRFD and ASD)

4-8.   Select the lightest MC12 section that will resist a total factored load of 372 kips and a total service load of 248 kips. The member is to be 20 ft long and is assumed to be welded on the end as well as to each flange for a distance of 6 in along the length of the channel. A36 steel is used.

4-9 to 4-16.   *Select the lightest section for each of the situations described in Table 4.1. Assume that bolts are 3 in on center (unless noted otherwise). Do not consider block shear. Determine U from Table 3.2 of this text (except if given).*

TABLE 4.1

Prob. no.	Section	P_D (kips)	P_L (kips)	Length (ft)	Steel	End Connection	Answer
4-9	W8	75	100	24	A992	Two lines of 5/8-in Ø bolts (3 in a line 2 1/2 in on center) each flange	W8 × 28 LRFD and ASD

(Continued)

TABLE 4.1 (*Continued*)

Prob. No.	Section	P_D (kips)	P_L (kips)	Length (ft)	Steel	End Connection	Answer
4-10	W10	120	220	30	A992	Two lines of 3/4-in Ø bolts (3 in a line) each flange	
4-11	W12	150	175	26	A36	Two lines of 7/8-in Ø bolts (2 in a line 4 in on center) each flange	W12 × 58 LRFD W12 × 65 ASD
4-12	W10	135	100	28	A36	Longitudinal weld to flanges only, 6 in long	
4-13	W8	100	80	30	A992	Transverse welds to flanges only	W8 × 24 LRFD W8 × 28 ASD
4-14	S	60	100	22	A36	One line of 3/4-in Ø bolts (3 in a line 4 in on center) each flange	
4-15	WT6	80	120	20	A992	Longitudinal weld to flange only, 6 in long	WT6 × 26.5 LRFD and ASD
4-16	WT4	30	50	18	A36	Transverse weld to flange only	

4-17. Using A36 steel select the lightest equal leg single angle member to resist a tensile load of $P_D = 45$ k, $P_L = 25$ k, and $P_W = 88$ k. The member will be connected through one leg with two lines of three 3/4-in Ø bolts 3 1/2 in on center. The member length is 24 ft. Neglect block shear. (*Ans.* L6 × 6 × 1/2 for LRFD and ASD)

4-18. Select a pair of C10 channels for a tension member subjected to a dead lead of 120 kips and a live load of 275 kips. The channels are placed back to back and connected to a 3/4-in gusset plate by 7/8-in Ø bolts. Assume A588 Grade 50 steel for the channels and assume the gusset plate is sufficient. The member is 25 ft long. The bolts are arranged in two lines parallel to the length of the member. There are two bolts in each line 4 in on center.

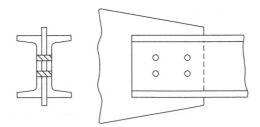

FIGURE P4-18

4-19. Select the lightest C6 channel shape to be used as a 12 ft long tension member to resist the following service loads, $P_D = 20$ k and $P_L = 32$ k. The member will be connected by a transverse weld at the end of the channel only. Use A36 Grade 36 steel with $F_u = 58$ ksi. (*Ans.* C6 × 10.5 LRFD and ASD)

4-20. Design member L_2L_3 of the truss shown in Fig. P4-20. It is to consist of a pair of angles with a 3/8-in gusset plate between the angles at each joint. Use A36 steel and assume two lines of three 3/4-in Ø bolts in each vertical angle leg, 4 in on center. Consider only the angles shown in the double-angle tables of the AISC Manual. For each load, $P_D = 60$ k and $P_L = 48$ k. Do not consider block shear.

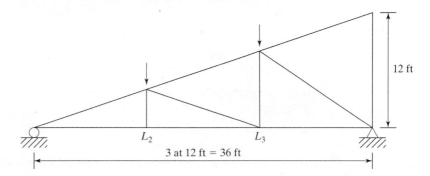

FIGURE P4-20

4-21. Select an ST shape to be used as a 20 ft long tension member that will safely support the service tensile loads: $P_D = 35$ k, $P_L = 115$ k, and $P_S = 65$ k (snow). The connection is through the flange with two lines of three 3/4-in Ø bolts 4 in on center. Use A572 Grade 50 steel. Neglect block shear. (*Ans.* ST10 × 33 LRFD and ASD)

4-22. Select the lightest WT4 shape to be used as a 20 ft long tension member to resist the following service loads; dead load, $D = 20$ k, live load, $L = 35$ k, snow load, $S = 25$ k, and earthquake, $E = 50$ k. The connection is two lines of bolts through the flange with three 3/4-in Ø bolts in each line spaced at 3 in on center. Use A992 Grade 50 steel. Neglect block shear.

4-23. A tension member is to consist of two C10 channels and two PL 1/2 × 11, arranged as shown in Fig. P4-23 to support the service loads, $P_D = 200$ k and $P_L = 320$ k. The member is assumed to be 30 ft long and is to have four lines of 3/4-in Ø bolts. Assume $U = 0.85$. All steel will be A36. Neglect block shear. (*Ans.* 2 − C10 × 25 LRFD and ASD)

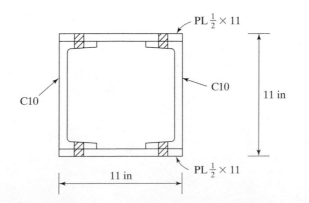

FIGURE P4-23

4-24. A pipe is supported at 25 ft intervals by a pipe strap hung from a threaded rod as shown. A 10-in Ø standard weight steel pipe full of water is used. What size round rod is required? Use A36 steel. Neglect weight of the pipe strap.

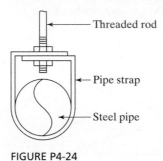

Threaded rod

Pipe strap

Steel pipe

FIGURE P4-24

4-25. Select a standard threaded round rod to support a factored tensile load of 72 kips (service tensile load = 50 kips) using A36 steel. (*Ans.* $1\frac{3}{4}$-in Ø rod LRFD and ASD)

4-26. What size threaded round rod is required for member AC shown in Fig. P4-26? The given load is a service live load. Use A36 steel.

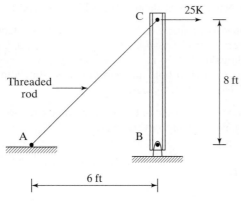

FIGURE P4-26

CHAPTER 5

Introduction to Axially Loaded Compression Members

5.1 GENERAL

There are several types of compression members, the column being the best known. Among the other types are the top chords of trusses and various bracing members. In addition, many other members have compression in some of their parts. These include the compression flanges of rolled beams and built-up beam sections, and members that are subjected simultaneously to bending and compressive loads. Columns are usually thought of as being straight vertical members whose lengths are considerably greater than their thicknesses. Short vertical members subjected to compressive loads are often called struts, or simply, compression members; however, the terms *column* and *compression member* will be used interchangeably in the pages that follow.

There are three general modes by which axially loaded columns can fail. These are flexural buckling, local buckling, and torsional buckling. These modes of buckling are briefly defined as follows:

1. *Flexural buckling* (also called Euler buckling) is the primary type of buckling discussed in this chapter. Members are subject to flexure, or bending, when they become unstable.

2. *Local buckling* occurs when some part or parts of the cross section of a column are so thin that they buckle locally in compression before the other modes of

buckling can occur. The susceptibility of a column to local buckling is measured by the width–thickness ratios of the parts of its cross section. This topic is addressed in Section 5.7.

3. *Flexural torsional buckling* may occur in columns that have certain cross-sectional configurations. These columns fail by twisting (torsion) or by a combination of torsional and flexural buckling. This topic is initially addressed in Section 6.10.

The longer a column becomes for the same cross section, the greater becomes its tendency to buckle and the smaller becomes the load it will support. The tendency of a member to buckle is usually measured by its *slenderness ratio*, which has previously been defined as the ratio of the length of the member to its least radius of gyration. This tendency to buckle is also affected by such factors as the types of end connections, eccentricity of load application, imperfection of column material, initial crookedness of columns, and residual stresses from manufacture.

The loads supported by a building column are applied by the column section above and by the connections of other members directly to the column. The ideal situation is for the loads to be applied uniformly across the column, with the center of gravity of the loads coinciding with the center of gravity of the column. Furthermore, it is desirable for the column to have no flaws, to consist of a homogeneous material, and to be perfectly straight, but these situations are obviously impossible to achieve.

Loads that are exactly centered over a column are referred to as *axial*, or *concentric loads*. The dead loads may or may not be concentrically placed over an interior building column, and the live loads may never be centered. For an outside column, the load situation is probably even more eccentric, as the center of gravity of the loads will often fall well on the inner side of the column. In other words, it is doubtful that a perfect axially loaded column will ever be encountered in practice.

The other desirable situations are also impossible to achieve because of the following conditions: imperfections of cross-sectional dimensions, residual stresses, holes punched for bolts, erection stresses, and transverse loads. It is difficult to take into account all of these imperfections in a formula.

Slight imperfections in tension members and beams can be safely disregarded, as they are of little consequence. On the other hand, slight defects in columns may be of major significance. A column that is slightly bent at the time it is put in place may have significant bending moments equal to the column load times the initial lateral deflection. Mill straightness tolerances, as taken from ASTM A6, are presented in Tables 1-22 through 1-28 of the AISC Manual.

Obviously, a column is a more critical member in a structure than is a beam or tension member, because minor imperfections in materials and dimensions mean a great deal. This fact can be illustrated by a bridge truss that has some of its members damaged by a truck. The bending of tension members probably will not be serious, as the tensile loads will tend to straighten those members; but the bending of any

Two International Place, Boston, MA. (Courtesy of Owen Steel Company, Inc.)

compression members is a serious matter, as compressive loads will tend to magnify the bending in those members.

The preceding discussion should clearly show that column imperfections cause them to bend, and the designer must consider stresses due to those moments as well as those due to axial loads. Chapters 5 to 7 are limited to a discussion of axially loaded columns, while members subjected to a combination of axial loads and bending loads are discussed in Chapter 11.

The spacing of columns in plan establishes what is called a *bay*. For instance, if the columns are 20 ft on center in one direction and 25 ft in the other direction, the bay size is 20 ft × 25 ft. Larger bay sizes increase the user's flexibility in space planning. As to economy, a detailed study by John Ruddy[1] indicates that when shallow spread footings are used, bays with length-to-width ratios of about 1.25 to 1.75, and areas of about 1000 sq ft, are the most cost efficient. When deep foundations are used, his study shows that larger bay areas are more economical.

5.2   RESIDUAL STRESSES

Research at Lehigh University has shown that residual stresses and their distribution are very important factors affecting the strength of axially loaded steel columns. These stresses are of particular importance for columns with slenderness ratios varying from approximately 40 to 120, a range that includes a very large percentage of practical columns. A major cause of residual stress is the uneven cooling of shapes after hot-rolling. For instance, in a W shape the outer tips of the flanges and the middle of the web cool quickly, while the areas at the intersection of the flange and web cool more slowly.

The quicker cooling parts of the sections, when solidified, resist further shortening, while those parts that are still hot tend to shorten further as they cool. The net result is that the areas that cooled more quickly have residual compressive stresses, and the slower cooling areas have residual tensile stresses. The magnitude of these stresses varies from about 10 to 15 ksi (69 to 103 MPa), although some values greater than 20 ksi (138 MPa), have been found.

When rolled-steel column sections with their residual stresses are tested, their proportional limits are reached at P/A values of only a little more than half of their yield stresses, and the stress–strain relationship is nonlinear from there up to the yield stress. Because of the early localized yielding occurring at some points of the column cross sections, buckling strengths are appreciably reduced. Reductions are greatest for columns with slenderness ratios varying from approximately 70 to 90 and may possibly be as high as 25 percent.[2]

As a column load is increased, some parts of the column will quickly reach the yield stress and go into the plastic range because of residual compression stresses. The stiffness of the column will be reduced and become a function of the part of the cross section that is still elastic. A column with residual stresses will behave as though it has a reduced cross section. This reduced section or elastic portion of the column will change as the applied stresses change. The buckling calculations for a particular column with residual stresses can be handled by using an effective moment of inertia I_e of the elastic portion of the cross section or by using the tangent modulus. For the usual sections used as columns, the two methods give fairly close results.

---

[1]J. L. Ruddy, "Economics of Low-Rise Steel-Framed Structures," *Engineering Journal,* AISC, vol. 20, no. 3 (3d quarter, 1983), pp. 107–118.

[2]L. S. Beedle and L. Tall, "Basic Column Strength," *Proc. ASCE 86* (July 1960), pp. 139–173.

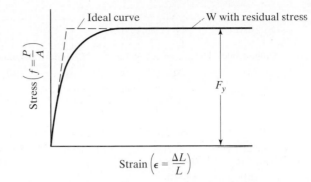

FIGURE 5.1

Effect of residual stresses on column stress–strain diagram.

In columns, welding can produce severe residual stresses that actually may approach the yield point in the vicinity of the weld. Another important fact is that columns may be appreciably bent by the welding process, which can decidedly affect their load-carrying ability. Figure 5.1 shows the effect of residual stresses (due to cooling and fabrication) on the stress–strain diagram for a hot-rolled W shape.

The welding together of built-up shapes frequently causes even higher residual stresses than those caused by the uneven cooling of hot-rolled I-shaped sections.

Residual stresses may also be caused during fabrication when *cambering* is performed by cold bending, and due to cooling after welding. Cambering is the bending of a member in a direction opposite to the direction of bending that will be caused by

the service loads. For instance, we may initially bend a beam upward 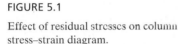 so that it will be approximately straight when its normal gravity loads are applied

.

5.3   SECTIONS USED FOR COLUMNS

Theoretically, an endless number of shapes can be selected to safely resist a compressive load in a given structure. From a practical viewpoint, however, the number of possible solutions is severely limited by such considerations as sections available, connection problems, and type of structure in which the section is to be used. The paragraphs that follow are intended to give a brief résumé of the sections which have proved to be satisfactory for certain conditions. These sections are shown in Fig. 5.2, and the letters in parentheses in the paragraphs to follow refer to the parts of this figure.

The sections used for compression members usually are similar to those used for tension members, with certain exceptions. The exceptions are caused by the fact that the strengths of compression members vary in some inverse relation to the slenderness ratios, and stiff members are required. Individual rods, bars, and plates usually are too slender to make satisfactory compression members, unless they are very short and lightly loaded.

Single-angle members (a) are satisfactory for use as bracing and compression members in light trusses. Equal-leg angles may be more economical than unequal-leg angles, because their least r values are greater for the same area of steel. The top chord

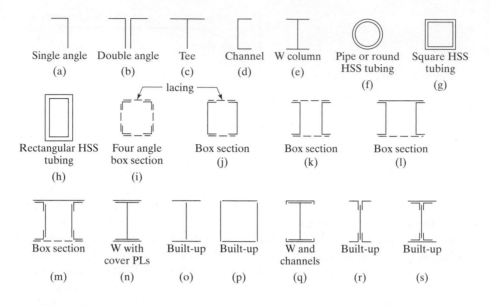

FIGURE 5.2

Types of compression members.

members of bolted roof trusses might consist of a pair of angles back to back (b). There will often be a space between them for the insertion of a gusset or connection plate at the joints necessary for connections to other members. An examination of this section will show that it is probably desirable to use unequal-leg angles with the long legs back to back to give a better balance between the r values about the x and y axes.

If roof trusses are welded, gusset plates may be unnecessary, and structural tees (c) might be used for the top chord compression members because the web members can be welded directly to the stems of the tees. Single channels (d) are not satisfactory for the average compression member because of their almost negligible r values about their web axes. They can be used if some method of providing extra lateral support in the weak direction is available. The W shapes (e) are the most common shapes used for building columns and for the compression members of highway bridges. Their r values, although far from being equal about the two axes, are much more nearly balanced than are the same values for channels.

Several famous bridges constructed during the nineteenth century (such as the Firth of Forth Bridge in Scotland and Ead's Bridge in St. Louis) made extensive use of tube-shaped members. Their use, however, declined due to connection problems and manufacturing costs, but with the development of economical welded tubing, the use of tube-shaped members is again increasing (although the tube shapes of today are very small compared with the giant ones used in those early steel bridges).

Hollow structural sections (square, rectangular, or round) and steel pipe are very valuable sections for buildings, bridges, and other structures. These clean, neat-looking sections are easily fabricated and erected. For small and medium loads, the round sections (f) are quite satisfactory. They are often used as columns in long series of windows, as short columns in warehouses, as columns for the roofs of covered walkways, in the basements and garages of residences, and in other applications. Round columns have the advantage of being equally rigid in all directions and are usually very economical, unless moments are too large for the sizes available. The AISC Manual furnishes the sizes of these sections and classifies them as being either round HSS sections or standard, extra strong, or double extra strong steel pipe.

Square and rectangular tubing (g) and (h) are being used more each year. For many years, only a few steel mills manufactured steel tubing for structural purposes. Perhaps the major reason tubing was not used to a great extent is the difficulty of making connections with rivets or bolts. This problem has been fairly well eliminated, however, by the advent of modern welding. The use of tubing for structural purposes by architects and engineers in the years to come will probably be greatly increased for several reasons:

1. The most efficient compression member is one that has a constant radius of gyration about its centroid, a property available in round HSS tubing and pipe sections. Square tubing is the next-most-efficient compression member.
2. Four-sided and round sections are much easier to paint than are the six-sided open W, S, and M sections. Furthermore, the rounded corners make it easier to apply paint or other coatings uniformly around the sections.
3. They have less surface area to paint or fireproof.
4. They have excellent torsional resistance.
5. The surfaces of tubing are quite attractive.
6. When exposed, the round sections have wind resistance of only about two-thirds of that of flat surfaces of the same width.
7. If cleanliness is important, hollow structural tubing is ideal, as it doesn't have the problem of dirt collecting between the flanges of open structural shapes.

A slight disadvantage that comes into play in certain cases is that the ends of tube and pipe sections that are subject to corrosive atmospheres may have to be sealed to protect their inaccessible inside surfaces from corrosion. Although making very attractive exposed members for beams, these sections are at a definite weight disadvantage compared with W sections, which have so much larger resisting moments for the same weights.

For many column situations, the weight of square or rectangular tube sections—usually referred to as *hollow structural sections* (HSS)—can be less than one-half the weights required for open-profile sections (W, M, S, channel, and angle sections). It is true that tube and pipe sections may cost perhaps 25 percent more per pound than open sections, but this still allows the possibility of up to 20 percent savings in many cases.[3]

Hollow structural sections are available with yield strengths up to 50 ksi and with enhanced atmospheric corrosion resistance. Today, they make up a small percentage of

---

[3]"High Design, Low Cost," *Modern Steel Construction* (Chicago: AISC, March–April 1990), pp. 32–34.

Administration Building Pensacola
Christian College, FL. (Courtesy
Britt, Peters and Associates.)

the structural steel fabricated for buildings and bridges in the United States. In Japan and Europe, the values are, respectively, 15 percent and 25 percent, and still growing. Thus, it is probable that their use will continue to rise in the United States in the years to come.[4] Detailed information, including various tables on hollow structural sections, can be obtained from the Steel Tube Institute (STI), 2000 Ponce de Leon, Suite 600, Coral Gables, Florida 33134.

Where compression members are designed for very large structures, it may be necessary to use built-up sections. Built-up sections are needed where the members are long and support very heavy loads and/or when there are connection advantages. Generally speaking, a single shape, such as a W section, is more economical than a built-up section having the same cross-sectional area. For heavy column loads, high-strength steels can

[4]Ibid., p. 34.

frequently be used with very economical results if their increased strength permits the use of W sections rather than built-up members.

When built-up sections are used, they must be connected on their open sides with some type of lacing (also called *lattice bars*) to hold the parts together in their proper positions and to assist them in acting together as a unit. The ends of these members are connected with *tie* plates (also called *batten* plates or *stay* plates). Several types of lacing for built-up compression members are shown in Fig. 6.9.

The dashed lines in Fig. 5.2 represent lacing or discontinuous parts, and the solid lines represent parts that are continuous for the full length of the members. Four angles are sometimes arranged as shown in (i) to produce large *r* values. This type of member may often be seen in towers and in crane booms. A pair of channels (j) is sometimes used as a building column or as a web member in a large truss. It will be noted that there is a certain spacing for each pair of channels at which their *r* values about the *x* and *y* axes are equal. Sometimes the channels may be turned out, as shown in (k).

A section well suited for the top chords of bridge trusses is a pair of channels with a cover plate on top (l) and with lacing on the bottom. The gusset or connection plates at joints are conveniently connected to the insides of the channels and may also be used as splices. When the largest channels available will not produce a top chord member of sufficient strength, a built-up section of the type shown in (m) may be used.

When the rolled shapes do not have sufficient strength to resist the column loads in a building or the loads in a very large bridge truss, their areas may be increased by plates added to the flanges (n). In recent years, it has been found that, for welded construction, a built-up column of the type shown in part (o) is a more satisfactory shape than a W with welded cover plates (n). It seems that in bending (as where a beam frames into the flange of a column) it is difficult to efficiently transfer tensile force through the cover plate to the column without pulling the plate away from the column. For very heavy column loads, a welded box section of the type shown in (p) has proved to be quite satisfactory. Some other built-up sections are shown in parts (q) through (s). The built-up sections shown in parts (n) through (q) have an advantage over those shown in parts (i) through (m) in that they do not require the expense of the lattice work necessary for some of the other built-up sections. Lateral shearing forces are negligible for the single column shapes and for the nonlatticed built-up sections, *but they are definitely not negligible for the built-up latticed columns.*

Today, *composite columns* are being increasingly used. These columns usually consist of steel pipe and structural tubing filled with concrete, or of W shapes encased in concrete, usually square or rectangular in cross section. (These columns are discussed in Chapter 17.)

5.4   DEVELOPMENT OF COLUMN FORMULAS

The use of columns dates to before the dawn of history, but it was not until 1729 that a paper was published on the subject, by Pieter van Musschenbroek, a Dutch mathematician.[5] He presented an empirical column formula for estimating the strength of

---

[5]L. S. Beedle et al., *Structural Steel Design* (New York: Ronald Press, 1964), p. 269.

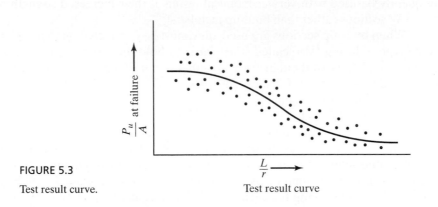

FIGURE 5.3

Test result curve.

Test result curve

rectangular columns. A few years later, in 1757, Leonhard Euler, a Swiss mathematician, wrote a paper of great value concerning the buckling of columns. He was probably the first person to realize the significance of buckling. The Euler formula, the most famous of all column expressions, is derived in Appendix A of this text. This formula, which is discussed in the next section, marked the real beginning of theoretical and experimental investigation of columns.

Engineering literature is filled with formulas developed for ideal column conditions, but these conditions are not encountered in actual practice. Consequently, practical column design is based primarily on formulas that have been developed to fit, with reasonable accuracy, test-result curves. The reasoning behind this procedure is simply that the independent derivation of column expressions does not produce formulas that give results which compare closely with test-result curves for all slenderness ratios.

The testing of columns with various slenderness ratios results in a scattered range of values, such as those shown by the broad band of dots in Fig. 5.3. The dots will not fall on a smooth curve, even if all of the testing is done in the same laboratory, because of the difficulty of exactly centering the loads, lack of perfect uniformity of the materials, varying dimensions of the sections, residual stresses, end restraint variations, and other such issues. The usual practice is to attempt to develop formulas that give results representative of an approximate average of the test results. The student should also realize that laboratory conditions are not field conditions, and column tests probably give the limiting values of column strengths.

The magnitudes of the yield stresses of the sections tested are quite important for short columns, as their failure stresses are close to those yield stresses. For columns with intermediate slenderness ratios, the yield stresses are of lesser importance in their effect on failure stresses, and they are of no significance for long slender columns. For intermediate range columns, residual stresses have more effect on the results, while the failure stresses for long slender columns are very sensitive to end support conditions. In addition to residual stresses and nonlinearity of material, another dominant factor in its effect on column strength is member out-of-straightness.

5.5   THE EULER FORMULA

Obviously, the stress at which a column buckles decreases as the column becomes longer. After it reaches a certain length, that stress will have fallen to the proportional limit of the steel. For that length and greater lengths, the buckling stress will be elastic.

For a column to buckle elastically, it will have to be long and slender. Its buckling load P can be computed with the Euler formula that follows:

$$P = \frac{\pi^2 EI}{L^2}.$$

This formula usually is written in a slightly different form that involves the column's slenderness ratio. Since $r = \sqrt{I/A}$, we can say that $I = Ar^2$. Substituting this value into the Euler formula and dividing both sides by the cross-sectional area, the Euler buckling stress is obtained:

$$\frac{P}{A} = \frac{\pi^2 E}{(L/r)^2} = F_e.$$

Example 5-1 illustrates the application of the Euler formula to a steel column. If the value obtained for a particular column exceeds the steel's proportional limit, the elastic Euler formula is not applicable.

Example 5-1

(a)   A W10 × 22 is used as a 15-ft long pin-connected column. Using the Euler expression, determine the column's critical or buckling load. Assume that the steel has a proportional limit of 36 ksi.

(b)   Repeat part (a) if the length is changed to 8 ft.

Solution

(a)   Using a 15-ft long W10 × 22 ($A = 6.49 \text{ in}^2$, $r_x = 4.27 \text{ in}$, $r_y = 1.33 \text{ in}$)

   Minimum $r = r_y = 1.33 \text{ in}$

$$\frac{L}{r} = \frac{(12 \text{ in/ft})(15 \text{ ft})}{1.33 \text{ in}} = 135.34$$

   Elastic or buckling stress $F_e = \dfrac{(\pi^2)(29 \times 10^3 \text{ ksi})}{(135.34)^2}$

$$= 15.63 \text{ ksi} < \text{the proportional limit of 36 ksi}$$

   OK column is in elastic range

   Elastic or buckling load $= (15.63 \text{ ksi})(6.49 \text{ in}^2) = 101.4 \text{ k}$

(b)   Using an 8-ft long W10 × 22,

$$\frac{L}{r} = \frac{(12 \text{ in/ft})(8 \text{ ft})}{1.33 \text{ in}} = 72.18$$

   Elastic or buckling stress $F_e = \dfrac{(\pi^2)(29 \times 10^3 \text{ ksi})}{(72.18)^2} = 54.94 \text{ ksi} > 36 \text{ ksi}$

   $\therefore$ column is in inelastic range and Euler equation is not applicable.

The student should carefully note that the buckling load determined from the Euler equation is independent of the strength of the steel used.

The Euler equation is useful only if the end support conditions are carefully considered. The results obtained by application of the formula to specific examples compare very well with test results for concentrically loaded, long, slender columns with pinned ends. Designers, however, do not encounter perfect columns of this type. The columns with which they work do not have pinned ends and are not free to rotate, because their ends are bolted or welded to other members. These practical columns have different amounts of restraint against rotation, varying from slight restraint to almost fixed conditions. For the actual cases encountered in practice where the ends are not free to rotate, different length values can be used in the formula, and more realistic buckling stresses will be obtained.

To successfully use the Euler equation for practical columns, the value of L should be the distance between points of inflection in the buckled shape. This distance is referred to as the *effective length* of the column. For a pinned-end column (whose ends can rotate, but cannot translate), the points of inflection, or zero moment, are located at the ends a distance L apart. For columns with different end conditions, the effective lengths may be entirely different. Effective lengths are discussed extensively in the next section.

450 Lexington Ave., New York City. (Courtesy of Owen Steel Company, Inc.)

5.6   END RESTRAINT AND EFFECTIVE LENGTHS OF COLUMNS

End restraint and its effect on the load-carrying capacity of columns is a very important subject indeed. Columns with appreciable rotational and translational end restraint can support considerably more load than can those with little rotational end restraint, as at hinged ends.

The effective length of a column is defined in the previous section as the distance between points of zero moment in the column, that is, the distance between its inflection points. In steel specifications, the effective length of a column is referred to as KL, where K is the *effective length factor*. K is the number that must be multiplied by the length of the column to find its effective length. Its magnitude depends on the rotational restraint supplied at the ends of the column and upon the resistance to lateral movement provided.

The concept of effective lengths is simply a mathematical method of taking a column, whatever its end and bracing conditions, and replacing it with an equivalent pinned-end braced column. A complex buckling analysis could be made for a frame to determine the critical stress in a particular column. The K factor is determined by finding the pinned-end column with an equivalent length that provides the same critical stress. The K factor procedure is a method of making simple solutions for complicated frame-buckling problems.

Columns with different end conditions have entirely different effective lengths. For this initial discussion, it is assumed that no sidesway or joint translation is possible between the member ends. Sidesway or joint translation means that one or both ends of a column can move laterally with respect to each other. Should a column be connected with frictionless hinges, as shown in Fig. 5.4(a), its effective length would be equal to the actual length of the column and K would equal 1.0. If there were such a thing as a perfectly fixed-ended column, its points of inflection (or points of zero moment) would occur at its one-fourth points and its effective length would equal $L/2$, as shown in Fig. 5.4(b). As a result, its K value would equal 0.50.

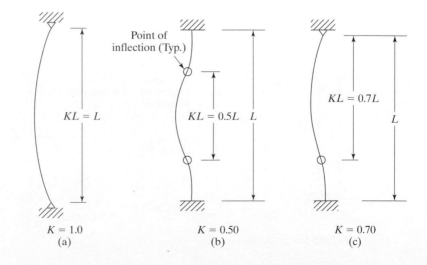

FIGURE 5.4

Effective length (KL) for columns in braced frames (sidesway prevented).

Obviously, the smaller the effective length of a particular column, the smaller its danger of lateral buckling and the greater its load-carrying capacity will be. In Fig. 5.4(c), a column is shown with one end fixed and one end pinned. The K value for this column is theoretically equal to 0.70.

This discussion would seem to indicate that column effective lengths always vary from an absolute minimum of $L/2$ to an absolute maximum of L, but there are many exceptions to this rule. An example is given in Fig. 5.5(a), where a simple bent is shown. The base of each of the columns is pinned, and the other end is free to rotate and move laterally (called *sidesway*). Examination of this figure will show that the effective length will exceed the actual length of the column as the elastic curve will theoretically take the shape of the curve of a pinned-end column of twice its length and K will theoretically equal 2.0. Notice, in part (b) of the figure, how much smaller the lateral deflection of column AB would be if it were pinned at both top and bottom so as to prevent sidesway.

Structural steel columns serve as parts of frames, and these frames are sometimes *braced* and sometimes *unbraced*. A braced frame is one for which sidesway or joint translation is prevented by means of bracing, shear walls, or lateral support from adjoining structures. An unbraced frame does not have any of these types of bracing supplied and must depend on the stiffness of its own members and the rotational rigidity of the joints between the frame members to prevent lateral buckling. For braced frames, K values can never be greater than 1.0, but for unbraced frames, the K values will always be greater than 1.0 because of sidesway.

Table C-C2.2 of the AISC Commentary on the Specification provides recommended effective length factors when ideal conditions are approximated. This table is reproduced here as Table 5.1 with the permission of the AISC. Two sets of K values are provided in the table, the theoretical values and the recommended design values, based on the fact that perfectly pinned and fixed conditions are not possible. If the ends of

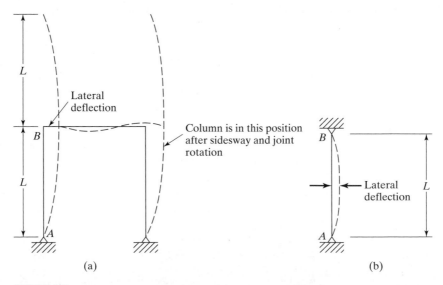

FIGURE 5.5

TABLE 5.1	Approximate Values of Effective Length Factor, *K*					

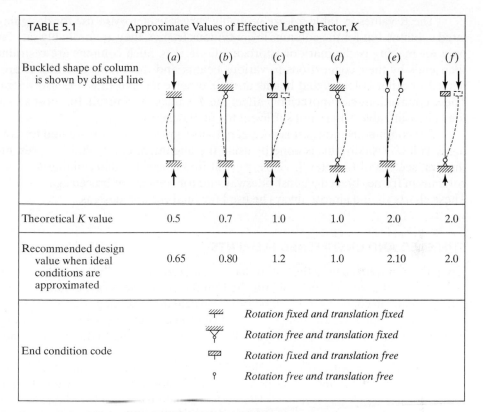

	(a)	(b)	(c)	(d)	(e)	(f)
Buckled shape of column is shown by dashed line						
Theoretical *K* value	0.5	0.7	1.0	1.0	2.0	2.0
Recommended design value when ideal conditions are approximated	0.65	0.80	1.2	1.0	2.10	2.0

End condition code	Rotation fixed and translation fixed
	Rotation free and translation fixed
	Rotation fixed and translation free
	Rotation free and translation free

Source: Commentary on the Specification, Appendix 7 – Table C-A-7.1, p. 16.1-511, June 22, 2010. Copyright © American Institute of Steel Construction. Reprinted with permission. All rights reserved.

the column of Fig. 5.4(b) were not quite fixed, the column would be a little freer to bend laterally, and its points of inflection would be farther apart. The recommended design *K* given in Table 5.1 is 0.65, while the theoretical value is 0.5. As no column ends are perfectly fixed or perfectly hinged, the designer may wish to interpolate between the values given in the table, basing the interpolation on his or her judgment of the actual restraint conditions.

The values in Table 5.1 are very useful for preliminary designs. When using this table, we will almost always apply the design values, and not the theoretical values. In fact, the theoretical values should be used only for those very rare situations where fixed ends are really almost perfectly fixed and/or when simple supports are almost perfectly frictionless. (*This means almost never.*)

You will note in the table that, for cases (a), (b), (c), and (e), the design values are greater than the theoretical values, but this is not the situation for cases (d) and (f), where the values are the same. The reason, in each of these two latter cases, is that if the pinned conditions are not perfectly frictionless, the *K* values will become smaller, not larger. Thus, by making the design values the same as the theoretical ones, we are staying on the safe side.

The *K* values in Table 5.1 are probably very satisfactory to use for designing isolated columns, but for the columns in continuous frames, they are probably satisfactory only for making preliminary or approximate designs. Such columns are restrained at their ends by their connections to various beams, and the beams themselves are connected to other columns and beams at their other ends and thus are also restrained. These connections can appreciably affect the *K* values. As a result, for most situations, the values in Table 5.1 are not sufficient for final designs.

For continuous frames, it is necessary to use a more accurate method for computing *K* values. Usually, this is done by using the alignment charts that are presented in the first section of Chapter 7. There, we will find charts for determining *K* values for columns in frames braced against sidesway and for frames not braced against sidesway. These charts should almost always be used for final column designs.

5.7   STIFFENED AND UNSTIFFENED ELEMENTS

Up to this point in the text, the author has considered only the overall stability of members, and yet it is entirely possible for the thin flanges or webs of a column or beam to buckle locally in compression well before the calculated buckling strength of the whole member is reached. When thin plates are used to carry compressive stresses, they are particularly susceptible to buckling about their weak axes due to the small moments of inertia in those directions.

The AISC Specification (Section B4) provides limiting values for the width–thickness ratios of the individual parts of compression members and for the parts of beams in their compression regions. The student should be well aware of the lack of stiffness of thin pieces of cardboard or metal or plastic with free edges. If, however, one of these elements is folded or restrained, its stiffness is appreciably increased. For this reason, two categories are listed in the AISC Manual: *stiffened elements* and *unstiffened elements*.

An unstiffened element is a projecting piece with one free edge parallel to the direction of the compression force, while a stiffened element is supported along the two edges in that direction. These two types of elements are illustrated in Fig. 5.6. In each case, the width, *b*, and the thickness, *t*, of the elements in question are shown.

Depending on the ranges of different width–thickness ratios for compression elements, and depending on whether the elements are stiffened or unstiffened, the elements will buckle at different stress situations.

For establishing width–thickness ratio limits for the elements of compression members, the AISC Specification divides members into three classifications, as follows: compact sections, noncompact sections, and slender compression elements. These classifications, which decidedly affect the design compression stresses to be used for columns, are discussed in the paragraphs to follow.

5.7.1   Classification of Compression Sections for Local Buckling

Compression sections are classified as either a nonslender element or a slender element. A nonslender element is one where the width-to-thickness of its compression elements does not exceed λ_r, from Table B4.1a of the AISC Specification. When the width-to-thickness ratio does exceed λ_r, the section is defined as a slender-element

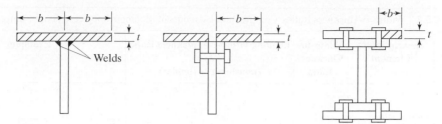

(a) Unstiffened elements

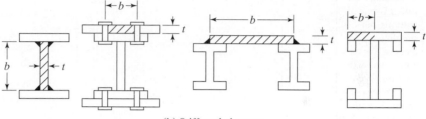

(b) Stiffened elements

FIGURE 5.6

section. The limiting values for λ_r are given in Table 5.2 of this text, which is Table B4.1a of the AISC Specification.

Almost all of the W and HP shapes listed in the Compression Member Section of the AISC Manual are nonslender for 50 ksi yield stress steels. A few of them are slender (and are so indicated in the column tables of the Manual). The values in the tables reflect the reduced design stresses available for slender sections.

If the member is defined as a nonslender element compression member, we should refer to Section E3 of the AISC Specification. The nominal compressive strength is then determined based only on the limit state of flexural buckling.

When the member is defined as a slender element compression member, the nominal compressive strength shall be taken as the lowest value based on the limit states of flexural buckling, torsional buckling, and flexural–torsional buckling. We should refer to Section E7 of the AISC Specification for this condition. Section 6.9 of this text presents an illustration of the determination of the design and allowable strengths of a column which contains slender elements.

5.8   LONG, SHORT, AND INTERMEDIATE COLUMNS

A column subject to an axial compression load will shorten in the direction of the load. If the load is increased until the column buckles, the shortening will stop and the column will suddenly bend or deform laterally and may at the same time twist in a direction perpendicular to its longitudinal axis.

TABLE 5.2 Width-to-Thickness Ratios: Compression Elements in Members Subject to Axial Compression

	Case	Description of Element	Width-to-Thickness Ratio	Limiting Width-to-Thickness Ratio λ_r (nonslender / slender)	Examples
Unstiffened Elements	1	Flanges of rolled I-shaped sections, plates projecting from rolled I-shaped sections, out-standing legs of pairs of angles connected with continuous con-tact, flanges of channels, and flanges of tees	b/t	$0.56\sqrt{\dfrac{E}{F_y}}$	
	2	Flanges of built-up I-shaped sec-tions and plates or angle legs projecting from built-up I-shaped sections	b/t	$0.64\sqrt{\dfrac{k_c E}{F_y}}^{[a]}$	
	3	Legs of single angles, legs of double angles with separators, and all other unstiffened elements	b/t	$0.45\sqrt{\dfrac{E}{F_y}}$	
	4	Stems of tees	d/t	$0.75\sqrt{\dfrac{E}{F_y}}$	

TABLE 5.2 Continued

	Case	Description of Element	Width-to-Thickness Ratio	Limiting Width-to-Thickness Ratio λ_r (nonslender / slender)	Examples
Stiffened Elements	5	Webs of doubly symmetric I-shaped sections and channels	h/t_w	$1.49\sqrt{\frac{E}{F_y}}$	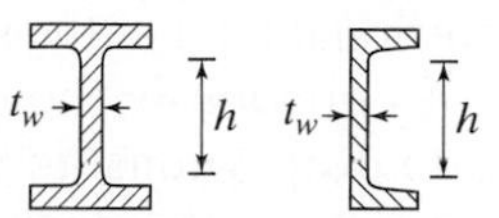
	6	Walls of rectangular HSS and boxes of uniform thickness	b/t	$1.40\sqrt{\frac{E}{F_y}}$	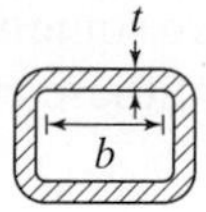
	7	Flange cover plates and diaphragm plates between lines of fasteners or welds	b/t	$1.40\sqrt{\frac{E}{F_y}}$	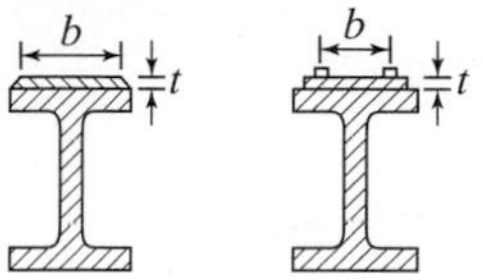
	8	All other stiffened elements	b/t	$1.49\sqrt{\frac{E}{F_y}}$	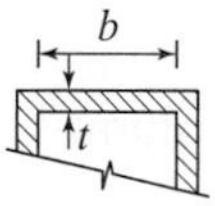
	9	Round HSS	D/t	$0.11\frac{E}{F_y}$	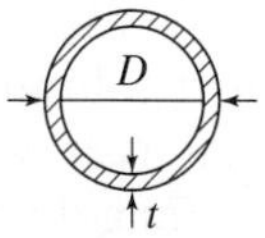

Source: AISC Specification, Table B4.1A, p. 16.1-16. June 22, 2010. Copyright © American Institute of Steel Construction. Reprinted with permission. All rights reserved.

The strength of a column and the manner in which it fails are greatly dependent on its effective length. A very short, stocky steel column may be loaded until the steel yields and perhaps on into the strain-hardening range. As a result, it can support about the same load in compression that it can in tension.

As the effective length of a column increases, its buckling stress will decrease. If the effective length exceeds a certain value, the buckling stress will be less than the proportional limit of the steel. Columns in this range are said to fail *elastically*.

As previously shown in Section 5.5, very long steel columns will fail at loads that are proportional to the bending rigidity of the column (EI) and independent of the strength of the steel. For instance, a long column constructed with a 36-ksi yield stress steel will fail at just about the same load as one constructed with a 100-ksi yield stress steel.

Columns are sometimes classed as being long, short, or intermediate. A brief discussion of each of these classifications is presented in the paragraphs to follow.

5.8.1   Long Columns

The Euler formula predicts very well the strength of long columns where the axial buckling stress remains below the proportional limit. Such columns will buckle *elastically*.

5.8.2   Short Columns

For very short columns, the failure stress will equal the yield stress and no buckling will occur. (For a column to fall into this class, it would have to be so short as to have no practical application. Thus, no further reference is made to them here.)

5.8.3   Intermediate Columns

For intermediate columns, some of the fibers will reach the yield stress and some will not. The members will fail by both yielding and buckling, and their behavior is said to be *inelastic*. Most columns fall into this range. (For the Euler formula to be applicable to such columns, it would have to be modified according to the reduced modulus concept or the tangent modulus concept to account for the presence of residual stresses.)

In Section 5.9, formulas are presented with which the AISC estimates the strength of columns in these different ranges.

5.9   COLUMN FORMULAS

The AISC Specification provides one equation (the Euler equation) for long columns with elastic buckling and an empirical parabolic equation for short and intermediate columns. With these equations, a flexural buckling stress, F_{cr}, is determined for a compression member. Once this stress is computed for a particular member, it is multiplied by the cross-sectional area of the member to obtain its nominal strength P_n. The LRFD design strength and ASD allowable strength of a column may be determined as follows:

$$P_n = F_{cr}A_g \qquad \text{(AISC Equation E3-1)}$$

$$\phi_c P_n = \phi_c F_{cr}A_g = \text{LRFD compression strength } (\phi_c = 0.90)$$

$$\frac{P_n}{\Omega_c} = \frac{F_{cr}A_g}{\Omega_c} = \text{ASD allowable compression strength } (\Omega_c = 1.67)$$

The following expressions show how F_{cr}, the flexural buckling stress of a column, may be determined for members without slender elements:

(a) If $\dfrac{KL}{r} \leq 4.71\sqrt{\dfrac{E}{F_y}} \left(\text{or } \dfrac{F_y}{F_e} \leq 2.25\right)$

$$F_{cr} = \left[0.658^{\frac{F_y}{F_e}}\right]F_y \qquad \text{(AISC Equation E3-2)}$$

(b) If $\dfrac{KL}{r} > 4.71\sqrt{\dfrac{E}{F_y}} \left(\text{or } \dfrac{F_y}{F_e} \leq 2.25\right)$

$$F_{cr} = 0.877F_e \qquad \text{(AISC Equation E3-3)}$$

In these expressions, F_e is the elastic critical buckling stress—that is, the Euler stress—calculated with the effective length of the column KL.

$$F_e = \frac{\pi^2 E}{\left(\dfrac{KL}{r}\right)^2} \qquad \text{(AISC Equation E3-4)}$$

These equations are represented graphically in Fig. 5.7.

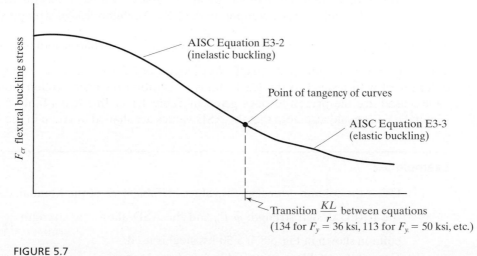

FIGURE 5.7

AISC column curve.

After looking at these column equations, the reader might think that their use would be tedious and time consuming with a pocket calculator. Such calculations, however, rarely have to be made, because the AISC Manual provides computed values of critical stresses $\phi_c F_{cr}$ and $\dfrac{F_{cr}}{\Omega_c}$ in their Table 4-22. The values are given for practical KL/r values (0 to 200) and for steels with F_y = 35, 36, 42, 46, and 50 ksi.

5.10   MAXIMUM SLENDERNESS RATIOS

The AISC Specification no longer provides a specific maximum slenderness ratio, as it formerly did and as is the practice of many other specifications. The AISC Commentary (E2) does indicate, however, that if KL/r is >200, the critical stress F_{cr} will be less than 6.3 ksi. In the past, the AISC maximum permitted KL/r was 200. That value was based on engineering judgment, practical economics, and the fact that special care had to be taken to keep from injuring such a slender member during fabrication, shipping, and erection. As a result of these important practical considerations, the engineer applying the 2010 AISC Specification will probably select compression members with slenderness values below 200, except in certain special situations. For those special cases, both the fabricators and the erectors will be on notice to use extra-special care in handling the members.

5.11   EXAMPLE PROBLEMS

In this section, four simple numerical column problems are presented. In each case, the design strength of a column is calculated. In Example 5-2(a), we determine the strength of a W section. The value of K is calculated as described in Section 5.6, the effective slenderness ratio is computed, and the available critical stresses $\phi_c F_{cr}$ and $\dfrac{F_{cr}}{\Omega}$ are obtained from Table 4-22 in the Manual.

It will be noted that the Manual, in Part 4, Tables 4-1 to 4-11, has further simplified the needed calculations by computing the LRFD column design strengths $(\phi_c P_n)$ and the ASD allowable column strengths $\left(\dfrac{P_n}{\Omega_c}\right)$ for each of the shapes normally employed as columns for the commonly used effective lengths or KL values. These strengths were determined with respect to the least radius of gyration for each section, and the F_y values used are the preferred ones given in Table 1.1 of this text (Table 2-3 in the Manual). You should also note that the ASD values are shaded in green in the Manual.

Example 5-2

(a) Using the column critical stress values in Table 4-22 of the Manual, determine the LRFD design strength $\phi_c P_n$ and the ASD allowable strength $\dfrac{P_n}{\Omega_c}$ for the column shown in Fig. 5.8, if a 50-ksi steel is used.

(b) Repeat the problem, using Table 4-1 of the Manual.

(c) Calculate $\phi_c P_n$ and $\dfrac{P_n}{\Omega_c}$, using the equations of AISC Section E3.

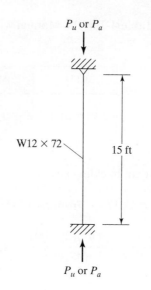

FIGURE 5.8

Solution

(a) Using a W12 × 72 ($A = 21.1$ in², $r_x = 5.31$ in, $r_y = 3.04$ in, $d = 12.3$ in, $b_f = 12.00$ in, $t_f = 0.670$ in, $k = 1.27$ in, $t_w = 0.430$ in)

$$\frac{b}{t} = \frac{12.00/2}{0.670} = 8.96 < 0.56\sqrt{\frac{E}{F_y}} = 0.56\sqrt{\frac{29,000}{50}} = 13.49$$

∴ Nonslender unstiffened flange element

$$\frac{h}{t_w} = \frac{d - 2k}{t_w} = \frac{12.3 - 2(1.27)}{0.430} = 22.70 < 1.49\sqrt{\frac{E}{F_y}} = 1.49\sqrt{\frac{29,000}{50}} = 35.88$$

∴ Nonslender stiffened web element

$K = 0.80$ from Table 5.1.

Obviously, $(KL/r)_y > (KL/r)_x$ and thus controls

$$\left(\frac{KL}{r}\right)_y = \frac{(0.80)(12 \times 15) \text{ in}}{3.04 \text{ in}} = 47.37$$

By straight-line interpolation, $\phi_c F_{cr} = 38.19$ ksi, and $\dfrac{F_{cr}}{\Omega_c} = 25.43$ ksi from Table 4-22 in the Manual using $F_y = 50$ ksi steel.

LRFD	ASD	
$\phi_c P_n = \phi_c F_{cr} A_g = (38.19)(21.1) = 805.8$ k	$\dfrac{P_n}{\Omega_c} = \dfrac{F_{cr} A_g}{\Omega_c} = (25.43)(21.1) = 536.6$ k	

(b) Entering Table 4-1 in the Manual with KL $(0.8)(15) = 12$ ft

LRFD	ASD
$\phi_t P_n = 807$ k	$\dfrac{P_n}{\Omega_c} = 537$ k

(c) Elastic critical buckling stress

$$\left(\frac{KL}{r}\right)_y = 47.37 \qquad \text{from part (a)}$$

$$F_e = \frac{\pi^2 E}{\left(\dfrac{KL}{r}\right)^2} = \frac{(\pi^2)(29{,}000)}{(47.37)^2} = 127.55 \text{ ksi} \qquad \text{(AISC Equation E3-4)}$$

Flexural buckling stress F_{cr}

$$4.71\sqrt{\frac{E}{F_y}} = 4.71\sqrt{\frac{29{,}000 \text{ ksi}}{50 \text{ ksi}}} = 113.43 > \left(\frac{KL}{r}\right)_y = 47.37$$

$$\therefore F_{cr} = \left[0.658^{\frac{F_y}{F_e}}\right] F_y = \left[0.658^{\frac{50}{127.55}}\right] 50 = 42.43 \; ksi \qquad \text{(AISC Equation E3-2)}$$

LRFD $\phi_c = 0.90$	ASD $\Omega_c = 1.67$
$\phi_c F_{cr} = (0.90)(42.43) = 38.19$ ksi	$\dfrac{F_{cr}}{\Omega_c} = \dfrac{42.43}{1.67} = 25.41$ ksi
$\phi_c P_n = \phi_c F_{cr} A = (38.19)(21.1)$	$\dfrac{P_n}{\Omega_c} = \dfrac{F_{cr}}{\Omega_c} A = (25.41)(21.1)$
$= 805.8$ k	$= 536.2$ k

Example 5-3

An HSS 16 × 16 × $\dfrac{1}{2}$ with $F_y = 46$ ksi is used for an 18-ft-long column with simple end supports.

(a) Determine $\phi_c P_n$ and $\dfrac{P_n}{\Omega_c}$ with the appropriate AISC equations.

(b) Repeat part (a), using Table 4-4 in the AISC Manual.

Solution

(a) Using an HSS

$$16 \times 16 \times \frac{1}{2}(A = 28.3 \text{ in}^2, t_{\text{wall}} = 0.465 \text{ in}, r_x = r_y = 6.31 \text{ in})$$

Calculate $\dfrac{b}{t}$ (AISC Table B4.1a, Case 6)

b is approximated as the tube size $-2 \times t_{\text{wall}}$

$$\frac{b}{t} = \frac{16 - 2(0.465)}{0.465} = 32.41 < 1.40\sqrt{\frac{E}{F_y}} = 1.40\sqrt{\frac{29{,}000}{46}}$$

$$= 35.15 \quad \therefore \text{ Section has no slender elements}$$

$\dfrac{b}{t}$ ratio also available from Table 1-12 of Manual

Calculate $\dfrac{KL}{r}$ and F_{cr}

$K = 1.0$

$$\left(\frac{KL}{r}\right)_x = \left(\frac{KL}{r}\right)_y = \frac{(1.0)(12 \times 18) \text{ in}}{6.31 \text{ in}} = 34.23$$

$$< 4.71\sqrt{\frac{E}{F_y}} = 4.71\sqrt{\frac{29{,}000}{46}} = 118.26$$

$\therefore$ Use AISC Equation E3-2 for F_{cr}

$$F_e = \frac{\pi^2 E}{\left(\dfrac{KL}{r}\right)^2} = \frac{(\pi^2)(29{,}000)}{(34.23)^2} = 244.28 \text{ ksi}$$

$$F_{cr} = \left[0.658^{\frac{F_y}{F_e}}\right]F_y = \left[0.658^{\frac{46}{244.28}}\right]46$$

$$= 42.51 \text{ ksi}$$

LRFD $\phi_c = 0.90$	ASD $\Omega_c = 1.67$
$\phi_c F_{cr} = (0.90)(42.51) = 38.26 \text{ ksi}$	$\dfrac{F_{cr}}{\Omega_c} = \dfrac{42.51}{1.67} = 25.46 \text{ ksi}$
$\phi_c P_n = \phi_c F_{cr} A = (38.26)(28.3)$	$\dfrac{P_n}{\Omega_c} = \dfrac{F_{cr}}{\Omega_c} A = (25.46)(28.3)$
$= 1082 \text{ k}$	$= 720 \text{ k}$

(b) From the Manual, Table 4-4

LRFD	ASD
$\phi_c P_n = 1080$ k	$\dfrac{P_n}{\Omega_c} = 720$ k

Though width–thickness ratios should be checked for all of the column and beam sections used in design, most of these necessary calculations are left out in this text to conserve space. To make this check in most situations, it is necessary only to refer to the Manual column tables where slender sections are clearly indicated for common shapes.

In Example 5-4, the author illustrates the computations necessary to determine the design strength of a built-up column section. Several special requirements for built-up column sections are described in Chapter 6.

Example 5-4

Determine the LRFD design strength $\phi_c P_n$ and the ASD allowable strength $\dfrac{P_n}{\Omega_c}$ for the axially loaded column shown in Fig. 5.9 if $KL = 19$ ft and 50-ksi steel is used.

Solution

$$A_g = (20)\left(\tfrac{1}{2}\right) + (2)(12.6) = 35.2 \text{ in}^2$$

$$\bar{y} \text{ from top} = \frac{(10)(0.25) + (2)(12.6)(9.50)}{35.2} = 6.87 \text{ in}$$

$$I_x = (2)(554) + (2)(12.6)(9.50 - 6.87)^2 + \left(\frac{1}{12}\right)(20)\left(\frac{1}{2}\right)^3 + (10)(6.87 - 0.25)^2$$

$$= 1721 \text{ in}^4$$

$$I_y = (2)(14.3) + (2)(12.6)(6.877)^2 + \left(\tfrac{1}{12}\right)\left(\tfrac{1}{2}\right)(20)^3 = 1554 \text{ in}^4$$

$$r_x = \sqrt{\frac{1721}{35.2}} = 6.99 \text{ in}$$

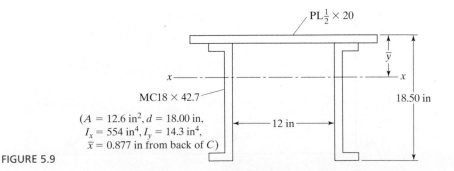

$\text{PL}\tfrac{1}{2} \times 20$

MC18 × 42.7

$(A = 12.6 \text{ in}^2, d = 18.00 \text{ in},$
$I_x = 554 \text{ in}^4, I_y = 14.3 \text{ in}^4,$
$\bar{x} = 0.877 \text{ in from back of } C)$

12 in

18.50 in

$\bar{y}$

FIGURE 5.9

$$r_y = \sqrt{\frac{1554}{35.2}} = 6.64 \text{ in}$$

$$\left(\frac{KL}{r}\right)_x = \frac{(12)(19)}{6.99} = 32.62$$

$$\left(\frac{KL}{r}\right)_y = \frac{(12)(19)}{6.64} = 34.34 \leftarrow$$

From the Manual, Table 4-22, we read for $\dfrac{KL}{r} = 34.34$ that $\phi_c F_{cr} = 41.33$ ksi and $\dfrac{F_{cr}}{\Omega_c} = 27.47$ ksi, for 50 ksi steel.

LRFD	ASD
$\phi_c P_n = \phi_c F_{cr} A_g = (41.33)(35.2) = 1455 \text{ k}$	$\dfrac{P_n}{\Omega_c} = \dfrac{F_{cr} A_g}{\Omega_c} = (27.47)(35.2) = 967 \text{ k}$

To determine the design compression stress needed for a particular column, it is theoretically necessary to compute both $(KL/r)_x$ and $(KL/r)_y$. The reader will notice, however, that for most of the steel sections used for columns, r_y will be much less than r_x. As a result, only $(KL/r)_y$ is calculated for most columns and used in the applicable column formulas.

For some columns, particularly the long ones, bracing is supplied perpendicular to the weak axis, thus reducing the slenderness or the length free to buckle in that direction. Bracing may be accomplished by framing braces or beams into the sides of a column. For instance, horizontal members or *girts* running parallel to the exterior walls of a building frame may be framed into the sides of columns. The result is stronger columns and ones for which the designer needs to calculate both $(KL/r)_x$ and $(KL/r)_y$. The larger ratio obtained for a particular column indicates the weaker direction and will be used for calculating the design stress $\phi_c F_{cr}$ and the allowable stress $\dfrac{F_{cr}}{\Omega_c}$ for that member.

Bracing members must be capable of providing the necessary lateral forces, without buckling themselves. The forces to be taken are quite small and are often conservatively estimated to equal 0.02 times the column design loads. These members can be selected as are other compression members. A bracing member must be connected to other members that can transfer the horizontal force by shear to the next restrained level. If this is not done, little lateral support will be provided for the original column in question.

If the lateral bracing were to consist of a single bar or rod (⊥), it would not prevent twisting and torsional buckling of the column. (See Chapter 6.) As torsional buckling is a difficult problem to handle, we should provide lateral bracing that prevents lateral movement and twist.[6]

---

[6]J. A. Yura, "Elements for Teaching Load and Resistance Factor Design" (New York: AISC, August 1987), p. 20.

Steel columns may also be built into heavy masonry walls in such a manner that they are substantially supported in the weaker direction. The designer, however, should be quite careful in assuming that there is complete lateral support parallel to the wall, because the condition of the wall may be unknown and a poorly built wall will not provide 100 percent lateral support.

Example 5-5 illustrates the calculations necessary to determine the LRFD design strength and the ASD allowable strength of a column with two unbraced lengths.

Example 5-5

(a) Determine the LRFD design strength $\phi_c P_n$ and the ASD allowable design strength $\dfrac{P_n}{\Omega_c}$ for the 50 ksi axially loaded W14 × 90 shown in Fig. 5.10.

Because of its considerable length, this column is braced perpendicular to its weak, or y, axis at the points shown in the figure. These connections are assumed to permit rotation of the member in a plane parallel to the plane of the flanges. At the same time, however, they are assumed to prevent translation or sidesway and twisting of the cross section about a longitudinal axis passing through the shear center of the cross section. (The shear center is the point in the cross section of a member through which the resultant of the transverse loads must pass so that no torsion will occur. See Chapter 10.)

(b) Repeat part (a), using the column tables of Part 4 of the Manual.

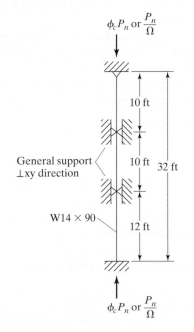

FIGURE 5.10

Solution

(a) Using W14 × 90 ($A = 26.5$ in², $r_x = 6.14$ in, $r_y = 3.70$ in)
Determining effective lengths

$$K_xL_x = (0.80)(32) = 25.6 \text{ ft}$$
$$K_yL_y = (1.0)(10) = 10 \text{ ft} \leftarrow \text{governs for } K_yL_y$$
$$K_yL_y = (0.80)(12) = 9.6 \text{ ft}$$

Computing slenderness ratios

$$\left(\frac{KL}{r}\right)_x = \frac{(12)(25.6)}{6.14} = 50.03 \leftarrow$$

$$\left(\frac{KL}{r}\right)_y = \frac{(12)(10)}{3.70} = 32.43$$

$$\left.\begin{array}{l} \phi_c F_{cr} = 37.49 \text{ ksi} \\ \dfrac{F_{cr}}{\Omega_c} = 24.90 \text{ ksi} \end{array}\right\} \begin{array}{l} \text{from Manual,} \\ \text{Table 4-22, } F_y = 50 \text{ ksi} \end{array}$$

LRFD	ASD
$\phi_c P_n = \phi_c F_{cr} A_g = (37.49)(26.5)$	$\dfrac{P_n}{\Omega_c} = \dfrac{F_{cr} A_g}{\Omega_c} = (24.90)(26.5) = 660 \text{ k}$
$= 993 \text{ k}$	

(b) Noting from part (a) solution that there are two different KL values

$$K_xL_x = 25.6 \text{ ft}$$
$$K_yL_y = 10 \text{ ft}$$

We would like to know which of these two values is going to control. This can easily be learned by determining a value of K_xL_x that is equivalent to K_yL_y. The slenderness ratio in the x direction is equated to an equivalent value in the y direction as follows:

$$\frac{K_xL_x}{r_x} = \text{Equivalent } \frac{K_yL_y}{r_y}$$

$$\text{Equivalent } K_yL_y = r_y\frac{K_xL_x}{r_x} = \frac{K_xL_x}{\dfrac{r_x}{r_y}}$$

Thus, the controlling K_yL_y for use in the tables is the larger of the real $K_yL_y = 10$ ft, or the equivalent K_yL_y.

$\dfrac{r_x}{r_y}$ for W14 × 90 (from the bottom of Table 4–1 of the Manual) = 1.66

$$\text{Equivalent } K_yL_y = \frac{25.6}{1.66} = 15.42 \text{ ft} > K_yL_y \text{ of 10 ft}$$

From column tables with K_yL_y = 15.42 ft, we find by interpolation that

$\phi_c P_n = 991 \text{ k and } \dfrac{P_n}{\Omega_c} = 660 \text{ k}.$

5.12   PROBLEMS FOR SOLUTION

5-1 to 5-4. *Determine the critical buckling load for each of the columns, using the Euler equation. E = 29,000 ksi. Proportional limit = 36,000 psi. Assume simple ends and maximum permissible L/r = 200.*

5-1. A solid round bar of 1¼ in diameter:

    a. $L = 4$ ft 0 in (*Ans.* 14.89 kips)

    b. $L = 2$ ft 3 in (*Ans.* Euler equation not applicable, F_e exceeds proportional limit)

    c. $L = 6$ ft 6 in (*Ans.* Euler equation not applicable, L/r exceeds 200)

5-2. The pipe section shown:

    a. $L = 21$ ft 0 in

    b. $L = 16$ ft 0 in

    c. $L = 10$ ft 0 in

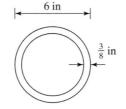

6 in

$\frac{3}{8}$ in

FIGURE P5-2

5-3. A W12 × 50, $L = 20$ ft 0 in. (*Ans.* 278.7 k)

5-4. The four − L4 × 4 × 1/4 shown for $L = 40$ ft 0 in.

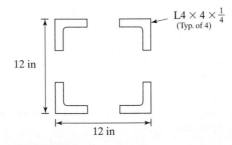

L4 × 4 × $\frac{1}{4}$
(Typ. of 4)

12 in

12 in

FIGURE P5-4

5-5 to 5-8. *Determine the LRFD design strength, $\Phi_c P_n$, and the ASD allowable strength, P_n/Ω_c, for each of the compression members shown. Use the AISC Specification and a steel with $F_y = 50$ ksi, except for Problem 5-8, $F_y = 46$ ksi.*

5-5. (*Ans.* 212 k LRFD; 141 k ASD)

5-6.

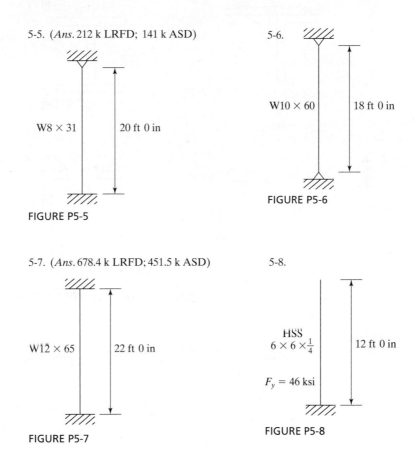

W8 × 31   20 ft 0 in

FIGURE P5-5

W10 × 60   18 ft 0 in

FIGURE P5-6

5-7. (*Ans.* 678.4 k LRFD; 451.5 k ASD)

5-8.

W12 × 65   22 ft 0 in

FIGURE P5-7

HSS
6 × 6 × $\frac{1}{4}$ 12 ft 0 in

$F_y = 46$ ksi

FIGURE P5-8

5-9 to 5-17. *Determine $\Phi_c P_n$, and P_n/Ω_c for each of the columns, using the AISC Specification and $F_y = 50$ ksi, unless noted otherwise.*

   5-9. a. W12 × 120 with $KL = 18$ ft. (*Ans.* 1120 k LRFD; 744 k ASD)
          b. HP10 × 42 with $KL = 15$ ft. (*Ans.* 371 k LRFD; 247 k ASD)
          c. WT8 × 50 with $KL = 20$ ft. (*Ans.* 294 k LRFD; 196 k ASD)

   5-10. Note that F_y is different for parts (c) to (e).

          a. A W8 × 24 with pinned ends, $L = 12$ ft.
          b. A W14 × 109 with fixed ends, $L = 20$ ft.
          c. An HSS 8 × 6 × 3/8, $F_y = 46$ ksi with pinned ends, $L = 15$ ft.
          d. A W12 × 152 with one end fixed and the other end pinned, $L = 25$ ft 0 in, $F_y = 36$ ksi.
          e. A Pipe 10 STD with pinned ends, $L = 18$ ft 6 in, $F_y = 35$ ksi.

5-11. A W10 × 39 with a 1/2 × 10 in cover plate welded to each flange is to be used for a column with KL = 14 ft. (*Ans.* 685 k LRFD; 455 k ASD)

5-12.

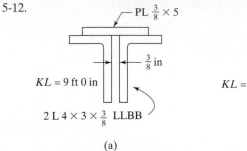

PL $\frac{3}{8}$ × 5

$\frac{3}{8}$ in

KL = 9 ft 0 in

2 L 4 × 3 × $\frac{3}{8}$ LLBB

(a)

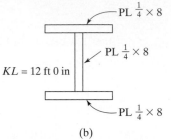

PL $\frac{1}{4}$ × 8

PL $\frac{1}{4}$ × 8

KL = 12 ft 0 in

PL $\frac{1}{4}$ × 8

(b)

5-13.

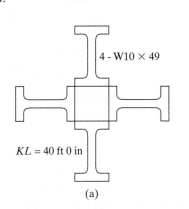

4 PL $\frac{3}{8}$ × 6

$6\frac{3}{4}$ in

KL = 12 ft 8 in   6 in

(a) (*Ans.* 297 k LRFD; 198 k ASD)

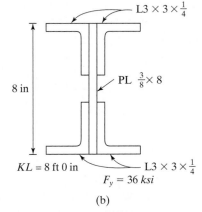

MC 10 × 28.5

KL = 18 ft 0 in

10 in

(b) (*Ans.* 601 k LRFD; 400 k ASD)

5-14.

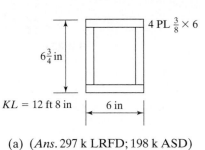

4 - W10 × 49

KL = 40 ft 0 in

(a)

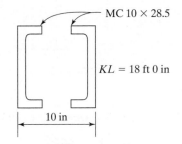

L3 × 3 × $\frac{1}{4}$

PL $\frac{3}{8}$× 8

8 in

KL = 8 ft 0 in   L3 × 3 × $\frac{1}{4}$

F_y = 36 *ksi*

(b)

5-15.

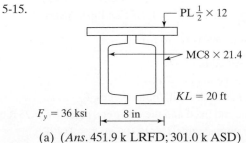

PL $\frac{1}{2}$ × 12

MC8 × 21.4

KL = 20 ft

F_y = 36 ksi   8 in

(a) (*Ans.* 451.9 k LRFD; 301.0 k ASD)

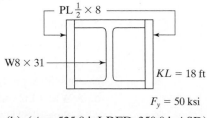

PL $\frac{1}{2}$ × 8

W8 × 31

KL = 18 ft

F_y = 50 ksi

(b) (*Ans.* 525.9 k LRFD; 350.0 k ASD)

5-16.

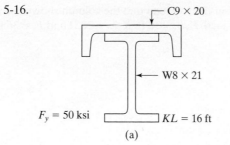

$F_y = 50$ ksi

C9 × 20

W8 × 21

$KL = 16$ ft

(a)

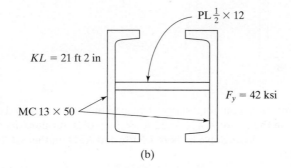

PL $\frac{1}{2}$ × 12

$KL = 21$ ft 2 in

MC 13 × 50

$F_y = 42$ ksi

(b)

5-17. A 24 ft axially loaded W12 x 96 column that has the bracing and end support conditions shown in the figure. (*Ans.* 1023.3 k LRFD; 680.4 k ASD)

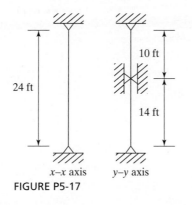

24 ft

10 ft

14 ft

x–x axis

y–y axis

FIGURE P5-17

5-18. Determine the maximum service live load that the column shown can support if the live load is twice the dead load. $K_xL_x = 18$ ft, $K_yL_y = 12$ ft and $F_y = 36$ ksi. Solve by LRFD and ASD methods.

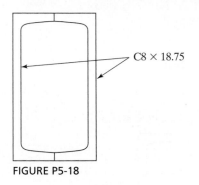

C8 × 18.75

FIGURE P5-18

5-19. Compute the maximum total service live load that can be applied to the A36 section shown in the figure, if $K_xL_x = 12$ ft., $K_yL_y = 10$ ft. Assume the load is 1/2 dead load and 1/2 live load. Solve by both LRFD and ASD methods. (*Ans.* 29.0 k LRFD; 27.0 k ASD)

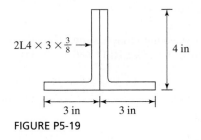

2L4 × 3 × $\frac{3}{8}$

4 in

3 in   3 in

FIGURE P5-19

C H A P T E R 6

Design of Axially Loaded Compression Members

6.1 INTRODUCTION

In this chapter, the designs of several axially loaded columns are presented. Included are the selections of single shapes, W sections with cover plates, and built-up sections constructed with channels. Also included are the designs of sections whose unbraced lengths in the x and y directions are different, as well as the sizing of lacing and tie plates for built-up sections with open sides. Another topic that is introduced is flexural torsional buckling of sections.

The design of columns by formulas involves a trial-and-error process. The LRFD design stress $\phi_c F_{cr}$ and the ASD allowable stress F_{cr}/Ω_c are not known until a column size is selected, and vice versa. A column size may be assumed, the r values for that section obtained from the Manual or calculated, and the design stress found by substituting into the appropriate column formula. It may then be necessary to try a larger or smaller section. Examples 6-1, 6-3, and 6-4 illustrate this procedure.

The designer may assume an LRFD design stress or an ASD allowable stress and divide that stress into the appropriate column load to give an estimated column area, select a column section with approximately that area, determine its design stress, and multiply that stress by the cross-sectional area of the section to obtain the member's design strength. Thus, the designer can see if the section selected is over- or underdesigned and—if it is appreciably so—try another size. The student may feel that he or she does not have sufficient background or knowledge to make reasonable initial design stress assumptions. If this student will read the information contained in the next few paragraphs, however, he or she will immediately be able to make excellent estimates.

The effective slenderness ratio (KL/r) for the average column of 10- to 15-ft length will generally fall between about 40 and 60. For a particular column, a KL/r

somewhere in this approximate range is assumed and substituted into the appropriate column equation to obtain the design stress. (To do this, you will first note that the AISC for KL/r values from 0 to 200 has substituted into the equations, with the results shown, in AISC Table 4-22. This greatly expedites our calculations.)

To estimate the effective slenderness ratio for a particular column, the designer may estimate a value a little higher than 40 to 60 if the column is appreciably longer than the 10- to 15-ft range, and vice versa. A very heavy factored column load—say, in the 750- to 1000-k range or higher—will require a rather large column for which the

Georgia Railroad Bank and Trust Company Building, Atlanta, GA. (Courtesy of Bethlehem Steel Corporation.)

radii of gyration will be larger, and the designer may estimate a little smaller value of KL/r. For lightly loaded bracing members, he or she may estimate high slenderness ratios of 100 or more.

In Example 6-1, a column is selected by the LRFD method. An effective slenderness ratio of 50 is assumed, and the corresponding design stress $\phi_c F_{cr}$ is picked from AISC Table 4-22. By dividing this value into the factored column load, an estimated required column area is derived and a trial section selected. After a trial section is selected with approximately that area, its actual slenderness ratio and design strength are determined. The first estimated size in Example 6-1, though quite close, is a little too small, but the next larger section in that series of shapes is satisfactory.

The author follows a similar procedure with the ASD formula as follows: Assume $KL/r = 50$, determine F_{cr}/Ω_c from AISC Table 4-22, divide this value into the ASD column load to obtain the estimated area required, and pick a trial section and determine its allowable load.

Example 6-1

Using $F_y = 50$ ksi, select the lightest W14 available for the service column loads $P_D = 130$ k and $P_L = 210$ k. $KL = 10$ ft.

Solution

LRFD	ASD
$P_u = (1.2)(130\,\text{k}) + (1.6)(210\,\text{k}) = 492\,\text{k}$	$P_a = 130\,\text{k} + 210\,\text{k} = 340\,\text{k}$
Assume $\dfrac{KL}{r} = 50$	Assume $\dfrac{KL}{r} = 50$
Using $F_y = 50$ ksi steel	Using $F_y = 50$ ksi steel
$\phi_c F_{cr}$ from AISC Table 4-22 $= 37.5$ ksi	$\dfrac{F_{cr}}{\Omega_c} = 24.9$ ksi (AISC Table 4-22)
A Reqd $= \dfrac{P_u}{\phi_c F_{cr}} = \dfrac{492\,\text{k}}{37.5\,\text{ksi}} = 13.12\,\text{in}^2$	A Reqd $= \dfrac{P_a}{F_{cr}/\Omega} = \dfrac{340\,\text{k}}{24.9\,\text{ksi}} = 13.65\,\text{in}^2$
Try W14 × 48 ($A = 14.1\,\text{in}^2$, $r_x = 5.85$ in, $r_y = 1.91$ in)	Try W14 × 48 ($A = 14.1\,\text{in}^2$, $r_x = 5.85$ in, $r_y = 1.91$ in)
$\left(\dfrac{KL}{r}\right)_y = \dfrac{(12\,\text{in/ft})(10\,\text{ft})}{1.91\,\text{in}} = 62.83$	$\left(\dfrac{KL}{r}\right)_y = \dfrac{(12\,\text{in/ft})(10\,\text{ft})}{1.91\,\text{in}} = 62.83$
$\phi_c F_{cr} = 33.75$ ksi from AISC Table 4-22	$\dfrac{F_{cr}}{\Omega_c} = 22.43$ ksi from AISC Table 4-22
$\phi_c P_n = (33.75\,\text{ksi})(14.1\,\text{in}^2)$ $= 476\,\text{k} < 492\,\text{k}$ N.G.	

Try next larger section W14 × 53 (A = 15.6 in², r_y = 1.92 in)

$$\frac{P_n}{\Omega_c} = (22.43 \text{ ksi})(14.1 \text{ in}^2) = 316 \text{ k} < 340 \text{ k N.G.}$$

$$\left(\frac{KL}{r}\right)_y = \frac{(12 \text{ in/ft})(10 \text{ ft})}{1.92 \text{ in}} = 62.5$$

Try next larger section W14 × 53 (A = 15.6 in², r_y = 1.92 in).

$$\phi_c F_{cr} = 33.85 \text{ ksi}$$

$$\left(\frac{KL}{r}\right)_y = \frac{(12 \text{ in/ft})(10 \text{ ft})}{1.92 \text{ in}} = 62.5$$

$$\frac{F_{cr}}{\Omega_c} = 22.5 \text{ ksi}$$

$$\phi_c P_n = (33.85 \text{ ksi})(15.6 \text{ in}^2)$$

$$\frac{P_n}{\Omega_c} = (22.5 \text{ ksi})(15.6 \text{ in}^2) = 351 \text{ k} > 340 \text{ k } \textbf{OK}$$

$$= 528 \text{ k} > 492 \text{ k} \quad \textbf{OK}$$

Use W14 × 53.

Use W14 × 53.

Note: Table 4-1 does not indicate that a W14 × 53 is a slender member for compression.

6.2   AISC DESIGN TABLES

For Example 6-2, Table 4-1 of the Manual is used to select various column sections from tables, without the necessity of using a trial-and-error process. These tables provide axial design strengths ($\phi_c P_n$) and allowable design loads (P_n/Ω_c) for various practical effective lengths of the steel sections commonly used as columns. The values are given with respect to the least radii of gyration, for Ws and WTs with 50 ksi steel. Other grade steels are commonly used for other types of sections, as shown in the Manual and listed here. These include 35 ksi for steel pipe, 36 ksi for Ls, 42 ksi for round HSS sections, and 46 ksi for square and rectangular HSS sections.

For most columns consisting of single steel shapes, the effective slenderness ratio with respect to the y axis $(KL/r)_y$ is larger than the effective slenderness ratio with respect to the x axis $(KL/r)_x$. As a result, the controlling, or smaller, design stress is for the y axis. Because of this, the AISC tables provide design strengths of columns with respect to their y axes. We will learn in the pages to follow how to handle situations in which $(KL/r)_x$ is larger than $(KL/r)_y$.

The resulting tables are very simple to use. The designer takes the KL value for the minor principal axis in feet, enters the table in question from the left-hand side, and moves horizontally across the table. Under each section is listed the design strength $\phi_c P_n$ and the allowable design strength P_n/Ω_c for that KL and steel yield stress. As an illustraion, assume that we have a factored design strength P_u = 1200 k, $K_y L_y$ = 12 ft, and we want to select the lightest available W14 section, using 50 ksi steel and the LRFD method. We enter with KL = 12 ft into the left-hand column of the first page of AISC Table 4-1 and read from left to right under the $\phi_c P_n$ columns. The values are successively 9030 k, 8220 k, 7440 k, and so on until a few pages later, where the consecutive values 1290 k and 1170 k are found. The 1170 k value is not sufficient, and we go back to the 1290 k value, which falls under the W14 × 109 section.

A similar procedure can be followed in the tables subsequent to AISC Table 4-1 for the selection of rectangular, square, and round HSS sections; pipe; WT shapes; angles; and so on.

Example 6-2 illustrates the selection of various possible sections to be used for a particular column. Among the sections chosen are the round HSS sections in AISC Table 4-5 and the steel pipe sections shown in Table 4-6 of the Manual. It is possible to support a given load with a standard pipe (labeled "std" in the table); with an extra strong pipe (XS), which has a smaller diameter, but thicker walls and thus is heavier and more expensive; or with a double extra-strong pipe (XXS), which has an even smaller diameter, even thicker walls, and greater weight. The XXS sizes are available only in certain sizes (pipes 4, 5, 6, and 8).

Example 6-2

Use the AISC column tables (both LRFD and ASD) for the designs to follow.

(a) Select the lightest W section available for the loads, steel, and KL of Example 6-1. $F_y = 50$ ksi.

(b) Select the lightest satisfactory rectangular or square HSS sections for the situation in part (a). $F_y = 46$ ksi.

(c) Select the lightest satisfactory round HSS section, $F_y = 42$ ksi for the situation in part (a).

(d) Select the lightest satisfactory pipe section, $F_y = 35$ ksi, for the situation in part (a).

Solution

Entering Tables with $K_y L_y = 10$ ft, $P_u = 492$ k for LRFD and $P_a = 340$ k for ASD from Example 6-1 solution.

LRFD	ASD
(a) W8 × 48 ($\phi_c P_n = 497$ k > 492 k)	(a) W10 × 49 $\left(\dfrac{P_n}{\Omega_c} = 366 \text{ k} > 340 \text{ k} \right)$
from Table 4-1	from Table 4-1
(b) Rectangular HSS	(b) Rectangular HSS
HSS 12 × 8 × $\dfrac{3}{8}$ @ 47.8 #/ft	HSS 12 × 10 × $\dfrac{3}{8}$ @ 52.9 #/ft
($\phi_c P_n = 499$ k > 492 k)	$\left(\dfrac{P_n}{\Omega_c} = 379 \text{ k} > 340 \text{ k} \right)$
from Table 4-3	from Table 4-3
Square HSS	Square HSS
HSS 10 × 10 × $\dfrac{3}{8}$ @ 47.8 #/ft	** HSS 12 × 12 × $\dfrac{5}{16}$ @ 48.8 #/ft
($\phi_c P_n = 513$ k > 492 k)	$\left(\dfrac{P_n}{\Omega_c} = 340 \text{ k} = 340 \text{ k} \right)$
from Table 4-4	from Table 4-4

(c) Round HSS 16.000 × 0.312

   @ 52.3 #/ft ($\phi_c P_n = 529$ k > 492 k)

   from Table 4-5

(d) XS Pipe 12 @ 65.5 #/ft

   ($\phi_c P_n = 530$ k > 492 k)

   from Table 4-6

(c) Round HSS 16.000 × 0.312

   @ 52.3 #/ft $\left(\dfrac{P_n}{\Omega_c} = 352 \text{ k} > 340 \text{ k}\right)$

   from Table 4-5

(d) XS Pipe 12 @ 65.5 #/ft

   $\left(\dfrac{P_n}{\Omega_c} = 353 \text{ k} > 340 \text{ k}\right)$

   from Table 4-6

**Note: The AISC Column Tables used in this problem indicate that only the HSS 12 × 12 × 5/16 from part (b)—ASD design method is a slender member for compression. The value of $P_n\Omega_c = 340$ k reflects the reduced design strength available for slender sections (per E7 AISC Specification).

An axially loaded column is laterally restrained in its weak direction, as shown in Fig. 6.1. Example 6-3 illustrates the design of such a column with its different unsupported lengths in the x and y directions. The student can easily solve this problem by trial and error. A trial section can be selected, as described in Section 6.1, the slenderness values $\left(\dfrac{KL}{r}\right)_x$ and $\left(\dfrac{KL}{r}\right)_y$ computed, and $\phi_c F_{cr}$ and $\dfrac{F_{cr}}{\Omega_c}$ determined for the larger value and multiplied, respectively, by A_g to determine $\phi_c P_n$ and $\dfrac{P_n}{\Omega_c}$. Then, if necessary, another size can be tried, and so on.

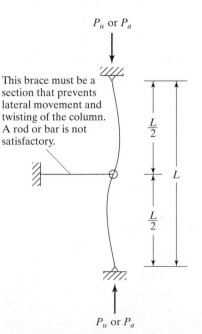

FIGURE 6.1
A column laterally restrained at
mid-depth in its weaker direction.

For this discussion, it is assumed that K is the same in both directions. Then, if we are to have equal strengths about the x and y axes, the following relation must hold:

$$\frac{L_x}{r_x} = \frac{L_y}{r_y}$$

For L_y to be equivalent to L_x we would have

$$L_x = L_y \frac{r_x}{r_y}$$

If $L_y(r_x/r_y)$ is less than L_x, then L_x controls; if greater than L_x, then L_y controls.

Consistent with the preceding information, the AISC Manual provides a method by which a section can be selected from its tables, with little trial and error, when the unbraced lengths are different. The designer enters the appropriate table with K_yL_y, selects a shape, takes the r_x/r_y value given in the table for that shape, and multiplies it by L_y. If the result is larger than K_xL_x, then K_yL_y controls and the shape initially selected is the correct one. If the result of the multiplication is less than K_xL_x, then K_xL_x controls and the designer will reenter the tables with a larger K_yL_y equal to $K_xL_x/(r_x/r_y)$ and select the final section.

Example 6-3 illustrates the two procedures described here for selecting a W section that has different effective lengths in the x and y directions.

Example 6-3

Select the lightest available W12 section, using both the LRFD and ASD methods for the following conditions. $F_y = 50$ ksi, $P_D = 250$ k, $P_L = 400$ k, $K_xL_x = 26$ ft and $K_yL_y = 13$ ft.

(a) By trial and error

(b) Using AISC tables

Solution

(a) Using trial and error to select a section, using the LRFD expressions, and then checking the section with both the LRFD and ASD methods

LRFD	ASD
$P_u = (1.2)(250 \text{ k}) + (1.6)(400 \text{ k}) = 940 \text{ k}$	$P = 250 \text{ k} + 400 \text{ k} = 650 \text{ k}$
Assume $\dfrac{KL}{r} = 50$	Assume $\dfrac{KL}{r} = 50$
Using $F_y = 50$ ksi steel $\phi_c F_{cr} = 37.5$ ksi (AISC Table 4-22)	Using $F_y = 50$ ksi steel $\dfrac{F_{cr}}{\Omega_c} = 24.9$ ksi (AISC Table 4-22)
$A \text{ Reqd} = \dfrac{940 \text{ k}}{37.5 \text{ ksi}} = 25.07 \text{ in}^2$	$A \text{ Reqd} = \dfrac{650 \text{ k}}{24.9 \text{ ksi}} = 26.10 \text{ in}^2$

Try W12 × 87 ($A = 25.6$ in^2, $r_x = 5.38$ in, $r_y = 3.07$ in)

$$\left(\frac{KL}{r}\right)_x = \frac{(12 \text{ in/ft})(26 \text{ ft})}{5.38 \text{ in}} = 57.99 \leftarrow \quad \therefore \left(\frac{KL}{r}\right)_x \text{ controls}$$

$$\left(\frac{KL}{r}\right)_y = \frac{(12 \text{ in/ft})(13 \text{ ft})}{3.07 \text{ in}} = 50.81$$

$$\phi_c F_{cr} = 35.2 \text{ ksi (Table 4-22)}$$

$$\phi_c P_n = (35.2 \text{ ksi})(25.6 \text{ in}^2)$$

$$= 901 \text{ k} < 940 \text{ k } \textbf{N.G.}$$

Try W12 × 87 ($A = 25.6$ in^2, $r_x = 5.38$ in, $r_y = 3.07$ in)

$$\left(\frac{KL}{r}\right)_x = \frac{(12 \text{ in/ft})(26 \text{ ft})}{5.38 \text{ in}} = 57.99 \leftarrow \quad \therefore \left(\frac{KL}{r}\right)_x \text{ controls}$$

$$\left(\frac{KL}{r}\right)_y = \frac{(12 \text{ in/ft})(13 \text{ ft})}{3.07 \text{ in}} = 50.81$$

$$\frac{F_{cr}}{\Omega_c} = 23.4 \text{ ksi (Table 4-22)}$$

$$\frac{P_n}{\Omega_c} = (23.4 \text{ ksi})(25.6 \text{ in}^2)$$

$$= 599 \text{ k} < 650 \text{ k } \textbf{N.G.}$$

Eversharp, Inc., Building at Milford, CN. (Courtesy of Bethlehem Steel Corporation.)

A subsequent check of the next-larger W12 section, a W12 × 96, shows that it will work for both the LRFD and ASD procedures.

(b) Using AISC Tables. Assuming $K_y L_y$ controls

Enter Table 4-1 with $K_y L_y = 13$ ft, $F_y = 50$ ksi and $P_u = 940$ k

LRFD

Try W12 × 87 $\left(\dfrac{r_x}{r_y} = 1.75 \right)$; $\phi P_n = 954$ k

Equivalent $K_y L_y = \dfrac{K_x L_x}{\dfrac{r_x}{r_y}}$

$= \dfrac{26}{1.75} = 14.86 > K_y L_y$ of 13 ft. $\therefore$ $K_x L_x$ controls

Use $K_y L_y = 14.86$ ft and reenter tables

LRFD	ASD
Use W12 × 96	Use W12 × 96
$\phi_c P_n = 994$ k > 940 k **OK**	$\dfrac{P_n}{\Omega_c} = 662$ k > 650 k **OK**

Note: Table 4-1 does not indicate that the W12 × 96 is a slender member for compression.

6.3   COLUMN SPLICES

For multistory buildings, column splices are desirably placed 4 feet above finished floors to permit the attachment of safety cables to the columns, as may be required at floor edges or openings. This offset also enables us to keep the splices from interfering with the beam and column connections.

Typical column splices are shown in Fig. 6.2. Many more, examples are shown in Table 14-3 of the AISC Manual. The column ends are usually milled so they can be placed firmly in contact with each other for purposes of load transfer. When the contact surfaces are milled, a large part of the axial compression (if not all) can be transferred through the contacting areas. It is obvious, however, that splice plates are necessary, even though full contact is made between the columns and only axial loads are involved. For instance, the two column sections need to be held together during erection and afterward. What is necessary to hold them together is decided primarily on the basis of the experience and judgment of the designer. Splice plates are even more necessary when consideration is given to the shears and moments existing in practical columns subjected to off-center loads, lateral forces, moments, and so on.

There is, obviously, a great deal of difference between tension splices and compression splices. In tension splices, all load has to be transferred through the splice, whereas in splices for compression members, a large part of the load can be transferred directly in bearing between the columns. The splice material is then needed to transfer only the remaining part of the load.

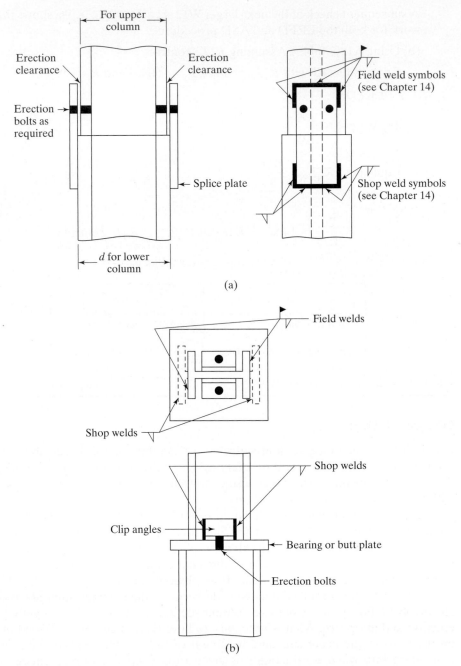

FIGURE 6.2

Column splices. (a) Columns from same W series with total depths close to each other ($d_{lower} <$ 2 in greater than d_{upper}). (b) Columns from different W series.

The amount of load to be carried by the splice plates is difficult to estimate. Should the column ends not be milled, the plates should be designed to carry 100 percent of the load. When the surfaces are milled and axial loads only are involved, the amount of load to be carried by the plates might be estimated to be from 25 to 50 percent of the total load. If bending is involved, perhaps 50 to 75 percent of the total load may have to be carried by the splice material.

The bridge specifications spell out very carefully splice requirements for compression members, but the AISC Specification does not. A few general requirements are given in Section J1.4 of the AISC Specification.

Figure 6.2(a) shows a splice that may be used for columns with substantially the same nominal depths. The student will notice in the AISC Manual that the W shapes of a given nominal size are divided into groups rolled with the same set of rolls. Because of the fixed dimensions of each set of rolls, the clear distances between the flanges are constant for each shape in that group, although their total depths may vary greatly. For instance, the inside distance for each of the 28 shapes (running from the W14 × 61 to the W14 × 730) is approximately 12.60 in, although their total depths vary from 13.9 in to 22.4 in. (Notice that the T values, which are the distances between the web toes of the fillets, are all 10 in for the W14 × 90 through the W14 × 730.)

It is most economical to employ the simple splices of Fig. 6.2(a). This can easily be accomplished by using one series of shapes for as many stories of a building as possible. For instance, we may select a particular W14 column section for the top story or top two stories of a building and then keep selecting heavier and heavier W14s for that column as we come down in the building. We may also switch to higher-strength steel columns as we move down in the building, thus enabling us to stay with the same W series for even more

Welded column splice for Colorado State
Building, Denver, CO. (Courtesy of
Lincoln Electric Company.)

floors. It will be necessary to use filler plates between the splice plates and the upper column if the upper column has a total depth significantly less than that of the lower column.

Figure 6.2(b) shows a type of splice that can be used for columns of equal or different nominal depths. For this type of splice, the butt plate is shop-welded to the lower column, and the clip angles used for field erection are shop-welded to the upper column. In the field, the erection bolts shown are installed, and the upper column is field-welded to the butt plate. The horizontal welds on this plate resist shears and moments in the columns.

Sometimes, splices are applied to all four sides of columns. The web splices are bolted in place in the field and field-welded to the column webs. The flange splices are shop-welded to the lower column and field-welded to the top column. The web plates may be referred to as *shear plates* and the flange plates as *moment plates*.

For multistory buildings, the columns may be fabricated for one or more stories. Theoretically, column sizes can be changed at each floor level so that the lightest total column weight is used. The splices needed at each floor will be quite expensive, however, and as a result it is usually more economical to use the same column sizes for at least two stories, even though the total steel weight will be higher. Seldom are the same sizes used for as many as three stories, because three-story columns are so difficult to erect. The two-story heights work out very well most of the time.

6.4   BUILT-UP COLUMNS

As previously described in Section 5.3, compression members may be constructed with two or more shapes built up into a single member. They may consist of parts in contact with each other, such as cover-plated sections $\perp$; or they may consist of parts in near contact with each other, such as pairs of angles $\top\hspace{-2pt}\top$ that may be separated from each other by a small distance equal to the thickness of the end connections or gusset plates between them. They may consist of parts that are spread well apart, such as pairs of channels $\lbrack\;\rbrack$ or four angles $\lbrack\;\lrcorner$, and so on.

Two-angle sections probably are the most common type of built-up member. (For example, they frequently are used as the members of light trusses.) When a pair of angles are used as a compression member, they need to be fastened together so that they will act as a unit. Welds may be used at intervals (with a spacer bar between the parts if the angles are separated), or they may be connected with *stitch bolts*. When the connections are bolted, washers, or *ring fills*, are placed between the parts to keep them at the proper spacing if the angles are to be separated.

For long columns, it may be convenient to use built-up sections where the parts of the columns are spread out or widely separated from each other. Before heavy W sections were made available, such sections were very commonly used in both buildings and bridges. Today, these types of built-up columns are commonly used for crane booms and for the compression members of various kinds of towers. The widely spaced parts of these types of built-up members must be carefully laced or tied together.

Sections 6.5 and 6.6 concern compression members that are built up from parts in direct contact (or nearly so) with each other. Section 6.7 addresses built-up compression members whose parts are spread widely apart.

6.5   BUILT-UP COLUMNS WITH COMPONENTS IN CONTACT WITH EACH OTHER

Should a column consist of two equal-size plates, as shown in Fig. 6.3, and should those plates not be connected together, each plate will act as a separate column, and each will resist approximately half of the total column load. In other words, the total moment of inertia of the column will equal two times the moment of inertia of one plate. The two "columns" will act the same and have equal deformations, as shown in part (b) of the figure.

Should the two plates be connected together sufficiently to prevent slippage on each other, as shown in Fig. 6.4, they will act as a unit. Their moment of inertia may be computed for the whole built-up section as shown in the figure, and will be four times as large as it was for the column of Fig. 6.3, where slipping between the plates was possible. The reader should also notice that the plates of the column in Fig. 6.4 will deform different amounts as the column bends laterally.

Should the plates be connected in a few places, it would appear that the strength of the resulting column would be somewhere in between the two cases just described.

Reference to Fig. 6.3(b) shows that the greatest displacement between the two plates tends to occur at the ends and the least displacement tends to occur at mid-depth. As a result, connections placed at column ends that will prevent slipping between the parts have the greatest strengthening effect, while those placed at mid-depth have the least effect.

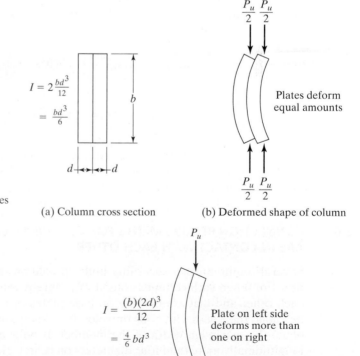

$$I = 2\frac{bd^3}{12}$$

$$= \frac{bd^3}{6}$$

FIGURE 6.3
Column consisting of two plates
not connected to each other.

(a) Column cross section

(b) Deformed shape of column

Plates deform
equal amounts

$$I = \frac{(b)(2d)^3}{12}$$

$$= \frac{4}{6}bd^3$$

Plate on left side
deforms more than
one on right

FIGURE 6.4
Column consisting of two plates fully
connected to each other.

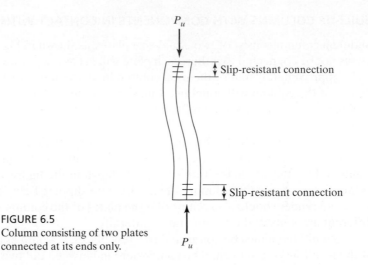

FIGURE 6.5
Column consisting of two plates
connected at its ends only.

Should the plates be fastened together at their ends with slip-resistant connectors, those ends will deform together and the column will take the shape shown in Fig. 6.5. As the plates are held together at the ends, the column will bend in an S shape, as shown in the figure.

If the column were to bend in the S shape shown, its K factor would theoretically equal 0.5 and its KL/r value would be the same as the one for the continuously connected column of Fig. 6.4.[1]

$$\frac{KL}{r} \text{ for the column of Fig. 6.4} = \frac{(1)(L)}{\sqrt{\frac{4}{6}bd^3/2bd}} = 1.732L$$

$$\frac{KL}{r} \text{ for the end-fastened column of Fig. 6.5} = \frac{(0.5)(L)}{\sqrt{\frac{1}{6}bd^3/2bd}} = 1.732L$$

Thus, the design stresses are equal for the two cases, and the columns would carry the same loads. This is true for the particular case described here, but is not applicable for the common case where the parts of Fig. 6.5 begin to separate.

6.6 CONNECTION REQUIREMENTS FOR BUILT-UP COLUMNS WHOSE COMPONENTS ARE IN CONTACT WITH EACH OTHER

Several requirements concerning built-up columns are presented in AISC Specification E6. When such columns consist of different components that are in contact with each other and that are bearing on base plates or milled surfaces, they must be connected at their ends with bolts or welds. If welds are used, the weld lengths must at least equal the maximum width of the member. If bolts are used, they may not be spaced longitudinally more than four diameters on center, and the connection must extend for a distance at least equal to $1\frac{1}{2}$ times the maximum width of the member.

---

[1]J. A. Yura, *Elements for Teaching Load and Resistance Factor Design* (Chicago: AISC, July 1987), pp. 17–19.

The AISC Specification also requires the use of welded or bolted connections between the components of the column end as described in the last paragraph. These connections must be sufficient to provide for the transfer of calculated stresses. If it is desired to have a close fit over the entire faying surfaces between the components, it may be necessary to place the connectors even closer than is required for shear transfer.

When the component of a built-up column consists of an outside plate, the AISC Specification provides specific maximum spacings for fastening. If intermittent welds are used along the edges of the components, or if bolts are provided along all gage lines at each section, their maximum spacing may not be greater than the thickness of the thinner outside plate times $0.75\sqrt{E/F_y}$, nor be greater than 12 in. Should the fasteners be staggered, the maximum spacing along each gage line shall not be greater than the thickness of the thinner outside plate times $1.12\sqrt{E/F_y}$, nor be greater than 18 in (AISC Specification Section E6.2).

In Chapter 12, high-strength bolts are referred to as being *snug-tight* or *slip-critical*. Snug-tight bolts are those that are tightened until all the plies of a connection are in firm contact with each other. This usually means the tightness obtained by the full manual effort of a worker with a spud wrench, or the tightness obtained after a few impacts with a pneumatic wrench.

Slip-critical bolts are tightened much more firmly than are snug-tight bolts. They are tightened until their bodies, or shanks, have very high tensile stresses (approaching the lower bound of their yield stress). Such bolts clamp the fastened parts of a connection so tightly together between the bolt and nut heads that loads are resisted by friction, and slippage is nil. (We will see in Chapter 12 that where slippage is potentially a problem, slip-critical bolts should be used. For example, they should be used if the working or service loads cause a large number of stress changes resulting in a possible fatigue situation in the bolts.)

In the discussion that follows, the letter a represents the distance between connectors, and r_i is the least radius of gyration of an individual component of the column. If compression members consisting of two or more shapes are used, they must be connected together at intervals such that the effective slenderness ratio Ka/r_i of each of the component shapes between the connectors is not larger than 3/4 times the governing or controlling slenderness ratio of the whole built-up member (AISC Commentary E6.1). The end connections must be made with welds or slip-critical bolts with clean mill scale, or blasted, cleaned faying surfaces, with Class A or B faying surfaces. (These surfaces are described in Section J3.8 of the AISC Specification.)

The design strength of compression members built up from two or more shapes in contact with each other is determined with the usual applicable AISC Sections E3, E4 or E7, with one exception. Should the column tend to buckle in such a manner that relative deformations in the different parts cause shear forces in the connectors between the parts, it is necessary to modify the KL/r value for that axis of buckling. This modification is required by Section E6 of the AISC Specification.

Reference is made here to the cover-plated column of Fig. 6.6. If this section tends to buckle about its y axis, the connectors between the W shape and the plates are not subjected to any calculated load. If, on the other hand, it tends to buckle about its x axis, the connectors are subjected to shearing forces. The flanges of the W section and the cover plates will have different stresses and thus different deformations. (In this

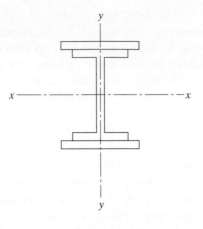

FIGURE 6.6

case, the cover plates and the W flanges to which they are attached bend in the same manner, and thus theoretically no shear or slippage occurs between them.) The result will be shear in the connection between these parts, and $(KL/r)_x$ will have to be modified by AISC Equations E6-1, E6-2a or E6-2b, as described next. (Equation E6-1 is based upon test results that supposedly account for shear deformations in the connectors. Equations E6-2a and E6-2b are based upon theory and was checked by means of tests.)

a.   For intermediate connectors that are snug-tight bolted,

$$\left(\frac{KL}{r}\right)_m = \sqrt{\left(\frac{KL}{r}\right)_o^2 + \left(\frac{a}{r_i}\right)^2}$$ (AISC Equation E6-1)

It is important to remember that the design strength of a built-up column will be reduced if the spacing of connectors is such that one of the components of the column can buckle before the whole column buckles.

b.   For intermediate connectors that are welded or have pretensioned bolts, as required for slip-critical joints,

when $\dfrac{a}{r_i} \leq 40$

$$\left(\frac{KL}{r}\right)_m = \left(\frac{KL}{r}\right)_o$$ (AISC Equation E6-2a)

when $\dfrac{a}{r_i} > 40$

$$\left(\frac{KL}{r}\right)_m = \sqrt{\left(\frac{KL}{r}\right)_o^2 + \left(\frac{K_i a}{r_i}\right)^2}$$ (AISC Equation E6-2b)

In these two equations,

$$\left(\frac{KL}{r}\right)_o$$ = column slenderness ratio of the whole built-up member acting as a unit in the buckling direction

$$\left(\frac{KL}{r}\right)_m$$ = modified slenderness ratio of built-up member because of shear

a = distance between connectors, in

r_i = minimum radius of gyration of individual component, in

K_i = 0.50 for angles back-to-back

= 0.75 for channels back-to-back

= 0.86 for all other cases

For the case in which the column tends to buckle about an axis such as to cause shear in the connection between the column parts, it will be necessary to compute a modified slenderness ratio $(KL/r)_m$ for that axis and to check to see whether that value will cause a change in the design strength of the member. If it does, it may be necessary to revise sizes and repeat the steps just described.

AISC Equation E6-1 is used to compute the modified slenderness ratio $(KL/r)_m$ about the major axis to find out whether it is greater than the slenderness ratio about the minor axis. If it is, that value should be used for determining the design strength of the member.

Section E6 of the AISC Commentary states that, on the basis of judgment and experience, the longitudinal spacing of the connectors for built-up compression members must be such that the slenderness ratios of the individual parts of the members do not exceed three-fourths of the slenderness ratio of the entire member.

Example 6-4 illustrates the design of a column consisting of a W section with cover plates bolted to its flanges, as shown in Fig. 6.7. Even though snug-tight bolts are used for this column, you should realize that AISC Specification E6 states that the end bolts must be pretensioned with Class A or B faying surfaces or the ends must be welded. This is required so that the parts of the built-up section will not slip with respect to each other and thus will act as a unit in resisting loads. (As a practical note, the typical steel company required to tighten the end bolts to a slip-critical condition will probably just go ahead and tighten them all to that condition.)

As this type of built-up section is not shown in the column tables of the AISC Manual, it is necessary to use a trial-and-error design procedure. An effective slenderness ratio is assumed. Then, $\phi_c F_{cr}$ or F_{cr}/Ω_c for that slenderness ratio is determined and divided into the column design load to estimate the column area required. The area of the W section is subtracted from the estimated total area to obtain the estimated cover plate area. Cover plate sizes are then selected to provide the required estimated area.

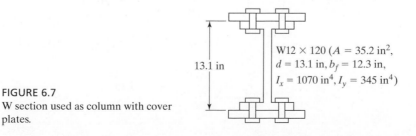

FIGURE 6.7
W section used as column with cover plates.

13.1 in

W12 × 120 (A = 35.2 in², d = 13.1 in, b_f = 12.3 in, I_x = 1070 in⁴, I_y = 345 in⁴)

Example 6-4

You are to design a column for $P_D = 750$ k and $P_L = 1000$ k, using $F_y = 50$ ksi and $KL = 14$ ft. A W12 × 120 (for which $\phi_c P_n = 1290$ k and $P_n/\Omega_c = 856$ k from AISC Manual, Table 4-1) is on hand. Design cover plates to be snug-tight bolted at 6-in spacings to the W section, as shown in Fig. 6.7, to enable the column to support the required load.

Solution

LRFD	ASD
$P_a = (1.2)(750) + (1.6)(1000) = 2500$ k	$P_a = 750 + 1000 = 1750$ k

Assume $\dfrac{KL}{r} = 50$

$$\phi_c F_{cr} = 37.50 \text{ ksi from AISC Table 4-22}$$

$$A \text{ reqd} = \frac{2500 \text{ k}}{37.50 \text{ ksi}} = 66.67 \text{ in}^2$$

$$-A \text{ of } W12 \times 120 = -35.30$$

Estimated A of 2 plates $= 31.37$ in^2 or 15.69 in^2 each

Try one PL1 × 16 each flange

$$A = 35.20 + (2)(1)(16) = 67.20 \text{ in}^2$$

$$I_x = 1070 + (2)(16)\left(\frac{13.1 + 1.00}{2}\right)^2 = 2660 \text{ in}^4$$

$$r_x = \sqrt{\frac{2660}{67.20}} = 6.29 \text{ in}$$

$$\left(\frac{KL}{r}\right)_x = \frac{(12 \text{ in/ft})(14 \text{ ft})}{6.29 \text{ in}} = 26.71$$

$$I_y = 345 + (2)\left(\frac{1}{12}\right)(1)(16)^3 = 1027.7 \text{ in}^4$$

$$r_y = \sqrt{\frac{1027.7}{67.20}} = 3.91 \text{ in}$$

$$\left(\frac{KL}{r}\right)_y = \frac{(12 \text{ in/ft})(14 \text{ ft})}{3.91 \text{ in}} = 42.97 \leftarrow$$

Computing the modified slenderness ratio yields

$$r_i = \sqrt{\frac{I}{A}} = \sqrt{\frac{\left(\frac{1}{12}\right)(16)(1)^3}{(1)(16)}} = 0.289 \text{ in}$$

$$\frac{a}{r_i} = \frac{6 \text{ in}}{0.289 \text{ in}} = 20.76$$

$$\left(\frac{KL}{r}\right)_x = \sqrt{\left(\frac{KL}{r}\right)_0^2 + \left(\frac{a}{r_i}\right)^2} = \sqrt{(26.71)^2 + (20.76)^2} \qquad \text{(AISC Equation E6-1)}$$

$$= 33.83 < 42.97 \therefore \text{ does not control}$$

Checking the slenderness ratio of the plates, we have

$$\frac{k_a}{r_i} = 20.76 < \frac{3}{4}\left(\frac{KL}{r}\right)_y = \left(\frac{3}{4}\right)(42.97) = 32.23$$

$$\text{For } \left(\frac{KL}{r}\right)_y = 42.97.$$

LRFD	ASD
$\phi_c F_{cr} = 39.31$ ksi from Table 4-22, $F_y = 50$ ksi	$\dfrac{F_{cr}}{\Omega_c} = 26.2$ ksi from Table 4-22, $F_y = 50$ ksi
$\phi_c P_n = (39.31)(67.30) = 2646 \text{ k} > 2500 \text{ k}$	$\dfrac{P_n}{\Omega_c} = (26.2)(67.30) = 1763 \text{ k} > 1750 \text{ k}$

Use W12 × 120 with one cover plate 1 × 16 each flange, $F_y = 50$ ksi. (Note: Many other plate sizes could have been selected.)

6.7 BUILT-UP COLUMNS WITH COMPONENTS NOT IN CONTACT WITH EACH OTHER

Example 6-5 presents the design of a member built up from two channels that are not in contact with each other. The parts of such members need to be connected or laced together across their open sides. The design of lacing is discussed immediately after this example and is illustrated in Example 6-6.

Example 6-5

Select a pair of 12-in standard channels for the column shown in Fig. 6.8, using $F_y = 50$ ksi. For connection purposes, the back-to-back distance of the channels is to be 12 in. $P_D = 100$ k and $P_L = 300$ k. Consider both LRFD and ASD procedures.

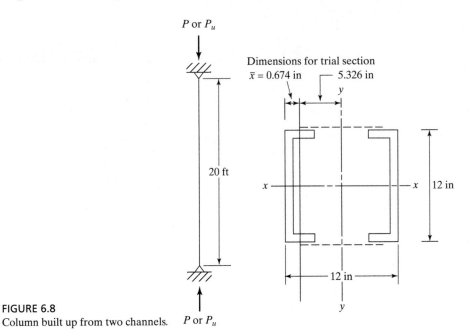

FIGURE 6.8
Column built up from two channels.

Solution

LRFD	ASD
$P_u = (1.2)(100) + (1.6)(300) = 600$ k	$P_a = 100 + 300 = 400$ k

Assume $\dfrac{KL}{r} = 50$

$\phi_c F_{cr} = 37.50$ ksi, from Table 4-22 ($F_y = 50$ ksi steel)

$A \text{ reqd} = \dfrac{600 \text{ k}}{37.50 \text{ ksi}} = 16.00 \text{ in}^2$

Try 2C12 × 30s. (For each channel, A = 8.81 in², I_x = 162 in⁴, I_y = 5.12 in⁴, $\bar{x}$ = 0.674 in.)

$$I_x = (2)(162) = 324 \text{ in}^4$$

$$I_y = (2)(5.12) + (2)(8.81)(5.326)^2 = 510 \text{ in}^4$$

$$r_x = \sqrt{\frac{324}{(2)(8.81)}} = 4.29 \text{ in controls}$$

$$KL = (1.0)(20 \text{ ft}) = 20 \text{ ft}$$

$$\frac{KL}{r} = \frac{(12 \text{ in/ft})(20 \text{ ft})}{4.29 \text{ in}} = 55.94$$

LRFD	ASD
$\phi_c F_{cr}$ = 35.82 ksi (AISC Table 4-22) F_y = 50 ksi	$\dfrac{F_{cr}}{\Omega_c}$ = 23.81 ksi (AISC Table 4-22) F_y = 50 ksi
$\phi_c P_n$ = (35.82)(2 × 8.81) = 631 k > 600 k **OK**	$\dfrac{P_n}{\Omega_c}$ = (23.81)(2 × 8.81) = 419.5 k > 400 k **OK**

Checking   width   thickness   ratios   of   channels   (d = 12.00 in, b_f = 3.17 in, t_f = 0.501 in, t_w = 0.510 in, $k = 1\frac{1}{8}$ in)

Flanges

$$\frac{b}{t} = \frac{3.17}{0.501} = 6.33 < 0.56\sqrt{\frac{29,000}{50}} = 13.49 \quad \text{OK (Case 1, AISC Table B4-1a)}$$

Webs

$$\frac{h}{t_w} = \frac{12.00 - (2)(1.125)}{0.510} = 19.12 < 1.49\sqrt{\frac{29,000}{50}}$$

$$= 35.88 \text{ (Case 5, AISC Table B4-1a)}$$

$\therefore$ Nonslender member

Use 2C12 × 30s.

The open sides of compression members that are built up from plates or shapes may be connected together with continuous cover plates with perforated holes for access purposes, or they may be connected together with lacing and tie plates. (The consideration of lacing is important because of retrofit work where it is used particularly for channels.)

The purposes of the perforated cover plates and the lacing, or latticework, are to hold the various parts parallel and the correct distance apart, and to equalize the stress distribution between the various parts. The student will understand the necessity for lacing if he or she considers a built-up member consisting of several sections (such as the four-angle member of Fig. 5.2(i)) that supports a heavy compressive load. Each of the parts will tend to individually buckle laterally, unless they are tied together to act as a unit in supporting the load. In addition to lacing, it is necessary to have *tie plates* (also called *stay plates* or *batten plates*) as near the ends of the member as possible, and at intermediate points if the lacing is interrupted. Parts (a) and (b) of Fig. 6.9 show arrangements of tie plates and lacing. Other possibilities are shown in parts (c) and (d) of the same figure.

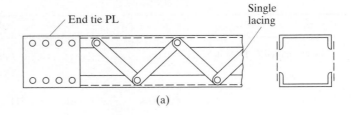

(a)

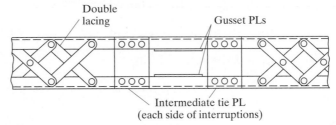

(b)

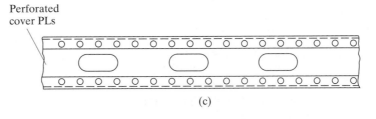

(c)

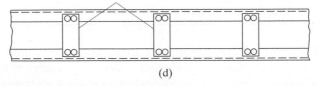

(d)

FIGURE 6.9
Lacing and perforated cover plates.

The failure of several structures in the past has been attributed to inadequate lacing of built-up compression members. Perhaps the best-known example was the failure of the Quebec Bridge in 1907. Following its collapse, the general opinion was that the latticework of the compression chords was too weak, and this resulted in failure.

If continuous cover plates perforated with access holes are used to tie the members together, the AISC Specification E6.2 states that (a) they must comply with the limiting width–thickness ratios specified for compression elements in Section B4.1 of the AISC Specification; (b) the ratio of the access hole length (in the direction of stress) to the hole width may not exceed 2; (c) the clear distance between the holes in the direction of stress may not be less than the transverse distance between the nearest lines of connecting fasteners or welds; and (d) the periphery of the holes at all points must have a radius no less than $1\frac{1}{2}$ in. Stress concentrations and secondary bending stresses are usually neglected, but lateral shearing forces must be checked as they are for other types of latticework. (The unsupported width of such plates at access holes is assumed to contribute to the design strength $\phi_c P_n$ of the member if the conditions as to sizes, width–thickness ratios, etc., described in AISC Specification E6 are met.) Perforated cover plates are attractive to many designers because of several advantages they possess:

1. They are easily fabricated with modern gas cutting methods.
2. Some specifications permit the inclusion of their net areas in the effective section of the main members, provided that the holes are made in accordance with empirical requirements, which have been developed on the basis of extensive research.
3. Painting of the members is probably simplified, compared with painting of ordinary lacing bars.

Dimensions of tie plates and lacing are usually controlled by specifications. Section E6 of the AISC Specification states that tie plates shall have a thickness at least equal to one-fiftieth of the distance between the connection lines of welds or other fasteners.

Lacing may consist of flat bars, angles, channels, or other rolled sections. These pieces must be so spaced that the individual parts being connected will not have L/r values between connections which exceed three-fourths of the governing value for the entire built-up member. (The governing value is KL/r for the whole built-up section.) Lacing is assumed to be subjected to a shearing force normal to the member, equal to not less than 2 percent of the compression design strength $\phi_c P_n$ of the member. The AISC column formulas are used to design the lacing in the usual manner. Slenderness ratios are limited to 140 for single lacing and 200 for double lacing. Double lacing or single lacing made with angles should preferably be used if the distance between connection lines is greater than 15 in.

Example 6-6 illustrates the design of lacing and end tie plates for the built-up column of Example 6-5. Bridge specifications are somewhat different in their lacing requirements from the AISC, but the design procedures are much the same.

Example 6-6

Using the AISC Specification and 36 ksi steel, design bolted single lacing for the column of Example 6-5. Reference is made to Fig. 6.10. Assume that 3/4-in bolts are used.

Solution. Distance between lines of bolts is 8.5 in $<$ 15 in; therefore, single lacing is OK.

Assume that lacing bars are inclined at 60° with axis of member. Length of channels between lacing connections is 8.5/cos 30° = 9.8 in, and *KL/r* of 1 channel between connections is 9.8/0.762 = 12.9 $<$ 3/4 $\times$ 55.94, which is *KL/r* of main member previously determined in Example 6-5. Only the LRFD solution is shown.

Force on lacing bar:

$$V_u = 0.02 \text{ times available design compressive strength of member}$$
$$\text{(from Example 6-5), } \phi P_n = 631 \text{ k}$$

$$V_u = (0.02)(631 \text{ k}) = 12.62 \text{ k}$$

$$\frac{1}{2}V_u = 6.31 \text{ k} = \text{shearing force on each plane of lacing}$$

Force in bar (with reference to bar dimensions in Fig. 6.10):

$$\left(\frac{9.8}{8.5}\right)(6.31) = 7.28 \text{ k}$$

Properties of flat bar:

$$I = \tfrac{1}{12}bt^3$$

$$A = bt$$

$$r = \sqrt{\frac{\frac{1}{12}bt^3}{bt}} = 0.289t$$

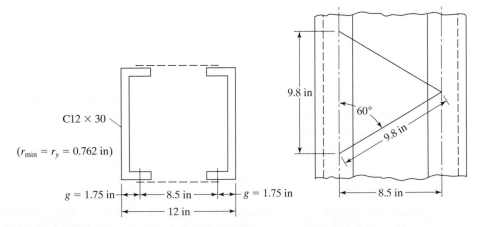

FIGURE 6.10
Two-channel column section with lacing.

Design of bar:

$$\text{Assume } \frac{KL}{r} = \text{maximum value of 140 and } K = 1.0$$

$$\frac{(1.0)\, 9.8 \text{ in}}{0.289t \text{ in}} = 140$$

$$t = 0.242 \text{ in } (\text{try } \tfrac{1}{4}\text{-in flat bar})$$

$$\frac{KL}{r} = \frac{(1.0)\, 9.8 \text{ in}}{(0.289)(0.250 \text{ in})} = 136$$

$$\phi_c F_{cr} = 12.2 \text{ ksi, from AISC Table 4-22, } F_y = 36 \text{ ksi}$$

$$\text{Area reqd} = \frac{7.28 \text{ k}}{12.2 \text{ ksi}} = 0.597 \text{ in}^2 \left(2.39 \times \tfrac{1}{4} \text{ needed}\right) \text{ Use } \tfrac{1}{4} \times 2\tfrac{1}{2} \text{ bar}$$

Minimum edge distance if $\tfrac{3}{4}$-in bolt used $= 1\tfrac{1}{4}$ in   AISC Table J3.4

$\therefore$ Minimum length of bar $= 9.8 + (2)\left(1\tfrac{1}{4}\right) = 12.3$ in, say, 14 in

Use $\tfrac{1}{4} \times 2\tfrac{1}{2} \times$ 1-ft 2-in bars, $F_y = 36$ ksi.

Design of end tie plates:

Minimum length $= 8.5$ in

Minimum $t = \left(\tfrac{1}{50}\right)(8.5) = 0.17$ in

Minimum width $= 8.5 + (2)\left(1\tfrac{1}{4}\right) = 11$ in

Use $\tfrac{3}{16} \times 8\tfrac{1}{2} \times$ 0-ft 12-in end tie plates.

6.8   SINGLE-ANGLE COMPRESSION MEMBERS

You will note that we have not discussed the design of single-angle compression members up to this point. The AISC has long been concerned about the problems involved in loading such members fairly concentrically. It can be done rather well if the ends of the angles are milled and if the loads are applied through bearing plates. In practice, however, single-angle columns are often used in such a manner that rather large eccentricities of load applications are present. The sad result is that it is somewhat easy to greatly underdesign such members.

In Section E5 of the AISC Specification, a special specification is provided for the design of single-angle compression members. Though this specification includes information for tensile, shear, compressive, flexural, and combined loadings, the present discussion is concerned only with the compression case.

In Table 4-11 of the Manual, the calculated strengths of concentrically loaded single angles are provided. The values shown are based on KL/r_z values. So often, though, single angles are connected at their ends by only one leg—an eccentric loading situation. Section E5 of the AISC Specification presents a method for handling such situations where eccentric compression loads are introduced to angles through one connected leg.

The writers of the specification assumed that connections to one leg of an angle provided considerable resistance to bending about the y axis of that angle or that is perpendicular to the connected leg. As a result, the angle was assumed to bend and buckle about the x axis of the member; thus, attention is given to the L/r_x ratio. To account for the eccentricity of loading larger L/r_x ratios for various situations are provided by AISC Equations E5-1 to E5-4 and are to be used for obtaining design stresses.

The first two of the equations apply to equal leg angles and to unequal leg angles connected through their longer legs. Further, the angles are to be used as members in two-dimensional, or planar, trusses, where the other members joining the ones in question are connected at their ends on the same side of gusset plates or on the same side of truss chord members. For these conditions, the following increased slenderness ratios are to be used for strength calculations:

When $L/r_x \leq 80$:

$$\frac{KL}{r} = 72 + 0.75\frac{L}{r_x} \qquad \text{(AISC Equation E5-1)}$$

When $L/r_x > 80$:

$$\frac{KL}{r} = 32 + 1.25\frac{L}{r_x} \leq 200 \qquad \text{(AISC Equation E5-2)}$$

Some variations are given in the specification for unequal leg angles if the leg length ratios are < 1.7 and if the shorter leg is connected. In addition, AISC Equations E5-3 and E5-4 are provided for cases where the single angles are members of box or space trusses. Example 6-7, which follows, illustrates the use of the first of these equations.

Example 6-7

Determine the $\phi_c P_n$ and P_n/Ω_c values for a 10-ft-long A36 angle $8 \times 8 \times 3/4$ with simple end connections, used in a planar truss. The other web members meeting at the ends of this member are connected on the same side of the gusset plates.

Solution

Using an L8 $\times$ 8 $\times$ $\dfrac{3}{4}(A = 11.5 \text{ in}^2, r_x = 2.46 \text{ in})$

$$\frac{L}{r_x} = \frac{(12 \text{ in/ft})(10 \text{ ft})}{2.46 \text{ in}} = 48.78 < 80$$

$$\therefore \quad \frac{KL}{r} = 72 + 0.75\frac{L}{r_x} \qquad \text{(AISC Equation E5-1)}$$

$$= 72 + (0.75)(48.78) = 108.6$$

LRFD	ASD
$\phi_c F_{cr}$ from AISC Table 4-22, $F_y = 36$ ksi	$\dfrac{F_{cr}}{\Omega_c}$ from AISC Table 4-22, $F_y = 36$ ksi
$= 17.38$ ksi	$= 11.58$ ksi
$\phi_c P_n = \phi_c F_{cr} A_g$	$\dfrac{P_n}{\Omega_c} = \dfrac{F_{cr}}{\Omega_c} A_g$
$= (17.38 \text{ ksi})(11.5 \text{ in}^2) = \mathbf{199.9\ k}$	$= (11.58 \text{ ksi})(11.5 \text{ in}^2) = \mathbf{133.2\ k}$

Table 4-12 in the Manual provides design values for angles eccentrically loaded in Example 6-7, because the AISC used some different conditions in solving the problem. The values in the table are the lower-bound axial compression strengths of single angles, with no consideration of end restraint. When the conditions described in AISC Specification E5 are not met, this table can be used. The values given were computed, considering biaxial bending about the principal axis of the angle, with the load applied at a given eccentricity, as described on page 4-8 in the Manual.

6.9   SECTIONS CONTAINING SLENDER ELEMENTS

A good many of the square and rectangular HSS have slender walls. The reader will be happy to learn that the effects of slender elements on column strengths have been included in the tables of Part 4 of the Manual. As a result, the designer rarely has to go through the calculations to take into account those items. Several equations are presented in AISC Section E7.1 and E7.2 for the consideration of members containing slender elements. Included are sections with stiffened elements and sections with unstiffened elements. Example 6-8 makes use of the appropriate equations for computing the strengths of such members.

The values obtained in the example problem to follow (Example 6-8) are smaller than the values given in Table 4-3 in the Manual for rectangular HSS sections, because f was assumed to equal F_y, whereas in the proper equations it is actually equal to $\dfrac{P_n}{A_e}$.

This conservative assumption will cause our hand calculations for design strengths, when slender elements are present, to be on the low, or safe, side. To use the correct value of f, it is necessary to use an iterative solution—a procedure for which the computer is ideally suited. In any case, the values calculated by hand, shown as follows, will be several percent on the conservative or low side.

Example 6-8

Determine the axial compressive design strength $\phi_c P_n$ and the allowable design strength $\dfrac{P_n}{\Omega_c}$ of a 24-ft HSS 14 × 10 × $\dfrac{1}{4}$ column section. The base of the column is considered to be fixed, while the upper end is assumed to be pinned. $F_y = 46$ ksi.

Solution

Using an HSS 14 × 10 × $\dfrac{1}{4}$ ($A = 10.8$ in², $r_x = 5.35$ in, $r_y = 4.14$ in, $t_w = 0.233$ in,

$$\frac{b}{t} = 39.9 \text{ and } \frac{h}{t} = 57.1) \text{ All values from Table 1-11}$$

Limiting width–thickness ratio (AISC Table B4.1a, Case 6)

$$\lambda_r = 1.40\sqrt{\frac{E}{F_y}} = 1.40\sqrt{\frac{29{,}000}{46}} = 35.15 < \frac{b}{t} \text{ and } \frac{h}{t}$$

∴ **Both the 10-in walls and the 14-in walls are slender elements.**

Computing b and h and noting that in the absence of the exact fillet dimensions, the AISC recommends that the widths and depths between the web toes of the fillets equal the outside dimensions $-3t_w$.

$$b = 10.00 - (3)(0.233) = 9.30 \text{ in}$$

$$h = 14.00 - (3)(0.233) = 13.30 \text{ in}$$

Computing the effective widths and heights of the walls by using AISC Equation E7-18 yields

$$b_e = 1.92\, t\sqrt{\frac{E}{f}}\left[1 - \frac{0.38}{(b/t)}\sqrt{\frac{E}{f}}\right] \le b$$

b_e for the 10-in wall

$$= (1.92)(0.233)\sqrt{\frac{29{,}000}{46}}\left[1 - \frac{0.38}{39.9}\sqrt{\frac{29{,}000}{46}}\right]$$

$$= 8.55 \text{ in} < 9.30 \text{ in}$$

Length that cannot be used $= 9.30 - 8.55 = 0.75$ in

b_e for the 14-in wall

$$= (1.92)(0.233)\sqrt{\frac{29{,}000}{46}}\left[1 - \frac{0.38}{57.1}\sqrt{\frac{29{,}000}{46}}\right]$$

$$= 9.36 \text{ in} < 13.30 \text{ in}$$

Length that cannot be used $= 13.30 - 9.36 = 3.94$ in

$$A_e = 10.8 - (2)(0.233)(0.75) - (2)(0.233)(3.94) = 8.61 \text{ in}^2$$

$$Q = Q_a = \frac{A_e}{A_g} = \frac{8.61}{10.8} = 0.7972$$

Determine equation to use for F_{cr}

$$\left(\frac{KL}{r}\right)_y = \frac{(0.8)(12 \text{ in/ft} \times 24 \text{ ft})}{4.14 \text{ in}} = 55.65$$

$$< 4.71\sqrt{\frac{29,000}{(0.7972)(46)}} = 132.45$$

$$F_e = \frac{\pi^2 E}{\left(\dfrac{KL}{r}\right)^2} = \frac{(\pi^2)(29,000)}{(55.65)^2} = 92.42 \text{ ksi} \qquad \text{(AISC Equation E3-4)}$$

$$F_{cr} = Q\left[0.658^{\frac{QF_y}{F_e}}\right]F_y \qquad \text{(AISC Equation E7-2)}$$

$$= 0.7972\left[0.658^{\frac{0.7972 \times 46}{92.42}}\right]46$$

$$= 31.06 \text{ ksi}$$

$$P_n = (10.8)(31.06) = 335.4 \text{ k} \qquad \text{(AISC Equation E7-1)}$$

LRFD $\phi_c = 0.90$	ASD $\Omega_c = 1.67$
$\phi_c P_n = (0.90)(335.4 \text{ k}) = \textbf{301.9 k}$	$\dfrac{P_n}{\Omega_c} = \dfrac{335.4 \text{ k}}{1.67} = \textbf{200.8 k}$

6.10   FLEXURAL-TORSIONAL BUCKLING OF COMPRESSION MEMBERS

Usually symmetrical members such as W sections are used as columns. Torsion will not occur in such sections if the lines of action of the lateral loads pass through their shear centers. The *shear center* is that point in the cross section of a member through which the resultant of the transverse loads must pass so that no torsion will occur. The calculations necessary to locate shear centers are presented in Chapter 10. The shear centers of the commonly used doubly symmetrical sections occur at their centroids. This is not necessarily the case for other sections such as channels and angles. Shear center locations for several types of sections are shown in Fig. 6.11. Also shown in the figure are the coordinates x_0 and y_0 for the shear center of each section with respect to its centroid. These values are needed to solve the flexural-torsional formulas, presented later in this section.

Even though loads pass through shear centers, torsional buckling still may occur. If you load any section through its shear center, no torsion will occur, but one still

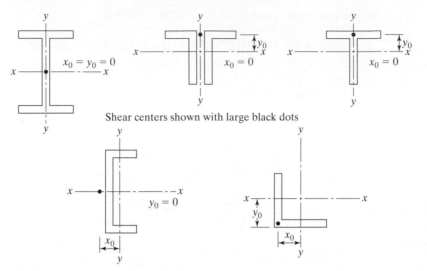

FIGURE 6.11
Shear center locations for some common column sections.

computes torsional buckling strength for these members—that is, buckling load does not depend on the nature of the axial or transverse loading; rather, it depends on the cross-section properties, column length, and support conditions.

Axially loaded compression members can theoretically fail in four different fashions: by local buckling of elements that form the cross section, by flexural buckling, by torsional buckling, or by flexural-torsional buckling.

Flexural buckling (also called *Euler buckling* when elastic behavior occurs) is the situation considered up to this point in our column discussions where we have computed slenderness ratios for the principal column axes and determined $\phi_c F_{cr}$ for the highest ratios so obtained. Doubly symmetrical column members (such as W sections) are subject only to local buckling, flexural buckling, and torsional buckling.

Because torsional buckling can be very complex, it is very desirable to prevent its occurrence. This may be done by careful arrangements of the members and by providing bracing to prevent lateral movement and twisting. If sufficient end supports and intermediate lateral bracing are provided, flexural buckling will always control over torsional buckling. The column design strengths given in the AISC column tables for W, M, S, tube, and pipe sections are based on flexural buckling.

Open sections such as Ws, Ms, and channels have little torsional strength, but box beams have a great deal. Thus, if a torsional situation is encountered, it may be well to use box sections or to make box sections out of W sections by adding welded side plates (). Another way in which torsional problems can be reduced is to shorten the lengths of members that are subject to torsion.

For a singly symmetrical section such as a tee or double angle, Euler buckling may occur about the x or y axis. For equal-leg single angles, Euler buckling may occur about the z axis. For all these sections, flexural-torsional buckling is definitely a possibility and may control. (It will always control for unequal-leg single-angle columns.) The values given in the AISC column load tables for double-angle and structural tee sections were computed for buckling about the weaker of the x or y axis and for flexural-torsional buckling.

The average designer does not consider the torsional buckling of symmetrical shapes or the flexural-torsional buckling of unsymmetrical shapes. The feeling is that these conditions don't control the critical column loads, or at least don't affect them very much. This assumption can be far from the truth. Should we have unsymmetrical columns or even symmetrical columns made up of thin plates, however, we will find that torsional buckling or flexural-torsional buckling may significantly reduce column capacities.

Section E4 of the AISC Specification is concerned with the torsional or flexural-torsional buckling of steel columns. Part (b) of the section presents a general method for handling the problem which is applicable to all shapes. Part (a) of the same section is a modification of the procedure presented in part (b) and is applicable specifically to double angles and tee sections used as columns.

The general approach of part (b) is presented here. The procedure involves using AISC Equation E4-9 for the determination of the elastic torsional buckling stress F_{ez} (which is analogous to the Euler buckling stress). After this value is determined, it is used in the appropriate one of AISC Equations E4-4, E4-5, and E4-6 to obtain F_e, the torsional or flexural-torsional elastic buckling stress. The critical stress, F_{cr}, is then determined according to Equation E3-2 or E3-3.

The procedure for part (a), which is for double-angle and tee-shaped compression members, is presented next. The critical stress, F_{cr}, is determined using AISC Equation E4-2. In this equation, F_{cry} is taken as F_{cr} from Equation E3-2 or E3-3, and F_{crz} and H are obtained from Equations E4-3 and E4-10 respectively.

For either procedure, the nominal compressive strength P_n, for the limit states of torsional and flexural-torsional buckling, is determined using AISC Equation E4-1. In the equation, the previously calculated F_{cr} is multiplied by A_g.

Usually it is unnecessary to consider torsional buckling for doubly symmetrical shapes. Furthermore we rarely have to consider the topic for shapes without an axis of symmetry because we probably will not use such members as columns. On some occasions, however, we probably will select sections with one axis of symmetry as columns and for them lateral torsional buckling must be considered.

A numerical example (Example 6-9) for flexural torsional buckling is presented in this section for a WT section used as a column. For such a shape the x axis will be subject to flexural buckling while there may be flexural buckling about the y axis (the axis of symmetry) as well as lateral torsional buckling.

There are four steps involved in solving this type of problem with the AISC Specification. These follow:

1. Determine the flexural buckling strength of the member for its x axis using AISC Equations E3-4, E3-2 or E3-3, as applicable, and E3-1.
2. Determine the flexural buckling strength of the member for its y axis using AISC Equations E3-4, E3-2 or E3-3, as applicable, and E3-1.
3. Determine the flexural torsional buckling strength of the member for its y axis using AISC Equations E4-11, E4-9, E4-10, E4-5, E3-2 or E3-3, as applicable, and E4-1.
4. Select the smallest P_n value determined in the preceding three steps.

Example 6-9

Determine the nominal compressive strength, P_n, of a WT10.5 $\times$ 66 with $KL_x = 25$ ft and $KL_y = KL_z = 20$ ft. Use the general approach given in part (b) of AISC Specification E4(b) and A992 steel.

Solution

Using a WT10.5 $\times$ 66 ($A = 19.4$ in^2, $t_f = 1.04$ in, $I_x = 181$ in^4, $r_x = 3.06$ in, $I_y = 166$ in^4, $r_y = 2.93$ in, $\bar{y} = 2.33$ in, $J = 5.62$ in^4, $C_w = 23.4$ in^6, and $G = 11,200$ ksi)

(1) Determine the flexural buckling strength for the x axis

$$\left(\frac{KL}{r}\right)_x = \frac{(12 \text{ in/ft})(25 \text{ ft})}{3.06 \text{ in}} = 98.04$$

$$F_{ex} = \frac{\pi^2 E}{\left(\dfrac{KL}{r}\right)_x^2} = \frac{(\pi^2)(29{,}000)}{(98.04)^2} = 29.78 \text{ ksi} \qquad \text{(AISC Equation E3-4)}$$

$$\left(\frac{KL}{r}\right)_x = 98.04 < 4.71\sqrt{\frac{29{,}000}{50}} = 113.43$$

$$\therefore F_{cr} = \left[0.658^{\frac{F_y}{F_e}}\right]F_y \qquad \text{(AISC Equation E3-2)}$$

$$= \left[0.658^{\frac{50}{29.78}}\right]50 = 24.76 \text{ ksi}$$

The nominal strength P_n for flexural buckling about x-axis is

$$P_n = F_{cr}A_g = (24.76)(19.4) = \mathbf{480.3 \ k} \qquad \text{(AISC Equation E3-1)}$$

(2) Determine the flexural buckling strength for the y axis

$$\left(\frac{KL}{r}\right)_y = \frac{12 \text{ in/ft} \times 20 \text{ ft}}{2.93 \text{ in}} = 81.91$$

$$F_{ey} = \frac{\pi^2 E}{\left(\dfrac{KL}{r}\right)_y^2} = \frac{(\pi^2)(29{,}000)}{(81.91)^2} = 42.66 \text{ ksi} \qquad \text{(AISC Equation E3-4)}$$

$$\left(\frac{KL}{r}\right)_y = 81.91 < 4.71\sqrt{\frac{29{,}000}{50}} = 113.43$$

$$\therefore F_{cr} = \left[0.658^{\frac{F_y}{F_e}}\right]F_y \qquad \text{(AISC Equation E3-2)}$$

$$= \left[0.658^{\frac{50}{42.66}}\right]50 = 30.61 \text{ ksi}$$

The nominal strength P_n for flexural buckling about y axis is

$$P_n = F_{cr} A_g = (30.61)(19.4) = \textbf{593.8 k} \qquad \text{(AISC Equation E3-1)}$$

(3) Determine the flexural–torsional buckling strength of the member about the y axis.

Note that x_0 and y_0 are the coordinates of the shear center with respect to the centroid of the section. Here x_0 equals 0 because the shear center of the WT is located on the y–y axis, while y_0 is $\overline{y} - \dfrac{t_f}{2}$ since the shear center is located at the intersection of the web and flange center lines as shown in Fig. 6.11.

$$x_0 = 0$$

$$y_0 = \overline{y} - \frac{t_f}{2} = 2.33 - \frac{1.04}{2} = 1.81$$

$\overline{r}_0 = $ the polar radius of gyration about the shear center

$$\overline{r}_0^2 = x_0^2 + y_0^2 + \frac{I_x + I_y}{A_g} \qquad \text{(AISC Equation E4-11)}$$

$$= 0^2 + 1.81^2 + \frac{181 + 166}{19.4} = 21.16 \text{ in}^2$$

$$F_{ez} = \left(\frac{\pi^2 E C_w}{(K_z L)^2} + GJ\right)\frac{1}{A_g \overline{r}_0^2} \qquad \text{(AISC Equation E4-9)}$$

$$= \left[\frac{\pi^2(29,000)(23.4)}{(12 \times 20)^2} + 11,200(5.62)\right]\frac{1}{19.4(21.16)} = 153.62 \text{ ksi}$$

$$H = 1 - \frac{x_0^2 + y_0^2}{\overline{r}_0^2} \qquad \text{(AISC Equation E4-10)}$$

$$= 1 - \frac{0^2 + 1.81^2}{21.16} = 0.84517$$

$$F_e = \left(\frac{F_{ey} + F_{ez}}{2H}\right)\left[1 - \sqrt{1 - \frac{4 F_{ey} F_{ez} H}{(F_{ey} + F_{ez})^2}}\right] \qquad \text{(AISC Equation E4-5)}$$

$$= \left(\frac{42.66 + 153.62}{(2)(0.84517)}\right)\left[1 - \sqrt{1 - \frac{(4)(42.66)(153.62)(0.84517)}{(42.66 + 153.62)^2}}\right]$$

$$= 40.42 \text{ ksi}$$

Now we need to go back to either AISC Equation E3-2 or E3-3 to determine the compressive strength of the member.

$$40.42 \text{ ksi} > \frac{F_y}{2.25} = 22.22 \text{ ksi}$$

$$\therefore \text{ Must use AISC Equation E3-2.}$$

$$F_{cr} = \left[0.658^{\frac{F_y}{F_e}}\right]F_y = \left(0.658^{\frac{50}{40.42}}\right)50 = 29.79 \text{ ksi}$$

The nominal strength is

$$P_n = F_{cr}A_g = (29.79)(19.4) = \mathbf{577.9 \text{ k}} \qquad \text{(AISC Equation E4-1)}$$

(4) The smallest one of the P_n values determined in (a), (b), and (c) is our nominal load.

$$\boldsymbol{P_n = 480.3 \text{ k}}$$

LRFD $\phi_c = 0.90$	ASD $\Omega_c = 1.67$
$\phi_c P_n = (0.90)(480.3 \text{ k}) = 432.3$	$\dfrac{P_n}{\Omega_c} = \dfrac{480.3 \text{ k}}{1.67} = 287.6 \text{ k}$

6.11   PROBLEMS FOR SOLUTION

All of the columns for these problems are assumed to be in frames that are braced against sidesway. Each problem is solved by both the LRFD and the ASD procedures.

6-1 to 6-3.   *Use a trial-and-error procedure in which a KL/r value is estimated, the stresses $\phi_c F_{cr}$ and F_{cr}/Ω_c determined from AISC Table 4-22, required areas calculated and trial sections selected–and checked and revised if necessary.*

6-1.   Select the lightest available W10 section to support the axial compression loads $P_D = 100$ k and $P_L = 160$ k if $KL = 15$ ft and A992, Grade 50 steel is used. (*Ans.* W10 × 49, LRFD and ASD)

6-2.   Select the lightest available W8 section to support the axial loads $P_D = 75$ k and $P_L = 125$ k if $KL = 13$ ft and $F_y = 50$ ksi.

6-3.   Repeat Prob. 6-2 if $F_y = 36$ ksi. (*Ans.* W8 × 48, LRFD and ASD)

6-4 to 6-17.   *Take advantage of all available column tables in the AISC Manual, particularly those in Part 4.*

6-4.   Repeat Prob. 6-1.

6-5.   Repeat Prob. 6-2. (*Ans.* W8 × 35, LRFD and ASD)

6-6.   Repeat Prob. 6-1 if $P_D = 150$ k and $P_L = 200$ k.

6-7.   Several building columns are to be designed, using A992 steel and the AISC Specification. Select the lightest available W sections and state the LRFD design strength, $\phi_c P_n$, and the ASD allowable strength, P_n/Ω_c, for these columns that are described as follows:

a. $P_D = 170$ k, $P_L = 80$ k, $L = 16$ ft, pinned end supports, W8.
   (*Ans.* W8 × 48, LRFD $\phi_c P_n = 340$ k $> P_u = 332$ k; W8 × 58, ASD $P_n/\Omega_c = 278$ k $>$ $P_a = 250$ k)

b. $P_D = 100$ k, $P_L = 220$ k, $L = 25$ ft, fixed at bottom, pinned at top, W14.
   (*Ans.* W14 × 74, LRFD $\phi_c P_n = 495$ k $> P_u = 472$ k; W14 × 74, ASD $P_n/\Omega_c = 329$ k $>$ $P_a = 320$ k)

c. $P_D = 120$ k, $P_L = 100$ k, $L = 25$ ft, fixed end supports, W12.
   (*Ans.* W12 × 50, LRFD $\phi_c P_n - 319$ k $> P_u = 304$ k; W12 × 53, ASD $P_n/\Omega_c = 297$ k $>$ $P_a = 220$ k)

d. $P_D = 250$ k, $P_L = 125$ k, $L = 18.5$ ft, pinned end supports, W14.
   (*Ans.* W14 × 74, LRFD $\phi_c P_n = 546$ k $> P_u = 500$ k; W14 × 82, ASD $P_n/\Omega_c = 400$ k $>$ $P_a = 375$ k)

6-8. Design a column with an effective length of 22 ft to support a dead load of 65 k, a live load of 110 k, and a wind load of 144 k. Select the lightest W12 of A992 steel.

6-9. A W10 section is to be selected to support the loads $P_D = 85$ k and $P_L = 140$ k. The member, which is to be 20 ft long, is fixed at the bottom and fixed against rotation but free to translate at the top. Use A992 steel. (*Ans.* W10 × 68, LRFD and ASD, $\phi_c P_n = 363$ k and $P_n/\Omega_c = 241$ k)

6-10. A W14 section is to be selected to support the loads $P_D = 500$ k and $P_L = 700$ k. The member is 24 ft long with pinned end supports and is laterally supported in the weak direction at the one-third points of the total column length. Use 50 ksi steel.

6-11. Repeat Prob. 6-10 if column length is 18 ft long and $P_D = 250$ k and $P_L = 350$ k. (*Ans.* W14 × 74, LRFD, $\phi_c P_n = 893$ k; W14 × 82, ASD, $P_n/\Omega_c = 655$ k)

6-12. A 28 ft long column is pinned at the top and fixed at the bottom, and has additional pinned support in the weak axis direction at a point 12 ft from the top. Assume the column is part of a braced frame. Axial gravity loads are $P_D = 220$ k and $P_L = 270$ k. Choose the lightest W12 column.

6-13. A 24 ft column in a braced frame building is to be built into a wall in such a manner that it will be continuously braced in its weak axis direction but not about its strong axis direction. If the member is to consist of 50 ksi and is assumed to have fixed ends, select the lightest satisfactory W10 section available using the AISC Specification. Loads are $P_D = 220$ k and $P_L = 370$ k. (*Ans.* W10 × 77 LRFD and ASD)

6-14. Repeat Prob. 6-13 if $P_D = 175$ k and $P_L = 130$ k. Select the lightest satisfactory W8 section available.

6-15. A W12 section of 50 ksi steel is to be selected to support the axial compressive loads of $P_D = 375$ k and $P_L = 535$ k. The member is 36 ft long, is to be pinned top and bottom and is to have lateral support at its one-quarter points, perpendicular to the y-axis (pinned). (*Ans.* W12 × 152 LRFD; W12 × 170 ASD)

6-16. Using the steels for which the column tables are provided in Part 4 of the Manual, select the lightest available rolled sections (W, HP, HSS Square, and HSS Round) that are adequate for the following conditions:
   a. $P_D = 150$ k, $P_L = 225$ k, $L = 25$ ft, one end pinned and the other fixed
   b. $P_D = 75$ k, $P_L = 225$ k, $L = 16$ ft, fixed ends
   c. $P_D = 50$ k, $P_L = 150$ k, $L = 30$ ft, pinned ends

6-17. Assuming axial loads only, select W10 sections for the interior column of the laterally braced frame shown in the accompanying illustration. Use $F_y = 50$ ksi and the LRFD method only. A column splice will be provided just above point B;

therefore, select a column section for column *AB* and a second different column section for column *BC* and *CD*. Miscellaneous data: Concrete weighs 150 lb/ft³. LL on roof = 30 psf. Roofing DL = 10 psf. LL on floors = 75 psf. Superimposed DL on floors = 12 psf. Partition load on floors = 15 psf. All joints are assumed to be pinned. Frames are 35 feet on center. (*Ans.* Column *AB* − W10×68, Column *BC* & *CD* − W10×39)

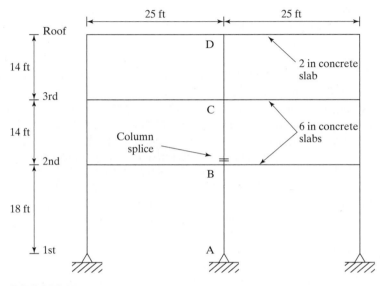

FIGURE P6-17

6-18.   You are requested to design a column for $P_D = 225$ k and $P_L = 400$ k, using A992 steel with $KL = 16$ ft. A W14 × 68 is available but may not provide sufficient capacity. If it does not, cover plates may be added to the section to increase the W14's capacity. Design the cover plates to be 12 inch wide and welded to the flanges of the section to enable the column to support the required load (see Fig. P6-18). Determine the minimum plate thickness required, assuming the plates are available in 1/16th-in increments.

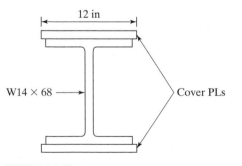

FIGURE P6-18

6-19.   Determine the LRFD design strength and the ASD allowable strength of the section shown if snug-tight bolts 3 ft on center are used to connect the A36 angles. The two angles, $5 \times 3\frac{1}{2} \times \frac{1}{2}$, are oriented with the long legs back-to-back (2L $5 \times 3\frac{1}{2} \times \frac{1}{2}$

LLBB) and separated by 3/8 inch. The effective length, $(KL)_x = (KL)_y = 15$ ft. (*Ans.* 101.9 k LRFD; 67.8 k ASD)

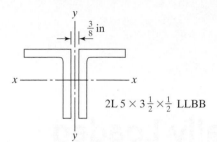

2L 5 × 3½ × ½ LLBB

FIGURE P6-19

6-20. Repeat Prob. 6-19 if the angles are welded together with their long legs back-to-back at intervals of 5 ft.

6-21. Four $3 \times 3 \times \frac{1}{4}$ angles are used to form the member shown in the accompanying illustration. The member is 24 ft long, has pinned ends, and consists of A36 steel. Determine the LRFD design strength and the ASD allowable strength of the member. Design single lacing and end tie plates, assuming connections are made to the angles with $\frac{3}{4}$-in diameter bolts. (*Ans.* 159.1 k LRFD; 106.0 k ASD)

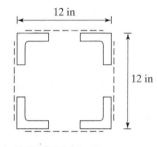

12 in

12 in

FIGURE P6-21

6-22. Select the lightest pair of C9 channels to support the loads $P_D = 50$ k and $P_L = 90$ k. The member is to be 20 ft long with both ends pinned and is to be arranged as shown in the accompanying illustration. Use A36 steel and design single lacing and end tie plates, assuming that $\frac{3}{4}$-in diameter bolts are to be used for connections. Assume that the bolts are located $1\frac{1}{4}$ in from the back of channels. Solve by LRFD and ASD procedures.

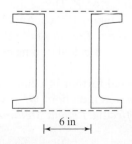

6 in

FIGURE P6-22

Design of Axially Loaded Compression Members (Continued) and Column Base Plates

7.1   INTRODUCTION

In this chapter, the available axial strengths of columns used in unbraced steel frames are considered. These frames are also referred to as moment frames or frames with sidesway uninhibited. As the column ends may move laterally the columns must be able to resist both axial loads and bending moments. As a consequence, they are generally referred to as beam–columns. Such members are discussed in detail in Chapter 11 of the text.

The AISC Specification provides several methods to deal with the stability analysis and design of beam–columns. One is the *Direct Analysis Method* (DM) that is specified in Chapter C of the Specification. This approach uses factors required to more accurately determine forces and moments during the analysis phase and eliminates the requirement for calculating the effective length factor, K. This is due to the fact that the effective length of compression members, KL, is taken as the actual length, L, that is, K is taken equal to 1.0. A second method, the *Effective Length Method* (ELM), is given in Appendix 7 of the Specification. In this method, K is calculated using one of the procedures discussed in this chapter.

These two design methods will be discussed further in Chapter 11 of the text. In this chapter, the available strength of compression members, ΦP_n, will be determined in building frames calculating KL using the *Effective Length Method*.

7.2   FURTHER DISCUSSION OF EFFECTIVE LENGTHS

The subject of effective lengths was introduced in Chapter 5, and some suggested K factors were presented in Table 5.1. These factors were developed for columns with certain idealized conditions of end restraint, which may be very different from practical design conditions. The table values are usually quite satisfactory for preliminary designs and for situations in which sidesway is prevented by bracing. Should the columns be part of a continuous frame subject to sidesway, however, it would often be advantageous to make a more detailed analysis, as described in this section. To a lesser extent, this is also desirable for columns in frames braced against sidesway.

Perhaps a few explanatory remarks should be made at this point, defining sidesway as it pertains to effective lengths. For this discussion, sidesway refers to a type of buckling. In statically indeterminate structures, sidesway occurs where the frames deflect laterally due to the presence of lateral loads or unsymmetrical vertical loads, or where the frames themselves are unsymmetrical. Sidesway also occurs in columns whose ends can move transversely when they are loaded to the point that buckling occurs.

Should frames with diagonal bracing or rigid shear walls be used, the columns will be prevented from sidesway and provided with some rotational restraint at their ends. For these situations, pictured in Fig. 7.1, the K factors will fall somewhere between cases (a) and (d) of Table 5.1.

The AISC Specification Appendix 7 (7.2.3(a)) states that $K = 1.0$ should be used for columns in frames with sidesway inhibited, unless an analysis shows that a smaller value can be used. A specification like $K = 1.0$ is often quite conservative, and an analysis made as described herein may result in some savings.

The true effective length of a column is a property of the whole structure, of which the column is a part. In many existing buildings, it is probable that the masonry walls provide sufficient lateral support to prevent sidesway. When light curtain walls are used, however, as they often are in modern buildings, there is probably little resistance to sidesway. Sidesway is also present in tall buildings in appreciable amounts, unless a definite diagonal bracing system or shear walls are used. For these cases, it seems logical to assume that resistance to sidesway is primarily provided by the lateral stiffness of the frame alone.

Theoretical mathematical analyses may be used to determine effective lengths, but such procedures are typically too lengthy and perhaps too difficult for the average designer. The usual procedure is to consult either Table 5.1, interpolating between the idealized values as the designer feels is appropriate, or the alignment charts that are described in this section.

The most common method for obtaining effective lengths is to employ the charts shown in Fig. 7.2. They were developed by O. G. Julian and L. S. Lawrence, and frequently are referred to as the Jackson and Moreland charts, after the firm where Julian and Lawrence worked.[1,2] The charts were developed from a slope-deflection analysis

---

[1]O. G. Julian and L. S. Lawrence, "Notes on J and L Monograms for Determination of Effective Lengths" (1959). Unpublished.
[2]Structural Stability Research Council, *Guide to Stability Design Criteria for Metal Structures*, 4th ed. T. V. Galambos, ed. (New York: Wiley, 1988).

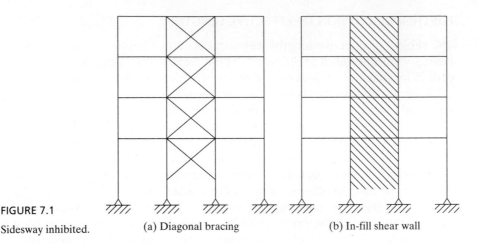

FIGURE 7.1

Sidesway inhibited.

(a) Diagonal bracing   (b) In-fill shear wall

of the frames that included the effect of column loads. One chart was developed for columns braced against sidesway and one for columns subject to sidesway. Their use enables the designer to obtain good K values without struggling through lengthy trial-and-error procedures with the buckling equations.

To use the alignment charts, it is necessary to have preliminary sizes for the girders and columns framing into the column in question before the K factor can be determined for that column. In other words, before the chart can be used, we have to either estimate some member sizes or carry out a preliminary design.

When we say sidesway is *inhibited*, we mean there is something present other than just columns and girders to prevent sidesway or the horizontal translation of the joints. That means we have a definite system of lateral bracing, or we have shear walls. If we say that sidesway is *uninhibited*, we are saying that resistance to horizontal translation is supplied only by the bending strength and stiffness of the girders and beams of the frame in question, with its continuous joints.

The resistance to rotation furnished by the beams and girders meeting at one end of a column is dependent on the rotational stiffnesses of those members. The moment needed to produce a unit rotation at one end of a member if the other end of the member is fixed is referred to as its *rotational stiffness*. From our structural analysis studies, this works out to be equal to $4EI/L$ for a homogeneous member of constant cross section. On the basis of the preceding, we can say that the rotational restraint at the end of a particular column is proportional to the ratio of the sum of the column stiffnesses to the girder stiffnesses meeting at that joint, or

$$G = \frac{\sum \dfrac{4EI}{L} \text{ for columns}}{\sum \dfrac{4EI}{L} \text{ for girders}} = \frac{\sum \dfrac{E_c I_c}{L_c}}{\sum \dfrac{E_g I_g}{L_g}}.$$

The subscripts A and B refer to the joints at the ends of the columns being considered. G is defined as:

$$G = \frac{\Sigma\left(\dfrac{E_c I_c}{L_c}\right)}{\Sigma\left(\dfrac{E_g I_g}{L_g}\right)} \quad \text{AISC Equation (C-A-7-2)}$$

The symbol Σ indicates a summation of all members rigidly connected to that joint and located in the plane in which buckling of the column is being considered. E_c is the elastic modulus of the column, I_c is the moment of inertia of the column, and L_c is the unsupported length of the column. E_g is the elastic modulus of the girder, I_g is the moment of inertia of the girder, and L_g is the unsupported length of the girder or other restraining member. I_c and I_g are taken about axes perpendicular to the plane of buckling being considered. The alignment charts are valid for different materials if an appropriate effective rigidity, EI, is used in the calculation of G.

Adjustments for Columns with Differing End Conditions. For column ends supported by, but not rigidly connected to, a footing or foundation, G is theoretically infinity but unless designed as a true friction-free pin, may be taken as 10 for practical designs. If the column is rigidly attached to a properly designed footing, G may be taken as 1.0. Smaller values may be used if justified by analysis.

From American Institute of Steel Construction Specification, ANSI/AISC 360-10, Commentary to Appendix 7. Fig. C-A-7.1 and C-A-7.2, p. 16.1–512 and 16.1–513 (Chicago: AISC, 2010) "Copyright © American Institute of Steel Construction. Reprinted with permission. All rights reserved."

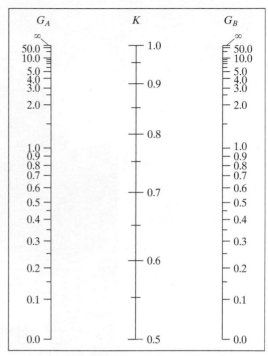

(a) Sidesway inhibited (Braced Frame)

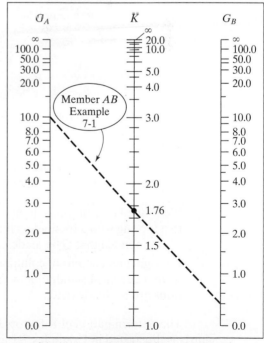

(b) Sidesway uninhibited (Moment Frame)

FIGURE 7.2

Jackson and Moreland alignment charts for effective lengths of columns in continous frames.

Shearson Lehman/American Express Information Services
Center, New York City. (Courtesy of Owen Steel Company, Inc.)

In applying the charts, the G factors at the column bases are quite variable. It is recommended that the following two rules be applied to obtain their values:

1. For pinned columns, G is theoretically infinite, such as when a column is connected to a footing with a frictionless hinge. Since such a connection is not frictionless, it is recommended that G be made equal to 10 where such nonrigid supports are used.
2. For rigid connections of columns to footings, G theoretically approaches zero, but from a practical standpoint, a value of 1.0 is recommended, because no connections are perfectly rigid.

The determination of K factors for the columns of a steel frame by the alignment charts is illustrated in Examples 7-1 and 7-2. The following steps are taken:

1. Select the appropriate chart (sidesway inhibited or sidesway uninhibited).

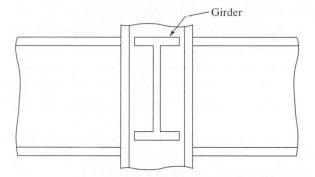

FIGURE 7.3

2. Compute G at each end of the column and label the values G_A and G_B, as desired.

3. Draw a straight line on the chart between the G_A and G_B values, and read K where the line hits the center K scale.

When G factors are being computed for a rigid frame structure (rigid in both directions), the torsional resistance of the perpendicular girders is generally neglected in the calculations. With reference to Fig. 7.3, it is assumed that we are calculating G for the joint shown for buckling in the plane of the paper. For such a case, the torsional resistance of the girder shown, which is perpendicular to the plane being considered, is probably neglected.

If the girders at a joint are very stiff (that is, they have very large EI/L values) the value of $G = \Sigma(E_c I_c/L_c)/\Sigma(E_g I_g/L_g)$ will approach zero and the K factors will be small. If G is very small, the column moments cannot rotate the joint very much; thus, the joint is close to a fixed-end situation. Usually, however, G is appreciably larger than zero, resulting in significantly larger values of K.

The effective lengths of each of the columns of a frame are estimated with the alignment charts in Example 7-1. (When sidesway is possible, it will be found that the effective lengths are always greater than the actual lengths, as is illustrated in this example. When frames are braced in such a manner that sidesway is not possible, K will be less than 1.0.) An initial design has provided preliminary sizes for each of the members in the frame. After the effective lengths are determined, each column can be redesigned. Should the sizes change appreciably, new effective lengths can be determined, the column designs repeated, and so on. Several tables are used in the solution of this example. These should be self-explanatory after the clear directions given on the alignment chart are examined.

7.3 FRAMES MEETING ALIGNMENT CHART ASSUMPTIONS

The Jackson and Moreland charts were developed on the basis of a certain set of assumptions, a complete list of which is given in Section 7.2 of the Commentary of Appendix 7 of the AISC Specification. Among these assumptions are the following:

1. The members are elastic, have constant cross sections, and are connected with rigid joints.

2. All columns buckle simultaneously.

3. For braced frames, the rotations at opposite ends of each beam are equal in magnitude, and each beam bends in single curvature.

4. For unbraced frames, the rotations at opposite ends of each beam are equal in magnitude, but each beam bends in double curvature.

5. Axial compression forces in the girders are neglibible.

The frame of Fig. 7.4 is assumed to meet all of the assumptions on which the alignment charts were developed. From the charts, the column effective length factors are determined, as shown in Example 7-1.

Example 7-1

Determine the effective length factor for each of the columns of the frame shown in Fig. 7.4 if the frame is not braced against sidesway. Use the alignment charts of Fig. 7.2(b).

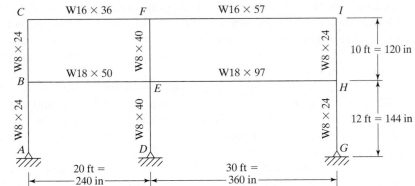

FIGURE 7.4

Solution. Stiffness factors: E is assumed to be 29,000 ksi for all members and is therefore neglected in the equation to calculate G.

	Member	Shape	I	L	I/L
Columns	AB	W8 × 24	82.7	144	0.574
	BC	W8 × 24	82.7	120	0.689
	DE	W8 × 40	146	144	1.014
	EF	W8 × 40	146	120	1.217
	GH	W8 × 24	82.7	144	0.574
	HI	W8 × 24	82.7	120	0.689
Girders	BE	W18 × 50	800	240	3.333
	CF	W16 × 36	448	240	1.867
	EH	W18 × 97	1750	360	4.861
	FI	W16 × 57	758	360	2.106

G factors for each joint:

Joint	$\Sigma(I_c/L_c)/\Sigma(I_g/L_g)$	G
A	Pinned Column, $G = 10$	10.0
B	$\dfrac{0.574 + 0.689}{3.333}$	0.379
C	$\dfrac{0.689}{1.867}$	0.369
D	Pinned Column, $G = 10$	10.0
E	$\dfrac{1.014 + 1.217}{(3.333 + 4.861)}$	0.272
F	$\dfrac{1.217}{(1.867 + 2.106)}$	0.306
G	Pinned Column, $G = 10$	10.0
H	$\dfrac{0.574 + 0.689}{4.861}$	0.260
I	$\dfrac{0.689}{2.106}$	0.327

Column K factors from chart [Fig. 7.2(b)]:

Column	G_A	G_B	K*
AB	10.0	0.379	1.76
BC	0.379	0.369	1.12
DE	10.0	0.272	1.74
EF	0.272	0.306	1.10
GH	10.0	0.260	1.73
HI	0.260	0.327	1.10

*It is a little difficult to read the charts to the three decimal places shown by the author. He has used a larger copy of Fig. 7.2 for his work. For all practical design purposes, the K values can be read to two places, which can easily be accomplished with this figure.

For most buildings, the values of K_x and K_y should be examined separately. The reason for such individual study lies in the different possible framing conditions in the two directions. Many multistory frames consist of rigid frames in one direction and conventionally connected frames with sway bracing in the other. In addition, the points of lateral support may often be entirely different in the two planes.

There is available a set of rather simple equations for computing effective length factors. On some occasions, the designer may find these expressions very convenient to use, compared with the alignment charts just described. Perhaps the most useful situation is for computer programs. You can see that it would be rather inconvenient to stop occasionally in the middle of a computer design to read K factors from the charts and input them to the computer. The equations, however, can easily be included in the programs, eliminating the necessity of using alignment charts.[3]

The alignment chart of Fig. 7.2(b) for frames with sidesway uninhibited always indicates that $K \geq 1.0$. In fact, calculated K factors of 2.0 to 3.0 are common, and even larger values are occasionally obtained. To many designers, such large factors seem completely unreasonable. If the designer derives seemingly high K factors, he or she should carefully review the numbers used to enter the chart (that is, the G values), as well as the basic assumptions made in preparing the charts. These assumptions are discussed in detail in Sections 7.4 and 7.5.

7.4 FRAMES NOT MEETING ALIGNMENT CHART ASSUMPTIONS AS TO JOINT ROTATIONS

In this section, a few comments are presented regarding frames whose joint rotations (and thus their beam stiffnesses) are not in agreement with the assumptions made for developing the charts.

It can be shown by structural analysis that the rotation at point B in the frame of Fig. 7.5 is twice as large as the rotation at B assumed in the construction of the nomographs. Therefore, beam BC in the figure is only one-half as stiff as the value assumed for the development of the alignment charts.

The Jackson and Moreland charts can be accurately used for situations in which the rotations are different from those assumed by making adjustments to the computed beam stiffnesses before the chart values are read. Relative stiffnesses for situations other than the one shown in Fig. 7.5 also can be determined by structural analysis. Table 7.1 presents correction factors to be multiplied by calculated beam stiffnesses, for situations where the beam end conditions are different from those assumed for the development of the charts.

Example 7-2 shows how the correction factors can be applied to a building frame where the rotations at the ends of some of the beams vary from the assumed conditions of the charts.

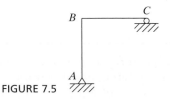

FIGURE 7.5

[3]P. Dumonteil, "Simple Equations for Effective Length Factors," *Engineering Journal*, AISC, vol. 29, no. 3 (3rd Quarter, 1992), pp. 111–115.

TABLE 7.1   Multipliers for Rigidly Attached Members

Condition at Far End of Girder	Sidesway Prevented, Multiply by:	Sidesway Uninhibited, Multiply by:
Pinned	1.5	0.5
Fixed against rotation	2.0	0.67

Example 7-2

Determine K factors for each of the columns of the frame shown in Fig. 7.6. Here, W sections have been tentatively selected for each of the members of the frame and their I/L values determined and shown in the figure.

Solution. First, the G factors are computed for each joint in the frame. In this calculation, the I/L values for members FI and GJ are multiplied by the appropriate factors from Table 7.1.

1. For member FI, the I/L value is multiplied by 2.0, because its far end is fixed and there is no sidesway on that level.

2. For member, GJ, I/L is multiplied by 1.5, because its far end is pinned and there is no sidesway on that level.

$$G_A = 10 \text{ as described in Section 7.2, Pinned Column}$$

$$G_B = \frac{23.2 + 23.2}{70} = 0.663$$

$$G_C = \frac{23.2 + 20.47}{70} = 0.624$$

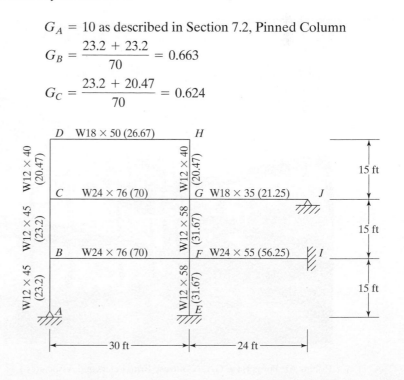

FIGURE 7.6

Steel shapes, including their I/L values.

$$G_D = \frac{20.47}{26.67} = 0.768$$

$G_E = 1.0$ as described in Section 7.2, Fixed Column

$$G_F = \frac{31.67 + 31.67}{70 + (2.0)(56.25)} = 0.347$$

$$G_G = \frac{31.67 + 20.47}{70 + (1.5)(21.25)} = 0.512$$

$$G_H = \frac{20.47}{26.67} = 0.768$$

Finally, the K factors are selected from the appropriate alignment chart of Fig. 7.2.

Column	G Factors	Chart used	K Factors
AB	10 and 0.663	7.2 (a) no sidesway	0.83
BC	0.663 and 0.624	7.2 (a) no sidesway	0.72
CD	0.624 and 0.768	7.2 (b) has sidesway	1.23
EF	1.0 and 0.347	7.2 (a) no sidesway	0.71
FG	0.347 and 0.512	7.2 (a) no sidesway	0.67
GH	0.512 and 0.768	7.2 (b) has sidesway	1.21

Robins Air Force Base, GA. (Courtesy Britt, Peters and Associates.)

7.5 STIFFNESS-REDUCTION FACTORS

As previously mentioned, the alignment charts were developed according to a set of idealized conditions that are seldom, if ever, completely met in a real structure. Included among those conditions are the following: The column behavior is purely elastic, all columns buckle simultaneously, all members have constant cross sections, all joints are rigid, and so on.

If the actual conditions are different from these assumptions, unrealistically high K factors may be obtained from the charts, and overconservative designs may result. A large percentage of columns will appear in the inelastic range, but the alignment charts were prepared with the assumption of elastic failure. This situation, previously discussed in Chapter 5, is illustrated in Fig. 7.7. For such cases, the chart K values are too conservative and should be corrected as described in this section.

In the elastic range, the stiffness of a column is proportional to EI, where $E = 29,000$ ksi; in the inelastic range, its stiffness is more accurately proportional to $E_T I$, where E_T is a reduced or tangent modulus.

The buckling strength of columns in framed structures is shown in the alignment charts to be related to

$$G = \frac{\text{column stiffness}}{\text{girder stiffness}} = \frac{\Sigma(EI/L) \text{ columns}}{\Sigma(EI/L) \text{ girders}}$$

If the columns behave elastically, the modulus of elasticity will be canceled from the preceding expression for G. If the column behavior is inelastic, however, the column stiffness factor will be smaller and will equal $E_T I/L$. As a result, the G factor used to enter the alignment chart will be smaller, and the K factor selected from the chart will be smaller.

Though the alignment charts were developed for elastic column action, they may be used for an inelastic column situation if the G value is multiplied by a correction factor, τ_b. This reduction factor is specified in Section C2-3 of the AISC Specification.

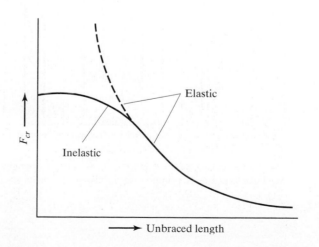

FIGURE 7.7

TABLE 7.2 Stiffness Reduction Factor, τ_b

ASD $\dfrac{P_a}{A_g}$ / LRFD $\dfrac{P_u}{A_g}$	F_y, ksi									
	35		36		42		46		50	
	ASD	LRFD	ASD	LRFD	ASD	LRFD	ASD	LRFD	ASD	LRFD
45	–	–	–	–	–	–	–	0.0851	–	0.360
44	–	–	–	–	–	–	–	0.166	–	0.422
43	–	–	–	–	–	–	–	0.244	–	0.482
42	–	–	–	–	–	–	–	0.318	–	0.538
41	–	–	–	–	–	0.0930	–	0.388	–	0.590
40	–	–	–	–	–	0.181	–	0.454	–	0.640
39	–	–	–	–	–	0.265	–	0.516	–	0.686
38	–	–	–	–	–	0.345	–	0.575	–	0.730
37	–	–	–	–	–	0.420	–	0.629	–	0.770
36	–	–	–	–	–	0.490	–	0.681	–	0.806
35	–	–	–	0.108	–	0.556	–	0.728	–	0.840
34	–	0.111	–	0.210	–	0.617	–	0.771	–	0.870
33	–	0.216	–	0.306	–	0.673	–	0.811	–	0.898
32	–	0.313	–	0.395	–	0.726	–	0.847	–	0.922
31	–	0.405	–	0.478	–	0.773	–	0.879	0.0317	0.942
30	–	0.490	–	0.556	–	0.816	–	0.907	0.154	0.960
29	–	0.568	–	0.627	–	0.855	–	0.932	0.267	0.974
28	–	0.640	–	0.691	–	0.889	0.102	0.953	0.373	0.986
27	–	0.705	–	0.750	–	0.918	0.229	0.970	0.470	0.994
26	–	0.764	–	0.802	0.0377	0.943	0.346	0.983	0.559	0.998
25	–	0.816	–	0.849	0.181	0.964	0.454	0.992	0.640	1.00
24	–	0.862	–	0.889	0.313	0.980	0.552	0.998	0.713	
23	–	0.901	–	0.923	0.434	0.991	0.640	1.00	0.777	
22	–	0.934	0.0869	0.951	0.543	0.998	0.719		0.834	
21	0.154	0.960	0.249	0.972	0.640	1.00	0.788		0.882	
20	0.313	0.980	0.395	0.988	0.726		0.847		0.922	
19	0.457	0.993	0.525	0.997	0.800		0.896		0.953	
18	0.583	0.999	0.640	1.00	0.862		0.936		0.977	
17	0.693	1.00	0.739		0.913		0.967		0.992	
16	0.786		0.822		0.952		0.987		0.999	
15	0.862		0.889		0.980		0.998		1.00	
14	0.922		0.940		0.996		1.00			
13	0.964		0.976		1.00					
12	0.991		0.996							
11	1.00		1.00							
10										
9										
8										
7										
6										
5										

– Indicates the stiffness reduction parameter is not applicable because the required strength exceeds the available strength for $KL/r = 0$.

Source: AISC Manual, Table 4-21, p. 4-321, 14th ed., 2011. "Copyright © American Institute of Steel Construction. Reprinted with permission. All rights reserved."

When $\alpha P_r/P_y$ is less than or equal to 0.5, then τ_b equals 1.0 per AISC Equation C2-2a. When $\alpha P_r/P_y$ is greater than 0.5, then $\tau_b = 4(\alpha P_r/P_y)[1 - (\alpha P_r/P_y)]$ per AISC Equation C2-2b. The factor, α, is taken as 1.0 for the LRFD method and 1.6 for the ASD design basis. P_r is the required axial compressive strength using LRFD or ASD load combinations, P_u or P_a respectively. P_y is the axial yield strength, F_y times the column gross area, A_g. Values of τ_b are shown for various P_u/A_g and P_a/A_g values in Table 7.2, which is Table 4-21 in the AISC Manual.

The τ_b factor is then used to reduce the column stiffness in the equation to calculate G, where $G_{(inelastic)} = \dfrac{\tau_b \sum (I_c/L_c)}{\sum (I_g/L_g)} = \tau_b G_{(elastic)}$. If the end of a column is pinned $(G = 10.0)$

or fixed $(G = 1.0)$, the value of G at that end should not be multiplied by a stiffness reduction factor.

Example 7-3 illustrates the steps used for the determination of the inelastic effective length factor for a column in a frame subject to sidesway. *In this example, note that the author has considered only in-plane behavior and only bending about the x axis.* As a result of inelastic behavior, the effective length factor is appreciably reduced.

Structures designed by inelastic analysis must meet the provisions of Appendix 1 of the AISC Specification.

Example 7-3

(a) Determine the effective length factor for column AB of the unbraced frame shown in Fig. 7.8, assuming that we have elastic behavior and that all of the other assumptions on which the alignment charts were developed are met. $P_D = 450$ k, $P_L = 700$ k, $F_y = 50$ ksi. Assume that column AB is a W12 × 170 and the columns above and below are as indicated on the figure.

(b) Repeat part (a) if inelastic column behavior is considered.

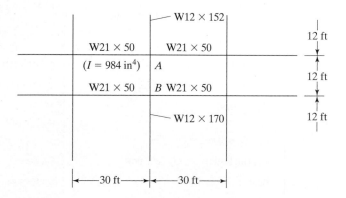

FIGURE 7.8

Solution

LRFD	ASD
$P_u = (1.2)(450) + (1.6)(700) = 1660$ k	$P_a = 450 + 700 = 1150$ k

(a) Assuming that the column is in the elastic range.

Using W12 × 170 ($A = 50$ in², $I_x = 1650$ in⁴) for column AB and the column below.

Using W12 × 152 ($A = 44.7$ in², $I_x = 1430$ in⁴) for column above.

$$G_A = \frac{\sum(I_c/L_c)}{\sum(I_g/L_g)} = \frac{\dfrac{1430}{12} + \dfrac{1650}{12}}{2\left(\dfrac{984}{30}\right)} = 3.91$$

$$G_B = \frac{\sum(I_c/L_c)}{\sum(I_g/L_g)} = \frac{2\left(\dfrac{1650}{12}\right)}{2\left(\dfrac{984}{30}\right)} = 4.19$$

From Fig. 7.2(b) alignment chart

$$K = 2.05$$

(b) Inelastic solution

LRFD	ASD
$\alpha = 1.0$	$\alpha = 1.6$
$P_r = P_u = 1660$ k	$P_r = P_a = 1150$ k
$P_y = F_y A_g = 50$ ksi (50 in²) = 2500 k	$P_y = F_y A_g = 50$ ksi (50 in²) = 2500 k
$\alpha\dfrac{P_r}{P_y} = \dfrac{1.0(1660)}{2500} = 0.664 > 0.5$	$\alpha\dfrac{P_r}{P_y} = \dfrac{1.6(1150)}{2500} = 0.736 > 0.5$
Use AISC Equation C2-2b	Use AISC Equation C2-2b
$\tau_b = 4\left(\alpha\dfrac{P_r}{P_y}\right)\left[1 - \left(\alpha\dfrac{P_r}{P_y}\right)\right]$	$\tau_b = 4\left(\alpha\dfrac{P_r}{P_y}\right)\left[1 - \left(\alpha\dfrac{P_r}{P_y}\right)\right]$
$\tau_b = 4(0.664)\,[1 - (0.664)]$	$\tau_b = 4(0.736)\,[1 - (0.736)]$
$\tau_b = 0.892$	$\tau_b = 0.777$
Determine τ_b from Table 7.2	Determine τ_b from Table 7.2
$\dfrac{P_u}{A_g} = \dfrac{1660}{50} = 33.2$	$\dfrac{P_a}{A_g} = \dfrac{1150}{50} = 23$
$\therefore\ \tau_b = 0.892$	$\therefore\ \tau_b = 0.777$
G_A (inelastic) $= \tau_b\, G_A$ (elastic)	G_A (inelastic) $= \tau_b\, G_A$ (elastic)
$0.892\,(3.91) = 3.49$	$0.777\,(3.91) = 3.04$
G_B (inelastic) $= \tau_b\, G_B$ (elastic)	G_B (inelastic) $= \tau_b\, G_A$ (elastic)
$0.892\,(4.19) = 3.74$	$0.777\,(4.19) = 3.26$
From Fig. 7.2(b) alignment chart	From Fig. 7.2 (b) alignment chart
$K = 1.96$	$K = 1.86$

7.6   COLUMNS LEANING ON EACH OTHER FOR IN-PLANE DESIGN

When we have an unbraced frame with beams rigidly attached to columns, it is safe to design each column individually, using the sidesway uninhibited alignment chart to obtain the K factors (which will probably be appreciably larger than 1.0).

A column cannot buckle by sidesway unless all of the columns on that story buckle by sidesway. One of the assumptions on which the alignment chart of Fig. 7.2(b) was prepared is that all of the columns in the story in question would buckle at the same time. If this assumption is correct, the columns cannot support or brace each other, because if one gets ready to buckle, they all supposedly are ready to buckle.

In some situations, however, certain columns in a frame have some excess buckling strength. If, for instance, the buckling loads of the exterior columns of the unbraced frame of Fig. 7.9 have not been reached when the buckling loads of the interior columns are reached, the frame will not buckle. The interior columns, in effect, will lean against the exterior columns; that is, the exterior columns will brace the interior ones. For this situation, shear resistance is provided in the exterior columns that resist the sidesway tendency.[4]

A pin-ended column that does not help provide lateral stability to a structure is referred to as a *leaning* column. Such a column depends on the other parts of the structure to provide lateral stability. AISC Commentary Appendix 7 – Section 7.2 states that the effects of gravity-loaded leaning columns shall be included in the design of moment frame columns.

There are many practical situations in which some columns have excessive buckling strength. This might happen when the designs of different columns on a particular story are controlled by different loading conditions. For such cases, failure of the frame will occur only when the gravity loads are increased sufficiently to offset the extra strength of the lightly loaded columns. As a result, the critical loads for the interior columns of Fig. 7.9 are increased, and, in effect, their effective lengths are decreased. In other words, if the exterior columns are bracing the interior ones against sidesway, the K factors for those interior columns are approaching 1.0. Yura[5] says that the effective length of some of the columns in a frame subject to sidesway can be reduced to 1.0 in this type of situation, even though there is no apparent bracing system present.

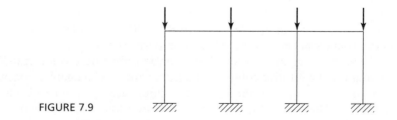

FIGURE 7.9

[4]J. A. Yura, "The Effective Length of Columns in Unbraced Frames," *Engineering Journal*, AISC, vol. 8, no. 2 (second quarter, 1971), pp. 37–4

[5]Ibid., pp. 39–40.

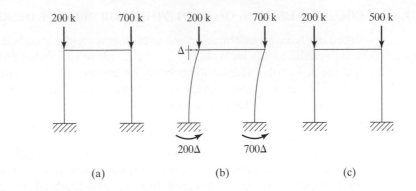

FIGURE 7.10   (a)   (b)   (c)

The net effect of the information presented here is that the total gravity load that an unbraced frame can support equals the sum of the strength of the individual columns. In other words, the total gravity load that will cause sidesway buckling in a frame can be split up among the columns in any proportion, so long as the maximum load applied to any one column does not exceed the maximum load that column could support if it were braced against sidesway, with $K = 1.0$.

For this discussion, the unbraced frame of Fig. 7.10(a) is considered. It is assumed that each column has $K = 2.0$ and will buckle under the loads shown.

When sidesway occurs, the frame will lean to one side, as shown in part (b) of the figure, and $P\Delta$ moments equal to 200Δ and 700Δ will be developed.

Suppose that we load the frame with 200 k on the left-hand column and 500 k on the right-hand column (or 200 k less than we had before). We know that for this situation, which is shown in part (c) of the figure, the frame will not buckle by sidesway until we reach a moment of 700Δ at the right-hand column base. This means that our right-hand column can take an additional moment of 200Δ. Thus, as Yura says, the right-hand column has a reserve of strength that can be used to brace the left-hand column and prevent its sidesway buckling.

Obviously, the left-hand column is now braced against sidesway, and sidesway buckling will not occur until the moment that its base reaches 200Δ. Therefore, it can be designed with a K factor less than 2.0 and can support an additional load of 200 k, giving it a total load of 400 k—*but this load must not be greater than its capacity would be if it were braced against sidesway with $K = 1.0$.* It should be mentioned that the total load the frame can carry is still 900 k, as in part (a) of the figure.

The advantage of the frame behavior described here is illustrated in Fig. 7.11. In this situation, the interior columns of a frame are braced against sidesway by the exterior columns. As a result, the interior columns are assumed to each have K factors equal to 1.0. They are designed for the factored loads shown (660 k each). Then the K factors for the exterior columns are determined with the sidesway uninhibited chart of Fig. 7.2, and they are each designed for column loads equal to $440 + 660 = 1100$ k. To fully understand the benefit of the *leaner column theory*, we must first realize that the frame is assumed to be braced against sidesway in the y or out of plane direction such that $K_y = 1.0$. Each of the end columns of Figure 7.11 will need to support

440 k + 660 k, but these loads are tending to buckle the end columns about their x axes. As a result the $\dfrac{KL}{r}$ value used to determine $\phi_c F_{cr}\left(\text{or } \dfrac{F_{cr}}{\Omega_c}\right)$ is $\left(\dfrac{KL}{r}\right)_x$ and not the much larger $\left(\dfrac{KL}{r}\right)_y$.

Example 7-4

For the frame of Fig. 7.11, which consists of 50 ksi steel, beams are rigidly connected to the exterior columns, while all other connections are simple. The columns are braced top and bottom against sidesway, out of the plane of the frame, so that $K_y = 1.0$ in that direction. Sidesway is possible in the plane of the frame. Using the LRFD method, design the interior columns assuming that $K_x = K_y = 1.0$, and design the exterior columns with K_x as determined from the alignment chart and $P_u = 1100$ k. (With this approach to column buckling, the interior columns could carry no load at all, since they appear to be unstable under sidesway conditions.) The end columns are assumed to have no bending moment at the top of the member.

Solution. *Design of interior columns:*

Assume $K_x = K_y = 1.0$, $KL = (1.0)(15) = 15$ ft, $P_u = 660$ k.

Use W14 × 74; $\phi P_n = 667$ k > $P_u = 660$ k

Design of exterior columns:

In plane $P_u = 440 + 660 = 1100$ k, K_x to be determined from alignment chart. Estimating a column size a little larger than would be required for $P_u = 1100$ k. Try W14 × 120 ($A = 35.3$ in², $I_x = 1380$ in⁴, $r_x = 6.24$ in, $r_y = 3.74$ in).

$$G_{\text{top}} = \frac{1380/15}{2100/30 \times 0.5} = 2.63$$

(noting that girder stiffness is multiplied by 0.5, since sidesway is permitted and far end of girder is hinged).

$$G_{\text{bottom}} = 10$$
$$K_x = 2.22 \text{ from Fig. 7.2(b)}$$

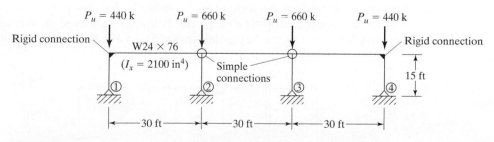

FIGURE 7.11

$$\frac{K_x L_x}{r_x} = \frac{(2.22)(12 \times 15)}{6.24} = 64.04$$

$$\phi_c F_{cr} = 33.38 \text{ ksi}$$

$$\phi_c P_n = (33.38)(35.3) = 1178 \text{ k} > P_u = 1100 \text{ k}$$

Out of plane: $K_y = 1.0, P_u = 440 \text{ k}$

$$\frac{K_y L_y}{r_y} = \frac{1.0 (12 \times 15)}{3.74} = 48.13$$

$$\phi F_{cr} = 37.96 \text{ ksi}$$

$$\phi_c P_n = (37.96) (35.3) = 1340 \text{ k} > P_u = 440 \text{ k}$$

Use W14 × 120.

It is rather frightening to think of additions to existing buildings and the leaner column theory. If we have a building (represented by the solid lines in Fig. 7.12) and we decide to add onto it (indicated by the dashed lines in the same figure), we may think that we can use the old frame to brace the new one and that we can keep expanding laterally with no effect on the existing building. Sadly, we may be in for quite a surprise. The leaning of the new columns may cause one of the old ones to fail.

7.7 BASE PLATES FOR CONCENTRICALLY LOADED COLUMNS

The design compressive stress in a concrete or other type of masonry footing is much smaller than it is in a steel column. When a steel column is supported by a footing, it is necessary for the column load to be spread over a sufficient area to keep the footing from being overstressed. Loads from steel columns are transferred through a steel base plate to a fairly large area of the footing below. (Note that a footing performs a related function, in that it spreads the load over an even larger area so that the underlying soil will not be overstressed.)

The base plates for steel columns can be welded directly to the columns, or they can be fastened by means of some type of bolted or welded lug angles. These connection methods are illustrated in Fig. 7.13. A base plate welded directly to the column is shown in part (a) of the figure. For small columns, these plates are probably shop-welded to the columns, but for larger columns it may be necessary to ship the plates separately and set them to the correct elevations. For this second case, the columns are connected to the footing with anchor bolts that pass through the lug angles which have been shop-welded to the columns. This type of arrangement is shown in part (b) of the figure. Some designers like to use lug angles on both flanges

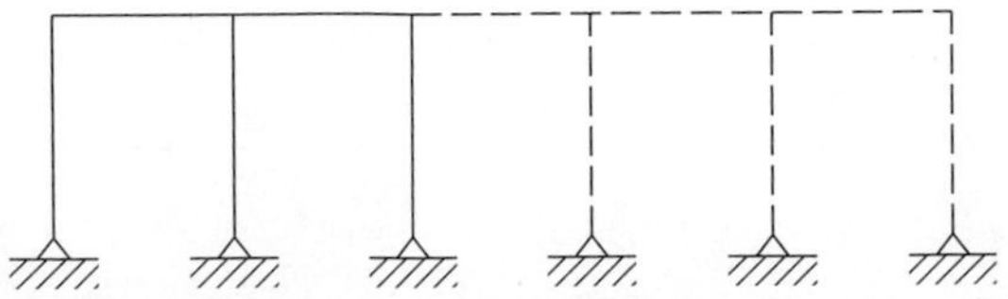

FIGURE 7.12

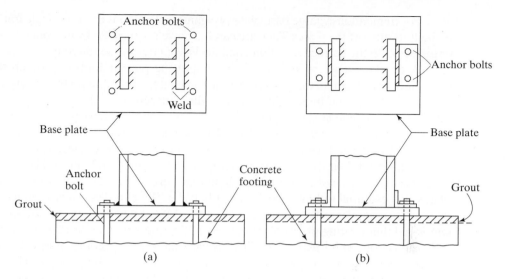

FIGURE 7.13

Column base plates.

and web. (The reader should be aware of OSHA regulations for the safe erection of structural steel, which require the use of no less than four anchor bolts for each column. These bolts will preferably be placed at base plate corners.)

A critical phase in the erection of a steel building is the proper positioning of column base plates. If they are not located at their correct elevations, serious stress changes may occur in the beams and columns of the steel frame. One of the following three methods is used for preparing the site for the erection of a column to its proper elevation: leveling plates, leveling nuts, or preset base plates. An article by Ricker[6] describes these procedures in considerable detail.

For small-to-medium base plates (up to 20 to 22 in), approximately 0.25-in-thick leveling plates with the same dimensions as the base plates (or a little larger) are shipped to the job and carefully grouted in place to the proper elevations. Then the columns with their attached base plates are set on the leveling plates.

As these leveling plates are very light and can be handled manually, they are set by the foundation contractor. This is also true for the lighter base plates. On the other hand, large base plates that have to be lifted with a derrick or crane are usually set by the steel erector.

For larger base plates, up to about 36 in, some types of leveling nuts are used to adjust the base plates up or down. To ensure stability during erection, these nuts must be used on at least four anchor bolts.

If the base plates are larger than about 36 in, the columns with the attached base plates are so heavy and cumbersome that it is difficult to ship them together. For such cases, the base plates are shipped to the job and placed in advance of the steel erection. They can be leveled with shims or wedges.

---

[6]D. T. Ricker, "Some Practical Aspects of Column Bases," *Engineering Journal*, AISC vol. 26, no. 3 (3d quarter, 1989), pp. 81–89.

For tremendously large base plates weighing several tons or more, angle frames may be built to support the plates. They are carefully leveled and filled with concrete, which is screeded off to the correct elevations, and the base plates are set directly on the concrete.

A column transfers its load to the supporting pier or footing through the base plate. Should the supporting concrete area A_2 be larger than the plate area A_1, the concrete strength will be higher. In that case, the concrete surrounding the contact area supplies appreciable lateral support to the directly loaded part, with the result that the loaded concrete can support more load. This fact is reflected in the design stresses.

The lengths and widths of column base plates are usually selected in multiples of even inches, and their thicknesses in multiples of $\frac{1}{8}$ in up to 1.25 in, and in multiples of $\frac{1}{4}$ in thereafter. To make sure that column loads are spread uniformly over their base plates, it is essential to have good contact between the two. The surface preparation of these plates is governed by Section M2.8 of the AISC Specification. In that section, it is stipulated that bearing plates of 2 in or less in thickness may be used without milling if satisfactory contact is obtained. (Milled surfaces have been accurately sawed or finished to a true plane.) Plates over 2 in thick, up through 4 in, may be pressed to be straightened, or they may be milled at the option of the steel fabricator. Plates thicker than 4 in must be milled if they are beyond the flatness tolerances specified in Table 1-29 of Part 1 of the AISC Manual, entitled "Rectangular Plates."

At least one hole should be provided near the center of large area base plates for placing grout. These holes will permit more even placement of grout under the plates,

Robins Air Force Base, GA. (Courtesy Britt Peters and Associates.)

which will tend to prevent air pockets. Grout holes are not needed if the grout is dry packed. Both anchor bolt holes and grout holes are usually flame cut, because their diameters are often too large for normal punching and drilling. Part 14 of the AISC Manual presents considerably more information concerning the installation of base plates.

If the bottom surfaces of the plates are to be in contact with cement grout, to ensure full bearing contact on the foundation, the plates do not have to be milled. Furthermore, the top surfaces of plates thicker than 4 in do not have to be milled if full-penetration welds (described in Chapter 14) are used. Notice that when finishing is required, as described here, the plates will have to be ordered a little thicker than is necessary for their final dimensions, to allow for the cuts.

Initially, columns will be considered that support average-size loads. Should the loads be very small, so that base plates are very small, the design procedure will have to be revised, as described later in this section.

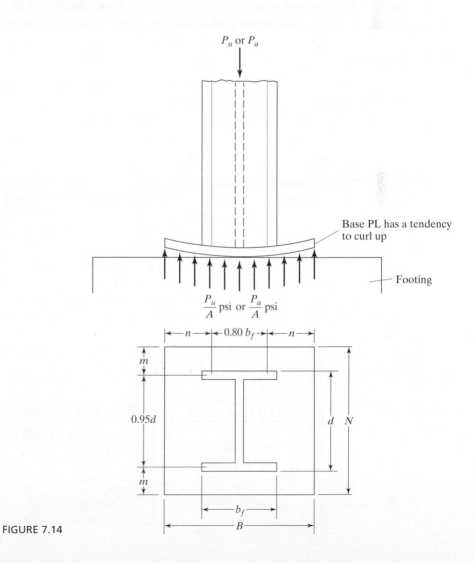

FIGURE 7.14

The AISC Specification does not stipulate a particular method for designing column base plates. The method presented here is based upon example problems presented in the CD accompanying the Manual.

To analyze the base plate shown in Fig. 7.14, note that the column is assumed to apply a total load to the base plate equal to P_u (for LRFD) or P_a (for ASD). Then the load is assumed to be transmitted uniformly through the plate to the footing below, with a pressure equal to P_u/A or P_a/A, where A is the area of the base plate. The footing will push back with an equal pressure and will tend to curl up the cantilevered parts of the base plate outside of the column, as shown in the figure. This pressure will also tend to push up the base plate between the flanges of the column.

With reference to Fig. 7.14, the AISC Manual suggests that maximum moments in a base plate occur at distances $0.80b_f$ and $0.95\,d$ apart. The bending moment can be calculated at each of these sections, and the larger value used to determine the plate thickness needed. This method of analysis is only a rough approximation of the true conditions, because the actual plate stresses are caused by a combination of bending in two directions.

7.7.1   Plate Area

The design strength of the concrete in bearing beneath the base plate must at least equal the load to be carried. When the base plate covers the entire area of the concrete, the nominal bearing strength of the concrete (P_p) is

$$P_p = 0.85f_c'A_1. \qquad \text{(AISC Equation J8-1)}$$

In this expression, f_c' is the 28-day compression strength of the concrete and A_1 is the area of the base plate. For LRFD design ϕ_c is 0.65, while for ASD design Ω_c is 2.31.

Should the full area of the concrete support not be covered by the plate, the concrete underneath the plate, surrounded by concrete outside, will be somewhat stronger. For this situation, the AISC Specification permits the nominal strength $0.85f_c'A_1$ to be increased by multiplying it by $\sqrt{A_2/A_1}$. In the resulting expression, A_2 is the maximum area of the portion of the supporting concrete, which is geometrically similar to and concentric with the loaded area. The quantity $\sqrt{A_2/A_1}$ is limited to a maximum value of 2, as shown in the expression that follows. You should note that A_1 may not be less than the depth of the column times its flange width. (Min $A_1 = b_f d$.)

$$P_p = (0.85f_c'A_1)\sqrt{\frac{A_2}{A_1}} \le 1.7f_c'A_1 \qquad \text{(LRFD Equation J8-2)}$$

LRFD with $\phi_c = 0.65$	ASD with $\Omega_c = 2.31$	
$P_u = \phi_c P_p = \phi_c(0.85f_c'A_1)\sqrt{\dfrac{A_2}{A_1}}$	$P_a = \dfrac{P_p}{\Omega_c} = \dfrac{0.85f_c'A_1\sqrt{\dfrac{A_2}{A_1}}}{\Omega_c}$	
$A_1 = \dfrac{P_u}{\phi_c(0.85f_c')\sqrt{\dfrac{A_2}{A_1}}}$	$A_1 = \dfrac{P_a\Omega_c}{(0.85f_c')\sqrt{\dfrac{A_2}{A_1}}}.$	

After the controlling value of A_1 is determined as just described, the plate dimensions B and N (shown in Fig. 7.14), are selected to the nearest 1 or 2 in so that the values of m and n shown in the figure are roughly equal. Such a procedure will make the cantilever moments in the two directions approximately the same. This will enable us to keep the plate thicknesses to a minimum. The condition $m = n$ can be approached if the following equation is satisfied:

$$N \approx \sqrt{A_1} + \Delta$$

Here,

$$A_1 = \text{area of plate} = BN$$

$$\Delta = 0.5\,(0.95\,d - 0.80\,b_f)$$

$$N = \sqrt{A_1} + \Delta$$

$$B \approx \frac{A_1}{N}$$

From a practical standpoint, designers will often use square base plates with anchor bolts arranged in a square pattern. Such a practice simplifies both field and shop work.

7.7.2 Plate Thickness

To determine the required plate thickness, t, moments are taken in the two directions as though the plate were cantilevered out by the dimensions m and n. Reference is again made here to Figure 7.14. In the expressions to follow, the load P is P_u for LRFD design and P_a for ASD design. The moments in the two directions are

$$\left(\frac{P_u}{BN}\right)(m)\left(\frac{m}{2}\right) = \frac{P_u m^2}{2BN} \text{ or } \left(\frac{P}{BN}\right)(n)\left(\frac{n}{2}\right) - \frac{P m^2}{2BN}, \text{ both computed for a 1-in width}$$

of plate.

If lightly loaded base plates, such as those for the columns of low-rise buildings and preengineered metal buildings, are designed by the procedure just described, they will have quite small areas. They will, as a result, extend very little outside the edges of the columns and the computed moments, and the resulting plate thicknesses will be very small, perhaps so small as to be impractical.

Several procedures for handling this problem have been proposed. In 1990, W. A. Thornton[7] combined three of the methods into a single procedure applicable to either heavily loaded or lightly loaded base plates. This modified method is used for the example base plate problems presented on the CD accompanying the AISC manual, as well as for the example problems in this chapter.

Thornton proposed that the thickness of the plates be determined by the largest of m, n, or $\lambda n'$. He called this largest value ℓ.

$$\ell = \max\,(m, n, \text{ or } \lambda n')$$

[7]W. A. Thornton, "Design of Base Plates for Wide Flange Columns—A Concatenation of Methods," *Engineering Journal*, AISC, vol. 27, no. 4 (4th quarter, 1990) pp. 173, 174.

To determine $\lambda n'$, it is necessary to substitute into the following expressions, which are developed in his paper:

$$\phi_c P_p = \phi_c 0.85 f'_c A_1 \qquad \text{for plates covering the full area of the concrete support}$$

$$\phi_c P_p = \phi_c 0.85 f'_c A_1 \sqrt{\frac{A_2}{A_1}}, \quad \text{where } \sqrt{\frac{A_2}{A_1}} \text{ must be } \leq 2 \text{ for plates not covering the}$$
entire area of concrete support

LRFD	ASD
$X = \left[\dfrac{4db_f}{(d + b_f)^2} \right] \dfrac{P_u}{\phi_c P_p}$	$\overline{X} = \dfrac{4db_f}{(d + b_f)^2} \dfrac{\Omega_c P_a}{P_p}$
$\lambda = \dfrac{2\sqrt{X}}{1 + \sqrt{1 - X}} \leq 1$	$\lambda = \dfrac{2\sqrt{\overline{X}}}{1 + \sqrt{1 - \overline{X}}} \leq 1$
$\lambda n' = \dfrac{\lambda\sqrt{db_f}}{4}$	$\lambda m' = \dfrac{\lambda\sqrt{db_f}}{4}$

According to Thornton it is permissible to conservatively assume λ equals 1.0 for all cases; this practice is followed in the examples to follow. As a result, it is unnecessary to substitute into the equations listed for $\overline{X}$, λ and $\lambda n'$. Thus the authors drop the λ from $\lambda n'$ and just uses n'.

Letting the largest value of $m, n,$ or $\lambda n'$ be referred to as ℓ, we find that the largest moment in the plate will equal $\left(\dfrac{P_u}{BN}\right)(\ell)\left(\dfrac{\ell}{2}\right) = \dfrac{P_u \ell^2}{2BN}$ for LRFD and $\dfrac{P_a \ell^2}{2BN}$ for ASD.

In the next few chapters of this text, the reader will learn how to calculate the resisting moments of plates (as well as the resisting moments for other steel sections). For plates, these values are $\dfrac{\phi_b F_y bt^2}{4}$ for LRFD, with $\phi_b = 0.9$, and $\dfrac{F_y bt^2}{4\Omega_b}$ for ASD, with $\Omega_b = 1.67$.

If these resisting moments are equated to the maximum bending moments, the resulting expressions may be solved for the required depth or thickness t with the following results, noting that $b = 1$ in:

LRFD with $\phi_b = 0.9$	ASD with $\Omega_b = 1.67$
$\dfrac{\phi_b F_y bt^2}{4} = \dfrac{P_u l^2}{2BN}$	$\dfrac{F_y bt^2}{4\Omega_b} = \dfrac{P_a l^2}{2BN}$
$t_{\text{reqd}} = \ell\sqrt{\dfrac{2P_u}{0.9F_y BN}}$	$t_{\text{reqd}} = \ell\sqrt{\dfrac{3.33P_a}{F_y BN}}$

Four example base plate designs are presented in the next few pages. Example 7-5 illustrates the design of a base plate supported by a large reinforced concrete footing, with A_2 many times as large as A_1. In Example 7-6, a base plate is designed that is supported by a concrete pedestal, where the plate covers the entire concrete area. In Example 7-7, a base plate is selected for a column that is to be supported on a pedestal 4 in wider on each side than the plate. This means that A_2 cannot be determined until the plate area is computed. Finally, Example 7-8 presents the design of a base plate for an HSS column.

Example 7-5

Design a base plate of A36 steel ($F_y = 36$ ksi) for a W12 × 65 column ($F_y = 50$ ksi) that supports the loads $P_D = 200$ k and $P_L = 300$ k. The concrete has a compressive strength $f'_c = 3$ ksi, and the footing has the dimensions 9 ft × 9 ft.

Solution. Using a W12 × 65 column ($d = 12.1$ in, $b_f = 12.0$ in)

LRFD	ASD
$P_u = (1.2)(200) + (1.6)(300) = 720$ k	$P = 200 + 300 = 500$ k
$A_2 = $ footing area $= \left(12\dfrac{\text{in}}{\text{ft}} \times 9 \text{ ft}\right)\left(12\dfrac{\text{in}}{\text{ft}} \times 9 \text{ ft}\right) = 11{,}664 \text{ in}^2$	$A_2 = 11{,}664 \text{ in}^2$

Determine required base plate area $A_1 = BN$. Note that the area of the supporting concrete is for greater than the base plate area, such that $\sqrt{\dfrac{A_2}{A_1}} = 2.0$.

LRFD $\phi_c = 0.65$	ASD $\Omega_c = 2.31$
$A_1 = \dfrac{P_u}{\phi_c(0.85f'_c)\sqrt{\dfrac{A_2}{A_1}}}$	$A_1 = \dfrac{P_a\Omega_c}{0.85f'_c\sqrt{\dfrac{A_2}{A_1}}} = \dfrac{(500)(2.31)}{(0.85)(3)(2)}$
$= \dfrac{720}{(0.65)(0.85)(3)(2)} = 217.2 \text{ in}^2$	$= 226.5 \text{ in}^2$

The base plate must be at least as large as the column $b_f d = (12.0)(12.1) = 145.2 \text{ in}^2 < 217.2 \text{ in}^2$ and 226.5 in^2 optimize base plate dimensions to make m and n approximately equal. Refer to Fig. 7.15.

LRFD	ASD
$\Delta = \dfrac{0.95d - 0.8b_f}{2}$	$\Delta = 0.947$ in
$= \dfrac{(0.95)(12.1) - (0.8)(12.0)}{2} = 0.947$ in	
$N = \sqrt{A_1} + \Delta = \sqrt{217.2} + 0.947 = 15.7$ in	$N = \sqrt{226.5} + 0.947 = 16.0$ in
Say 16 in	Say 16 in
$B = \dfrac{A_1}{N} = \dfrac{217.2}{16} = 13.6$ in	$B = \dfrac{226.5}{16} = 14.2$ in

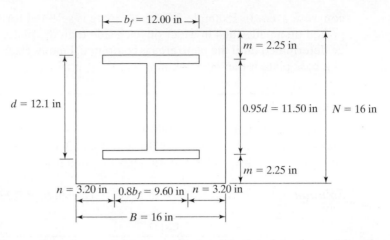

FIGURE 7.15

As previously mentioned, we might very well simplify the plates by making them square—say, 16 in × 16 in.

Check the bearing strength of the concrete

LRFD $\phi_c = 0.65$	ASD $\Omega_c = 2.31$
$\phi_c P_p = \phi_c 0.85 f'_c A_1 \sqrt{\dfrac{A_2}{A_1}}$	$\dfrac{P_p}{\Omega_c} = \dfrac{0.85 f'_c A_1}{\Omega_c} \sqrt{\dfrac{A_2}{A_1}}$
$= (0.65)(0.85)(3)(16 \times 16)(2)$	$= \dfrac{(0.85)(3)(16 \times 16)(2)}{2.31} = 565.2\text{ k} > 500\text{ k } \textbf{OK}$
$= 848.6\text{ k} > 720\text{ k}\quad \textbf{OK}$	

Computing required base plate thickness

$$m = \frac{N - 0.95d}{2} = \frac{16 - (0.95)(12.1)}{2} = 2.25 \text{ in}$$

$$n = \frac{B - 0.8b_f}{2} = \frac{16 - (0.8)(12.0)}{2} = 3.20 \text{ in}$$

$$n' = \frac{\sqrt{db_f}}{4} = \frac{\sqrt{(12.1)(12.0)}}{4} = 3.01 \text{ in}$$

$$\ell = \text{largest of } m, n, \text{ or } n' = 3.20 \text{ in}$$

LRFD	ASD
$t_{reqd} = \ell \sqrt{\dfrac{2P_u}{0.9 F_y BN}}$	$t_{reqd} = \ell \sqrt{\dfrac{3.33 P_a}{F_y BN}}$
$= 3.20 \sqrt{\dfrac{(2)(720)}{(0.9)(36)(16 \times 16)}} = 1.33 \text{ in}$	$= 3.20 \sqrt{\dfrac{(3.33)(500)}{(36)(16)(16)}} = 1.36 \text{ in}$
Use PL $1\frac{1}{2} \times 16 \times 1$ ft 4 in A36.	Use PL $1\frac{1}{2} \times 16 \times 1$ ft 4 in A36.

Example 7-6

A base plate is to be designed for a W12 × 152 column (F_y = 50 ksi) that supports the loads P_D = 200 k and P_L = 450 k. Select an A36 plate (F_y = 36 ksi) to cover the entire area of the 3 ksi concrete pedestal underneath.

Solution. Using a W12 × 152 column (d = 13.7 in, b_f = 12.5 in)

LRFD	ASD
$P_u = (1.2)(200) + (1.6)(450) = 960$ k	$P_a = 200 + 450 = 650$ k

Determine the required base plate area, noting that the term $\sqrt{\dfrac{A_2}{A_1}}$ is equal to 1.0, since $A_1 = A_2$.

LRFD ϕ_c = 0.65	ASD Ω_c = 2.31
$A_1 = \dfrac{P_u}{\phi_c(0.85f'_c)\sqrt{\dfrac{A_2}{A_1}}}$	$A_1 = \dfrac{P_a\Omega_c}{0.85f'_c\sqrt{\dfrac{A_2}{A_1}}}$
$= \dfrac{960}{(0.65)(0.85 \times 3)(1)}$	$= \dfrac{(650)(2.31)}{(0.85)(3)(1)}$
$= 579.2$ in^2 ←	$= 588.8$ in^2 ←
A_1 min $= db_f = (13.7)(12.5)$	A_1 min $= db_f = (13.7)(12.5)$
$= 171.2$ in^2	$= 171.2$ in^2

Optimizing base plate dimensions $n \sim m$

LRFD	ASD
$\Delta = \dfrac{0.95d - 0.8b_f}{2}$	
$= \dfrac{(0.95)(13.7) - (0.8)(12.5)}{2} = 1.51$ in	$\Delta = 1.51$ in
$N = \sqrt{A_1} + \Delta = \sqrt{579.2} + 1.51$	$N = \sqrt{588.8} + 1.51$
$= 25.6$ in **Say 26 in**	$= 25.8$ in **Say 26 in**
$B = \dfrac{A_1}{N} = \dfrac{579.2}{26} = 22.3$ in	$B = \dfrac{588.8}{26} = 22.6$ in
Say 23 in	**Say 23 in**

Check the bearing strength of the concrete

LRFD $\phi_c = 0.65$	ASD $\Omega_c = 2.31$
$\phi_c P_p = \phi_c 0.85 f'_c A_1 \sqrt{\dfrac{A_2}{A_1}}$	$\dfrac{P_p}{\Omega_c} = \dfrac{0.85 f'_c A_1}{\Omega_c} \sqrt{\dfrac{A_2}{A_1}}$
$= (0.65)(0.85)(3)(23 \times 26)(1.0)$	$= \dfrac{(0.85)(3)(23 \times 26)}{2.31}(1.0)$
$= 991.2 \text{ k} > 960 \text{ k}$ **OK**	$= 660.1 \text{ k} > 650 \text{ k}$ **OK**

Computing required base plate thickness

$$m = \frac{N - 0.95d}{2} = \frac{26 - (0.95)(13.7)}{2} = 6.49 \text{ in}$$

$$n = \frac{B - 0.8b_f}{2} = \frac{23 - (0.8)(12.5)}{2} = 6.50 \text{ in}$$

$$n' = \frac{\sqrt{db_f}}{4} = \frac{\sqrt{(13.7)(12.5)}}{4} = 3.27 \text{ in}$$

$$\ell = \text{maximum of } m, n \text{ or } n' = 6.50 \text{ in}$$

LRFD	ASD
$t_{\text{reqd}} = \ell \sqrt{\dfrac{2P_u}{0.9 F_y BN}}$	$t_{\text{reqd}} = \ell \sqrt{\dfrac{3.33 P_a}{F_y BN}}$
$= 6.50 \sqrt{\dfrac{(2)(960)}{(0.9)(36)(26 \times 23)}}$	$= 6.50 \sqrt{\dfrac{(3.33)(650)}{(36)(23 \times 26)}}$
$= 2.05 \text{ in}$	$= 2.06 \text{ in}$

Use $2\frac{1}{8} \times 23 \times 2$ ft 2 in A36 base plate with 23×26 concrete pedestal ($f'_c = 3$ ksi).

Example 7-7

Repeat Example 7-6 if the column is to be supported by a concrete pedestal 2 in wider on each side than the base plate.

Solution. Using a W12 × 152 ($d = 13.7$ in, $b_f = 12.5$ in)

LRFD	ASD
$P_u = (1.2)(200) + (1.6)(450) = 960 \text{ k}$	$P_a = 200 + 450 = 650 \text{ k}$
A_1 required from Example 7-6 solution was 579.2 in^2	A_1 required from Example 7-6 solution was 588.8 in^2

If we try a plate 24×25 ($A_1 = 600$ in^2), the pedestal area will equal $(24 + 4)(25 + 4) = 812$ in^2, and $\sqrt{\dfrac{A_2}{A_1}}$ will equal $\sqrt{\dfrac{812}{600}} = 1.16$. Recalculating the A_1 values gives

LRFD $\phi_c = 0.65$	ASD $\Omega_c = 2.31$
$A_1 = \dfrac{P_u}{\phi_c(0.85f'_c)\sqrt{\dfrac{A_2}{A_1}}}$	$A_1 = \dfrac{P_a\Omega_c}{0.85f'_c\sqrt{\dfrac{A_2}{A_1}}}$
$= \dfrac{960}{(0.65)(0.85)(3)(1.16)} = 499.3$ in^2	$= \dfrac{(650)(2.31)}{(0.85)(3)(1.16)} = 507.6$ in^2

Trying a 22×23 plate (506 in^2), the pedestal area will be $(22 + 4)(23 + 4) = 702$ in^2, and $\sqrt{\dfrac{A_2}{A_1}} = \sqrt{\dfrac{702}{506}} = 1.18$. Thus, A_1 (LRFD) will be 490.8 in^2 and A_1 (ASD) will be 499.0 in^2.

Optimizing base plate dimensions $n \sim m$

LRFD	ASD
$\Delta = \dfrac{0.95d - 0.8b_f}{2}$	
$= \dfrac{(0.95)(13.7) - (0.8)(12.5)}{2} = 1.51$ in	$\Delta = 1.51$ in
$N = \sqrt{A_1} + \Delta = \sqrt{490.8} + 1.51$	$N = \sqrt{A_1} + \Delta = \sqrt{499.0} + 1.51$
$= 23.66$ in **Say, 24 in**	$= 23.85$ in. **Say, 24 in**
$B = \dfrac{A_1}{N} = \dfrac{490.8}{24} = 20.45$ in	$B = \dfrac{A_1}{N} = \dfrac{499}{24} = 20.79$ in
Say, 21 in	**Say, 21 in**
Use pedestal 25 $\times$ 28	
$\sqrt{\dfrac{A_2}{A_1}} = \sqrt{\dfrac{(25)(28)}{(21)(24)}} = 1.18$	Same.

Check the bearing strength of the concrete

LRFD $\phi_c = 0.65$	ASD $\Omega_c = 2.31$
$\phi_c P_p = \phi_c 0.85f'_c A_1 \sqrt{\dfrac{A_2}{A_1}}$	$\dfrac{P_p}{\Omega_c} = \dfrac{0.85f'_c A_1}{\Omega_c}\sqrt{\dfrac{A_2}{A_1}}$
$= (0.65)(0.85)(3)(21 \times 24)(1.18)$	$= \dfrac{(0.85)(3)(21 \times 24)}{2.31}(1.18)$
$= 985.7$ k > 960 k **OK**	$= 656.5$ k > 650 k **OK**

Computing required base plate thickness

$$m = \frac{N - 0.95d}{2} = \frac{24 - (0.95)(13.7)}{2} = 5.49 \text{ in}$$

$$n = \frac{B - 0.8b_f}{2} = \frac{21 - (0.8)(12.5)}{2} = 5.50 \text{ in}$$

$$n' = \frac{\sqrt{db_f}}{4} = \frac{\sqrt{(13.7)(12.5)}}{4} = 3.27 \text{ in}$$

$$\ell = \text{maximum of } m, n \text{ or } n' = 5.50 \text{ in}$$

LRFD	ASD
$t_{reqd} = \ell\sqrt{\dfrac{2P_u}{0.9F_yBN}}$	$t_{reqd} = \ell\sqrt{\dfrac{3.33P_a}{F_yBN}}$
$= 5.50\sqrt{\dfrac{(2)(960)}{(0.9)(36)(21)(24)}}$	$= 5.50\sqrt{\dfrac{(3.33)(650)}{(36)(21)(24)}}$
$= 1.89 \text{ in}$	$= 1.90 \text{ in}$

Use 2 × 21 × 2 ft 0 in A36 base plate with 25 × 28 concrete pedestal (f'_c = 3 ksi).

Example 7-8

A HSS 10 × 10 × $\dfrac{5}{16}$ with F_y = 46 ksi is used to support the service loads P_D = 100 k and P_L = 150 k. A spread footing underneath is 9 ft-0 in × 9 ft-0 in and consists of reinforced concrete with f'_c = 4000 psi. Design a base plate for this column with A36 steel (F_y = 36 ksi and F_u = 58 ksi).

Solution. Required strength

LRFD	ASD
$P_u = (1.2)(100) + (1.6)(150) = 360 \text{ k}$	$P_a = 100 + 150 = 250 \text{ k}$

Try a base plate extending 4 in from the face of the column in each direction—that is, an 18 in × 18 in plate.

Determine the available strength of the concrete footing.

$$A_1 = (18)(18) = 324 \text{ in}^2$$
$$A_2 = (12 \times 9)(12 \times 9) = 11{,}664 \text{ in}^2$$

$$P_p = 0.85f'_cA_1\sqrt{\frac{A_2}{A_1}} = (0.85)(4)(324)\sqrt{\frac{11{,}664}{324}} = 6609.6 \text{ K}$$

$$\text{since } \sqrt{\frac{11{,}664}{324}} = 6.0 > 2.0 \therefore P_p = 1.7f_c' A_1$$

$$P_p = 1.7f_c' A_1 = 1.7(4)(324) = 2203.2 \text{ k}$$

LRFD $\phi_c = 0.65$	ASD $\Omega_c = 2.31$
$\phi_c P_p = (0.65)(2203.2)$	$\dfrac{P_p}{\Omega_c} = \dfrac{2203.2}{2.31}$
$= 1432.1 \text{ k} > 360 \text{ k}$ **OK**	$= 953.8 \text{ k} > 250 \text{ k}$ **OK**

Determine plate thickness.

$$m = n = \frac{N - (0.95)(\text{outside dimension of HSS})}{2}$$

$$= \frac{18 - (0.95)(10)}{2} = 4.25 \text{ in}$$

Notice that these values for m and n are both less than the distance from the center of the base plate to the center of the HSS walls. However, the moment in the plate outside the walls is greater than the moment in the plate between the walls. You can verify this statement by drawing the moment diagrams for the situation shown in Fig. 7.16.

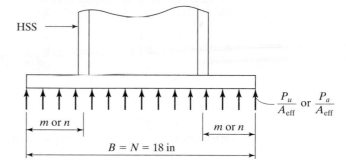

FIGURE 7.16

LRFD	ASD
$f_{pu} = \dfrac{P_u}{A_{\text{eff}}} = \dfrac{360}{(18)(18)} = 1.11 \text{ ksi}$	$f_{pa} = \dfrac{P_a}{A_{\text{eff}}} = \dfrac{250}{324} = 0.772 \text{ ksi}$
$t_{\text{reqd}} = \ell \sqrt{\dfrac{2P_u}{0.9 F_y BN}}$	$t_{\text{reqd}} = \ell \sqrt{\dfrac{3.33 P_a}{F_y BN}}$
$= 4.25 \sqrt{\dfrac{(2)(360)}{(0.9)(36)(18)(18)}} = 1.11 \text{ in}$	$= 4.25 \sqrt{\dfrac{(3.33)(250)}{(36)(18)(18)}} = 1.14 \text{ in}$

Use $1\frac{1}{4} \times 18 \times 1$ ft 6 in A36 base plate for both LRFD and ASD.

7.7.3   Moment Resisting Column Bases

The designer will often be faced with the need for moment resisting column bases. Before such a topic is introduced, however, the student needs to be familiar with the design of welds (Chapter 14) and moment resisting connections between members (Chapters 14 and 15). For this reason, the subject of moment resisting base plates has been placed in Appendix D.

7.8   PROBLEMS FOR SOLUTION

7-1.   Using the alignment chart from the AISC Specification, determine the effective length factors for columns *IJ, FG,* and *GH* of the frame shown in the accompanying figure, assuming that the frame is subject to sidesway and that all of the assumptions on which the alignment charts were developed are met. (*Ans.* 1.27, 1.20, and 1.17)

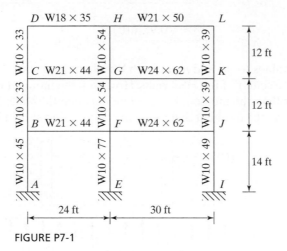

FIGURE P7-1

7-2.   Determine the effective length factors for all of the columns of the frame shown in the accompanying figure. Note that columns *CD* and *FG* are subject to sidesway, while columns *BC* and *EF* are braced against sidesway. Assume that all of the assumptions on which the alignment charts were developed are met.

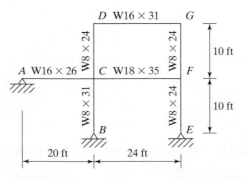

FIGURE P7-2

7-3. to 7-6.   *Use both LRFD and ASD methods.*

7-3.   a. Determine the available column strength for column AB in the frame shown if $F_y = 50$ ksi, and only in-plane behavior is considered. Furthermore, assume that the column immediately above or below AB are the same size as AB, and also that all the other assumptions on which the alignment charts were developed are met. (*Ans.* 825 k, LRFD; 549 k, ASD)

b. Repeat part (a) if inelastic behavior is considered and $P_D = 200$ k and $P_L = 340$ k. (*Ans.* 838 k, LRFD; 563 k, ASD)

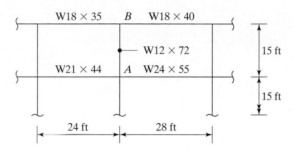

FIGURE P7-3

7-4.   Repeat Prob. 7-3 if $P_D = 250$ k and $P_L = 400$ k and a W14 × 90 section is used.

7-5.   Determine the available column strength for column AB in the frame shown for which $F_y = 50$ ksi. Otherwise, the conditions are exactly as those described for Prob. 7-3.
a. Assume elastic behavior. (*Ans.* 1095 k, LRFD; 729 k, ASD)
b. Assume inelastic behavior and $P_D = 240$ k and $P_L = 450$ k. (*Ans.* 1098 k, LRFD; 735 k, ASD)

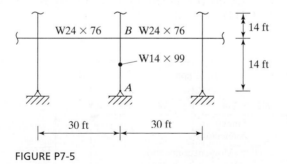

FIGURE P7-5

7-6.   Repeat Prob. 7-5 if $P_D = 225$ k and $P_L = 375$ k and a W12 × 87 section is used.

7-7. to 7-13.   *Use the Effective Length Method, assume elastic behavior, and use both the LRFD and ASD methods. The columns are assumed to have no bending moments.*

7-7. Design W12 columns for the bent shown in the accompanying figure, with 50 ksi steel. The columns are braced top and bottom against sidesway out of the plane of the frame so that $K_y = 1.0$ in that direction. Sidesway is possible in the plane of the frame, the x-x axis. Design the right-hand column as a leaning column, $K_x = K_y = 1.0$ and the left-hand column as a moment frame column, K_x determined from the alignment chart. $P_D = 350$ k and $P_L = 240$ k for each column. The beam has a moment connection to the left column, and has a simple or pinned connection to the right column. (*Ans.* (Right) W12 × 79, LRFD; W12 × 87, ASD – (Left) W12 × 170, LRFD; W12 × 190, ASD)

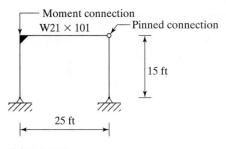

FIGURE P7-7

7-8. Repeat Prob. 7-7 if the loads on each column are $P_D = 120$ k and $P_L = 220$ k, and the girder is a W21 × 68.

7-9. Design W14 columns for the bent shown in the accompanying figure, with 50 ksi steel. The columns are braced top and bottom against sidesway out of the plane of the frame so that $K_y = 1.0$ in that direction. Sidesway is possible in the plane of the frame, the x-x axis. Design the interior column as a leaning column, $K_x = K_y = 1.0$ and the exterior columns as a moment frame columns, K_x determined from the alignment chart. (*Ans.* (Interior) W14 × 176, LRFD; W14 × 193, ASD – (Exterior) W14 × 211, LRFD and ASD)

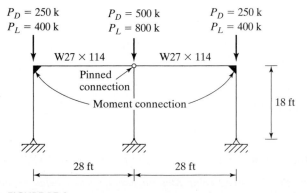

FIGURE P7-9

7-10. Repeat Prob. 7-9, assuming that the outside columns are fixed at the bottom.

7-11. The frame shown in the accompanying figure is unbraced against sidesway about the x-x axis. Determine K_x for column AB. Support conditions in the direction perpendicular to the frame are such that $K_y = 1.0$. Determine if the W14 × 109 column for member AB is capable of resisting a dead load of 250 k and a live load of 500 k. A992 steel is used. (*Ans.* LRFD W14 × 109, OK, $\Phi P_n = 1205\text{ k} > P_u = 1100\text{ k}$; ASD W14 × 109, OK, $P_n/\Omega = 803\text{ k} > P_a = 750\text{ k}$)

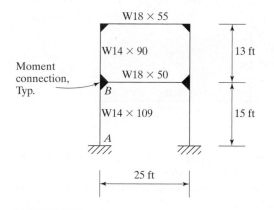

W18 × 55

W14 × 90

13 ft

Moment connection, Typ.

W18 × 50

B

W14 × 109

15 ft

A

25 ft

FIGURE P7-11

7-12. The frame shown in the accompanying figure is unbraced against sidesway about the x-x axis. The columns are W8 and the beams are W12 × 16. ASTM A572 steel is used for the columns and beams. The beams and columns are oriented so that bending is about the x-x axis. Assume that $K_y = 1.0$, and for column AB the service load is 175 k, in which 25 percent is dead load and 75 percent is live load. Select the lightest W8 shape for column AB.

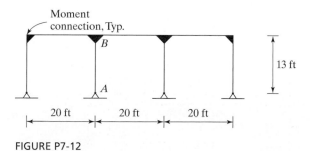

Moment connection, Typ.

B

13 ft

A

20 ft 20 ft 20 ft

FIGURE P7-12

7-13. Select the lightest W12 shape for column AB of the pinned-base unbraced-moment frame shown in the figure. All steel is ASTM A992. The horizontal girder is a W18 × 76. The girder and columns are oriented so that bending is about the x-x axis. In the plane perpendicular to the frame, $K_y = 1.0$ and bracing is provided to the y-y axis of the column at the top and mid-height using pinned end connections. The loads on each are $P_D = 150$ k and $P_L = 200$ k. (*Ans.* W12 × 53, LRFD; W12 × 58, ASD)

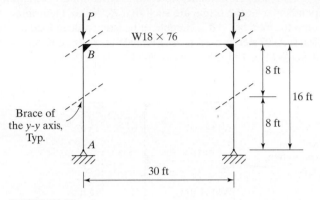

FIGURE P7-13

7-14. Design a square base plate with A36 steel for a W10 × 60 column with a service dead load of 175 k and a service live load of 275 k. The concrete 28-day strength, f'_c, is 3000 psi. The base plate rests on a 12 ft 0 in × 12 ft 0 in concrete footing. Use the LRFD and ASD design methods.

7-15. Repeat Prob. 7-14 if the column is supported by a 24 in × 24 in concrete pedestal. (*Ans.* B PL – 1¾ × 18 × 1 ft 6 in A36 LRFD and ASD)

7-16. Design a rectangular base plate for a W8 × 28 column with $P_D = 80$ k and $P_L = 150$ k if A36 steel is used and $f'_c = 3$ ksi for the concrete. Assume that the column is supported by a 7 ft 0 in × 7 ft 0 in concrete footing. Use the LRFD and ASD design methods.

CHAPTER 8

Introduction to Beams

8.1 TYPES OF BEAMS

Beams are usually said to be members that support transverse loads. They are probably thought of as being used in horizontal positions and subjected to gravity or vertical loads, but there are frequent exceptions—roof rafters, for example.

Among the many types of beams are joists, lintels, spandrels, stringers, and floor beams. *Joists* are the closely spaced beams supporting the floors and roofs of buildings, while *lintels* are the beams over openings in masonry walls, such as windows and doors. *Spandrel beams* support the exterior walls of buildings and perhaps part of the floor and hallway loads. The discovery that steel beams as a part of a structural frame could support masonry walls (together with the development of passenger elevators) is said to have permitted the construction of today's high-rise buildings. *Stringers* are the beams in bridge floors running parallel to the roadway, whereas *floor beams* are the larger beams in many bridge floors, which are perpendicular to the roadway of the bridge and are used to transfer the floor loads from the stringers to the supporting girders or trusses. The term *girder* is rather loosely used, but usually indicates a large beam and perhaps one into which smaller beams are framed. These and other types of beams are discussed in the sections to follow.

8.2 SECTIONS USED AS BEAMS

The W shapes will normally prove to be the most economical beam section, and they have largely replaced channels and S sections for beam usage. Channels are sometimes used for beams subjected to light loads, such as purlins, and in places where clearances available require narrow flanges. They have very little resistance to lateral forces and need to be braced, as illustrated by the sag rod problem in Chapter 4. The W shapes have more steel concentrated in their flanges than do S beams and thus have larger moments of inertia and resisting moments for the same weights. They are relatively wide and have appreciable lateral stiffness. (The small amount of space devoted to

Harrison Avenue Bridge, Beaumont, TX. (Courtesy of Bethlehem Steel Corporation.)

S beams in the AISC Manual clearly shows how much their use has decreased from former years. Today, they are used primarily for special situations, such as when narrow flange widths are desirable, where shearing forces are very high, or when the greater flange thickness next to the web may be desirable where lateral bending occurs, as perhaps with crane rails or monorails.)

Another common type of beam section is the open-web steel joist, or bar joist, which is discussed at length in Chapter 19. This type of section, which is commonly used to support floor and roof slabs, is actually a light shop-fabricated parallel chord truss. It is particularly economical for long spans and light loads.

8.3   BENDING STRESSES

For an introduction to bending stresses, the rectangular beam and stress diagrams of Fig. 8.1 are considered. (For this initial discussion, the beam's compression flange is assumed to be fully braced against lateral buckling. Lateral buckling is discussed at length in Chapter 9). If the beam is subjected to some bending moment, the stress at any point may be computed with the usual flexure formula $f_b = Mc/I$. *It is to be remembered, however, that this expression is applicable only when the maximum computed stress in the beam is below the elastic limit.* The formula is based on the usual elastic assumptions: Stress is proportional to strain, a plane section before bending remains a plane section after bending, etc. The value of I/c is a constant for a particular section and is known as the *section modulus* (S). The flexure formula may then be written as follows:

$$f_b = \frac{Mc}{I} = \frac{M}{S}$$

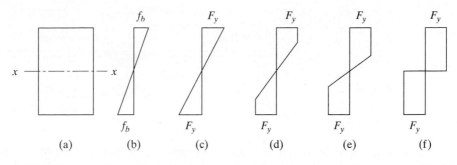

FIGURE 8.1

Variations in bending stresses due to increasing moment about x axis.

Initially, when the moment is applied to the beam, the stress will vary linearly from the neutral axis to the extreme fibers. This situation is shown in part (b) of Fig. 8.1. If the moment is increased, there will continue to be a linear variation of stress until the yield stress is reached in the outermost fibers, as shown in part (c) of the figure. The *yield moment* of a cross section is defined as the moment that will just produce the yield stress in the outermost fiber of the section.

If the moment in a ductile steel beam is increased beyond the yield moment, the outermost fibers that had previously been stressed to their yield stress will continue to have the same stress, but will yield, and the duty of providing the necessary additional resisting moment will fall on the fibers nearer to the neutral axis. This process will continue, with more and more parts of the beam cross section stressed to the yield stress (as shown by the stress diagrams of parts (d) and (e) of the figure), until finally a full plastic distribution is approached, as shown in part (f). Note that the variation of strain from the neutral axis to the outer fibers remains linear for all of these cases. When the stress distribution has reached this stage, a *plastic hinge* is said to have formed, because no additional moment can be resisted at the section. Any additional moment applied at the section will cause the beam to rotate, with little increase in stress.

The *plastic moment* is the moment that will produce full plasticity in a member cross section and create a plastic hinge. The ratio of the plastic moment M_p to the yield moment M_y is called the *shape factor*. The shape factor equals 1.50 for rectangular sections and varies from about 1.10 to 1.20 for standard rolled-beam sections.

8.4 PLASTIC HINGES

This section is devoted to a description of the development of a plastic hinge as in the simple beam shown in Fig. 8.2. The load shown is applied to the beam and increased in magnitude until the yield moment is reached and the outermost fiber is stressed to the yield stress. The magnitude of the load is further increased, with the result that the outer fibers begin to yield. The yielding spreads out to the other fibers, away from the section of maximum moment, as indicated in the figure. The distance in which this yielding occurs away from the section in question is dependent on the loading conditions and the member cross section. For a concentrated load applied at the center line of a simple beam with a rectangular cross section, yielding in the extreme fibers at the time the plastic hinge is formed will extend for one-third of the span. For a W shape in similar

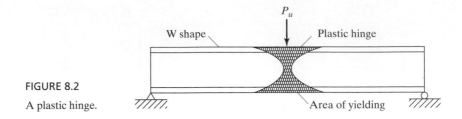

FIGURE 8.2

A plastic hinge.

circumstances, yielding will extend for approximately one-eighth of the span. During this same period, the interior fibers at the section of maximum moment yield gradually, until nearly all of them have yielded and a plastic hinge is formed, as shown in Fig. 8.2.

Although the effect of a plastic hinge may extend for some distance along the beam, for analysis purposes it is assumed to be concentrated at one section. For the calculation of deflections and for the design of bracing, the length over which yielding extends is quite important.

For plastic hinges to form, the sections must be compact. This term was previously introduced in Section 5.7.1. There compact sections were defined as being those which have sufficiently stocky profiles such that they are capable of developing fully plastic stress distributions before they buckle locally. This topic is continued in some detail in Section 9.9.

The student must realize that for plastic hinges to develop the members must not only be compact but also must be braced in such a fashion that lateral buckling is prevented. Such bracing is discussed in Section 9.4.

Finally the effects of shear, torsion, and axial loads must be considered. They may be sufficiently large as to cause the members to fail before plastic hinges can form. In the study of plastic behavior, strain hardening is not considered.

When steel frames are loaded to failure, the points where rotation is concentrated (plastic hinges) become quite visible to the observer before collapse occurs.

8.5   ELASTIC DESIGN

Until recent years, almost all steel beams were designed on the basis of the elastic theory. The maximum load that a structure could support was assumed to equal the load that first caused a stress somewhere in the structure to equal the yield stress of the material. The members were designed so that computed bending stresses for service loads did not exceed the yield stress divided by a safety factor (e.g., 1.5 to 2.0). Engineering structures have been designed for many decades by this method, with satisfactory results. The design profession, however, has long been aware that ductile members do not fail until a great deal of yielding occurs after the yield stress is first reached. This means that such members have greater margins of safety against collapse than the elastic theory would seem to indicate.

8.6   THE PLASTIC MODULUS

The yield moment M_y equals the yield stress times the elastic modulus. The elastic modulus equals I/c or $bd^2/6$ for a rectangular section, and the yield moment equals $F_y bd^2/6$. This same value can be obtained by considering the resisting internal couple shown in Fig. 8.3.

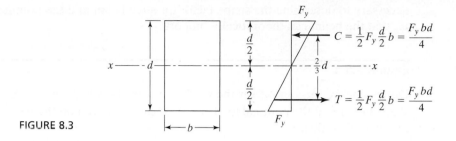

FIGURE 8.3

The resisting moment equals T or C times the lever arm between them, as follows:

$$M_y = \left(\frac{F_y bd}{4}\right)\left(\frac{2}{3}d\right) = \frac{F_y bd^2}{6}$$

The elastic section modulus can again be seen to equal $bd^2/6$ for a rectangular beam.

The resisting moment at full plasticity can be determined in a similar manner. The result is the so-called plastic moment, M_p. *It is also the nominal moment of the section*, M_n. This plastic, or nominal, moment equals T or C times the lever arm between them. For the rectangular beam of Fig. 8.4, we have

$$M_p = M_n = T\frac{d}{2} = C\frac{d}{2} = \left(F_y\frac{bd}{2}\right)\left(\frac{d}{2}\right) = F_y\frac{bd^2}{4}.$$

The plastic moment is said to equal the yield stress times the plastic section modulus. From the foregoing expression for a rectangular section, the plastic section modulus Z can be seen to equal $bd^2/4$. The shape factor, which equals $M_p/M_y = F_y Z/F_y S$, or Z/S, is $(bd^2/4)/(bd^2/6) = 1.50$ for a rectangular section. *A study of the plastic section modulus determined here shows that it equals the statical moment of the tension and compression areas about the plastic neutral axis.* Unless the section is symmetrical, the neutral axis for the plastic condition will not be in the same location as for the elastic condition. The total internal compression must equal the total internal tension. As all fibers are considered to have the same stress (F_y) in the plastic condition, the areas above and below the plastic neutral axis must be equal. This situation does not hold for unsymmetrical sections in the elastic condition. Example 8-1 illustrates the calculations

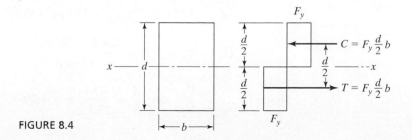

FIGURE 8.4

necessary to determine the shape factor for a tee beam and the nominal uniform load w_n that the beam can theoretically support.

___

Example 8-1

Determine M_y, M_n, and Z for the steel tee beam shown in Fig. 8.5. Also, calculate the shape factor and the nominal load (w_n) that can be placed on the beam for a 12-ft simple span. $F_y = 50$ ksi.

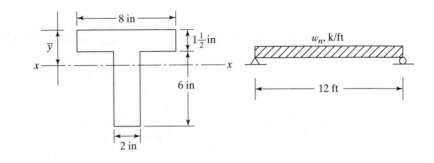

FIGURE 8.5

Solution. Elastic calculations:

$$A = (8\text{ in})\left(1\frac{1}{2}\text{ in}\right) + (6\text{ in})(2\text{ in}) = 24\text{ in}^2$$

$$\bar{y} = \frac{(12\text{ in})(0.75\text{ in}) + (12\text{ in})(4.5\text{ in})}{24\text{ in}^2} = 2.625\text{ in from top of flange}$$

$$I = \frac{1}{12}(8\text{ in})(1.5\text{ in})^3 + (8\text{ in})(1.5\text{ in})(1.875\text{ in})^2 + \frac{1}{12}(2\text{ in})(6\text{ in})^3$$
$$+ (2\text{ in})(6\text{ in})(1.875\text{ in})^2$$
$$= 122.6\text{ in}^4$$

$$S = \frac{I}{c} = \frac{122.6\text{ in}^4}{4.875\text{ in}} = 25.1\text{ in}^3$$

$$M_y = F_y S = \frac{(50\text{ ksi})(25.1\text{ in}^3)}{12\text{ in/ft}} = 104.6\text{ ft-k}$$

Plastic calculations (plastic neutral axis is at base of flange):

$$Z = (12\text{ in}^2)(0.75\text{ in}) + (12\text{ in}^2)(3\text{ in}) = 45\text{ in}^3$$

$$M_n = M_p = F_y Z = \frac{(50\text{ ksi})(45\text{ in}^3)}{12\text{ in/ft}} = 187.5\text{ ft-k}$$

$$\text{Shape factor} = \frac{M_p}{M_y} \quad \text{or} \quad \frac{Z}{S} = \frac{45\text{ in}^3}{25.1\text{ in}^3} = 1.79$$

$$M_n = \frac{w_n L^2}{8}$$

$$\therefore \quad w_n = \frac{(8)(187.5 \text{ ft-k})}{(12 \text{ ft})^2} = 10.4 \text{ k/ft}$$

The values of the plastic section moduli for the standard steel beam sections are tabulated in Table 3-2 of the AISC Manual, entitled "W Shapes Selection by Z_x", and are listed for each shape in the "Dimensions and Properties" section of the Handbook (Part 1). These Z values will be used continuously throughout the text.

8.7   THEORY OF PLASTIC ANALYSIS

The basic plastic theory has been shown to be a major change in the distribution of stresses after the stresses at certain points in a structure reach the yield stress. The theory is that those parts of the structure that have been stressed to the yield stress cannot resist additional stresses. They instead will yield the amount required to permit the extra load or stresses to be transferred to other parts of the structure where the stresses are below the yield stress, and thus in the elastic range and able to resist increased stress. Plasticity can be said to serve the purpose of equalizing stresses in cases of overload.

As early as 1914, Dr. Gabor Kazinczy, a Hungarian, recognized that the ductility of steel permitted a redistribution of stresses in an overloaded, statically indeterminate structure.[1] In the United States, Prof. J. A. Van den Broek introduced his plastic theory, which he called "limit design." This theory was published in a paper entitled "Theory of Limit Design" in February 1939, in the *Proceedings of the ASCE*.

For this discussion, the stress–strain diagram is assumed to have the idealized shape shown in Fig. 8.6. The yield stress and the proportional limit are assumed to occur at the same point for this steel, and the stress–strain diagram is assumed to be a

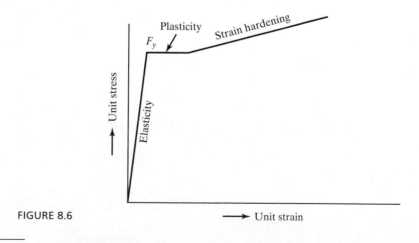

FIGURE 8.6

[1] Lynn S. Beedle, *Plastic Design of Steel Frames* (New York: Wiley, 1958), p. 3.

perfectly straight line in the plastic range. Beyond the plastic range there is a range of strain hardening. This latter range could theoretically permit steel members to withstand additional stress, but from a practical standpoint the strains which arise are so large that they cannot be considered. Furthermore, inelastic buckling will limit the ability of a section to develop a moment greater than M_p, even if strain hardening is significant.

8.8 THE COLLAPSE MECHANISM

A statically determinate beam will fail if one plastic hinge develops. To illustrate this fact, the simple beam of constant cross section loaded with a concentrated load at midspan, shown in Fig. 8.7(a), is considered. Should the load be increased until a plastic hinge is developed at the point of maximum moment (underneath the load in this case), an unstable structure will have been created, as shown in part (b) of the figure. Any further increase in load will cause collapse. P_n represents the nominal, or theoretical, maximum load that the beam can support.

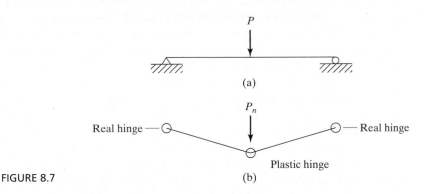

FIGURE 8.7

For a statically indeterminate structure to fail, it is necessary for more than one plastic hinge to form. The number of plastic hinges required for failure of statically indeterminate structures will be shown to vary from structure to structure, but may never be less than two. The fixed-end beam of Fig. 8.8, part (a), cannot fail unless the three plastic hinges shown in part (b) of the figure are developed.

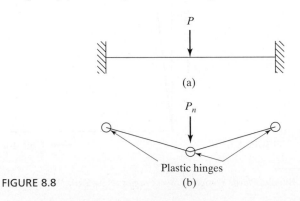

FIGURE 8.8

Although a plastic hinge may have formed in a statically indeterminate structure, the load can still be increased without causing failure if the geometry of the structure permits. The plastic hinge will act like a real hinge insofar as increased loading is concerned. As the load is increased, there is a redistribution of moment, because the plastic hinge can resist no more moment. As more plastic hinges are formed in the structure, there will eventually be a sufficient number of them to cause collapse. Actually, some additional load can be carried after this time, before collapse occurs, as the stresses go into the strain hardening range, but the deflections that would occur are too large to be permissible.

The propped beam of Fig. 8.9, part (a), is an example of a structure that will fail after two plastic hinges develop. Three hinges are required for collapse, but there is a real hinge on the right end. In this beam, the largest elastic moment caused by the design concentrated load is at the fixed end. As the magnitude of the load is increased, a plastic hinge will form at that point.

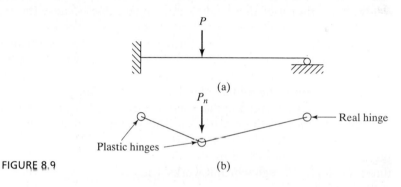

FIGURE 8.9

The load may be further increased until the moment at some other point (here it will be at the concentrated load) reaches the plastic moment. Additional load will cause the beam to collapse. The arrangement of plastic hinges and perhaps real hinges that permit collapse in a structure is called the *mechanism*. Parts (b) of Figs. 8.7, 8.8, and 8.9 show mechanisms for various beams.

After observing the large number of fixed-end and propped beams used for illustration in this text, the student may form the mistaken idea that he or she will frequently encounter such beams in engineering practice. These types of beams are difficult to find in actual structures, but are very convenient to use in illustrative examples. They are particularly convenient for introducing plastic analysis before continuous beams and frames are considered.

8.9 THE VIRTUAL-WORK METHOD

One very satisfactory method used for the plastic analysis of structures is the *virtual-work method*. The structure in question is assumed to be loaded to its nominal capacity, M_n, and is then assumed to deflect through a small additional displacement after the ultimate load is reached. The work performed by the external loads during this displacement is equated to the internal work absorbed by the hinges. For this discussion,

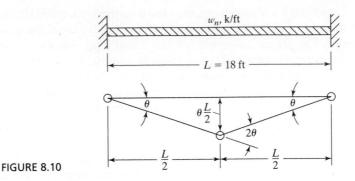

FIGURE 8.10

the *small-angle theory* is used. By this theory, the sine of a small angle equals the tangent of that angle and also equals the same angle expressed in radians. In the pages to follow, the author uses these values interchangeably because the small displacements considered here produce extremely small rotations or angles.

As a first illustration, the uniformly loaded fixed-ended beam of Fig. 8.10 is considered. This beam and its collapse mechanism are shown. Owing to symmetry, the rotations at the end plastic hinges are equal, and they are represented by θ in the figure; thus, the rotation at the middle plastic hinge will be 2θ.

The work performed by the total external load $(w_n L)$ is equal to $w_n L$ times the average deflection of the mechanism. The average deflection equals one-half the deflection at the center plastic hinge $(1/2 \times \theta \times L/2)$. The external work is equated to the internal work absorbed by the hinges, or to the sum of M_n at each plastic hinge times the angle through which it works. The resulting expression can be solved for M_n and w_n as follows:

$$M_n(\theta + 2\theta + \theta) = w_n L\left(\frac{1}{2} \times \theta \times \frac{L}{2}\right)$$

$$M_n = \frac{w_n L^2}{16}$$

$$w_n = \frac{16 M_n}{L^2}.$$

For the 18-ft span used in Fig. 8.10, these values become

$$M_n = \frac{(w_n)(18)^2}{16} = 20.25\, w_n$$

$$w_n = \frac{M_n}{20.25}.$$

Plastic analysis can be handled in a similar manner for the propped beam of Fig. 8.11. There, the collapse mechanism is shown, and the end rotations (which are equal to each other) are assumed to equal θ.

The work performed by the external load P_n as it moves through the distance $\theta \times L/2$ is equated to the internal work performed by the plastic moments at the hinges; note that there is no moment at the real hinge on the right end of the beam.

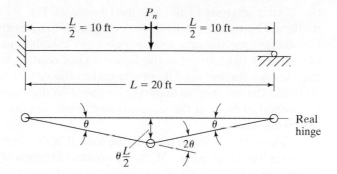

FIGURE 8.11

$$M_n(\theta + 2\theta) = P_n\left(\theta\frac{L}{2}\right)$$

$$M_n = \frac{P_n L}{6} \quad \text{(or } 3.33P_n \text{ for the 20-ft beam shown)}$$

$$P_n = \frac{6M_n}{L} \quad \text{(or } 0.3M_n \text{ for the 20-ft beam shown)}$$

The fixed-end beam of Fig. 8.12, together with its collapse mechanism and as-sumed angle rotations, is considered next. From this figure, the values of M_n and P_n can be determined by virtual work as follows:

$$M_n(2\theta + 3\theta + \theta) = P_n\left(2\theta \times \frac{L}{3}\right)$$

$$M_n = \frac{P_n L}{9} \quad \text{(or } 3.33P_n \text{ for this beam)}$$

$$P_n = \frac{9M_n}{L} \quad \text{(or } 0.3M_n \text{ for this beam).}$$

A person beginning the study of plastic analysis needs to learn to think of all the possible ways in which a particular structure might collapse. Such a habit is of the greatest importance when one begins to analyze more complex structures. In this light,

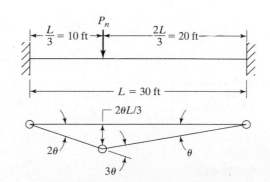

FIGURE 8.12

the plastic analysis of the propped beam of Fig. 8.13 is done by the virtual-work method. The beam with its two concentrated loads is shown, together with four possible collapse mechanisms and the necessary calculations. It is true that the mechanisms of parts (b), (d), and (e) of the figure do not control, but such a fact is not obvious to the average student until he or she makes the virtual-work calculations for each case. Actually, the mechanism of part (e) is based on the assumption that the plastic moment is reached at both of the concentrated loads simultaneously (a situation that might very well occur).

The value for which the collapse load P_n is the smallest in terms of M_n is the correct value (or the value where M_n is the greatest in terms of P_n). For this beam, the second plastic hinge forms at the P_n concentrated load, and P_n equals $0.154 M_n$.

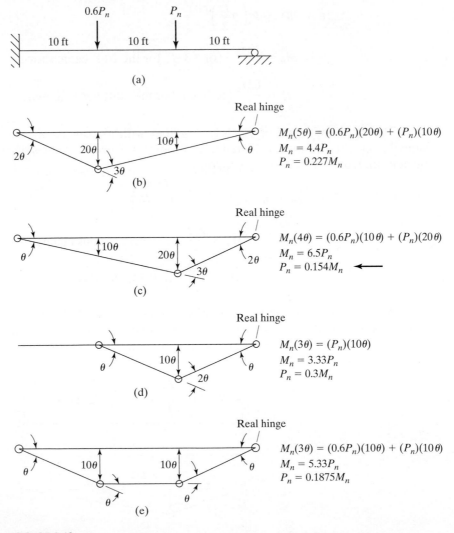

FIGURE 8.13

8.10   LOCATION OF PLASTIC HINGE FOR UNIFORM LOADINGS

There was no difficulty in locating the plastic hinge for the uniformly loaded fixed-end beam, but for other beams with uniform loads, such as propped or continuous beams, the problem may be rather difficult. For this discussion, the uniformly loaded propped beam of Fig. 8.14(a) is considered.

The elastic moment diagram for this beam is shown as the solid line in part (b) of the figure. As the uniform load is increased in magnitude, a plastic hinge will first form at the fixed end. At this time, the beam will, in effect, be a "simple" beam (so far as increased loads are concerned) with a plastic hinge on one end and a real hinge on the other. Subsequent increases in the load will cause the moment to change, as represented by the dashed line in part (b) of the figure. This process will continue until the moment at some other point (a distance x from the right support in the figure) reaches M_n and creates another plastic hinge.

The virtual-work expression for the collapse mechanism of the beam shown in part (c) of Fig. 8.14 is written as follows:

$$M_n\left(\theta + \theta + \frac{L - x}{x}\theta\right) = (w_nL)(\theta)(L - x)\left(\frac{1}{2}\right).$$

Solving this equation for M_n, taking $dM_n/dx = 0$, the value of x can be calculated to equal $0.414L$. This value is also applicable to uniformly loaded end spans of continuous beams with simple end supports, as will be illustrated in the next section.

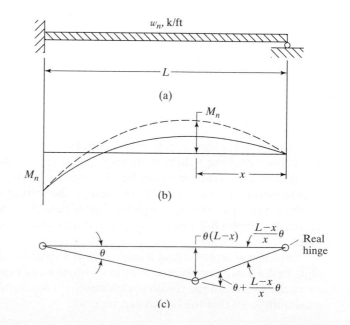

FIGURE 8.14

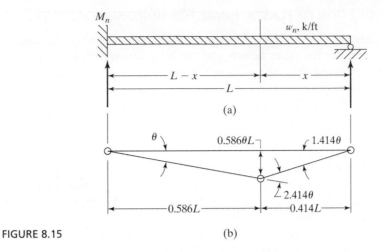

FIGURE 8.15 (b)

The beam and its collapse mechanism are redrawn in Fig. 8.15, and the following expression for the plastic moment and uniform load are written by the virtual-work procedure:

$$M_n(\theta + 2.414\theta) = (w_nL)(0.586\theta L)\left(\frac{1}{2}\right)$$

$$M_n = 0.0858w_nL^2$$

$$w_n = 11.65\frac{M_n}{L^2}$$

8.11 CONTINUOUS BEAMS

Continuous beams are very common in engineering structures. Their continuity causes analysis to be rather complicated in the elastic theory, and even though one of the complex "exact" methods is used for analysis, the resulting stress distribution is not nearly so accurate as is usually assumed.

Plastic analysis is applicable to continuous structures, as it is to one-span structures. The resulting values definitely give a more realistic picture of the limiting strength of a structure than can be obtained by elastic analysis. Continuous, statically indeterminate beams can be handled by the virtual-work procedure as they were for the single-span statically indeterminate beams. As an introduction to continuous beams, Examples 8-2 and 8-3 are presented to illustrate two of the more elementary cases.

Here, it is assumed that if any or all of a structure collapses, failure has occurred. Thus, in the continuous beams to follow, virtual-work expressions are written separately for each span. From the resulting expressions, it is possible to determine the limiting or maximum loads that the beams can support.

Example 8-2

A W18 × 55 ($Z_x = 112$ in^3) has been selected for the beam shown in Fig. 8.16. Using 50 ksi steel and assuming full lateral support, determine the value of w_n.

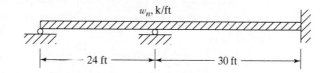

FIGURE 8.16

Solution

$$M_n = F_y Z = \frac{(50 \text{ ksi})(112 \text{ in}^3)}{12 \text{ in/ft}} = 466.7 \text{ ft-k}$$

Drawing the (collapse) mechanisms for the two spans:

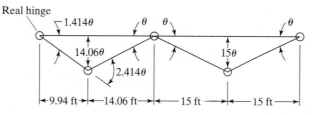

Left-hand span:

$$(M_n)(3.414\theta) = (24w_n)\left(\frac{1}{2}\right)(14.06\theta)$$

$$w_n = 0.0202 \, M_n = (0.0202)(466.7) = 9.43 \text{ k/ft}$$

Right-hand span:

$$(M_n)(4\theta) = (30w_n)\left(\frac{1}{2}\right)(15\theta)$$

$$w_n = 0.0178 \, M_n = (0.0178)(466.7) = 8.31 \text{ k/ft} \leftarrow$$

Additional spans have little effect on the amount of work involved in the plastic analysis procedure. The same cannot be said for elastic analysis. Example 8-3 illustrates the analysis of a three-span beam that is loaded with a concentrated load on each span. The student, from his or her knowledge of elastic analysis, can see that plastic hinges will initially form at the first interior supports and then at the center lines of the end spans, at which time each end span will have a collapse mechanism.

Example 8-3

Using a W21 × 44 ($Z_x = 95.4$ in^3) consisting of A992 steel, determine the value of P_n for the beam of Fig. 8.17.

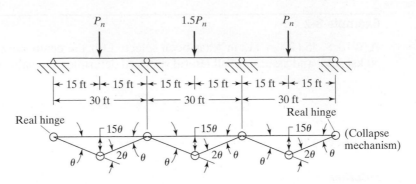

FIGURE 8.17

Solution

$$M_n = F_y Z = \frac{(50\ \text{ksi})(95.4\ \text{in}^3)}{12\ \text{in/ft}} = 397.5\ \text{ft-k}$$

For first and third spans:

$$M_m(3\theta) = (P_n)(15\theta)$$

$$P_n = 0.2\ M_n = (0.2)(397.5) = 79.5\ \text{k}$$

For center span:

$$(M_n)(4\theta) = (1.5\ P_n)(15\theta)$$

$$P_n = 0.178\ M_n = (0.178)(397.5) = 70.8\ \text{k} \leftarrow$$

8.12   BUILDING FRAMES

In this section, plastic analysis is applied to a small building frame. It is not the author's purpose to consider frames at length in this chapter. Rather, he wants to show the reader that the virtual-work method is applicable to frames as well as to beams, and that there are other types of mechanisms besides the beam types.

For the frame considered, it is assumed that the same W section is used for both the beam and the columns. If these members differed in size, it would be necessary to take that into account in the analysis.

The pin-supported frame of Fig. 8.18 is statically indeterminate to the first degree. The development of one plastic hinge will cause it to become statically determinate, while the forming of a second hinge can create a mechanism. There are, however, several types of mechanisms that might feasibly occur in this frame. A possible beam mechanism is shown in part (b), a sidesway mechanism is shown in part (c), and a combined beam and sidesway mechanism is shown in part (d). The critical condition is the one that will result in the smallest value of P_n.

Example 8-4 presents the plastic analysis of the frame of Fig. 8.18. The distances through which the loads tend to move in the various mechanisms should be carefully

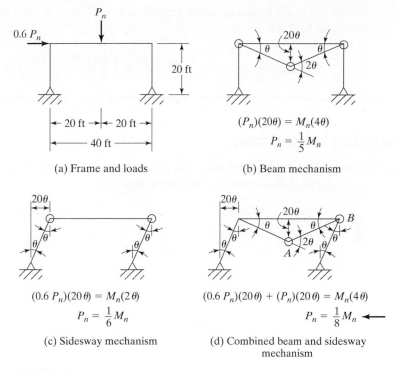

(a) Frame and loads

(b) Beam mechanism

$$(P_n)(20\theta) = M_n(4\theta)$$
$$P_n = \frac{1}{5}M_n$$

(c) Sidesway mechanism

$$(0.6\,P_n)(20\,\theta) = M_n(2\,\theta)$$
$$P_n = \frac{1}{6}M_n$$

(d) Combined beam and sidesway mechanism

$$(0.6\,P_n)(20\,\theta) + (P_n)(20\,\theta) = M_n(4\theta)$$
$$P_n = \frac{1}{8}M_n$$

FIGURE 8.18

Possible mechanisms for a frame.

studied. The solution of this problem brings out a point of major significance: *Superposition does not apply to plastic analysis.* You can easily see this by studying the virtual-work expressions for parts (b), (c), and (d) of the figure. The values of P_n obtained for the separate beam and sidesway mechanisms do not add up to the value obtained for the combined beam and sidesway mechanism.

For each mechanism, we want to consider the situation in which we have the fewest possible number of plastic hinges that will cause collapse. If you look at one of the virtual-work expressions, you will note that P_n becomes smaller as the number of plastic hinges decreases. With this in mind, look at part (d) of Fig. 8.18. The frame can feasibly sway to the right without the formation of a plastic hinge at the top of the left column. The two plastic hinges labeled A and B are sufficient for collapse to occur.

Example 8-4

A W12 × 72 ($Z_x = 108$ in) is used for the beam and columns of the frame shown in Fig. 8.18. If $F_y = 50$ ksi, determine the value of P_n.

Solution

The virtual-work expressions are written for parts (b), (c), and (d) of Fig. 8.18 and shown with the respective parts of the figure. The combined beam and sidesway case is found to be the critical case, and from it, the value of P_n is determined as follows:

$$P_n = \frac{1}{8} M_n = \left(\frac{1}{8}\right)(F_y Z) = \left(\frac{1}{8}\right)\left(\frac{50 \times 108}{12}\right) = 56.25 \text{ k}$$

8.13   PROBLEMS FOR SOLUTION

8-1 to 8-10.   *Find the values of S and Z and the shape factor about the horizontal x axes, for the sections shown in the accompanying figures.*

8-1.   (*Ans.* 446.3, 560, 1.25)

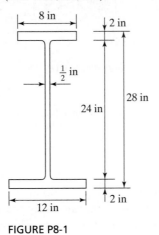

FIGURE P8-1

8-2.

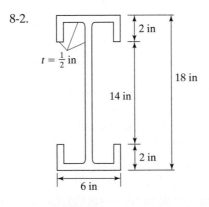

FIGURE P8-2

8-3. (*Ans.* 4.21, 7.15, 1.70)

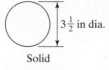

Solid

FIGURE P8-3

8-4.

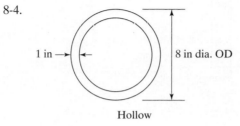

Hollow

FIGURE P8-4

8-5. (*Ans.* 4.33, 7.78, 1.80)

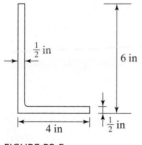

FIGURE P8-5

8-6.

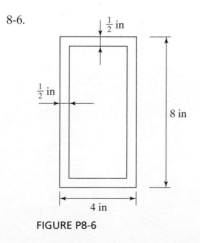

FIGURE P8-6

8-7. (*Ans.* 40.0, 45.8, 1.15)

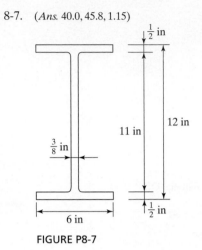

FIGURE P8-7

8-8.

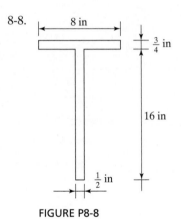

FIGURE P8-8

8-9. (*Ans.* 33.18, 43.0, 1.30)

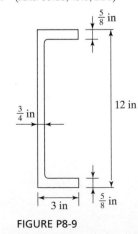

FIGURE P8-9

8-10.

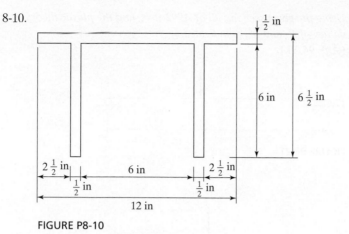

FIGURE P8-10

8-11 to 8-20. *Determine the values of S and Z and the shape factor about the horizontal x axes, unless otherwise directed. Use the web and flange dimensions given in the AISC Manual for making these calculations.*

8-11. A W21 × 122 (*Ans.* 271.8, 305.6, 1.12)

8-12. A W14 × 34 with a cover plate on each flange. The plate is 3/8 × 8 in.

8-13. Two 5 × 3 × 3/8 in Ls long legs vertical (LLV) and back to back. (*Ans.* 4.47, 7.95, 1.78)

8-14. Two C8 × 11.5 channels back to back.

8-15. Four 3 × 3 × 3/8 in Ls arranged as shown in Fig. P8-15. (*Ans.* 4.56, 7.49, 1.64)

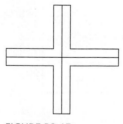

FIGURE P8-15

8-16. A W16 × 31.

8-17. The section of Prob. 8-7, considering the *y* axis. (*Ans.* 6.02, 9.39, 1.56)

8-18. Rework Prob. 8-9, considering the *y* axis.

8-19. Rework Prob. 8-12 considering the *y* axis. (*Ans.* 13.8, 22.6, 1.64)

8-20. Rework Prob. 8-14 considering the *y* axis.

8-21 to 8-39. *Using the given sections, all of A992 steel, and the plastic theory, determine the values of P_n and w_n as indicated.*

8-21. (*Ans.* 94.3 k)

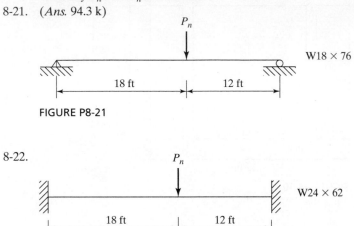

FIGURE P8-21

8-22.

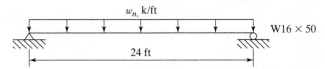

FIGURE P8-22

8-23. (*Ans.* 10.65 k/ft)

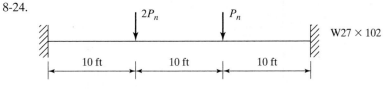

FIGURE P8-23

8-24.

FIGURE P8-24

8-25. (*Ans.* 189.5 k)

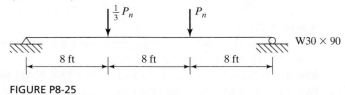

FIGURE P8-25

8-26.

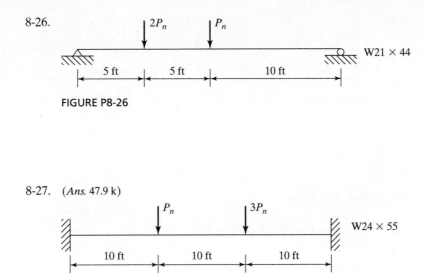

FIGURE P8-26

8-27. (*Ans.* 47.9 k)

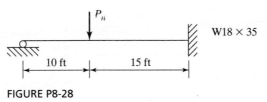

FIGURE P8-27

8-28.

FIGURE P8-28

8-29. (*Ans.* 49.3 k)

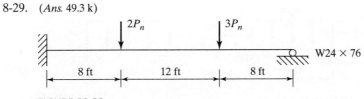

FIGURE P8-29

8-30.

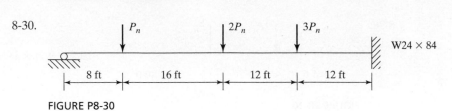

FIGURE P8-30

8-31. (*Ans.* 10.95 k/ft)

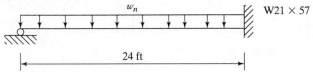

FIGURE P8-31

8-32.

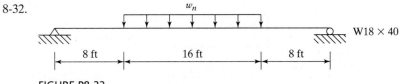

FIGURE P8-32

8-33. (*Ans.* 4.20 k/ft)

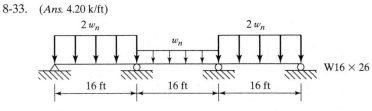

FIGURE P8-33

8-34.

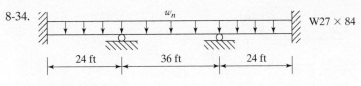

W27 × 84

| 24 ft | 36 ft | 24 ft |

FIGURE P8-34

8-35. (*Ans.* 9.56 k/ft)

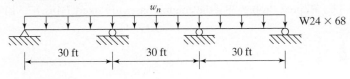

W24 × 68

| 30 ft | 30 ft | 30 ft |

FIGURE P8-35

8-36.

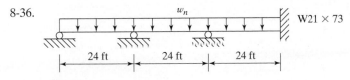

W21 × 73

| 24 ft | 24 ft | 24 ft |

FIGURE P8-36

8-37. (*Ans.* 88.2 k)

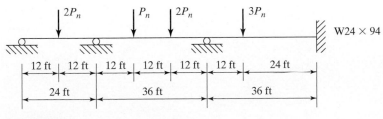

W24 × 94

| 12 ft | 12 ft | 12 ft | 12 ft | 12 ft | 12 ft | 24 ft |

| 24 ft | 36 ft | 36 ft |

FIGURE P8-37

8-38.

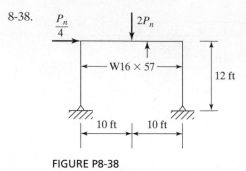

$\frac{P_n}{4}$

$2P_n$

W16 × 57

12 ft

10 ft   10 ft

FIGURE P8-38

8-39. Repeat Prob. 8-38 if the column bases are fixed. (*Ans*. 87.5 k)

C H A P T E R 9

Design of Beams for Moments

9.1 INTRODUCTION

If gravity loads are applied to a fairly long, simply supported beam, the beam will bend downward, and its upper part will be placed in compression and will act as a compression member. The cross section of this "column" will consist of the portion of the beam cross section above the neutral axis. For the usual beam, the "column" will have a much smaller moment of inertia about its y or vertical axis than about its x axis. If nothing is done to brace it perpendicular to the y axis, it will buckle laterally at a much smaller load than would otherwise have been required to produce a vertical failure. (You can verify these statements by trying to bend vertically a magazine held in a vertical position. The magazine will, just as a steel beam, always tend to buckle laterally unless it is braced in that direction.)

Lateral buckling will not occur if the compression flange of a member is braced laterally or if twisting of the beam is prevented at frequent intervals. In this chapter, the buckling moments of a series of compact ductile steel beams with different lateral or torsional bracing situations are considered. (As previously defined, a *compact section* is one that has a sufficiently stocky profile so that it is capable of developing a fully plastic stress distribution before buckling.)

In this chapter, we will look at beams as follows:

1. First, the beams will be assumed to have continuous lateral bracing for their compression flanges.
2. Next, the beams will be assumed to be braced laterally at short intervals.
3. Finally, the beams will be assumed to be braced laterally at larger and larger intervals.

In Fig. 9.1, a typical curve showing the nominal resisting or buckling moments of one of these beams with varying unbraced lengths is presented.

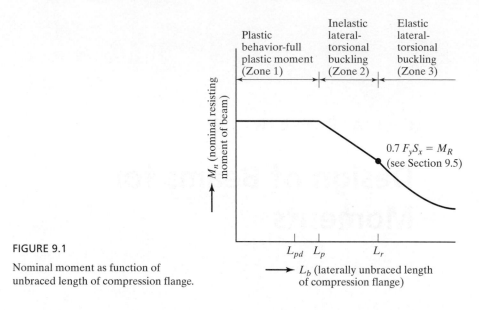

FIGURE 9.1

Nominal moment as function of
unbraced length of compression flange.

An examination of Fig. 9.1 will show that beams have three distinct ranges, or zones, of behavior, depending on their lateral bracing situation. If we have continuous or closely spaced lateral bracing, the beams will experience yielding of the entire cross section and fall into what is classified as Zone 1. As the distance between lateral bracing is increased further, the beams will begin to fail inelastically at smaller moments and fall into Zone 2. Finally, with even larger unbraced lengths, the beams will fail elastically and fall into Zone 3. A brief discussion of these three types of behavior is presented in this section, while the remainder of the chapter is devoted to a detailed discussion of each type, together with a series of numerical examples.

9.1.1   Plastic Behavior (Zone 1)

If we were to take a compact beam whose compression flange is continuously braced laterally, we would find that we could load it until its full plastic moment M_p is reached at some point or points; further loading then produces a redistribution of moments, as was described in Chapter 8. In other words, the moments in these beams can reach M_p and then develop a rotation capacity sufficient for moment redistribution.

If we now take one of these compact beams and provide closely spaced intermittent lateral bracing for its compression flanges, we will find that we can still load it until the plastic moment plus moment redistribution is achieved if the spacing between the bracing does not exceed a certain value, called L_p herein. (The value of L_p is dependent on the dimensions of the beam cross section and on its yield stress.) *Most beams fall in Zone 1.*

9.1.2   Inelastic Buckling (Zone 2)

If we now further increase the spacing between points of lateral or torsional bracing, the section may be loaded until some, but not all, of the compression fibers are stressed to F_y. The section will have insufficient rotation capacity to permit full moment

Bridge over Allegheny River at Kittanning, PA. (Courtesy of the American Bridge Company.)

redistribution and thus will not permit plastic analysis. In other words, in this zone we can bend the member until the yield strain is reached in some, but not all, of its compression elements before lateral buckling occurs. This is referred to as *inelastic buckling*.

As we increase the unbraced length, we will find that the moment the section resists will decrease, until finally it will buckle before the yield stress is reached anywhere in the cross section. The maximum unbraced length at which we can still reach F_y at one point is the end of the inelastic range. It's shown as L_r in Fig. 9.1; its value is dependent upon the properties of the beam cross section, as well as on the yield and residual stresses of the beam. At this point, as soon as we have a moment that theoretically causes the yield stress to be reached at some point in the cross section (actually, it's less than F_y because of residual stresses), the section will buckle.

9.1.3 Elastic Buckling (Zone 3)

If the unbraced length is greater than L_r, the section will buckle elastically before the yield stress is reached anywhere. As the unbraced length is further increased, the buckling moment becomes smaller and smaller. As the moment is increased in such a beam, the beam will deflect more and more transversely until a critical moment value M_{cr} is reached. At this time, the beam cross section will twist and the compression flange will move laterally. The moment M_{cr} is provided by the torsional resistance and the warping resistance of the beam, as will be discussed in Section 9.7.

150 Federal Street, Boston, MA. (Courtesy Owen Steel Company, Inc.)

9.2 YIELDING BEHAVIOR—FULL PLASTIC MOMENT, ZONE 1

In this section and the next two, beam formulas for yielding behavior (Zone 1) are presented, while in Sections 9.5 through 9.7, formulas are presented for inelastic buckling (Zone 2) and elastic buckling (Zone 3). After seeing some of these latter expressions, the reader may become quite concerned that he or she is going to spend an enormous amount of time in formula substitution. This is not generally true, however, as the values needed are tabulated and graphed in simple form in Part 3 of the AISC Manual.

If the unbraced length L_b of the compression flange of a compact I- or C-shaped section, including hybrid members, does not exceed L_p (if elastic analysis is being used) or L_{pd} (if plastic analysis is being used), then the member's bending strength about its major axis may be determined as follows:

$$M_n = M_p = F_y Z \qquad \text{(LRFD Equation F2-1)}$$

$$\phi_b M_n = \phi_b F_y Z \ (\phi_b = 0.90)$$

$$\frac{M_n}{\Omega_b} = \frac{F_y Z}{\Omega_b} \ (\Omega_b = 1.67)$$

When a conventional elastic analysis approach is used to establish member forces, L_b may not exceed the value L_p to follow if M_n is to equal $F_y Z$.

$$L_p = 1.76 \, r_y \sqrt{\frac{E}{F_y}} \qquad \text{(AISC Equation F2-5)}$$

When a plastic analysis approach is used to establish member forces for doubly symmetric and singly symmetric I-shaped members, with the compression flanges larger than their tension flanges (including hybrid members) loaded in the plane of the web, L_b (which is defined as the laterally unbraced length of the compression flange at plastic hinge locations associated with failure mechanisms) may not exceed the value L_{pd} to follow if M_n is to equal $F_y Z$.

$$L_{pd} = \left[0.12 - 0.076 \left(\frac{M_1'}{M_2'} \right) \right] \left(\frac{E}{F_y} \right) r_y \quad \text{(AISC Appendix Equation A-1-5)}$$

In this expression, M_1 is the smaller moment at the end of the unbraced length of the beam and M_2 is the larger moment at the end of the unbraced length, and the ratio M_1/M_2 is positive when the moments cause the member to be bent in double curvature (⌒⌣) and negative if they bend it in single curvature (⌣⌒). Only steels with F_y values (F_y is the specified minimum yield stress of the compression flange) of 65 ksi or less may be considered. Higher-strength steels may not be ductile.

There is no limit of the unbraced length for circular or square cross sections or for I-shaped beams bent about their minor axes. (If I-shaped sections are bent about their minor or y axes, they will not buckle before the full plastic moment M_p about the y axis is developed, as long as the flange element is compact.) Appendix Equation A1-8 of the AISC Specification also provides a value of L_{pd} for solid rectangular bars and symmetrical box beams.

9.3   DESIGN OF BEAMS, ZONE 1

Included in the items that need to be considered in beam design are moments, shears, deflections, crippling, lateral bracing for the compression flanges, fatigue, and others. Beams will probably be selected that provide sufficient design moment capacities

($\phi_b M_n$) and then checked to see if any of the other items are critical. The factored moment will be computed, and a section having that much design moment capacity will be initially selected from the AISC Manual, Part 3, Table 3-2, entitled "W Shapes Selection by Z_x." From this table, steel shapes having sufficient plastic moduli to resist certain moments can quickly be selected. Two important items should be remembered in selecting shapes:

1. These steel sections cost so many cents per pound, and it is therefore desirable to select the lightest possible shape having the required plastic modulus (assuming that the resulting section is one that will reasonably fit into the structure). The table has the sections arranged in various groups having certain ranges of plastic moduli. The heavily typed section at the top of each group is the lightest section in that group, and the others are arranged successively in the order of their plastic moduli. Normally, the deeper sections will have the lightest weights giving the required plastic moduli, and they will be generally selected, unless their depth causes a problem in obtaining the desired headroom, in which case a shallower but heavier section will be selected.

2. The plastic moduli values in the table are given about the horizontal axes for beams in their upright positions. If a beam is to be turned on its side, the proper plastic modulus about the y axis can be found in Table 3-4 of the Manual or in the tables giving dimensions and properties of shapes in Part 1 of the AISC Manual. A W shape turned on its side may be only from 10 to 30 percent as strong as one in the upright position when subjected to gravity loads. In the same manner, the strength of a wood joist with the actual dimensions 2×10 in turned on its side would be only 20 percent as strong as in the upright position.

The examples to follow illustrate the analysis and design of compact steel beams whose compression flanges have full lateral support or bracing, thus permitting plastic analysis. For the selection of such sections, the designer may enter the tables either with the required plastic modulus or with the factored design moment (if $F_y = 50$ ksi).

Example 9-1

Is the compact and laterally braced section shown in Fig. 9.2 sufficiently strong to support the given loads if $F_y = 50$ ksi? Check the beam with both the LRFD and ASD methods.

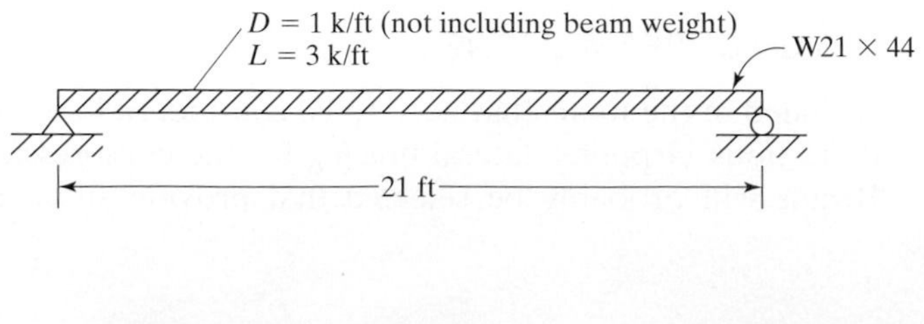

FIGURE 9.2

Solution. Using a W21 × 44 ($Z_x = 95.4 \text{ in}^3$)

LRFD $\phi_b = 0.9$	ASD $\Omega_b = 1.67$
Given beam wt = 0.044 k/ft	Given beam wt = 0.044 k/ft
$w_u = (1.2)(1 + 0.044) + (1.6)(3) = 6.05 \text{ k/ft}$	$w_a = (1 + 0.044) + 3 = 4.044 \text{ k/ft}$
$M_u = \dfrac{(6.05)(21)^2}{8} = 333.5 \text{ ft-k}$	$M_a = \dfrac{(4.044)(21)^2}{8} = 222.9 \text{ ft-k}$
$M_n \text{ of section} = \dfrac{F_y Z}{12} = M_{px}$	$M_n = 397.5 \text{ ft-k from LRFD solution}$
$= \dfrac{(50 \text{ ksi})(95.4 \text{ in}^3)}{12 \text{ in/ft}} = 397.5 \text{ ft-k}$	$\dfrac{M_n}{\Omega_2} = \dfrac{397.5}{1.67} = 238 \text{ ft-k} > 222.9 \text{ ft-k } \mathbf{OK}$
$M_u = \phi_b M_{px} = (0.9)(397.5)$	
$- 358 \text{ ft-k} > 333.5 \text{ ft-k } \mathbf{OK}$	

Note: Instead of using Z_x and $F_y Z_x$, we will usually find it easier to use the moment columns $\phi_b M_{px}$ and M_{px}/Ω_b in AISC Table 3-2. There, the term M_{px} represents the plastic moment of a section about its x axis. Following this procedure for a W21 × 44, we find the values $\phi_b M_{px} = 358$ ft-k and $M_{px}/\Omega_b = 238$ ft-k. These values agree with the preceding calculations.

9.3.1   Beam Weight Estimates

In each of the examples to follow, the weight of the beam is included in the calculation of the bending moment to be resisted, as the beam must support itself as well as the external loads. The estimates of beam weight are very close here, because the authors were able to perform a little preliminary paperwork before making his estimates. The beginning student is not expected to be able to glance at a problem and estimate exactly the weight of the beam required. A very simple method is available, however, with which the student can quickly and accurately estimate beam weights. He or she can calculate the maximum bending moment, not counting the effect of the beam weight, and pick a section from AISC Table 3-2. Then the weight of that section or a little bit more (since the beam's weight will increase the moment somewhat) can be used as the estimated beam weight. The resulting beam weights will almost always be very close to the weights of the member selected for the final designs. For future example problems in this text, the author does not show his calculations for estimated beam weights. The weight estimates used for those problems, however, were obtained in exactly the same manner as they are in Example 9-2, which follows.

Example 9-2

Select a beam section by using both the LRFD and ASD methods for the span and loading shown in Fig. 9.3, assuming full lateral support is provided for the compression flange by the floor slab above (that is, $L_b = 0$) and $F_y = 50$ ksi.

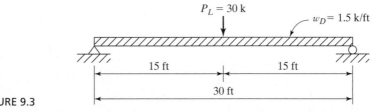

FIGURE 9.3

Solution

Estimate beam weights.

LRFD	ASD
w_u not including beam weight	w_a not including beam weight
$\quad = (1.2)(1.5) = 1.8$ k/ft	$\quad = 1.5$ k/ft
$P_u = (1.6)(30) = 48$ k	$P_a = 30$ k
$M_u = \dfrac{(1.8)(30)^2}{8} + \dfrac{(48)(30)}{4}$	$M_a = \dfrac{(1.5)(30)^2}{8} + \dfrac{(30)(30)}{4}$
$\quad = 562.5$ ft-k	$\quad = 393.8$ ft-k
From AISC Table 3-2 and the LRFD moment column ($\phi_b M_{px}$), a W24 × 62 is required.	From AISC Table 3-2 and the ASD moment column (M_{px}/Ω_b), a W21 × 68 is required.
$\phi_b M_{px} = 574$ ft-k	$\dfrac{M_{px}}{\Omega_b} = 399$ ft-k
Assume beam wt = 62 lb/ft.	**Assume beam wt = 68 lb/ft.**

Select beam section.

LRFD	ASD
$w_u = (1.2)(1.5 + 0.062) = 1.874$ k/ft	$w_a = 1.5 + 0.068 = 1.568$ k/ft
$P_u = (1.6)(30) = 48$ k	$P_a = 30$ k
$M_u = \dfrac{(1.874)(30)^2}{8} + \dfrac{(48)(30)}{4}$	$M_a = \dfrac{(1.568)(30)^2}{8} + \dfrac{(30)(30)}{4}$
$\quad = 570.8$ ft-k	$\quad = 401.4$ ft-k

(Continued)

LRFD	ASD
From AISC Table 3-2	From AISC Table 3-2
Use W24 × 62.	**Use W24 × 68.**
$(\phi_b M_{px} = 574 \text{ ft-k} > 570.8 \text{ ft-k})$	$(M_{px}/\Omega_b = 442 \text{ ft-k} > 401.4 \text{ ft k})$
OK	**OK**

Example 9-3

The 5-in reinforced-concrete slab shown in Fig. 9.4 is to be supported with steel W sections 8 ft 0 in on centers. The beams, which will span 20 ft, are assumed to be simply supported. If the concrete slab is designed to support a live load of 100 psf, determine the lightest steel sections required to support the slab by the LRFD and ASD procedures. It is assumed that the compression flange of the beam will be fully supported laterally by the concrete slab. The concrete weighs 150 lb/ft³. $F_y = 50$ ksi.

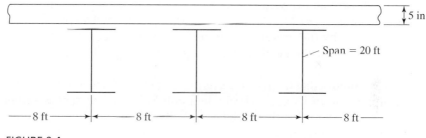

5 in

Span = 20 ft

— 8 ft — | — 8 ft — | — 8 ft — | — 8 ft —

FIGURE 9.4

Solution

LRFD	ASD
Assume beam wt $= 22$ lb/ft	Assume beam wt $= 22$ lb/ft
Slab wt $= \left(\dfrac{5}{12}\right)(150)(8) = 500$ lb/ft	Slab wt $= 500$ lb/ft
$w_D = 522$ lb/ft	$w_D = 522$ lb/ft
$w_L = (8)(100) = 800$ lb/ft	$w_L = 800$ lb/ft
$w_u = (1.2)(522) + (1.6)(800)$	$w_a = 522 + 800 = 1322$ lb/ft
$= 1906$ lb/ft $= 1.906$ k/ft	$= 1.322$ k/ft
$M_u = \dfrac{(1.906)(20)^2}{8} = 95.3$ ft-k	$M_a = \dfrac{(1.322)(20)^2}{8} = 66.1$ ft-k
From AISC Table 3-2	From AISC Table 3-2
Use W10 × 22.	**Use W12 × 22.**
$(\phi_b M_{px} = 97.5 \text{ ft-k} > 95.3 \text{ ft-k})$	$\left(\dfrac{M_{px}}{\Omega_b} = 73.1 \text{ ft-k} > 66.1 \text{ ft-k}\right)$

9.3.2   Holes in Beams

It is often necessary to have holes in steel beams. They are obviously required for the installation of bolts and sometimes for pipes, conduits, ducts, etc. If at all possible, these latter types of holes should be completely avoided. When absolutely necessary, they should be placed through the web if the shear is small and through the flange if the moment is small and the shear is large. Cutting a hole through the web of a beam does not reduce its section modulus greatly or its resisting moment; but, as will be described in Section 10.2, a large hole in the web tremendously reduces the shearing strength of a steel section. When large holes are put in beam webs, extra plates are sometimes connected to the webs around the holes to serve as reinforcing against possible web buckling.

When large holes are placed in beam webs, the strength limit states of the beams—such as local buckling of the compression flange, the web, or the tee-shaped compression zone above or below the opening—or the moment–shear interaction or serviceability limit states may control the size of the member. A general procedure for estimating these effects and the design of any required reinforcement is available for both steel and composite beams.[1,2]

The presence of holes of any type in a beam certainly does not make it stronger and in all probability weakens it somewhat. The effect of holes has been a subject that has been argued back and forth for many years. The questions, "Is the neutral axis affected by the presence of holes?" and "Is it necessary to subtract holes from the compression flange that are going to be plugged with bolts?" are frequently asked.

The theory that the neutral axis might move from its normal position to the theoretical position of its net section when bolt holes are present is rather questionable. Tests seem to show that flange holes for bolts do not appreciably change the location of the neutral axis. It is logical to assume that the location of the neutral axis will not follow the exact theoretical variation with its abrupt changes in position at bolt holes, as shown in part (b) of Fig. 9.5. A more reasonable change in neutral axis location is shown in part (c) of this figure, where it is assumed to have a more gradual variation in position.

It is interesting to note that flexure tests of steel beams seem to show that their failure is based on the strength of the compression flange, even though there may be bolt holes in the tension flange. The presence of these holes does not seem to be as serious as might be thought, particularly when compared with holes in a pure tension member. These tests show little difference in the strengths of beams with no holes and in beams with an appreciable number of bolt holes in either flange.

Bolt holes in the webs of beams are generally considered to be insignificant, as they have almost no effect on Z calculations.

---

[1]D. Darwin, "*Steel and Composite Beams with Web Openings*," AISC Design Guide Series No. 2 (Chicago: American Institute of Steel Construction, 1990).

[2]ASCE Task Committee on Design Criteria for Composite Structures in Steel and Concrete, "Proposed Specification for Structural Steel Beams with Web Openings," D. Darwin, Chairman, *Journal of Structural Engineering*, ASCE, vol. 118 (New York: ASCE, December, 1992).

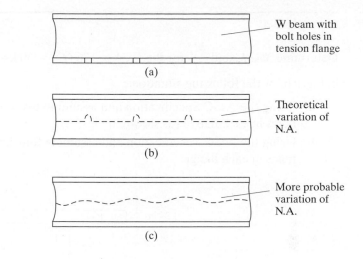

W beam with
bolt holes in
tension flange

(a)

Theoretical
variation of
N.A.

(b)

More probable
variation of
N.A.

FIGURE 9.5   (c)

Some specifications, notably the bridge ones, and some designers have not adopted the idea of ignoring the presence of all or part of the holes in tension flanges. As a result, they follow the more conservative practice of deducting 100 percent of all holes. For such a case, the reduction in Z_x will equal the statical moment of the holes (in both flanges) taken about the neutral axis. If we have bolt holes in the compression flanges only and they are filled with bolts, we forget the whole thing. This is because it is felt that the fasteners can adequately transfer compression through the holes by means of the bolts.

The flexural strengths of beams with holes in their tension flanges are predicted by comparing the value $F_y A_{fg}$ with $F_u A_{fn}$. In these expressions, A_{fg} is the gross area of the tension flange while A_{fn} is the net tension flange area after the holes are subtracted. In the expressions given herein for computing M_n, there is a term Y_t, which is called the hole reduction coefficient. Its value is taken as 1.0 if $F_y/F_u \leq 0.8$. For cases when the ratio of F_y/F_u is >0.8, Y_t is taken as 1.1.[3,4]

a. If $F_u A_{fn} \geq Y_t F_y A_{fg}$, the limit state of tensile rupture does not apply and there is no reduction in M_n because of the holes.

b. If $F_u A_{fn} < Y_t F_y A_{fg}$, the nominal flexural strength of the member at the holes is to be determined by the following expression, in which S_x is the section modulus of the member:

$$M_n = \frac{F_u A_{fn}}{A_{fg}} S_x \qquad \text{(AISC Equation F13-1)}$$

[3]R.J. Dexter and S.A. Altstadt, "Strength and Ductility of Tension Flanges in Girders," Proceedings of the Second New York City Bridge Conference (New York, 2003).
[4]Q. Yuan, J. Swanson and G.A. Rassati, "An Investigation of Hole Making Practices in the Fabrication of Structural Steel" (University of Cincinnati, Cincinnati, OH, 2004).

Example 9-4

Determine $\phi_b M_n$ and $\dfrac{M_n}{\Omega_b}$ for the W24 × 176 (F_y = 50 ksi, F_u = 65 ksi) beam shown in Fig. 9.6 for the following situations:

a. Using the AISC Specification and assuming two lines of 1-in bolts in standard holes in each flange (as shown in Fig. 9.6).

b. Using the AISC Specification and assuming four lines of 1-in bolts in standard holes in each flange.

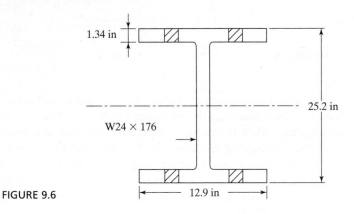

FIGURE 9.6

Solution. Using a W24 × 176 (b_f = 12.9 in, t_f = 1.34 in and S_x = 450 in³)

a.
$$A_{fg} = b_f t_f = (12.9 \text{ in})(1.34 \text{ in}) = 17.29 \text{ in}^2$$

$$A_{fn} = 17.29 \text{ in}^2 - (2)\left(1\tfrac{1}{8}\text{in}\right)(1.34 \text{ in}) = 14.27 \text{ in}^2$$

$$F_u A_{fn} = (65 \text{ ksi})(14.27 \text{ in}^2) = 927.6 \text{ k}$$

$$\frac{F_y}{F_u} = \frac{50}{65} = 0.77 < 0.8 \quad \therefore Y_t = 1.0$$

$$927.6 \text{ k} > Y_t F_y A_{fg} = (1.0)(50 \text{ ksi})(17.29 \text{ in}^2) = 864.5 \text{ k}$$

∴ **Tensile rupture does not apply and** $\phi_b M_{px}$ = **1920 ft-k and** $\dfrac{M_{px}}{\Omega_b}$ = **1270 ft-k**

from AISC Table 3-2.

b.
$$A_{fn} = 17.29 \text{ in}^2 - (4)\left(1\tfrac{1}{8}\text{in}\right)(1.34 \text{ in}) = 11.26 \text{ in}^2$$

$$\frac{F_y}{F_u} = \frac{50}{65} = 0.77 \quad \therefore Y_t = 1.0$$

$$F_u A_{fn} = (65 \text{ ksi})(11.26 \text{ in}^2) = 731.9 \text{ k}$$
$$< Y_t F_y A_{fg} = (1.0)(50 \text{ ksi})(17.29 \text{ in}^2) = 864.5 \text{ k}$$

∴ Tensile rupture expression does apply.

$$M_n = \frac{F_u A_{fn}}{A_{fg}} S_x = \frac{(65 \text{ ksi})(11.26 \text{ in}^2)(450 \text{ in}^3)}{(17.29 \text{ in}^2)} = 19{,}048 \text{ in-k} = 1587.4 \text{ ft-k}$$

LRFD $\phi_b = 0.9$	ASD $\Omega_b = 1.67$
$\phi_b M_n = (0.9)(1587.4)$	$\dfrac{M_n}{\Omega_b} = \dfrac{1587.4}{1.67}$
= 1429 ft-k	**= 951 ft-k**

Should a hole be present in only one side of a flange of a W section, there will be no axis of symmetry for the net section of the shape. A correct theoretical solution of the problem would be very complex. Rather than going through such a lengthy process over a fairly minor point, it seems logical to assume that there are holes in both sides of the flange. The results obtained will probably be just as satisfactory as those obtained by a laborious theoretical method.

9.4 LATERAL SUPPORT OF BEAMS

Probably most steel beams are used in such a manner that their compression flanges are restrained against lateral buckling. (Unfortunately, however, the percentage has not been quite as high as the design profession has assumed.) The upper flanges of

An ironworker awaits the lifting of a steel beam that will be field bolted to the column. (Courtesy of CMC South Carolina Steel.)

beams used to support concrete building and bridge floors are often incorporated in these concrete floors. For situations of this type, where the compression flanges are restrained against lateral buckling, the beams will fall into Zone 1.

Should the compression flange of a beam be without lateral support for some distance, it will have a stress situation similar to that existing in columns. As is well known, the longer and slenderer a column becomes, the greater becomes the danger of its buckling for the same loading condition. When the compression flange of a beam is long enough and slender enough, it may quite possibly buckle, unless lateral support is provided.

There are many factors affecting the amount of stress that will cause buckling in the compression flange of a beam. Some of these factors are properties of the material, the spacing and types of lateral support provided, residual stresses in the sections, the types of end support or restraints, the loading conditions, etc.

The tension in the other flange of a beam tends to keep that flange straight and restrain the compression flange from buckling; but as the bending moment is increased, the tendency of the compression flange to buckle may become large enough to overcome the tensile restraint. When the compression flange does begin to buckle, twisting or torsion will occur, and the smaller the torsional strength of the beam the more rapid will be the failure. The W, S, and channel shapes so frequently used for beam sections do not have a great deal of resistance to lateral buckling and the resulting torsion. Some other shapes—notably, the built-up box shapes—are tremendously stronger. These types of members have a great deal more torsional resistance than the W, S, and plate girder sections. Tests have shown that they will not buckle laterally until the strains developed are well in the plastic range.

Some judgment needs to be used in deciding what does and what does not constitute satisfactory lateral support for a steel beam. Perhaps the most common question asked by practicing steel designers is, "What is lateral support?" A beam that is wholly encased in concrete or that has its compression flange incorporated in a concrete slab is certainly well supported laterally. When a concrete slab rests on the top flange of a beam, the engineer must study the situation carefully before he or she counts on friction to provide full lateral support. Perhaps if the loads on the slab are fairly well fixed in position, they will contribute to the friction and it may be reasonable to assume full lateral support. If, on the other hand, there is much movement of the loads and appreciable vibration, the friction may well be reduced and full lateral support cannot be assumed. Such situations occur in bridges due to traffic, and in buildings with vibrating machinery such as printing presses.

Should lateral support of the compression flange not be provided by a floor slab, it is possible that such support may be provided with connecting beams or with special members inserted for that purpose. Beams that frame into the sides of the beam or girder in question and are connected to the compression flange can usually be counted on to provide full lateral support at the connection. If the connection is made primarily to the tensile flange, little lateral support is provided to the compression flange. Before support is assumed from these beams, the designer should note whether the beams themselves are prevented from moving. The beams represented with horizontal dashed lines in Fig. 9.7 provide questionable lateral support for the main beams between columns. For a situation of this type, some system of x-bracing may be desirable in one of the bays. Such a system is shown in Fig. 9.7. This one system will provide sufficient lateral support for the beams for several bays.

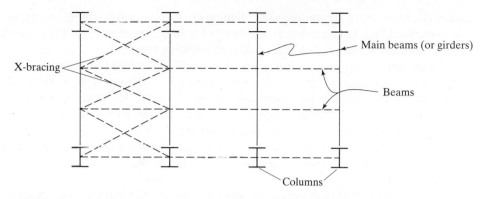

FIGURE 9.7

X-bracing for floor system.

The intermittent welding of metal roof or floor decks to the compression flanges of beams will probably provide sufficient lateral bracing. The corrugated sheet-metal roofs that are usually connected to the purlins with metal straps probably furnish only partial lateral support. A similar situation exists when wood flooring is bolted to supporting steel beams. At this time, the student quite naturally asks, "If only partial support is available, what am I to consider to be the distance between points of lateral support?" The answer to this question is for the student to use his or her judgment. As an illustration, a wood floor is assumed to be bolted every 4 ft to the supporting steel beams in such a manner that, it is thought, only partial lateral support is provided at those points. After studying the situation, the engineer might well decide that the equivalent of full lateral support at 8-ft intervals is provided. Such a decision seems to be within the spirit of the Specification.

Should there be doubt in the designer's mind as to the degree of lateral support provided, he or she should probably be advised to assume that there is none.

The reader should carefully study the provisions of Section C1 and Appendix 6 of the AISC Specification regarding stability bracing for beams and columns. In this Appendix, values are provided for calculating necessary bracing strengths and stiffnesses, and design formulas are given for obtaining those values. Included are various types of bracing for columns as well as torsional bracing for flexural members.

Two categories of bracing are considered in the Appendix: relative bracing and nodal bracing. With relative bracing a particular point is restrained in relation to another point or points. In other words, relative bracing is connected not only to the member to be braced but to other members as well (diagonal cross bracing, for example). Nodal bracing is used to prevent lateral movement or twisting of a member independently of other braces.

In Section 10.9 of this text, lateral bracing for the ends of beams supported on bearing plates is discussed.

9.5   INTRODUCTION TO INELASTIC BUCKLING, ZONE 2

If intermittent lateral bracing is supplied for the compression flange of a beam section, or if intermittent torsional bracing is supplied to prevent twisting of the cross section at the bracing points such that the member can be bent until the yield strain is reached in some (but not all) of its compression elements before lateral buckling occurs, we have inelastic buckling. In other words, the bracing is insufficient to permit the member to reach a full plastic strain distribution before buckling occurs.

Because of the presence of residual stresses (discussed in Section 5.2), yielding will begin in a section at applied stresses equal to $F_y - F_r$, where F_y is the yield stress of the web and F_r equals the compressive residual stress. The AISC Specification estimates this value ($F_y - F_r$) to be equal to about $0.7F_y$, and we will see that value in the AISC equations. It should be noted that the definition of plastic moment F_yZ in Zone 1 is not affected by residual stresses, because the sum of the compressive residual stresses equals the sum of the tensile residual stresses in the section and the net effect is, theoretically, zero.

When a constant moment occurs along the unbraced length, L_b, of a compact I- or C-shaped section and L_b is larger than L_p, the beam will fail inelastically, unless L_b is greater than a distance L_r (to be discussed) beyond which the beam will fail elastically before F_y is reached (thus falling into Zone 3).

9.5.1 Bending Coefficients

In the formulas presented in the next few sections for inelastic and elastic buckling, we will use a term C_b, called the *lateral-torsional buckling modification factor* for nonuniform moment diagrams, when both ends of the unsupported segment are braced. This is a moment coefficient that is included in the formulas to account for the effect of different moment gradients on lateral-torsional buckling. In other words, lateral buckling may be appreciably affected by the end restraint and loading conditions of the member.

As an illustration, the reader can see that the moment in the unbraced beam of part (a) of Fig. 9.8 causes a worse compression flange situation than does the moment in the unbraced beam of part (b). For one reason, the upper flange of the beam in part (a) is in compression for its entire length, while in (b) the length of the "column"—that is, the length of the upper flange that is in compression—is much less (thus, in effect, a much shorter "column").

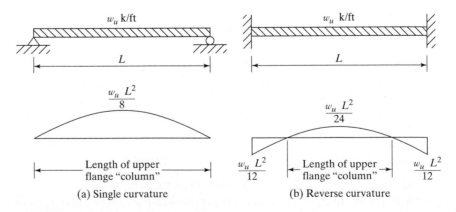

(a) Single curvature   (b) Reverse curvature

FIGURE 9.8

For the simply supported beam of part (a) of the figure, we will find that C_b is 1.14, while for the beam of part (b) it is 2.38 (see Example 9-5). The basic moment capacity equations for Zones 2 and 3 were developed for laterally unbraced beams subject to single curvature, with $C_b = 1.0$. Frequently, beams are not bent in single curvature, with the result that they can resist more moment. We have seen this in Fig. 9.8. To handle this situation, the AISC Specification provides moment or C_b coefficients larger than 1.0 that are to be multiplied by the computed M_n values. The results are higher moment capacities. The designer who conservatively says, "I'll always use $C_b = 1.0$ in is missing out on the possibility of significant savings in steel weight for some situations. *When using C_b values, the designer should clearly understand that the moment capacity obtained by multiplying M_n by C_b may not be larger than the plastic M_n of Zone 1, which is M_p and is equal to $F_y Z$.* This situation is illustrated in Fig. 9.9.

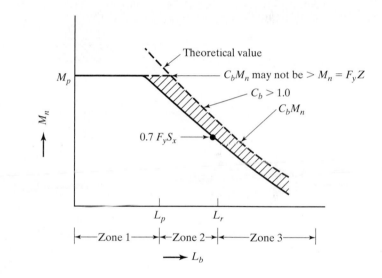

FIGURE 9.9

The value of C_b for singly symmetric members in single curvature and all doubly symmetric members is determined from the expression to follow, in which M_{max} is the largest moment in an unbraced segment of a beam, while M_A, M_B, and M_C are, respectively, the moments at the 1/4 point, 1/2 point, and 3/4 point in the segment:

$$C_b = \frac{12.5\,M_{max}}{2.5M_{max} + 3M_A + 4M_B + 3M_C} \qquad \text{(AISC Equation F1-1)}$$

In singly symmetric members subject to reverse curvature bending, the lateral-torsional buckling strength shall be checked for both top and bottom flanges. A more detailed analysis for C_b of singly symmetric members is presented in the AISC Commentary, F1. General Provisions.

C_b is equal to 1.0 for cantilevers or overhangs where the free end is unbraced. Some typical values of C_b, calculated with the previous AISC Equation F1-1, are shown in Fig. 9.10 for various beam and moment situations. Some of these values are also given in Table 3-1 of the AISC Manual.

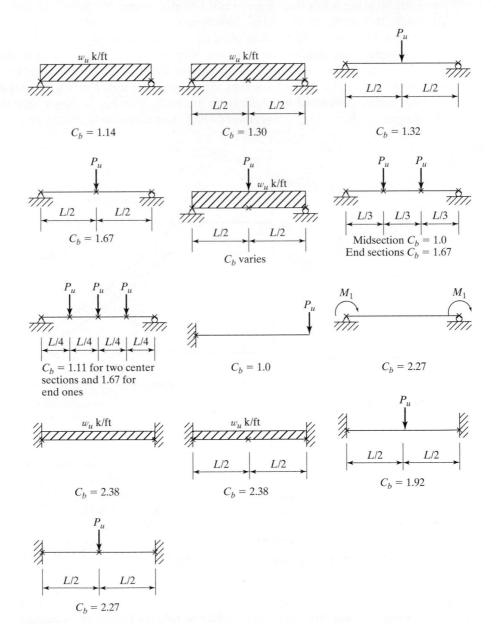

FIGURE 9.10

Sample C_b values for doubly symmetric members. (The X marks represent points of lateral bracing of the compression flange.)

Example 9-5

Determine C_b for the beam shown in Fig. 9.8 parts (a) and (b). Assume the beam is a doubly symmetric member.

a.

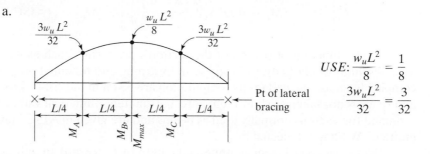

$USE: \dfrac{w_u L^2}{8} = \dfrac{1}{8}$

$\dfrac{3w_u L^2}{32} = \dfrac{3}{32}$

$$C_b = \frac{12.5 M_{max}}{2.5 M_{max} + 3M_A + 4M_B + 3M_C}$$

$$C_b = \frac{12.5\left(\dfrac{1}{8}\right)}{2.5\left(\dfrac{1}{8}\right) + 3\left(\dfrac{3}{32}\right) + 4\left(\dfrac{1}{8}\right) + 3\left(\dfrac{3}{32}\right)} = 1.14$$

b.

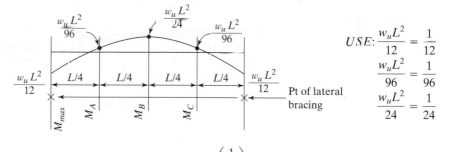

$USE: \dfrac{w_u L^2}{12} = \dfrac{1}{12}$

$\dfrac{w_u L^2}{96} = \dfrac{1}{96}$

$\dfrac{w_u L^2}{24} = \dfrac{1}{24}$

$$C_b = \frac{12.5\left(\dfrac{1}{12}\right)}{2.5\left(\dfrac{1}{12}\right) + 3\left(\dfrac{1}{96}\right) + 4\left(\dfrac{1}{24}\right) + 3\left(\dfrac{1}{96}\right)} = 2.38$$

9.6 MOMENT CAPACITIES, ZONE 2

When constant moment occurs along the unbraced length, or as the unbraced length of the compression flange of a beam or the distance between points of torsional bracing is increased beyond L_p, the moment capacity of the section will become smaller and smaller. Finally, at an unbraced length L_r, the section will buckle elastically as soon as the yield stress is reached. Owing to the rolling operation, however, there is a residual stress in the section equal to F_r. Thus, the elastically computed stress caused by bending can reach only $F_y - F_r = 0.7F_y$. The nominal moment strengths for unbraced lengths between L_p and L_r are calculated with the equation to follow:

$$M_n = C_b\left[M_p - (M_p - 0.7F_yS_x)\left(\frac{L_b - L_p}{L_r - L_p}\right)\right] \le M_p \qquad \text{(AISC Equation F2-2)}$$

L_r is a function of several of the section's properties, such as its cross-sectional area, modulus of elasticity, yield stress, and warping and torsional properties. The very complex formulas needed for its computation are given in the AISC Specification (F1), and space is not taken to show them here. Fortunately, numerical values have been determined for sections normally used as beams and are given in AISC Manual Table 3-2, entitled "W Shapes Selected by Z_x."

Going backward from an unbraced length of L_r toward an unbraced length L_p, we can see that buckling does not occur when the yield stress is first reached. We are in the inelastic range (Zone 2), where there is some penetration of the yield stress into the section from the extreme fibers. For these cases when the unbraced length falls between L_p and L_r, the nominal moment strength will fall approximately on a straight line between $M_{nx} = F_yZ_x$ at L_p and $0.7F_yS_x$ at L_r. For intermediate values of the unbraced length between L_p and L_r, we may interpolate between the end values that fall on a straight line. It probably is simpler, however, to use the expressions given at the end of this paragraph to perform the interpolation. Should C_b be larger than 1.0, the nominal moment strength will be larger, but not more than, $M_p = F_yZ_x$.

The interpolation expressions that follow are presented on page 3-8 of the AISC Manual. The bending factors (BFs) represent part of AISC Equation F2-2, as can be seen by comparing the equations to follow with that equation. Their numerical values in kips are given in Manual Table 3-2 for W shapes.

For LRFD $\phi_b M_n = C_b[\phi_b M_{px} - BF(L_b - L_p)] \le \phi_b M_{px}$

For ASD $\dfrac{M_n}{\Omega_b} = C_b\left[\dfrac{M_{px}}{\Omega_b} - BF(L_b - L_p)\right] \le \dfrac{M_{px}}{\Omega_b}$

Example 9-6

Determine the LRFD design moment capacity and the ASD allowable moment capacity of a W24 × 62 with $F_y = 50$ ksi, $L_b = 8.0$ ft, and $C_b = 1.0$.

Solution

Using a W24 × 62 (from AISC Table 3-2: $\phi_b M_{px} = 574$ ft-k, $M_{px}/\Omega_b = 382$ ft-k, $\phi_b M_{rx} = 344$ ft-k, $M_{rx}/\Omega_b = 229$ ft-k, $L_p = 4.87$ ft, $L_r = 14.4$ ft, BF for LRFD = 24.1 k, and BF for ASD = 16.1 k)

Noting $L_b > L_p < L_r$ ∴ falls in Zone 2, Fig. 9.1 in text.

LRFD	ASD
$\phi_b M_{nx} = C_b[\phi_b M_{px} - BF(L_b - L_p)]$	$\dfrac{M_{nx}}{\Omega_b} = C_b\left[\dfrac{M_{px}}{\Omega_b} - BF(L_b - L_p)\right]$
$\leq \phi_b M_{px}$	$\leq \dfrac{M_{px}}{\Omega_b}$
$\phi_b M_{nx} = 1.0[574 - 24.1(8.0 - 4.87)]$	$\dfrac{M_{nx}}{\Omega_b} = 1.0[382 - 16.1(8.0 - 4.87)]$
$= $ **499 ft-k** < 574 ft-k	$= $ **332 ft-k** < 382 ft-k
$\therefore \phi_b M_{nx} = 499$ ft-k	$\therefore \dfrac{M_{nx}}{\Omega} = 332$ ft-k

9.7   ELASTIC BUCKLING, ZONE 3

When the unbraced length of a beam is greater than L_r, the beam will fall in Zone 3. Such a member may fail due to buckling of the compression portion of the cross section laterally about the weaker axis, with twisting of the entire cross section about the beam's longitudinal axis between the points of lateral bracing. This will occur even though the beam is loaded so that it supposedly will bend about the stronger axis. The beam will bend initially about the stronger axis until a certain critical moment M_{cr} is reached. At that time, it will buckle laterally about its weaker axis. As it bends laterally, the tension in the other flange will try to keep the beam straight. As a result, the buckling of the beam will be a combination of lateral bending and a twisting (or torsion) of the beam cross section. A sketch of this situation is shown in Fig. 9.11.

The critical moment, or flexural-torsional moment M_{cr} in a beam will be made up of the torsional resistance (commonly called *St-Venant torsion*) and the warping resistance of the section.

If the unbraced length of the compression flange of a beam section or the distance between points that prevent twisting of the entire cross section is greater than L_r,

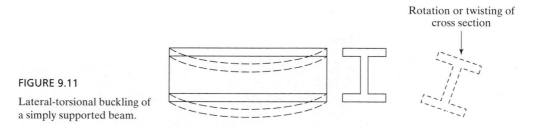

FIGURE 9.11

Lateral-torsional buckling of a simply supported beam.

Rotation or twisting of cross section

the section will buckle elastically before the yield stress is reached anywhere in the section. In Section F2.2 of the AISC Specification, the buckling stress for doubly symmetric I-shaped members is calculated with the following expression:

$$M_n = F_{cr}S_x \leq M_p \qquad \text{(AISC Equation F2-3)}$$

$$F_{cr} = \frac{C_b \pi^2 E}{\left(\dfrac{L_b}{r_{ts}}\right)^2} \sqrt{1 + 0.078 \frac{Jc}{S_x h_o}\left(\frac{L_b}{r_{ts}}\right)^2} \qquad \text{(AISC Equation F2-4)}$$

In this calculation,

r_{ts} = effective radius of gyration, in (provided in AISC Table 1-1)

J = torsional constant, in^4 (AISC Table 1-1)

c = 1.0 for doubly symmetric I-shapes

h_o = distance between flange centroids, in (AISC Table 1-1)

It is not possible for lateral-torsional buckling to occur if the moment of inertia of the section about the bending axis is equal to or less than the moment of inertia out of plane. For this reason the limit state of lateral-torsional buckling is not applicable for shapes bent about their minor axes, for shapes with $I_x \leq I_y$, or for circular or square shapes. Furthermore, yielding controls if the section is noncompact.

Example 9-7

Using AISC Equation F2-4, determine the values of $\phi_b M_{nx}$ and M_{nx}/Ω_b for a W18 × 97 with F_y = 50 ksi and an unbraced length L_b = 38 ft. Assume that C_b = 1.0.

Solution

Using a W18 × 97 (L_r = 30.4 ft, r_{ts} = 3.08 in, J = 5.86 in^4, c = 1.0 for doubly symmetric I section, S_x = 188 in^3, h_o = 17.7 in and Z_x = 211 in^3)

Noting L_b = 38 ft > L_r = 30.4 ft (from AISC Table 3-2), section is in Zone 3.

$$F_{cr} = \frac{(1.0)(\pi)^2(29 \times 10^3)}{\left(\dfrac{12 \times 38}{3.08}\right)^2} \sqrt{1 + (0.078)\frac{(5.86)(1.0)}{(188)(17.7)}\left(\frac{12 \times 38}{3.08}\right)^2}$$

$$= 26.2 \text{ ksi}$$

$$M_{nx} = F_{cr}S_x = \frac{(26.2)(188)}{12} = 410 \text{ ft-k} < M_p = \frac{(50)(211)}{12} = 879 \text{ ft-k}$$

LFRD ϕ_b = 0.9	ASD Ω_b = 1.67
$\phi_b M_{nx} = (0.9)(410)$	$\dfrac{M_{nx}}{\Omega_b} = \dfrac{410}{1.67}$
= **369 ft-k**	= **246 ft-k**

9.8   DESIGN CHARTS

Fortunately, the values of $\phi_b M_n$ and M_n/Ω_b for sections normally used as beams have been computed by the AISC, plotted for a wide range of unbraced lengths, and shown as Table 3-10 in the AISC Manual. These diagrams enable us to solve any of the problems previously considered in this chapter in just a few seconds.

The values provided cover unbraced lengths in the plastic range, in the inelastic range, and on into the elastic buckling range (Zones 1–3). They are plotted for $F_y = 50$ ksi and $C_b = 1.0$.

The LRFD curve for a typical W section is shown in Fig. 9.12. For each of the shapes, L_p is indicated with a solid circle ($\bullet$), while L_r is shown with a hollow circle ($\bigcirc$).

The charts were developed without regard to such things as shear, deflection, etc.—items that may occasionally control the design, as described in Chapter 10. They cover almost all of the unbraced lengths encountered in practice. If C_b is greater than 1.0, the values given will be magnified somewhat, as illustrated in Fig. 9.9.

To select a member, it is necessary to enter the chart only with the unbraced length L_b and the factored design moment M_u or the ASD moment M_a. For an illustration, let's assume that $C_b = 1.0$, $F_y = 50$ ksi and that we wish to select a beam with $L_b = 18$ ft, $M_u = 544$ ft-k (or $M_a = 362.7$ ft-k). For this problem, the appropriate page from AISC Table 3-10 is shown in Fig. 9.13, with the permission of the AISC.

First, for the LRFD solution, we proceed up from the bottom of the chart for an unbraced length $L_b = 18$ ft until we intersect a horizontal line from the ϕM_n column for $M_u = 544$ ft-k. Any section to the right and above this intersection point ($\nearrow$) will have a greater unbraced length and a greater design moment capacity.

Moving up and to the right, we first encounter the W16 × 89 and W14 × 90 sections. In this area of the charts, these sections are shown with dashed lines. The dashed lines indicate that the sections will provide the necessary moment capacities, but are in an uneconomical range. If we proceed further upward and to the right, the first solid line encountered will represent the lightest satisfactory section. In this case, it is a W24 × 84. For an ASD solution of the same problem, we enter the chart with $L_b = 18$ ft and $M_a = 362.7$ ft-k and use the left column entitled M_n/Ω. The result, again, is a W24 × 84. Other illustrations of these charts are presented in Examples 9-8 to 9-10.

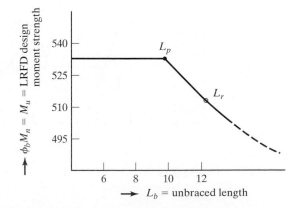

FIGURE 9.12

LRFD design moment for a beam plotted versus unbraced length, L_b.

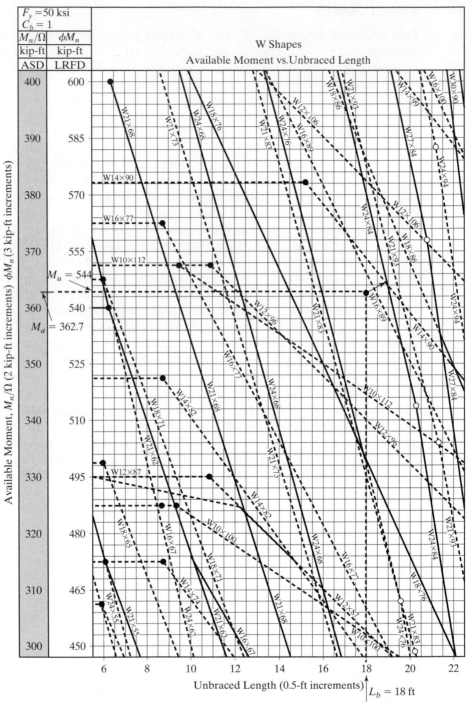

FIGURE 9.13

Example 9-8

Using 50 ksi steel, select the lightest available section for the beam of Fig. 9.14, which has lateral bracing provided for its compression flange, only at its ends. Assume that $C_b = 1.00$ for this example. (It's actually 1.14.) Use both LRFD and ASD methods.

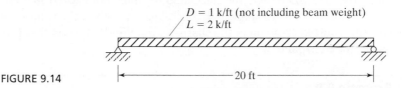

FIGURE 9.14

Solution

LRFD	ASD
Neglect beam wt initially then check after member selection is made	
$w_u = 1.2\,(1.0\text{ k/ft}) + 1.6\,(2.0\text{ k/ft})$ $= 4.4\text{ k/ft}$	$w_a = 1.0\text{ k/ft} + 2.0\text{ k/ft} = 3.0\text{ k/ft}$
$M_u = \dfrac{(4.4\text{ k/ft})\,(20\text{ ft})^2}{8} = 220\text{ ft-k}$	$M_a = \dfrac{(3.0\text{ k/ft})\,(20\text{ ft})^2}{8} = 150\text{ ft-k}$
Enter AISC Table 3-10 with $L_b = 20$ ft and $M_u = 220$ ft-k	Enter AISC Table 3-10 with $L_b = 20$ ft and $M_a = 150$ ft-k
Try W12 × 53	Try W12 × 53
Add self wt of 53 lb/ft	Add self wt of 53 lb/ft
$w_u = 1.2\,(1.053\text{ k/ft}) + 1.6\,(2.0\text{ k/ft})$ $= 4.46\text{ k/ft}$	$w_a = 1.053\text{ k/ft} + 2.0\text{ k/ft} = 3.05\text{ k/ft}$
$M_u = \dfrac{(4.46\text{ k/ft})\,(20\text{ ft})^2}{8} = 223\text{ ft-k}$	$M_a = \dfrac{(3.05\text{ k/ft})\,(20\text{ ft})^2}{8} = 153\text{ ft-k}$
Re-enter AISC Table 3-10	Re-enter AISC Table 3-10
Use W12 × 53.	**Use W12 × 53.**
$\phi M_n = 230.5$ ft-k $\geq M_u = 223$ ft-k **OK**	$\dfrac{M_n}{\Omega} = 153.6$ ft-k $\geq M_a = 153$ ft-k **OK**

Note: ϕM_n and M_n/Ω may be calculated from AISC Equations or more conveniently read from Table 3-10. To obtain the value of ϕM_n or M_n/Ω, proceed up from the bottom of the chart for an $L_b = 20$ ft until we intersect the line for the W12 × 53 member. Turn left and proceed with a horizontal line and read the value of either ϕM_n or M_n/Ω from the vertical axis.

For the example problem that follows, C_b is greater than 1.0. For such a situation, the reader should look back to Fig. 9.9. There, he or she will see that the design moment strength of a section can go to $\phi_b C_b M_n$ when $C_b > 1.0$, **but may under no circumstances exceed $\phi_b M_p = \phi_b F_y Z$, nor may M_a exceed M_p/Ω_b for the section.**

To handle such a problem, we calculate an effective moment, as shown next. (The numbers are taken from Example 9-9, which follows. Notice $C_b = 1.67$.

$$M_{u\,effective} = \frac{850}{1.67} = 509 \text{ ft-k} \quad \text{and} \quad M_{a\,effective} = \frac{595}{1.67} = 356 \text{ ft-k}$$

Then we enter the charts with $L_b = 17$ ft, and with M_u effective $= 509$ ft-k or M_a effective $= 356$ ft-k, and select a section. We must, however, make sure that, for LRFD design, the calculated M_u, does not exceed $\phi_b M_n = \phi F_y Z$ for the section selected. Similarly, for ASD design, the calculated M_a must not exceed $M_n/\Omega_b = F_y Z/\Omega_b$ for the section selected. For Example 9-9, which follows, both of the values are exceeded. As a result, one must keep going in the charts until we find a section that provides the necessary values for $\phi_b M_n$ and M_n/Ω_b.

Example 9-9

Using 50 ksi steel and both the LRFD and ASD methods, select the lightest available section for the situation shown in Fig. 9.15. Bracing is provided only at the ends and center line of the member, and thus, $L_b = 17$ ft.

Using Fig. 9.10, C_b is 1.67 if the only uniform load is the member self-weight and it is neglected. If the self-weight is considered then C_b will be between 1.67 and 1.30. Since the self-weight is a small portion of the design moment, C_b is near the value of 1.67 and using it would be a reasonable assumption.

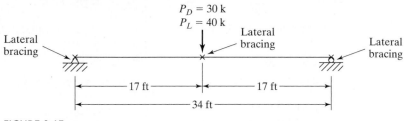

FIGURE 9.15

Solution

LRFD	ASD
Neglect beam wt initially then check after member selection is made	
$P_u = 1.2\,(30\text{ k}) + 1.6\,(40\text{ k}) = 100$ k	$P_a = 30\text{ k} + 40\text{ k} = 70$ k
$M_u = \dfrac{100\text{ k}\,(34\text{ ft})}{4} = 850$ ft-k	$M_a = \dfrac{70\text{ k}\,(34\text{ ft})}{4} = 595$ ft-k
Enter AISC Table 3-10 with $L_b = 17$ ft and M_u effective $= \dfrac{850}{1.67} = 509$ ft-k	Enter AISC Table 3-10 with $L_b = 17$ ft and M_a effective $= \dfrac{595}{1.67} = 356$ ft-k
Try W24 × 76 ($\phi_b M_p$ from AISC Table 3-2 = 750 ft-k < M_u = 850 ft-k **N.G.**)	Try W24 × 84 $\left(\dfrac{M_p}{\Omega_b} = 559\text{ ft-k from}\right.$ AISC Table 3-2 < M_a = 595 ft-k **N.G.**$\Big)$
Try W27 × 84 ($\phi_b M_p$ = 915 ft-k)	Try W27 × 84 $\left(\dfrac{M_p}{\Omega_b} = 609\text{ ft-k}\right)$
Add self-weight of 84 lb/ft	Add self-weight of 84 lb/ft

$w_u = 1.2\,(0.084\text{ k/ft}) = 0.101\text{ k/ft}$	$w_a = 0.084\text{ k/ft}$
$M_u = \dfrac{(0.101\text{ k/ft})(34\text{ ft})^2}{8} + \dfrac{100\text{ k}(34\text{ ft})}{4}$	$M_a = \dfrac{(0.084\text{ k/ft})(34\text{ ft})^2}{8} + \dfrac{70\text{ k}(34\text{ ft})}{4}$
$M_u = 865\text{ ft-k} < \phi_b M_p = 915\text{ ft-k}$ **OK**	$M_a = 607\text{ ft-k} < \dfrac{M_p}{\Omega_b} = 609\text{ ft-k}$ **OK**
Use W27 × 84.	**Use W27 × 84.**

Example 9-10

Using 50 ksi steel and both the LRFD and ASD methods, select the lightest available section for the situation shown in Fig. 9.16. Bracing is provided only at the ends and at midspan.

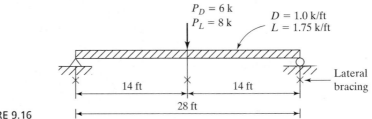

FIGURE 9.16

Solution

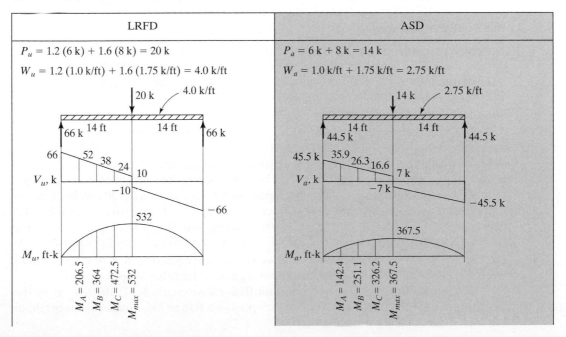

LRFD	ASD
$P_u = 1.2\,(6\text{ k}) + 1.6\,(8\text{ k}) = 20\text{ k}$	$P_a = 6\text{ k} + 8\text{ k} = 14\text{ k}$
$W_u = 1.2\,(1.0\text{ k/ft}) + 1.6\,(1.75\text{ k/ft}) = 4.0\text{ k/ft}$	$W_a = 1.0\text{ k/ft} + 1.75\text{ k/ft} = 2.75\text{ k/ft}$

$$C_b = \frac{12.5\, M_{max}}{2.5\, M_{max} + 3M_A + 4M_B + 3M_C}$$

$$C_b = \frac{12.5(532)}{2.5(532) + 3(206.5) + 4(364) + 3(472.5)}$$

$$= 1.38$$

Enter AISC Table 3-10 with $L_b = 14$ ft and

$$M_u \text{ effective} = \frac{532}{1.38} = 386 \text{ ft-k}$$

Try W21 × 62 ($\phi_b M_n = 405$ ft-k from Table 3-10, $\phi_b M_p$ = 540 ft-k from Table 3-2)

$C_b\, \phi_b\, M_n = 1.38\,(405) = 559$ ft-k $\nleq \phi M_p$

= 540 ft-k ← controls

$M_u = 532$ ft-k $\le \phi_b M_p = 540$ ft-k **OK**

Use W21 × 62.

$$C_b = \frac{12.5\, M_{max}}{2.5\, M_{max} + 3M_A + 4M_B + 3M_C}$$

$$C_b = \frac{12.5(367.5)}{2.5\,(367.5) + 3(142.4) + 4(251.1) + 3(326.2)}$$

$$= 1.38$$

Enter AISC Table 3-10 with $L_b = 14$ ft and

$$M_a \text{ effective} = \frac{367.5}{1.38} = 267 \text{ ft-k}$$

Try W21 × 62 ($M_n/\Omega_b = 270$ ft-k from Table 3.10,

$M_p/\Omega_b = 359$ ft-k $< M_u = 367.5$ ft-k **N.G.**

Try W24 × 62 ($M_p/\Omega_b = 382$ ft-k from Table 3-2)

$M_n/\Omega_b = 236$ ft-k from Table 3-10 with $C_b = 1.0$

$$\therefore \frac{C_b M_n}{\Omega_b} = 1.38(236) = 326 \text{ ft-k}$$

$< M_a = 367.5$ ft-k **N.G.**

Try W21 × 68 ($M_p/\Omega_b = 399$ ft-k from Table 3-2)

$M_n/\Omega = 304$ ft-k from Table 3-10 with $C_b = 1.0$

$$\therefore \frac{C_b M_n}{\Omega_b} = 1.38\,(304) = 420 \text{ ft-k} > M_p/\Omega_b$$

= 399 ft-k ← controls

$$M_a = 367.5 \text{ ft-k} \le \frac{M_p}{\Omega_b} = 399 \text{ ft-k} \quad \textbf{OK}$$

Use W21 × 68.

9.9 NONCOMPACT SECTIONS

A compact section is a section that has a sufficiently stocky profile so that it is capable of developing a fully plastic stress distribution before buckling locally (web or flange). The term *plastic* means stressed throughout to the yield stress and is discussed at length in Chapter 8. For a section to be compact, the width thickness ratio of the flanges of W- or other I-shaped rolled sections must not exceed a b/t value $\lambda_p = 0.38\sqrt{E/F_y}$. Similarly, the webs in flexural compression must not exceed an h/t_w value $\lambda_p = 3.76\sqrt{E/F_y}$. The values of b, t, h, and t_w are shown in Fig. 9.17.

A noncompact section is one for which the yield stress can be reached in some, but not all, of its compression elements before buckling occurs. It is not capable of reaching a fully plastic stress distribution. The noncompact sections are those that have web–thickness ratios greater than λ_p, but not greater than λ_r. The λ_r values are provided in Table 9.2, which is Table B4.1b of the AISC Specification. For the noncompact range, the width–thickness ratios of the flanges or W- or other I-shaped rolled sections must not exceed $\lambda_r = 1.0\sqrt{E/F_y}$, while those for the webs must not exceed $\lambda_r = 5.70\sqrt{E/F_y}$. Other values are provided in AISC Table B4.1b for λ_p and λ_r for other shapes.

For noncompact beams, the nominal flexural strength M_n is the lowest of the lateral-torsional buckling strength, the compression flange local buckling strength, or the web local buckling strength.

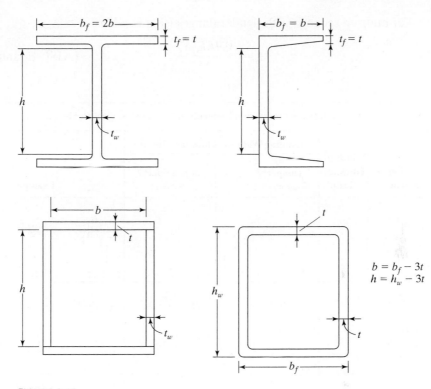

FIGURE 9.17

Values of h, b, t, and t_w to be used for computing λ = width–thickness ratios.

If we have a section with noncompact flanges—that is, one where $\lambda_p < \lambda \leq \lambda_r$ the value of M_n is given by the equation to follow, in which $k_c = 4/\sqrt{h/t_w} \geq 0.35 \leq 0.76$:

$$M_n = \left[M_p - (M_p - 0.7F_y\,S_x)\left(\frac{\lambda - \lambda_{pf}}{\lambda_{rf} - \lambda_{pf}}\right)\right] \qquad \text{(AISC Equation F3-1)}$$

Almost all of the standard hot-rolled W, M, S, and C shapes listed in the AISC Manual are compact, and none of them fall into the slender classification. All of these shapes have compact webs, but a few of them have noncompact flanges. We particularly must be careful when working with built-up sections as they may very well be noncompact or slender.

In this section the author considers a shape which has a noncompact flange. If a standard shape has a noncompact flange, it will be indicated in the Manual with an "f" footnote. The numerical values shown in the tables are based on the reduced stresses caused by noncompactness.

As indicated in AISC Specification F3 the flange of a member is noncompact if $\lambda_p < \lambda \leq \lambda_r$ and the member will buckle inelastically. These values are given for different shapes in AISC Specification Table B4.1b.

For built-up sections with slender flanges (that is, where $\lambda > \lambda_r$),

$$M_n = \frac{0.9Ek_c\,S_x}{\lambda^2}.$$ (AISC Equation F3-2)

TABLE 9.1 Width-to-Thickness Ratios: Compression Elements in Members Subject to Flexure

	Case	Description of Element	Width-to-Thickness Ratio	Limiting Width-to-Thickness Ratios		Example
				λ_r compact / noncompact)	λ_r noncompact / slender)	
Unstiffened Elements	10	Flanges of rolled I-shaped sections, channels, and tees	b/t	$0.38\sqrt{\dfrac{E}{F_y}}$	$1.0\sqrt{\dfrac{E}{F_y}}$	
	11	Flanges of doubly and singly symmetric I-shaped built-up sections	b/t	$0.38\sqrt{\dfrac{E}{F_y}}$	$0.95\sqrt{\dfrac{K_cE}{F_L}}$ [a][b]	
	12	Legs of single angles	b/t	$0.54\sqrt{\dfrac{E}{F_y}}$	$0.91\sqrt{\dfrac{E}{F_y}}$	
	13	Flanges of all I-shaped sections and channels in flexure about the weak axis	b/t	$0.38\sqrt{\dfrac{E}{F_y}}$	$1.0\sqrt{\dfrac{E}{F_y}}$	
	14	Stems of tees	d/t	$0.84\sqrt{\dfrac{E}{F_y}}$	$1.03\sqrt{\dfrac{E}{F_y}}$	
Stiffened Elements	15	Webs of doubly-symmetric I-shaped sections and channels	h/t_w	$3.76\sqrt{\dfrac{E}{F_y}}$	$5.70\sqrt{\dfrac{E}{F_y}}$	
	16	Webs of singly-symmetric I-shaped sections	h_c/t_w	$\dfrac{\dfrac{h_e}{h_p}\sqrt{\dfrac{E}{F_y}}}{\left(0.54\dfrac{M_p}{M_y} - 0.09\right)^2} \le \lambda_t$ [c]	$5.70\sqrt{\dfrac{E}{F_y}}$	

(Continued)

TABLE 9.1   (Continued)

Case	Description of Element	Width-to-Thickness Ratio	Limiting Width-to-Thickness Ratios		Example
			λ_r compact/ noncompact)	λ_r noncompact/ slender)	
17	Flanges of rectangular HSS and boxes of uniform thickness	b/t	$1.12\sqrt{\dfrac{E}{F_y}}$	$1.40\sqrt{\dfrac{E}{F_y}}$	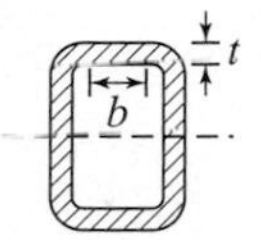
18	Flange cover plates and diaphragm plates between lines of fasteners or welds	b/t	$1.12\sqrt{\dfrac{E}{F_y}}$	$1.40\sqrt{\dfrac{E}{F_y}}$	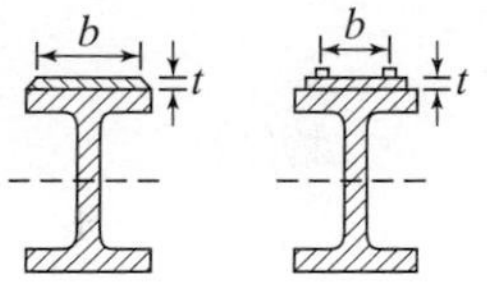
19	Webs of rectangular HSS and boxes	h/t	$2.42\sqrt{\dfrac{E}{F_y}}$	$5.70\sqrt{\dfrac{E}{F_y}}$	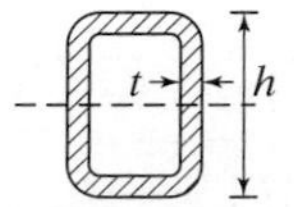
20	Round HSS	D/t	$0.07\dfrac{E}{F_y}$	$0.31\dfrac{E}{F_y}$	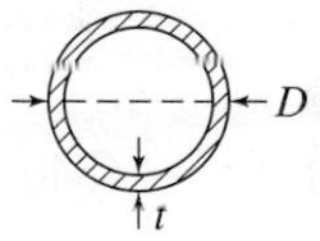

Stiffened Elements

[a] $K_c = \dfrac{4}{\sqrt{h/t_w}}$ but shall not be taken less than 0.35 nor greater than 0.76 for calculation purposes.

[b] $F_L = 0.7F_y$ for major axis bending of compact and noncompact web built-up I-shaped members with $S_{xy}/S_{xc} \geq 0.7$, $F_L = F_y S_{xy}/S_{xc} > 0.5F_y$ for major-axis bending of compact and noncompact web built-up I-shaped members with $S_{xy}/S_{xc} < 0.7$.

[c] M_y is the moment at yielding of the extreme fiber. M_p = plastic bending moment, kip-in. (N-mm)

E = modulus of elasticity of steel = 29,000 ksi (200 000 MPa)

F_y = specified minimum yield stress, ksi (MPa)

Example 9-11

Determine the LRFD flexural design stress and the ASD allowable flexural stress for a 50 ksi W12 × 65 section which has full lateral bracing.

Solution

Using a W12 × 65 (b_f = 12.00 in, t_f = 0.605 in, S_x = 87.9 in³, Z_x = 96.8 in³)

Is the flange noncompact?

$$\lambda_p = 0.38\sqrt{\frac{E}{F_y}} = 0.38\sqrt{\frac{29 \times 10^3}{50}} = 9.15$$

$$\lambda = \frac{b_f}{2t_f} = \frac{12.00}{(2)(0.605)} = 9.92$$

$$\lambda_r = 1.0\sqrt{\frac{E}{F_y}} = 1.0\sqrt{\frac{29 \times 10^3}{50}} = 24.08$$

$$\lambda_p = 9.15 < \lambda = 9.92 < \lambda_r = 24.08$$

$$\therefore \text{ The flange is noncompact.}$$

Calculate the nominal flexural stress.

$$M_p = F_y Z = (50)(96.8) = 4840 \text{ in-k}$$

$$M_n = \left[M_p - (M_p - 0.7F_y S_x)\left(\frac{\lambda - \lambda_p}{\lambda_r - \lambda_p}\right)\right] \qquad \text{(AISC Eq F3-1)}$$

$$M_n = \left[4840 - (4840 - 0.7 \times 50 \times 87.9)\left(\frac{9.92 - 9.15}{24.08 - 9.15}\right)\right]$$

$$= 4749 \text{ in-k} = 395.7 \text{ ft-k}$$

Determine $\phi_b M_n$ and M_n/Ω.

LRFD ϕ_b = 0.9	ASD Ω_b = 1.67
$\phi_b M_n$ = (0.9)(395.7) = 356 ft-k	$\dfrac{M_n}{\Omega_b} = \dfrac{395.7}{1.67}$ = 237 ft-k

Note: These values correspond to the values given in AISC Table 3-2.

The equations mentioned here were used, where applicable, to obtain the values used for the charts plotted in AISC Table 3-10. The designer will have little trouble with noncompact sections when F_y is no more than 50 ksi. He or she, however, will have to use the formulas presented in this section for shapes with larger F_y values.

9.10   PROBLEMS FOR SOLUTION

9-1 to 9-8.   *Using both LRFD and ASD, select the most economical sections, with $F_y = 50$ ksi, unless otherwise specified, and assuming full lateral bracing for the compression flanges. Working or service loads are given for each case, and beam weights are not included.*

9-1.

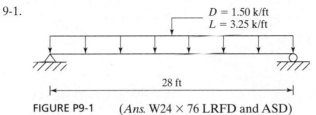

FIGURE P9-1   (*Ans.* W24 × 76 LRFD and ASD)

9-2.

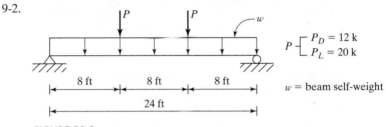

FIGURE P9-2

9-3.

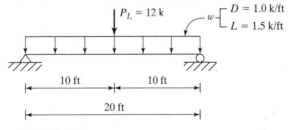

FIGURE P9-3   (*Ans.* W18 × 40 LRFD and ASD)

9-4.   Repeat Prob. 9-3, using $P_L = 20$ k.

9-5.

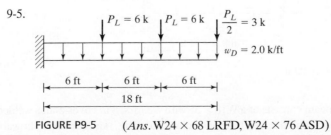

FIGURE P9-5   (*Ans.* W24 × 68 LRFD, W24 × 76 ASD)

9-6.

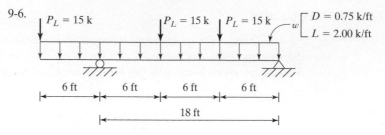

FIGURE P9-6

9-7.

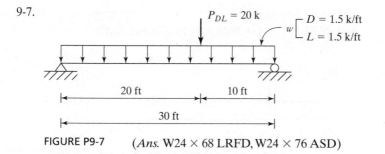

FIGURE P9-7   (*Ans.* W24 × 68 LRFD, W24 × 76 ASD)

9-8.   The accompanying figure shows the arrangement of beams and girders that are used to support a 5 in reinforced concrete floor for a small industrial building. Design the beams and girders assuming that they are simply supported. Assume full lateral support of the compression flange and a live load of 80 psf. Concrete weight is 150 lb/ft³.

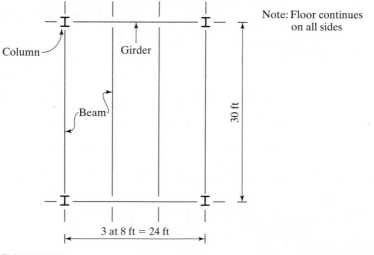

FIGURE P9-8

9-9.   A beam consists of a W18 × 35 with 3/8 in × 8 in cover plates welded to each flange. Determine the LRFD design uniform load, w_u, and the ASD allowable uniform load,

w_a, that the member can support in addition to its own weight for a 28 ft simple span. (*Ans.* 2.85 k/ft LRFD, 3.02 k/ft ASD)

9-10.   The member shown is made with 36 ksi steel. Determine the maximum service live load that can be placed on the beam if, in addition to its own weight, it is supporting a service dead load of 0.80 klf. The member is used for a 20 ft simple span. Use both LRFD and ASD methods.

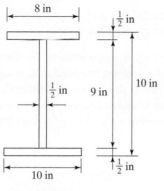

FIGURE P9-10

9-11 to 9-14.   *Use both LRFD and ASD methods for these beams for which full lateral bracing of the compression flange is provided.*

9-11.   Select a W section for a 24 ft simple span to support a service dead uniform load of 1.5 k/ft and a live service load of 1.0 k/ft if two holes for 3/4-in ϕ bolts are assumed present in each flange at the section of maximum moment. Use AISC Specification and A36 steel. Use both LRFD and ASD methods. (*Ans.* W21 × 44 LRFD, W21 × 48 ASD)

9-12.   Rework Prob. 9-11, assuming that four holes for 3/4-in ϕ bolts pass through each flange at the point of maximum moment. Use A992 steel.

9-13.   The section shown in Fig. P9-13 has two 3/4-in ϕ bolts passing though each flange and cover plate. Find the design load, w_a, and factored load, w_u, that the section can support, in addition to its own weight, for a 22 ft simple span if it consists of a steel with $F_y = 50$ ksi. Deduct all holes for calculating section properties. (*Ans.* Net w_u = 4.74 k/ft, Net w_a = 3.14 k/ft)

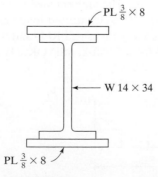

FIGURE P9-13

9-14.   A 36 ft simple span beam is to support two movable service 20 kip loads a distance of 12 ft apart. Assuming a dead load of 1.0 k/ft including the beam self-weight, select a 50 ksi steel section to resist the largest possible moment. Use LRFD method only.

9-15 to 9-28.   *For these problems, different values of L_b are given. Dead loads do not include beam weights. Use both LRFD and ASD methods.*

9-15.   Determine ΦM_n and M_n/Ω for a W18 × 46 used as a beam with an unbraced length of the compression flange of 4 ft and 12 ft. Use A992 steel and $C_b = 1.0$.

(*Ans. L_b* = 4 ft, 340 ft-k LRFD; 226 ft-k ASD)
(*Ans. L_b* = 12 ft, 231.4 ft-k LRFD; 154.3 ft-k ASD)

9-16.   Determine the lightest satisfactory W shape to carry a uniform dead load of 4.0 k/ft plus the beam self-weight and a uniform live load of 2.75 k/ft on a simple span of 12 ft. Assume bracing is provided at the ends only. Obtain C_b from Fig. 9.10 in text.

9-17.   Select the lightest satisfactory W-shape section if $F_y = 50$ ksi. Lateral bracing is provided at the ends only. Determine C_b. (*Ans.* W14 × 61 LRFD, W12 × 65 ASD)

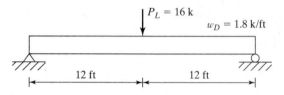

FIGURE P9-17

9-18.   Repeat Prob. 9-17 if lateral bracing is provided at the concentrated load as well as at the ends of the span. Determine C_b.

9-19.   A W18 × 55 of A992 steel is used on a simple span of 15 ft and has lateral support of compression flange at its ends only. If the only dead load present is the beam self-weight, what is the largest service concentrated live load that can be placed at the 1/3 points of the beam? Determine C_b. (*Ans.* 41.7 k LRFD, 44.3 k ASD)

9-20.   Repeat Prob. 9-19 if lateral bracing is supplied at the beam ends and at the concentrated loads. Determine C_b.

9-21.   The cantilever beam shown in Fig. P9-21 is a W18 × 55 of A992 steel. Lateral bracing is supplied at the fixed end only. The uniform load is a service dead load and includes the beam self-weight. The concentrated loads are service live loads. Determine whether the beam is adequate using LRFD and ASD methods. Assume $C_b = 1.0$. (*Ans.* LRFD OK, 363 ft-k > 335 ft-k; ASD OK, 241 ft-k > 212.5 ft-k)

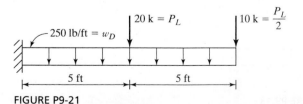

FIGURE P9-21

9-22. The given beam in Fig. P9-22 is A992 steel. If the live load is twice the dead load, what is the maximum total service load in k/ft that can be supported when (a) the compression flange is braced laterally for its full length, and (b) lateral bracing is supplied at the ends and centerline only?

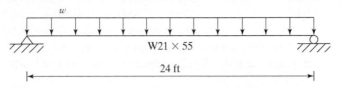

W21 × 55

24 ft

FIGURE P9-22

9-23. A W21 × 68 beam of A992 steel carries a uniformly distributed service dead load of 1.75 k/ft plus its self-weight and two concentrated service live loads at the third points of a 33 ft simply supported span. If lateral bracing is provided at the ends and the concentrated loads, determine the maximum service live load, P_L. Assume the concentrated loads are equal in value, determine C_b. (*Ans.* 12.26 k LRFD, 8.50 k ASD)

9-24. A beam of $F_y = 50$ ksi steel is used to support the loads shown in Fig. P9-24. Neglecting the beam self-weight, determine the lightest W shape to carry the loads if full lateral bracing is provided.

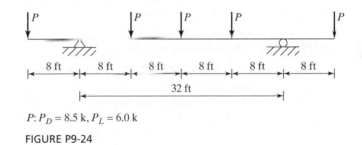

8 ft   8 ft   8 ft   8 ft   8 ft   8 ft

32 ft

$P: P_D = 8.5$ k, $P_L = 6.0$ k

FIGURE P9-24

9-25. Redesign the beam of Prob. 9-24 if lateral bracing is only provided at the supports and at the concentrated loads. Determine C_b. (*Ans.* W16 × 26 LRFD, W14 × 30 ASD)

9-26. Design the lightest W shape beam of 50 ksi steel to support the loads shown in Fig. P9-26. Neglect the beam self-weight. The beam has continuous lateral bracing between A and B, but is laterally unbraced between B and C. Determine C_b.

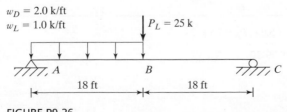

$w_D = 2.0$ k/ft
$w_L = 1.0$ k/ft

$P_L = 25$ k

A   B   C

18 ft   18 ft

FIGURE P9-26

9-27. A W16 × 36 beam of A992 steel is fixed at one support and simply supported at the other end. A concentrated load of dead load of 9.25 k and live load of 6.50 k is applied at the center of the 32 ft span. Assume lateral bracing of the compression flange is provided at the pinned support, the load point, and the fixed support. You may neglect the beam self-weight and assume that C_b = 1.0. Is the W16 adequate? (*Ans.* LRFD OK, 129.0 ft-k ≤ 136.6 ft-k; ASD N.G., 94.5 ft-k ≥ 90.9 ft-k)

9-28. A W24 × 104 beam is used to support the loads shown in Fig. P9-28. Lateral bracing of the compression flange is supplied only at the ends. Determine C_b. If F_y = 50 ksi, determine if the W24 is adequate to support these loads.

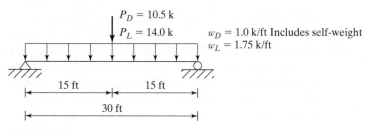

FIGURE P9-28

9-29. An A992, W18 × 60 steel beam is used on a 36 ft simple span to carry a uniformly distributed load. Determine the location of the lateral support, L_b, in order to provide just enough strength to carry a design moment. Use M_u = 416.8 ft-k for LRFD method and M_a = 277.5 ft-k for ASD method. Assume C_b = 1.0. (*Ans.* 9 ft LRFD and ASD)

9-30. The two steel beams shown in Fig. P9-30 are part of a two-span beam framing system with a pin (hinge) located 4.5 ft left of the interior support, making the system statically determinate. Determine the sizes (lightest) of the two W shape beams. Assume A992 steel and continuous lateral support of the compression flanges. The beam self-weight may be neglected. Use LRFD and ASD methods.

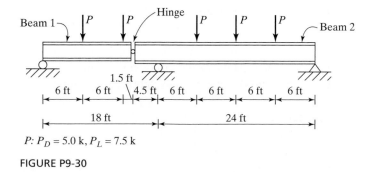

FIGURE P9-30

9-31.   A built-up shape steel beam consists of a ¼ in × 12 in web, and ⅜ in × 4 in top and bottom flanges. The member has the compression flange fully braced, therefore the moment capacity, ΦM_n, using LRFD method was calculated to be 103.4 ft-k using $F_y = 50$ ksi steel. During design it was thought that the factored moment, M_u, was 100 ft-k, but after the member was fabricated it was found that the actual design moment, M_u, should have been 130 ft-k. A brilliant and resourceful young engineer suggested adding a ¼ in × 6 in cover plate to the bottom flange of the member to increase its capacity. Compute the new moment capacity, ΦM_n, and state whether or not it will safely support the design moment, $M_u = 130$ ft-k. (*Ans.* Yes, $\Phi M_n = 131.5$ ft-k $> M_u = 130$ ft-k)

9-32.   A W21 × 93 has been specified for use on your design project. By mistake, a W21 × 73 was shipped to the field. This beam must be erected today. Assuming that ½ in thick plates are obtainable immediately, select cover plates to be welded to the top and bottom flanges to obtain the necessary section capacity. Use $F_y = 50$ ksi steel for all materials and assume that full bracing is supplied for the compression flange. Use LRFD and ASD methods.

CHAPTER 10

Design of Beams— Miscellaneous Topics (Shear, Deflection, etc.)

10.1 DESIGN OF CONTINUOUS BEAMS

Section B3 of the AISC Specification states that beams may be designed according to the provisions of the LRFD or ASD methods. Analysis of the members to determine their required strengths may be made by elastic, inelastic, or plastic analysis procedures. Design based on plastic analysis is permitted only for sections with yield stresses no greater than 65 ksi and is subject to some special requirements in Appendix 1 of the Specification Commentary.

Both theory and tests show clearly that continuous ductile steel members meeting the requirements for compact sections with sufficient lateral bracing supplied for their compression flanges have the desirable ability of being able to redistribute moments caused by overloads. If plastic analysis is used, this advantage is automatically included in the analysis.

If elastic analysis is used, the AISC handles the redistribution by a rule of thumb that approximates the real plastic behavior. AISC Appendix 1 Commentary, Section A1.3, states that for continuous compact sections, the design *may* be made on the basis of nine-tenths of the maximum negative moments caused by gravity loads that are maximum at points of support if the positive moments are increased by one-tenth of the average negative moments at the adjacent supports. *(The 0.9 factor is applicable only to gravity loads and not to lateral loads such as those caused by wind and earthquake.)* The factor can also be applied to columns that have axial forces not exceeding $0.15\phi_c F_y A_g$ for LRFD or

$0.15 F_y A_g / \Omega_c$ for ASD. This moment reduction does not apply to moments produced by loading on cantilevers nor for designs made according to Sections 1.4 through 1.8 of Appendix 1 of the AISC Specification Commentary.

Example 10-1

The beam shown in Fig. 10.1 is assumed to consist of 50 ksi steel. (a) Select the lightest W section available, using plastic analysis and assuming that full lateral support is provided for its compression flanges. (b) Design the beam, using elastic analysis with the service loads and the 0.9 rule, and assuming that full lateral support is provided for both the flanges.

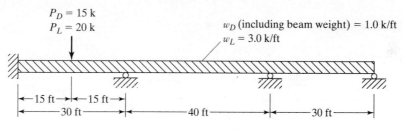

FIGURE 10.1

Solution

a, Plastic analysis and LRFD design

$$w_u = (1.2)(1.0) + (1.6)(3) = 6.0 \text{ klf}$$
$$P_u = (1.2)(15) + (1.6)(20) = 50 \text{ k}$$

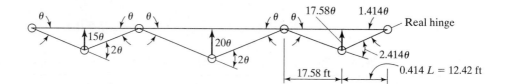

$$
\text{Span 1}
\begin{cases}
M_u 4\theta = (30 w_u)\left(\frac{1}{2}\right)(15\theta) + (P_u)(15\theta) \\
M_u = 56.25 w_u + 3.75 P_u \\
M_u = (56.25)(6.0) + (3.75)(50) \\
M_u = 525 \text{ ft-k}
\end{cases}
$$

$$
\text{Span 2}
\begin{cases}
M_u(4\theta) = (40 w_u)\left(\frac{1}{2}\right)(20\theta) \\
M_u = 100 w_u = (100)(6.0) \\
M_u = 600 \text{ ft-k} \leftarrow \text{controls}
\end{cases}
$$

$$\text{Span 3} \begin{cases} M_u(3.414\theta) = (30w_u)\left(\dfrac{1}{2}\right)(17.58\theta) \\ M_u = 77.24w_u = (77.24)(6.0) \\ \qquad = 463.4 \text{ ft-k} \end{cases}$$

Use **W21 × 68** (AISC Table 3-2). $\phi_b M_p = 600$ ft-k $= M_u = 600$ ft-k

b. Analyzing beam of Fig. 10.1 for service loads

$$w_a = 1.0 + 3.0 = 4 \text{ k/ft}$$

$$P_a = 15 + 20 = 35 \text{ k}$$

Drawing moment diagram, ft-k

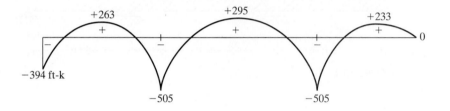

Maximum negative moment for design

$$= (0.9)(-505) = -454.5 \text{ ft-k} \leftarrow \text{controls}$$

Maximum positive moment for design

$$= +295 + \left(\frac{1}{10}\right)\left(\frac{505 + 505}{2}\right) = +345.5 \text{ ft-k}$$

Use **W24 × 76** (AISC Table 3-2). $M_p/\Omega = 499$ ft-k $> M_a = 454.5$ ft-k

Important Note: Should the lower flange of this W24 × 76 not be braced laterally, we must check the lengths of the span where negative moments are present, because L_b values may very well exceed L_p for the section and the design may have to be revised.

10.2 SHEAR

For this discussion, the beam of Fig. 10.2(a) is considered. As the member bends, shear stresses occur because of the changes in length of its longitudinal fibers. For positive bending the lower fibers are stretched and the upper fibers are shortened, while somewhere in between there is a neutral axis where the fibers do not change in length. Owing to these varying deformations, a particular fiber has a tendency to slip on the fiber above or below it.

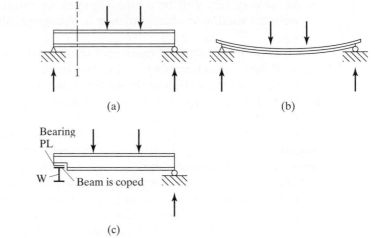

FIGURE 10.2

If a wooden beam was made by stacking boards on top of each other and not connecting them, they obviously would tend to take the shape shown in part (b) of the figure. The student may have observed short, heavily loaded timber beams with large transverse shears that split along horizontal planes.

This presentation may be entirely misleading in seeming to completely separate horizontal and vertical shears. In reality, horizontal and vertical shears at any point are the same, as long as the critical section at which the shear stress is evaluated is taken parallel to the axis of symmetry. Furthermore, one cannot occur without the other.

Generally, shear is not a problem in steel beams, because the webs of rolled shapes are capable of resisting rather large shearing forces. Perhaps it is well, however, to list here the most common situations where shear might be excessive:

1. Should large concentrated loads be placed near beam supports, they will cause large internal forces without corresponding increases in bending moments. A fairly common example of this type of loading occurs in tall buildings where, on a particular floor, the upper columns are offset with respect to the columns below. The loads from the upper columns applied to the beams on the floor level in question will be quite large if there are many stories above.

2. Probably the most common shear problem occurs where two members (as a beam and a column) are rigidly connected together so that their webs lie in a common plane. This situation frequently occurs at the junction of columns and beams (or rafters) in rigid frame structures.

3. Where beams are notched or coped, as shown in Fig. 10.2(c), shear can be a problem. For this case, shear forces must be calculated for the remaining beam depth. A similar discussion can be made where holes are cut in beam webs for ductwork or other items.

4. Theoretically, very heavily loaded short beams can have excessive shears, but practically, this does not occur too often unless it is like Case 1.

5. Shear may very well be a problem even for ordinary loadings when very thin webs are used, as in plate girders or in light-gage cold-formed steel members.

From his or her study of mechanics of materials, the student is familiar with the horizontal shear stress formula $f_v = VQ/Ib$, where V is the external shear; Q is the statical moment of that portion of the section lying outside (either above or below) the line on which f_v is desired, taken about the neutral axis; and b is the width of the section where the unit shearing stress is desired.

Figure 10.3(a) shows the variation in shear stresses across the cross section of an I-shaped member, while part (b) of the same figure shows the shear stress variation in a member with a rectangular cross section. It can be seen in part (a) of the figure that the shear in I-shaped sections is primarily resisted by the web.

If the load is increased on an I-shaped section until the bending yield stress is reached in the flange, the flange will be unable to resist shear stress and it will be carried in the web. If the moment is further increased, the bending yield stress will penetrate farther down into the web and the area of web that can resist shear will be further reduced. Rather than assuming the nominal shear stress is resisted by part of the web, the AISC Specification assumes that a reduced shear stress is resisted by the entire web area. This web area, A_w, is equal to the overall depth of the member, d, times the web thickness, t_w.

Shear strength expressions are given in AISC Specification G2. In these expressions, h is the clear distance between the web toes of the fillets for rolled shapes, while for built-up welded sections it is the clear distance between flanges. For bolted built-up sections, h is the distance between adjacent lines of bolts in the web. Different expressions are given for different h/t_w ratios, depending on whether shear failures would be plastic, inelastic, or elastic.

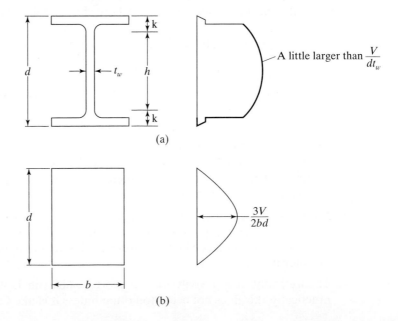

FIGURE 10.3

Combined welded and bolted joint, Transamerica Pyramid, San Francisco, CA.
(Courtesy of Kaiser Steel Corporation.)

The nominal shear strength of unstiffered or stiffened webs is specified as

$$V_n = 0.6F_y A_w C_v \qquad \text{(AISC Equation G2-1)}$$

Using this equation for the webs of I-shaped members when $h/t_w \le 2.24\sqrt{E/F_y}$, we find that $C_v = 1.0$, $\phi_v = 1.00$, and $\Omega_v = 1.50$. (Almost all current W, S, and HP shapes fall into this class. The exceptions are listed in Section G2 of the AISC Specification.)

For the webs of all doubly symmetric shapes, singly symmetric shapes, and channels, except round HSS, $\phi_v = 0.90$ and $\Omega_v = 1.67$ are used to determine the design shear strength, $\phi_v V_n$, and the allowable shear strength V_n/Ω. C_v, the web shear coefficient, is determined from the following situations and is substituted into AISC Equation G2-1:

a. For $\dfrac{h}{t_w} \le 1.10\sqrt{\dfrac{k_v E}{F_y}}$

$$C_v = 1.0 \qquad \text{(AISC Equation G2-3)}$$

b. For $1.10\sqrt{\dfrac{k_v E}{F_y}} < \dfrac{h}{t_w} \le 1.37\sqrt{\dfrac{k_v E}{F_y}}$

$$C_v = \dfrac{1.10\sqrt{\dfrac{k_v E}{F_y}}}{\dfrac{h}{t_w}} \qquad \text{(AISC Equation G2-4)}$$

c. For $\dfrac{h}{t_w} > 1.37\sqrt{\dfrac{k_v E}{F_y}}$

$$C_v = \frac{1.51 E k_v}{\left(\dfrac{h}{t_w}\right)^2 F_y}$$ (AISC Equation G2-5)

The web plate shear buckling coefficient, k_v, is specified in the AISC Specification G2.1b, parts (i) and (ii). For webs without transverse stiffeners and with $h/t_w < 260$: $k_v = 5$. This is the case for most rolled I-shaped members designed by engineers.

Example 10-2

A W21 × 55 with $F_y = 50$ ksi is used for the beam and loads of Fig. 10.4. Check its adequacy in shear,

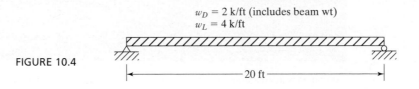

$w_D = 2$ k/ft (includes beam wt)
$w_L = 4$ k/ft

FIGURE 10.4

|← —————————— 20 ft —————————— →|

Solution

Using a W21 × 55 ($A = 16.2$ in^2, $d = 20.8$ in, $t_w = 0.375$ in, and $k_{des} = 1.02$ in)

$$h = 20.8 - 2k_{des} = 20.8 - (2)(1.02) = 18.76 \text{ in}$$

$$\frac{h}{t_w} = \frac{18.76}{0.375} = 50.03 < 2.24\sqrt{\frac{29{,}000}{50}} = 53.95$$

$$\therefore C_v = 1.0, \ \phi_v = 1.0 \text{ and } \Omega_v = 1.50$$

$$A_w = d \, t_w = (20.8 \text{ in})(0.375 \text{ in}) = 7.80 \text{ in}^2$$

$$\therefore V_n = 0.6 \, F_y \, A_w \, C_v = 0.6 \, (50 \text{ ksi})(7.80 \text{ in}^2)(1.0) = 234 \text{ k}$$

LRFD $\phi_v = 1.00$	ASD $\Omega_v = 1.50$
$w_u = (1.2)(2) + (1.6)(4) = 8.8$ k/ft	$w_a = 2 + 4 = 6$ k/ft
$V_u = \dfrac{8.8 \text{ k/ft } (20 \text{ ft})}{2} = 88 \text{ k}$	$V_a = \dfrac{6.0 \text{ k/ft } (20 \text{ ft})}{2} = 60 \text{ k}$
$\phi_v V_n = (1.00)(234) = 234$ k	$\dfrac{V_n}{\Omega_v} = \dfrac{234}{1.50} = 156$ k
> 88 k **OK**	> 60 k **OK**

Notes

1. The values of $\phi_v V_{nx}$ and V_{nx}/Ω_v with $F_y = 50$ ksi are given for W shapes in the manual, Table 3-2.

2. Two values are given in the AISC Manual for k. One is given in decimal form and is to be used for design calculations, while the other one is given in fractions and is to be used for detailing. These two values are based, respectively, on the minimum and maximum radii of the fillets and will usually be quite different from each other.

3. A very useful table (3-6) is provided in Part 3 of the AISC Manual for determining the maximum uniform load each W shape can support for various spans. The values given are for $F_y = 50$ ksi and are controlled by maximum moments or shears, as specified by LRFD or ASD.

Should V_u for a particular beam exceed the AISC specified shear strengths of the member, the usual procedure will be to select a slightly heavier section. If it is necessary, however, to use a much heavier section than required for moment, doubler plates (Fig. 10.5) may be welded to the beam web, or stiffeners may be connected to the webs in zones of high shear. Doubler plates must meet the width–thickness requirements for compact stiffened elements, as prescribed in Section B4 of the AISC Specification. In addition, they must be welded sufficiently to the member webs to develop their proportionate share of the load.

The AISC specified shear strengths of a beam or girder are based on the entire area of the web. Sometimes, however, a connection is made to only a small portion or depth of the web. For such a case, the designer may decide to assume that the shear is spread over only part of the web depth, for purposes of computing shear strength. Thus, he or she may compute A_w as being equal to t_w times the smaller depth, for use in the shear strength expression.

When beams that have their top flanges at the same elevations (the usual situation) are connected to each other, it is frequently necessary to cope one of them, as shown in Fig. 10.6. For such cases, there is a distinct possibility of a block shear failure along the broken lines shown. This subject was previously discussed in Section 3.7 and is continued in Chapter 15.

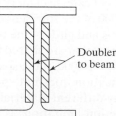

Doubler plates welded to beam web

FIGURE 10.5

Increasing shear strength of beam by using doubler plates.

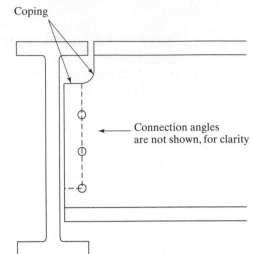

Coping

Connection angles
are not shown, for clarity

FIGURE 10.6

Block shear failure possible along dashed line.

10.3   DEFLECTIONS

The deflections of steel beams are usually limited to certain maximum values. Among the several excellent reasons for deflection limitations are the following:

1. Excessive deflections may damage other materials attached to or supported by the beam in question. Plaster cracks caused by large ceiling joist deflections are one example.
2. The appearance of structures is often damaged by excessive deflections.
3. Extreme deflections do not inspire confidence in the persons using a structure, although the structure may be completely safe from a strength standpoint.
4. It may be necessary for several different beams supporting the same loads to deflect equal amounts.

Standard American practice for buildings has been to limit service live-load deflections to approximately 1/360 of the span length. This deflection is supposedly the largest value that ceiling joists can deflect without causing cracks in underlying plaster. The 1/360 deflection is only one of many maximum deflection values in use because of different loading situations, different engineers, and different specifications. For situations where precise and delicate machinery is supported, maximum deflections may be limited to 1/1500 or 1/2000 of the span lengths. The 2010 AASHTO Specifications limit deflections in steel beams and girders due to live load and impact to 1/800 of the span. (For bridges in urban areas that are shared by pedestrians, the AASHTO *recommends* a maximum value equal to 1/1000 of the span lengths.)

The AISC Specification does not specify exact maximum permissible deflections. There are so many different materials, types of structures, and loadings that no one single set of deflection limitations is acceptable for all cases. Thus, limitations must be set by the individual designer on the basis of his or her experience and judgment.

The reader should note that deflection limitations fall in the serviceability area. Therefore, deflections are determined for service loads, and thus the calculations are identical for both LRFD and ASD designs.

Before substituting blindly into a formula that will give the deflection of a beam for a certain loading condition, the student should thoroughly understand the theoretical methods of calculating deflections. These methods include the moment area, conjugate beam, and virtual-work procedures. From these methods, various expressions can be determined, such as the following common one for the center line deflection of a uniformly loaded simple beam:

$$\Delta_{\mathfrak{C}} = \frac{5wL^4}{384EI}$$

To use deflection expressions such as this one, the reader must be very careful to apply consistent units. Example 10-3 illustrates the application of the preceding expression. The author has changed all units to pounds and inches. Thus, the uniform load given in the problem as so many kips per foot is changed to so many lb/in.

Example 10-3

A W24 × 55 ($I_x = 1350$ in) has been selected for a 21-ft simple span to support a total service live load of 3 k/ft (including beam weight). Is the center line deflection of this section satisfactory for the service live load if the maximum permissible value is 1/360 of the span?

Solution. Use $E = 29 \times 10^6$ lb/in^2

$$\Delta_{\mathfrak{C}} = \frac{5wL^4}{384EI} = \frac{(5)(3000/12)(12 \times 21)^4}{(384)(29 \times 10^6)(1350)} = 0.335 \text{ in total load deflection}$$

$$< \left(\frac{1}{360}\right)(12 \times 21) = 0.70 \text{ in} \qquad\qquad\qquad \text{OK}$$

Another way many engineers use to account for units in the deflection calculation is to keep the uniform load, w, in units of k/ft and span, L, in units of ft, and then convert units from ft to in by multiplying by 1728 (i.e. 12 × 12 × 12). In this method, E has units of k/in^2 (i.e. 29,000).

On page 3-7 in the AISC Manual, the following simple formula for determining maximum beam deflections for W, M, HP, S, C, and MC sections for several different loading conditions is presented:

$$\Delta = \frac{ML^2}{C_1 I_x}$$

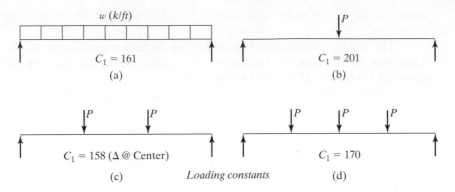

FIGURE 10.7

Values of constant C_1 for use in deflection expression. (Figure 3-2 in AISC Manual.)

In this expression, M is the maximum service load moment in ft-k, and it can be based on the 4 different loading conditions from Fig. 10.7; C_1 is a constant whose value can be determined from Fig. 10.7; L is the span length (ft); and I_x is the moment of inertia (in^4).

If we want to use this expression for the beam of Example 10-3, C_1 from part (a) of Fig. 10.7 would be 161, the center line bending moment would be $wL^2/8 = (3)(21)^2/8 = 165.375$ ft-k, and the center line deflection for the live service load would be

$$\Delta_{\substack{\text{\large C}}} = \frac{(165.375)(21)^2}{(161)(1350)} = 0.336 \text{ in}$$

Table 1604.3 of IBC 2009 presents maximum permissible deflections for quite a few types of members and loading conditions. Several of these are presented in Table 10.1. These values are not applicable to ponding situations.

Some specifications handle the deflection problem by requiring certain minimum depth–span ratios. For example, the AASHTO suggests the depth–span ratio be limited to a minimum value of 1/25. A shallower section is permitted, but it should have sufficient stiffness to prevent a deflection greater than would have occurred if the 1/25 ratio had been used.

A steel beam can be cold-bent, or cambered, an amount equal to the deflection caused by dead load, or the deflection caused by dead load plus some percentage of the live load. Approximately 25 percent of the camber so produced is elastic and will disappear when the cambering operation is completed. It should be remembered that a beam which is bent upward looks much stronger and safer than one that sags downward (even a very small distance).

A camber requirement is quite common for longer steel beams. In fact, a rather large percentage of the beams used in composite construction today (see Chapter 16) are cambered. On many occasions, however, it is more economical to select heavier

TABLE 10.1   Deflection Limits from IBC 2009

Members	Loading conditions		
	L	D + L	S or W
For floor members	$\dfrac{L}{360}$	$\dfrac{L}{240}$	—
For roof members supporting plaster ceiling*	$\dfrac{L}{360}$	$\dfrac{L}{240}$	$\dfrac{L}{360}$
For roof members supporting nonplaster ceilings*	$\dfrac{L}{240}$	$\dfrac{L}{180}$	$\dfrac{L}{240}$
For roof members not supporting ceilings*	$\dfrac{L}{180}$	$\dfrac{L}{120}$	$\dfrac{L}{180}$
*All roof members should be investigated for ponding.			

beams with their larger moments of inertia so as to reduce deflections and avoid the labor costs involved in cambering. One commonly used rule of thumb is that it takes about one extra man-work-hour to camber each beam.

Cambering is something of a nuisance to many fabricators, and it may very well introduce some additional problems. For instance, when beams are cambered, it may be necessary to adjust the connection detail so that proper fitting of members is achieved. The end of a cambered beam will be rotated, and thus it may be necessary to rotate the connection details by the same amount to insure proper fitting.

If we can move up one or two sections in weight, thus reducing deflections so that cambering is not needed, we may have a very desirable solution. Similarly, if a higher-strength steel is being used, it may be be desirable to switch the beams that need cambering to a lower yield stress steel. The results will be larger beams, but smaller deflections, and perhaps some economy will be achieved.

The most economical way to camber beams is with a mechanical press. However, beams less than about 24 ft in length may not fit into a standard press. Such a situation may require the use of heat to carry out the cambering, but expenses will increase by two or three or more times. For this reason, it is almost always wise for the sizes of such short members to be increased until cambering is not needed.

If necessary for a particular project, a member may be bent or cambered about its horizontal axis. This is called *sweep*.

Deflections may very well control the sizes of beams for longer spans, or for short ones where deflection limitations are severe. To assist the designer in selecting sections where deflections may control, the AISC Manual includes a set of tables numbered 3-3 and entitled "W-shapes Selection by I_x," in which the I_x values are given in numerically descending order for the sections normally used as beams. In this table, the sections are arranged in groups, with the lightest section in each group printed in roman type. Example 10-4 presents the design of a beam where deflections control the design.

Example 10-4

Using the LRFD and ASD methods, select the lightest available section with $F_y = 50$ ksi to support a service dead load of 1.2 k/ft and a service live load of 3 k/ft for a 30-ft simple span. The section is to have full lateral bracing for its compression flange, and the maximum total service load deflection is not to exceed 1/1500 the span length.

Solution. After some scratch work, assume that beam wt = 167 lb/ft

LRFD	ASD
$w_u = 1.2(1.2 + 0.167) + (1.6)(3) = 6.44$ klf	$w_a = (1.2 + 0.167) + 3 = 4.37$ k/ft
$M_u = \dfrac{(6.44 \text{ k/ft})(30 \text{ ft})^2}{8} = 724.5$ ft-k	$M_a = \dfrac{(4.37 \text{ k/ft})(30 \text{ ft})^2}{8} = 491.6$ ft-k
From AISC Table 3-2, try W24 × 76 ($I_x = 2100$ in⁴)	From AISC Table 3-2, try W24 × 76
Maximum permissible $\Delta = \left(\dfrac{1}{1500}\right)(12 \times 30) = 0.24$ in	
Actual $\Delta = \dfrac{ML^2}{C_1 I_x}$	All other calculations same as LRFD
$M = M_a = M_{\text{service}} = \dfrac{(4.37 \text{ k/ft})(30 \text{ ft})^2}{8}$	
$= 491.6$ ft-k	
$\Delta = \dfrac{(491.6)(30)^2}{(161)(2100)} = 1.31$ in > 0.24 in **N.G.**	
Min I_x required to limit	
Δ to 0.24 in	
$= \left(\dfrac{1.31}{0.24}\right)(2100) = 11{,}463$ in⁴	
From AISC Table 3-3	
Use W40 × 167. ($I_x = 11{,}600$ in⁴)	**Use W40 × 167.**

10.3.1 Vibrations

Though steel members may be selected that are satisfactory as to moment, shear, deflections, and so on, some very annoying floor vibrations may still occur. This is perhaps the most common serviceability problem faced by designers. Objectionable vibrations will frequently occur where long spans and large open floors without partitions or other items that might provide suitable damping are used. The reader may often have noticed this situation in the floors of large malls.

Damping of vibrations may be achieved by using framed-in-place partitions, each attached to the floor system in at least three places, or by installing "false" sheetrock partitions between ceilings and the underside of floor slabs. Further damping may be achieved by the thickness of the floor slabs, the stability of partitions, and by the weight of office furniture and perhaps the equipment used in the building. A better procedure is to control the stiffness of the structural system.[1, 2] (A rather common practice in the past has been to try to limit vibrations by selecting beams no shallower than 1/20 times span lengths.)

If the occupants of a building feel uneasy about or are annoyed by vibrations, the design is unsuccessful. It is rather difficult to correct a situation of this type in an existing structure. On the other hand, the situation can be easily predicted and corrected in the design stage. Several good procedures have been developed that enable the structural designer to estimate the acceptability of a given system by its users.

F. J. Hatfield[3] has prepared a useful chart for estimating the perceptibility of vibrations of steel beams and concrete slabs for both office and residential buildings.

10.3.2   Ponding

If water on a flat roof accumulates faster than it runs off, the increased load causes the roof to deflect into a dish shape that can hold more water, which causes greater deflections, and so on. This process of *ponding* continues until equilibrium is reached or until collapse occurs. Ponding is a serious matter, as illustrated by the large annual number of flat-roof failures in the United States.

Ponding will occur on almost any flat roof to a certain degree, even though roof drains are present. Drains may be inadequate during severe storms, or they may become stopped up. Furthermore, they are often placed along the beam lines, which are actually the high points of the roof. The best method of preventing ponding is to have an appreciable slope on the roof (1/4 in/ft or more), together with good drainage facilities. It has been estimated that probably two-thirds of the flat roofs in the United States have slopes less than this value, which is the minimum recommended by the National Roofing Contractors Association (NRCA). It costs approximately 3 to 6 percent more to construct a roof with this desired slope than to build with no slope.[4]

When a very large flat roof (perhaps an acre or more) is being considered, the effect of wind on water depth may be quite important. A heavy rainstorm will frequently be accompanied by heavy winds. When a large quantity of water is present on the roof, a strong wind may very well push a great deal of water to one end, creating a dangerous depth of water in terms of the load in pounds per square foot applied to the roof.

[1]T.M. Murray, "Controlling Floor Movement." *Modern Steel Construction* (AISC, Chicago, IL, June 1991) pp. 17–19.

[2]T.M. Murray, "Acceptability Criterion for Occupant-Induced Floor Vibrations," *Engineering Journal*, AISC, 18, 2 (2nd Quarter, 1981), pp. 62–70.

[3]F. J. Hatfield, "Design Chart for Vibration of Office and Residential Floors," *Engineering Journal*, AISC, 29, 4 (4th Quarter, 1992), pp. 141–144.

[4]Gary Van Ryzin, "Roof Design: Avoid Ponding by Sloping to Drain," *Civil Engineering* (New York: ASCE, January 1980), pp. 77–81.

For such situations, *scuppers* are sometimes used. These are large holes, or tubes, in the walls or parapets that enable water above a certain depth to quickly drain off the roof.

Ponding failures will be prevented if the roof system (consisting of the roof deck and supporting beams and girders) has sufficient stiffness. The AISC Specification (Appendix 2) describes a minimum stiffness to be achieved if ponding failures are to be prevented. If this minimum stiffness is not provided, it is necessary to make other investigations to be sure that a ponding failure is not possible.

Theoretical calculations for ponding are very complicated. The AISC requirements are based on work by F. J. Marino,[5] in which he considered the interaction of a two-way system of secondary members, or submembers, supported by a primary system of main members, or girders. A numerical ponding example is presented in Appendix E of this book.

10.4   WEBS AND FLANGES WITH CONCENTRATED LOADS

When steel members have concentrated loads applied that are perpendicular to one flange and symmetric to the web, their flanges and webs must have sufficient flange and web design strength in the areas of flange bending, web yielding, web crippling, and sidesway web buckling. Should a member have concentrated loads applied to both flanges, it must have a sufficient web design strength in the areas of web yielding, web crippling, and column web buckling. In this section, formulas for determining strengths in these areas are presented.

If flange and web strengths do not satisfy the requirements of AISC Specification Section J.10, it will be necessary to use transverse stiffeners at the concentrated loads. Should web design strengths not satisfy the requirements of AISC Specification J.10 it will be necessary to use doubler plates or diagonal stiffeners, as described herein. These situations are discussed in the paragraphs to follow.

10.4.1   Local Flange Bending

The flange must be sufficiently rigid so that it will not deform and cause a zone of high stress concentrated in the weld in line with the web. The nominal tensile load that may be applied through a plate welded to the flange of a W section is to be determined by the expression to follow, in which F_{yf} is the specified minimum yield stress of the flange (ksi) and t_f is the flange thickness (in):

$$R_n = 6.25t_f^2 F_{yf} \qquad \text{(AISC Equation J10-1)}$$

$$\phi = 0.90\,(\text{LRFD}) \quad \Omega = 1.67\,(\text{ASD})$$

It is not necessary to check this formula if the length of loading across the beam flange is less than 0.15 times the flange width b_f or if a pair of half-depth or deeper web stiffeners are provided. Fig. 10.9(a) shows a beam with local flange bending.

---

[5]F. J. Marino, "Ponding of Two-Way Roof System," *Engineering Journal* (New York: AISC, July 1966), pp. 93–100.

10.4.2 Local Web Yielding

The subject of local web yielding applies to all concentrated forces, tensile or compressive. Here we will try to limit the stress in the web of a member in which a force is being transmitted. Local web yielding is illustrated in part (b) of Fig. 10.9.

The nominal strength of the web of a beam at the web toe of the fillet when a concentrated load or reaction is applied is to be determined by one of the following two expressions, in which k is the distance from the outer edge of the flange to the web toe of the fillet, l_b is the length of bearing (in) of the force parallel to the plane of the web, F_{yw} is the specified minimum yield stress (ksi) of the web, and t_w is the thickness of the web:

If the force is a concentrated load or reaction that causes tension or compression and is applied at a distance greater than the member depth, d, from the end of the member, then

$$R_n = (5k + l_b)F_{yw}t_w \qquad \text{(AISC Equation J10-2)}$$

$$\phi = 1.00 \,(\text{LRFD}) \quad \Omega = 1.50 \,(\text{ASD})$$

If the force is a concentrated load or reaction applied at a distance d or less from the member end, then

$$R_n = (2.5k + l_b)F_{yw}t_w \qquad \text{(AISC Equation J10-3)}$$

$$\phi = 1.00 \,(\text{LRFD}) \quad \Omega = 1.50 \,(\text{ASD})$$

This beam is to be subjected to heavy concentrated loads. Full depth web stiffeners are used to prevent web crippling and to keep the top flange from twisting or warping at the load. (Courtesy of CMC South Carolina Steel.)

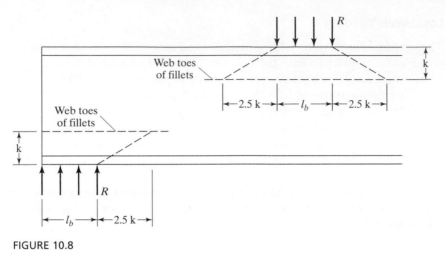

FIGURE 10.8

Local web yielding.

Reference to Fig. 10.8 clearly shows where these expressions were obtained. The nominal strength R_n equals the length over which the force is assumed to be spread when it reaches the web toe of the fillet times the web thickness times the yield stress of the web. Should a stiffener extending for at least half the member depth or a doubler plate be provided on each side of the web at the concentrated force, it is not necessary to check for web yielding.

10.4.3 Web Crippling

Should concentrated compressive loads be applied to a member with an unstiffened web (the load being applied in the plane of the web), the nominal web crippling strength of the web is to be determined by the appropriate equation of the two that follow (in which d is the overall depth of the member). If one or two web stiffeners or one or two doubler plates are provided and extend for at least half of the web depth, web crippling will not have to be checked. Research has shown that when web crippling occurs, it is located in the part of the web adjacent to the loaded flange. Thus, it is thought that stiffening the web in this area for half its depth will prevent the problem. Web crippling is illustrated in part (c) of Fig. 10.9.

If the concentrated load is applied at a distance greater than or equal to $d/2$ from the end of the member, then

$$R_n = 0.80t_w^2 \left[1 + 3\left(\frac{l_b}{d}\right)\left(\frac{t_w}{t_f}\right)^{1.5} \right] \sqrt{\frac{EF_{yw}t_f}{t_w}} \quad \text{(AISC Equation J10-4)}$$

$$\phi = 0.75 \text{ (LRFD)} \quad \Omega = 2.00 \text{ (ASD)}$$

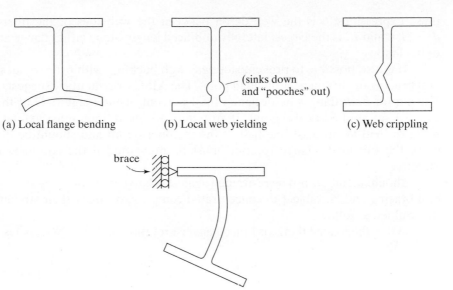

(a) Local flange bending   (b) Local web yielding   (c) Web crippling

(sinks down
and "pooches" out)

(d) Sidesway web buckling

FIGURE 10.9

If the concentrated load is applied at a distance less than $d/2$ from the end of the member, then

for $\dfrac{l_b}{d} \leq 0.2,$

$$R_n = 0.40t_w^2\left[1 + 3\left(\frac{l_b}{d}\right)\left(\frac{t_w}{t_f}\right)^{1.5}\right]\sqrt{\frac{EF_{yw}t_f}{t_w}}$$ (AISC Equation J10-5a)

$\phi = 0.75$ (LRFD) $\Omega = 2.00$ (ASD)

For $\dfrac{l_b}{d} > 0.2,$

$$R_n = 0.40t_w^2\left[1 + \left(4\frac{l_b}{d} - 0.2\right)\left(\frac{t_w}{t_f}\right)^{1.5}\right]\sqrt{\frac{EF_{yw}t_f}{t_w}}$$ (AISC Equation J10-5b)

$\phi = 0.75$ (LRFD) $\Omega = 2.00$ (ASD)

10.4.4 Sidesway Web Buckling

Should compressive loads be applied to laterally braced compression flanges, the web will be put in compression and the tension flange may buckle, as shown in Fig. 10-9(d).

It has been found that sidesway web buckling will not occur if the compression flange is restrained against rotation, with $(h/t_w)/(L_b/b_f) > 2.3$, or if $(h/t_w)/(L_b/b_f) > 1.7$ when the compression flange rotation is not restrained about its longitudinal axis. In these expressions, h is the web depth between the web toes of the fillets—that is, $d - 2k$—and L_b is the largest laterally unbraced length along either flange at the point of the load.

It is also possible to prevent sidesway web buckling with properly designed lateral bracing or stiffeners at the load point. The AISC Commentary suggests that local bracing for *both* flanges be designed for 1 percent of the magnitude of the concentrated load applied at the point. If stiffeners are used, they must extend from the point of load for at least one-half of the member depth and should be designed to carry the full load. Flange rotation must be prevented if the stiffeners are to be effective.

Should members not be restrained against relative movement by stiffeners or lateral bracing and be subject to concentrated compressive loads, their strength may be determined as follows:

When the loaded flange is braced against rotation and $(h/t_w)/(L_b/b_f)$ is ≤ 2.3,

$$R_n = \frac{C_r t_w^3 t_f}{h^2}\left[1 + 0.4\left(\frac{h/t_w}{L_b/b_f}\right)^3\right] \qquad \text{(AISC Equation J10-6)}$$

$$\phi = 0.85 \text{ (LRFD)} \quad \Omega = 1.76 \text{ (ASD)}$$

C&W Warehouse, Spartanburg, SC. (Courtesy of Britt, Peters and Associates.)

When the loaded flange is not restrained against rotation and $\dfrac{h/t_w}{L_b/b_f} \leq 1.7$,

$$R_n = \frac{C_r t_w^3 t_f}{h^2}\left[0.4\left(\frac{h/t_w}{L_b/b_f}\right)^3\right] \qquad \text{(AISC Equation J10-7)}$$

$$\phi = 0.85 \text{ (LRFD)} \quad \Omega = 1.76 \text{ (ASD)}$$

It is not necessary to check Equations J10-6 and J10-7 if the webs are subject to distributed load. Furthermore, these equations were developed for bearing connections and *do not apply to moment connections*. In these expressions,

$C_r = 960{,}000$ ksi when $M_u < M_y$ (LRFD) or $1.5M_a < M_y$ (ASD) at the location of the force, ksi.

$C_r = 480{,}000$ ksi when $M_u \geq M_y$ (LRFD) or $1.5M_a \geq M_y$ (ASD) at the location of the force, ksi.

10.4.5   Compression Buckling of the Web

This limit state relates to concentrated compression loads applied to both flanges of a member, such as, moment connections applied to both ends of a column. For such a situation, it is necessary to limit the slenderness ratio of the web to avoid the possibility of buckling. Should the concentrated loads be larger than the value of ϕR_n given in the next equation, it will be necessary to provide either one stiffener, a pair of stiffeners, or a doubler plate, extending for the full depth of the web and meeting the requirements of AISC Specification J10.8. (The equation to follow is applicable to moment connections, but not to bearing ones.)

$$R_n = \frac{24\, t_w^3 \sqrt{EF_{yw}}}{h} \qquad \text{(AISC Equation J10-8)}$$

$$\phi = 0.90 \text{ (LRFD)} \quad \Omega = 1.67 \text{ (ASD)}$$

If the concentrated forces to be resisted are applied at a distance from the member end that is less than $d/2$, then the value of R_n shall be reduced by 50 percent.

Example 10-5, which follows, presents the review of a beam for the applicable items discussed in this section.

Example 10-5

A W21 × 44 has been selected for moment in the beam shown in Fig. 10.10. Lateral bracing is provided for both flanges at beam ends and at concentrated loads. If the end bearing length is 3.50 in and the concentrated load bearing lengths are each 3.00 in, check the beam for web yielding, web crippling, and sidesway web buckling.

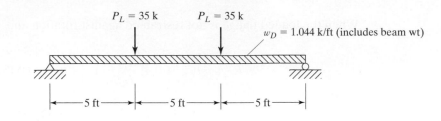

FIGURE 10.10

Solution

Using a W21 × 44 (d = 20.7 in, b_f = 6.50 in, t_w = 0.350 in, t_f = 0.450 in, k = 0.950 in)

LRFD	ASD
End reaction	End reaction
$R_u = (1.2)(1.044 \text{ k/ft})\left(\dfrac{15 \text{ ft}}{2}\right) + (1.6)(35 \text{ k})$	$R_a = (1.044 \text{ k/ft})\left(\dfrac{15 \text{ ft}}{2}\right) + 35 \text{ k}$
$= 65.4 \text{ k}$	$= 42.83 \text{ k}$
Concentrated load	Concentrated load
$P_u = (1.6)(35 \text{ k}) = 56 \text{ k}$	$P_a = 35 \text{ k}$

Local web yielding

(l_b = bearing length of reactions = 3.50 in, for concentrated loads l_b = 3.00 in)

At end reactions (AISC Equation J10-3)

$$R_n = (2.5 k + l_b)F_{yw}t_w = (2.5 \times 0.950 \text{ in} + 3.50 \text{ in})(50 \text{ ksi})(0.350 \text{ in}) = 102.8 \text{ k}$$

LRFD ϕ = 1.00	ASD Ω = 1.50
$\phi R_n = (1.00)(102.8) = 102.8 \text{ k}$	$\dfrac{R_n}{\Omega} = \dfrac{102.8}{1.50} = 68.5 \text{ k}$
> 65.4 k **OK**	> 42.83 k **OK**

At concentrated loads (AISC Equation J10-2)

$$R_n = (5\,k + l_b)F_{yw}t_w = (5 \times 0.950\text{ in} + 3.00\text{ in})(50\text{ ksi})(0.350\text{ in}) = 135.6\text{ k}$$

LRFD $\phi = 1.00$	ASD $\Omega = 1.50$
$\phi R_n = (1.00)(135.6) = 135.6\text{ k}$	$\dfrac{R_n}{\Omega} = \dfrac{135.6}{1.50} = 90.4\text{ k}$
$> 56\text{ k}$ **OK**	$> 35\text{ k}$ **OK**

Web crippling

At end reactions (AISC Equation J10-5a) since $\dfrac{l_b}{d} \leq 0.20$

$$\frac{l_b}{d} = \frac{3.5}{20.7} = 0.169 < 0.20$$

$$R_n = 0.40t_w^2\left[1 + 3\left(\frac{l_b}{d}\right)\left(\frac{t_w}{t_f}\right)^{1.5}\right]\sqrt{\frac{EF_{yw}t_f}{t_w}}$$

$$= (0.40)(0.350\text{ in})^2\left[1 + 3\left(\frac{3.5\text{ in}}{20.7\text{ in}}\right)\left(\frac{0.350\text{ in}}{0.450\text{ in}}\right)^{1.5}\right]$$

$$\sqrt{\frac{(29 \times 10^3\text{ ksi})(50\text{ ksi})(0.450\text{ in})}{0.350\text{ in}}}$$

$$= 90.3\text{ k}$$

LRFD $\phi = 0.75$	ASD $\Omega = 2.00$
$\phi R_n = (0.75)(90.3) = 67.7\text{ k}$	$\dfrac{R_n}{\Omega} = \dfrac{90.3}{2.00} = 45.1\text{ k}$
$> 65.4\text{ k}$ **OK**	$> 42.83\text{ k}$ **OK**

At concentrated loads (AISC Equation J10-4)

$$R_n = 0.80\,t_w^2\left[1 + 3\left(\frac{l_b}{d}\right)\left(\frac{t_w}{t_f}\right)^{1.5}\right]\sqrt{\frac{EF_{yw}t_f}{t_w}}$$

$$= (0.80)(0.350)^2\left[1 + 3\left(\frac{3.0}{20.7}\right)\left(\frac{0.350}{0.450}\right)^{1.5}\right]\sqrt{\frac{(29 \times 10^3)(50)(0.450)}{0.350}}$$

$$= 173.7\text{ k}$$

LRFD $\phi = 0.75$	ASD $\Omega = 2.00$
$\phi R_n = (0.75)(173.7)$	$\dfrac{R_n}{\Omega} = \dfrac{173.7}{2.00} = 86.8 \text{ k}$
$= 130.3 \text{ k} > 56 \text{ k}$ **OK**	$> 35 \text{ k}$ **OK**

Sidesway web buckling

The compression flange is restrained against rotation.

$$\frac{h}{t_w} \Big/ \frac{L_b}{b_f} = \frac{20.7 \text{ in} - 2 \times 0.950 \text{ in}}{0.350 \text{ in}} \Big/ \left(\frac{12 \text{ in/ft} \times 5 \text{ ft}}{6.50 \text{ in}} \right) = 5.82 > 2.3$$

$\therefore$ **Sidesway web buckling does not have to be checked.**

The preceding calculations can be appreciably shortened if use is made of the Manual tables numbered 9-4 and entitled "Beam Bearing Constants." In those tables, values are shown for ϕR_1, ϕR_2, ϕR_3, R_1/Ω, R_2/Ω, R_3/Ω, and so on. The values given represent parts of the equations used for checking web yielding and web crippling and are defined on page 9-19 in the Manual.

Instructions for use of the tables are provided on pages 9-19 and 9-20 of the Manual. The expressions for local web yielding at beam ends of a W21 × 44 are written next making use of the table values. Then, those expressions and the table values are used to check the previous calculations.

LRFD	ASD
$\phi R_n = \phi R_1 + \phi l_b R_2$	$\dfrac{R_n}{\Omega} = \left(\dfrac{R_1}{\Omega} \right) + l_b \left(\dfrac{R_2}{\Omega} \right)$
$= 41.6 + (3.5)(17.5)$	$= 27.7 + (3.5)(11.7)$
$= 102.8 \text{ k}$ **OK**	$= 68.6 \text{ k}$ **OK**

10.5   UNSYMMETRICAL BENDING

From mechanics of materials, it should be remembered that each beam cross section has a pair of mutually perpendicular axes known as the principal axes for which the product of inertia is zero. Bending that occurs about any axis other than one of the principal axes is said to be unsymmetrical bending. When the external loads are not in a plane with either of the principal axes, or when loads are simultaneously applied to the beam from two or more directions, unsymmetrical bending results.

If a load is not perpendicular to one of the principal axes, it may be broken into components that are perpendicular to those axes and to the moments about each axis, M_{ux} and M_{uy}, or M_{ax} and M_{ay}, determined as shown in Fig. 10.11.

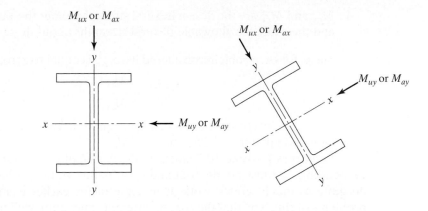

FIGURE 10.11

When a section has one axis of symmetry, that axis is one of the principal axes, and the calculations necessary for determining the moments are quite simple. For this reason, unsymmetrical bending is not difficult to handle in the usual beam section, which is probably a W, S, M, or C. Each of these sections has at least one axis of symmetry, and the calculations are appreciably reduced. A further simplifying factor is that the loads are usually gravity loads and probably perpendicular to the x axis.

Among the beams that must resist unsymmetrical bending are crane girders in industrial buildings and purlins for ordinary roof trusses. The x axes of purlins are parallel to the sloping roof surface, while the large percentage of their loads (roofing, snow, etc.) are gravity loads. These loads do not lie in a plane with either of the principal axes of the inclined purlins, and the result is unsymmetrical bending. Wind loads are generally considered to act perpendicular to the roof surface and thus perpendicular to the x axes of the purlins, with the result that they are not considered to cause unsymmetrical bending. The x axes of crane girders are usually horizontal, but the girders are subjected to lateral thrust loads from the moving cranes, as well as to gravity loads.

To check the adequacy of members bent about both axes simultaneously, the AISC provides an equation in Section H1 of their Specification. The equation that follows is for combined bending and axial force, if $P_r/P_c < 0.2$:

$$\frac{P_r}{2P_c} + \left(\frac{M_{rx}}{M_{cx}} + \frac{M_{ry}}{M_{cy}} \right) \leq 1.0 \qquad \text{(AISC Equation H1-1b)}$$

Here, the values are defined as follows:

1. P_r is the required axial strength under LRFD or the required allowable axial strength with ASD.
2. P_c is the available axial strength with LRFD or the available allowable axial strength with ASD.
3. M_{rx} and M_{ry} are the required design flexural strengths about the x and y axes under LRFD and the required allowable flexural strengths with ASD.

4. M_{cx} and M_{cy} are the design flexural strengths about the x and y axes using LRFD and the available allowable flexural strengths about those axes using ASD.

Since, for the problem considered here, P_r is equal to zero, the equation becomes

$$\frac{M_{rx}}{M_{cx}} + \frac{M_{ry}}{M_{cy}} \leq 1.0$$

This is an interaction, or percentage, equation. If M_{rx} is 75 percent of M_{cx}, then M_{ry} can be no greater than 25 percent of M_{cy}.

Examples 10-6 and 10-7 illustrate the design of beams subjected to unsymmetrical bending. To illustrate the trial-and-error nature of the problem, the author did not do quite as much scratch work as in some of his earlier examples. The first design problems of this type that the student attempts may quite well take several trials. Consideration needs to be given to the question of lateral support for the compression flange. Should the lateral support be of questionable nature, the engineer should reduce the design moment resistance by means of one of the expressions previously given for that purpose.

Example 10-6

A steel beam in its upright position must resist the following service moments: $M_{Dx} = 60$ ft-k, $M_{Lx} = 100$ ft-k, $M_{Dy} = 15$ ft-k, and $M_{Ly} = 25$ ft-k. These moments include the effects of the estimated beam weight. The loads are assumed to pass through the centroid of the section. Select a W24 shape of 50 ksi steel that can resist these moments, assuming full lateral support for the compression flange.

Solution

Try a W24 × 62 ($\phi_b M_{px} = 574$ ft-k, $\dfrac{M_{px}}{\Omega_b} = 382$ ft-k, $Z_y = 15.7$ in^3)

$$\phi_b M_{py} = \phi_b F_y Z_y = \frac{(0.9)(50 \text{ ksi})(15.7 \text{ in}^3)}{12 \text{ in/ft}} = 58.8 \text{ ft-k}, \quad \frac{M_{py}}{\Omega_b} = \frac{F_y Z_y}{\Omega_b}$$

$$= \frac{(50 \text{ ksi})(15.7 \text{ in}^3)}{(12 \text{ in/ft})(1.67)} = 39.1 \text{ ft-k}$$

LRFD	ASD
$M_{ux} = (1.2)(60) + (1.6)(100) = 232$ ft-k	$M_{ax} = 60 + 100 = 160$ ft-k
$M_{uy} = (1.2)(15) + (1.6)(25) = 58$ ft-k	$M_{ay} = 15 + 25 = 40$ ft-k
$\dfrac{M_{rx}}{M_{cx}} + \dfrac{M_{ry}}{M_{cy}} \leq 1.0$	$\dfrac{M_{rx}}{M_{cx}} + \dfrac{M_{ry}}{M_{cy}} \leq 1.0$
$\dfrac{232}{574} + \dfrac{58}{58.8} = 1.39 > 1.0$ **N.G.**	$\dfrac{160}{382} + \dfrac{40}{39.1} = 1.44 > 1.0$ **N.G.**

Try a W24 × 68 $(\phi_b M_{px} = 664 \text{ ft-k}, \dfrac{M_{px}}{\Omega_b} = 442 \text{ ft-k}, Z_y = 24.5 \text{ in}^3)$

$$\phi_b M_{py} = \frac{(0.9)(50 \text{ ksi})(24.5 \text{ in}^3)}{12 \text{ in/ft}} = 91.9 \text{ ft-k}, \quad \frac{M_{py}}{\Omega_b} = \frac{(50 \text{ ksi})(24.5 \text{ in}^3)}{(12 \text{ in/ft})(1.67)} = 61.1 \text{ ft-k}$$

LRFD	ASD
$\dfrac{232}{664} + \dfrac{58}{91.9} = 0.98$	$\dfrac{160}{442} + \dfrac{40}{61.1} = 1.02$
< 1.0 **OK**	> 1.00 **N.G.**
Use W24 × 68.	**Try a larger section, W24 × 76.**

It should be noted in the solution for Example 10-6 that, although the procedure used will yield a section that will adequately support the moments given, the selection of the absolutely lightest section listed in the AISC Manual could be quite lengthy because of the two variables Z_x and Z_y, which affect the size. If we have a large M_{ux} and a small M_{uy}, the most economical section will probably be quite deep and rather narrow, whereas if we have a large M_{uy} in proportion to M_{ux}, the most economical section may be rather wide and shallow.

10.6   DESIGN OF PURLINS

To avoid bending in the top chords of roof trusses, it is theoretically desirable to place purlins only at panel points. For large trusses, however, it is more economical to space them at closer intervals. If this practice is not followed for large trusses, the purlin sizes may become so large as to be impractical. When intermediate purlins are used, the top chords of the truss should be designed for bending as well as for axial stress, as described in Chapter 11. Purlins are usually spaced from 2 to 6 ft apart, depending on loading conditions, while their most desirable depth-to-span ratios are probably in the neighborhood of 1/24. Channels or S sections are the most frequently used sections, but on some occasions other shapes may be convenient.

As previously mentioned, the channel and S sections are very weak about their web axes, and sag rods may be necessary to reduce the span lengths for bending about those axes. Sag rods, in effect, make the purlins continuous sections for their y axes, and the moments about these axes are greatly reduced, as shown in Fig. 10.12. These moment diagrams were developed on the assumption that the changes in length of the sag rods are negligible. It is further assumed that the purlins are simply supported at the trusses. This assumption is on the conservative side, since they are often continuous over two or more trusses and appreciable continuity may be achieved at their splices. The student can easily reproduce these diagrams from his or her knowledge of moment distribution or by other analysis methods. In the diagrams, L is the distance between

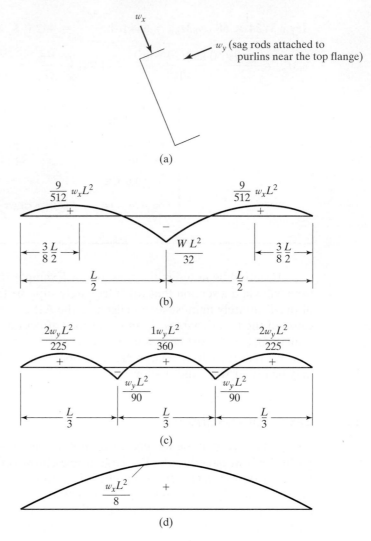

FIGURE 10.12

(a) Channel purlin. (b) Moment about web axis of purlins acting on the upper half of purlin—sag rods at midspan. (c) Moment about web axis of purlins acting on the upper half of purlin—sag rods at one-third points. (d) Moment about x axis of purlin.

trusses, w_{uy} is the load component *perpendicular* to the web axis of the purlin, and w_{ux} is the load component *parallel* to the web axis.

If sag rods were not used, the maximum moment about the web axis of a purlin would be $w_y L^2/8$. When sag rods are used at midspan, this moment is reduced to a maximum of $w_y L^2/32$ (a 75 percent reduction), and when used at one-third points is reduced to a maximum of $w_y L^2/90$ (a 91 percent reduction). In Example 10-7, sag rods are used at the midpoints, and the purlins are designed for a moment of $w_x L^2/8$ parallel to the web axis and $w_y L^2/32$ perpendicular to the web axis.

In addition to being advantageous in reducing moments about the web axes of purlins, sag rods can serve other useful purposes. First, they can provide lateral support for the purlins; second, they are useful in keeping the purlins in proper alignment during erection until the roof deck is installed and connected to the purlins.

Example 10-7

Using both the LRFD and ASD methods, select a W6 purlin for the roof shown in Fig. 10.13. The trusses are 18 ft 6 in on center, and sag rods are used at the midpoints between trusses. Full lateral support is assumed to be supplied from the roof above. Use 50 ksi steel and the AISC Specification. Loads are as follows in terms of pounds per square foot of roof surface:

$$\text{Snow} = 30 \text{ psf}$$
$$\text{Roofing} = 6 \text{ psf}$$
$$\text{Estimated purlin weight} = 3 \text{ psf}$$
$$\text{Wind pressure} = 15 \text{ psf} \perp \text{ to roof surface}$$

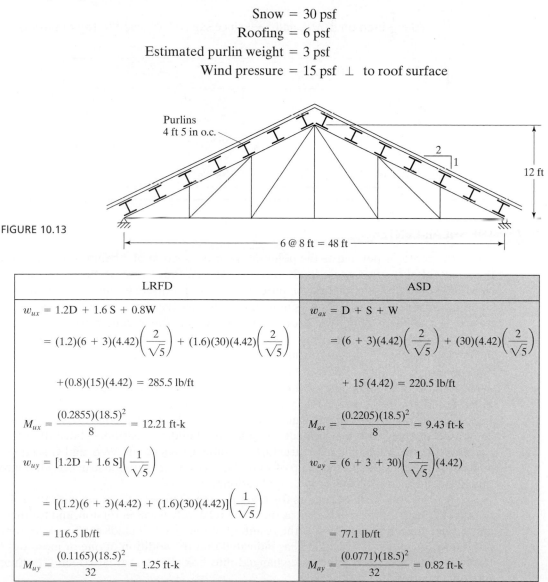

FIGURE 10.13

LRFD	ASD
$w_{ux} = 1.2D + 1.6\,S + 0.8W$	$w_{ax} = D + S + W$
$= (1.2)(6 + 3)(4.42)\left(\dfrac{2}{\sqrt{5}}\right) + (1.6)(30)(4.42)\left(\dfrac{2}{\sqrt{5}}\right)$	$= (6 + 3)(4.42)\left(\dfrac{2}{\sqrt{5}}\right) + (30)(4.42)\left(\dfrac{2}{\sqrt{5}}\right)$
$+ (0.8)(15)(4.42) = 285.5 \text{ lb/ft}$	$+ 15\,(4.42) = 220.5 \text{ lb/ft}$
$M_{ux} = \dfrac{(0.2855)(18.5)^2}{8} = 12.21 \text{ ft-k}$	$M_{ax} = \dfrac{(0.2205)(18.5)^2}{8} = 9.43 \text{ ft-k}$
$w_{uy} = [1.2D + 1.6\,S]\left(\dfrac{1}{\sqrt{5}}\right)$	$w_{ay} = (6 + 3 + 30)\left(\dfrac{1}{\sqrt{5}}\right)(4.42)$
$= [(1.2)(6 + 3)(4.42) + (1.6)(30)(4.42)]\left(\dfrac{1}{\sqrt{5}}\right)$	
$= 116.5 \text{ lb/ft}$	$= 77.1 \text{ lb/ft}$
$M_{uy} = \dfrac{(0.1165)(18.5)^2}{32} = 1.25 \text{ ft-k}$	$M_{ay} = \dfrac{(0.0771)(18.5)^2}{32} = 0.82 \text{ ft-k}$

Solution

$$\text{Try W6} \times 9 \ (Z_x = 6.23 \text{ in}^3, \ \phi_b M_{px} = \frac{(0.9)(50)(6.23)}{12} = 23.36 \text{ ft-k},$$

$$Z_y = 1.72 \text{ in}^3, \ \phi_b M_{py} = \frac{(0.9)(50)(1.72)}{12}\left(\frac{1}{2}\right) = 3.23 \text{ ft-k},$$

$$\frac{M_{px}}{\Omega_b} = \frac{(50)(6.23)}{(12)(1.67)} = 15.54 \text{ ft-k}, \quad \frac{M_{py}}{\Omega_b} = \frac{(50)(1.72)}{(12)(1.67)}\left(\frac{1}{2}\right) = 2.15 \text{ ft-k}$$

$$\text{(the } \frac{1}{2} \text{ used on } \phi_b M_{py} \text{ and } \frac{M_{py}}{\Omega_b} \text{ since sag rod attached to top of purlin).}$$

LRFD	ASD
$\dfrac{M_{rx}}{M_{cx}} + \dfrac{M_{ry}}{M_{cy}} \leq 1.0$	$\dfrac{M_{rx}}{M_{cx}} + \dfrac{M_{ry}}{M_{cy}} \leq 1.0$
$\dfrac{12.21}{23.36} + \dfrac{1.25}{3.23} = 0.910 < 1.00 \ \textbf{OK}$	$\dfrac{9.43}{15.54} + \dfrac{0.82}{2.15} = 0.988 < 1.00 \ \textbf{OK}$
Use W6 × 9.	**Use W6 × 9.**

10.7   THE SHEAR CENTER

The shear center is defined as the point on the cross section of a beam through which the resultant of the transverse loads must pass so that the stresses in the beam may be calculated only from the theories of pure bending and transverse shear. Should the resultant pass through this point, it is unnecessary to analyze the beam for torsional moments. For a beam with two axes of symmetry, the shear center will fall at the intersection of the two axes, thus coinciding with the centroid of the section. For a beam with one axis of symmetry, the shear center will fall somewhere on that axis, but not necessarily at the centroid of the section. This surprising statement means that to avoid torsion in some beams the lines of action of the applied loads and beam reactions should not pass through the centroids of the sections.

The shear center is of particular importance for beams whose cross sections are composed of thin parts that provide considerable bending resistance, but little resistance to torsion. Many common structural members, such as the W, S, and C sections, angles, and various beams made up of thin plates (as in aircraft construction) fall into this class, and the problem has wide application.

The location of shear centers for several open sections are shown with the solid dots (•) in Fig. 10.14. Sections such as these are relatively weak in torsion, and for them and similar shapes, the location of the resultant of the external loads can be a very serious matter. A previous discussion has indicated that the addition of one or more webs to these sections—so that they are changed into box shapes—greatly increases their torsional resistance.

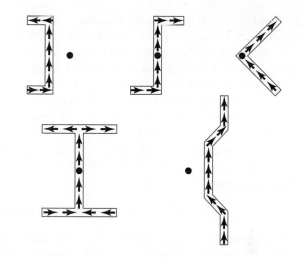

FIGURE 10.14

The average designer probably does not take the time to go through the some-times tedious computations involved in locating the shear center and calculating the effect of twisting. He or she may instead simply ignore the situation, or may make a rough estimate of the effect of torsion on the bending stresses. One very poor, but conservative, estimate practiced occasionally is to reduce $\phi_b M_{ny}$ by 50 percent for use in the interaction equation.

Shear centers can be located quickly for beams with open cross sections and relatively thin webs. For other beams, the shear centers can probably be found, but only with considerable difficulty. The term *shear flow* is often used when reference is made to thin-wall members, although there is really no flowing involved. It refers to the shear per inch of the cross section and equals the unit shearing stress times the thickness of the member. (The unit shearing stress F_v has been determined by the expression VQ/bI, and the shear flow q_v can be determined by VQ/I if the shearing stress is assumed to be constant across the thickness of the section.) The shear flow acts parallel to the sides of each element of a member.

The channel section of Fig. 10.15(a) will be considered for this discussion. In this figure, the shear flow is shown with the small arrows, and in part (b) the values are

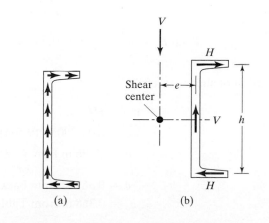

FIGURE 10.15   (a)   (b)

totaled for each component of the shape and labeled H and V. The two H values are in equilibrium horizontally, and the internal V value balances the external shear at the section. Although the horizontal and vertical forces are in equilibrium, the same cannot be said for the moment forces, unless the lines of action of the resultant of the external forces pass through a certain point called the *shear center*. The horizontal H forces in part (b) of the figure can be seen to form a couple. The moment produced by this couple must be opposed by an equal and opposite moment, which can be produced only by the two V values. The location of the shear center is a problem in equilibrium; therefore, moments should be taken about a point that eliminates the largest number of forces possible.

With this information, the following equation can be written from which the shear center can be located:

$$Ve = Hh \text{ (moments taken about c.g. of web)}$$

Examples 10-8 and 10-9 illustrate the calculations involved in locating the shear center for two shapes. Note that the location of the shear center is independent of the value of the external shear. (*Shear center locations are provided for channels in* Table 1-5 *of Part I of the AISC Manual.* Their location is given by e_o in those tables.)

Example 10-8

The C10 × 30 channel section shown in Fig. 10.16(a) is subjected to an external shear of V in the vertical plane. Locate the shear center.

Solution

Properties of section:

For this channel, the flanges are idealized as rectangular elements with a thickness equal to the average flange thickness.

$$I_x = \left(\frac{1}{12}\right)(3)(10)^3 - \left(\frac{1}{12}\right)(2.327)(9.128)^3 = 102.52 \text{ in}^4$$

$$q_v \text{ at } B = \frac{(V)(2.694 \times 0.436 \times 4.782)}{102.52} = 0.05479 \ V/\text{in}$$

$$\text{Total } H = \left(\frac{1}{2}\right)(2.694)(0.05479 \ V) = 0.07380 \ V$$

Location of shear center:

$$Ve = Hh$$
$$Ve = (0.07380 \ V)(9.564)$$
$$e = 0.706 \text{ in from } \mathcal{L} \text{ of web}$$
$$\text{or}$$
$$e = 0.369 \text{ in from back of web}$$
$$e_o = 0.368 \text{ in from Table 1-5 (C10 × 30)}$$

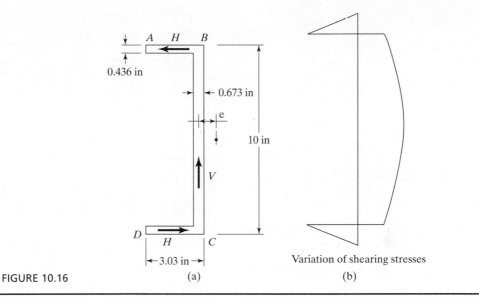

FIGURE 10.16
(a)   (b)

Variation of shearing stresses

The student should clearly understand that shear stress variation across the corners, where the webs and flanges join, cannot be determined correctly with the mechanics of materials expression (VQ/bI or VQ/I for shear flow) and cannot be determined too well, even after a complicated study with the theory of elasticity. As an approximation, in the two examples presented here, we have assumed that shear flow continues up to the middle of the corners on the same pattern (straight-line variation for horizontal members and parabolic for others). The values of Q are computed for the corresponding dimensions. Other assumptions could have been made, such as assuming that a shear flow variation continues vertically for the full depth of webs, and only for the protruding parts of flanges horizontally, or vice versa. It is rather disturbing to find that, whichever assumption is made, the values do not check out perfectly. For instance, in Example 10-9, which follows, the sum of the vertical shear flow values does not check out very well with the external shear.

Example 10-9

The open section of Fig. 10.17 is subjected to an external shear of V in a vertical plane. Locate the shear center.

Solution

Properties of section:

$$I_x = \left(\frac{1}{12}\right)(0.25)(16)^3 + (2)(3.75 \times 0.25)(4.87)^2 = 129.8 \text{ in}^4$$

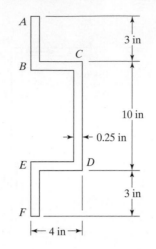

FIGURE 10.17

Values of shear flow (labeled q):

$$q_A = 0$$

$$q_B = \frac{(V)(3.12 \times 0.25 \times 6.44)}{129.8} = 0.0387 \, V/\text{in}$$

$$q_C = q_B + \frac{(V)(3.75 \times 0.25 \times 4.87)}{129.8} = 0.0739 \, V/\text{in}$$

$$q_{\mathcal{L}} = q_C + \frac{(V)(4.87 \times 0.25 \times 2.44)}{129.8} = 0.0968 \, V/\text{in}$$

These shear flow values are shown in Fig. 10.18(a), and the summation for each part of the member is given in part (b) of the figure.

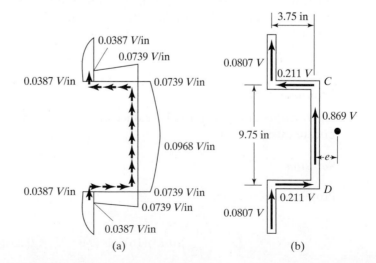

FIGURE 10.18   (a)   (b)

Taking moments about the center line of *CD*

$$-(0.211\ V)(9.75) + (2)(0.0807\ V)(3.75) + Ve = 0$$

$$e = 1.45\text{ in}$$

The theory of the shear center is a very useful one in design, but it has certain limitations that should be clearly understood. For instance, the approximate analysis given in this section is valid only for thin sections. In addition, steel beams often have variable cross sections along their spans, with the result that the loci of the shear centers are not straight lines along those spans. Thus, if the resultant of the loads passes through the shear center at one cross section, it might not do so at other cross sections along the beam.

When designers are faced with the application of loads to thin-walled sections, such that twisting of those sections will be a problem, they will usually provide some means by which the twisting can be constrained. They may specify special bracing at close intervals, or attachments to flooring or roofing, or other similar devices. Should such solutions not be feasible, designers will probably consider selecting sections with greater torsional stiffnesses. Two references on this topic are given here.[6,7]

10.8   BEAM-BEARING PLATES

When the ends of beams are supported by direct bearing on concrete or other masonry construction, it is frequently necessary to distribute the beam reactions over the masonry by means of beam-bearing plates. The reaction is assumed to be spread uniformly through the bearing plate to the masonry, and the masonry is assumed to push up against the plate with a uniform pressure equal to the reaction R_u or R_a over the area of the plate A_1. This pressure tends to curl up the plate and the bottom flange of the beam. The AISC Manual recommends that the bearing plate be considered to take the entire bending moment produced and that the critical section for moment be assumed to be a distance *k* from the center line of the beam (see Fig. 10.19). The distance *k* is the same as the distance from the outer face of the flange to the web toe of the fillet, given in the tables for each section (or it equals the flange thickness plus the fillet radius).

The determination of the true pressure distribution in a beam-bearing plate is a very formidable task, and the uniform pressure distribution assumption is usually made. This assumption is probably on the conservative side, as the pressure is typically larger at the center of the beam than at the edges. The outer edges of the plate and flange tend to bend upward, and the center of the beam tends to go down, concentrating the pressure there.

The required thickness of a 1-in-wide strip of plate can be determined as follows, with reference being made to Fig. 10.19:

$$Z \text{ of a 1-in-wide piece of plate of } t \text{ thickness} = (1)\left(\frac{t}{2}\right)\left(\frac{t}{4}\right)(2) = \frac{t^2}{4}$$

---

[6]C. G. Salmon and J. E. Johnson, *Steel Structures, Design and Behavior*, 4th ed. (New York: Harper & Row, 1996), pp. 430–443.
[7]AISC Engineering Staff, "Torsional Analysis of Steel Members" (Chicago: AISC, 1983).

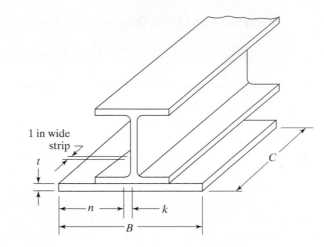

1 in wide
strip

t

n

k

B

C

FIGURE 10.19

The moments M_u and M_a are computed at a distance k from the web center line and are equated, respectively, to $\phi_b F_y Z$ and $F_y Z/\Omega_b$; the resulting equations are then solved for the required plate thickness.

LRFD $\phi_b = 0.90$	ASD $\Omega_b = 1.67$
$M_u = \dfrac{R_u}{A_1} n \left(\dfrac{n}{2} \right) = \dfrac{R_u n^2}{2 A_1}$	$M_a = \dfrac{R_a}{A_1} n \left(\dfrac{n}{2} \right) = \dfrac{R_a n^2}{2 A_1}$
$\dfrac{R_u n^2}{2 A_1} = \phi_b F_y \dfrac{t^2}{4}$	$\dfrac{R_a n^2}{2 A_1} = \dfrac{F_y \dfrac{t^2}{4}}{\Omega_b}$
From which $t_{\text{reqd}} = \sqrt{\dfrac{2 R_u n^2}{\phi_b A_1 F_y}}$	From which $t_{\text{reqd}} = \sqrt{\dfrac{2 R_a n^2 \Omega_b}{A_1 F_y}}$

In the absence of code regulations specifying different values, the design strength for bearing on concrete is to be taken equal to $\phi_c P_p$ or P_p/Ω_a according to AISC Specification J8. This specification states that when a bearing plate extends for the full area of a concrete support, the bearing strength of the concrete can be determined as follows:

$$P_p = 0.85 f'_c A_1 \qquad \text{(LRFD Equation J8-1)}$$

(The bearing values provided are the same as those given in Section 10.17 of the 2005 *ACI Building Code Requirements for Structural Concrete*.)

Should the bearing load be applied to an area less than the full area of the concrete support, $\phi_c P_p$ is to be determined with the following equation, in which A_2 is the maximum area of the supporting surface that is geometrically similar to and concentric with the loaded area, with $\sqrt{A_2/A_1}$ having a maximum value of 2:

$$P_p = 0.85 f'_c A_1 \sqrt{\dfrac{A_2}{A_1}} \leq 1.7 f'_c A_1 \qquad \text{(AISC Equation J8-2)}$$

In this expression, f'_c is the compression strength of the concrete in psi and A_1 is the area of the plate (in^2) bearing concentrically on the concrete. For the design of such a plate, its required area A_1 can be determined by dividing the factored reaction R_u by $\phi_c 0.85 f'_c$ for LRFD, or by dividing R_a by $0.85 f'_c / \Omega_c$ for ASD.

$$A_1 = \frac{R_u}{\phi_c 0.85 f'_c} \text{ with } \phi_c = 0.65 \quad \text{or } A_1 = \frac{\Omega_c R_a}{0.85 f'_c} \text{ with } \Omega_c = 2.31$$

After A_1 is determined, its length (parallel to the beam) and its width are selected. The length may not be less than the N required to prevent web yielding or web crippling of the beam, nor may it be less than about 3 1/2 or 4 in for practical construction reasons. It may not be greater than the thickness of the wall or other support, and actually, it may have to be less than that thickness, particularly at exterior walls, to prevent the steel from being exposed.

Example 10-10 illustrates the calculations involved in designing a beam-bearing plate. Notice that the width and the length of the plate are desirably taken to the nearest full inch.

Example 10-10

A W18 × 71 beam ($d = 18.5$ in, $t_w = 0.495$ in, $b_f = 7.64$ in, $t_f = 0.810$ in, $k = 1.21$ in) has one of its ends supported by a reinforced-concrete wall with $f'_c = 3$ ksi. Design a bearing plate for the beam with A36 steel, for the service loads $R_D = 30$ k and $R_L = 50$ k. The maximum length of end bearing ⊥ to the wall is the full wall thickness = 8.0 in.

Solution

Compute plate area A_1.

LRFD $\phi_c = 0.65$	ASD $\Omega_c = 2.31$
$R_u = (1.2)(30) + (1.6)(50) = 116$ k	$R_a = 30 + 50 = 80$ k
$A_1 = \dfrac{R_u}{\phi_c 0.85 f'_c} = \dfrac{116}{(0.65)(0.85)(3)}$	$A_1 = \dfrac{\Omega_c R_a}{0.85 f'_c} = \dfrac{(2.31)(80)}{(0.85)(3)}$
$= 70.0$ in^2	$= 72.5$ in^2
Try PL 8 × 10 (80 in^2).	Try PL 8 × 10 (80 in^2).

Check web local yielding.

$$R_n = (2.5k + l_b)F_{yw}t_w \qquad \text{(AISC Equation J10-3)}$$
$$= (2.5 \times 1.21 + 8)(36)(0.495) = 196.5 \text{ k}$$

LRFD $\phi = 1.00$	ASD $\Omega = 1.50$
$R_u = \phi R_n = (1.00)(196.5)$	$R_a = \dfrac{R_n}{\Omega} = \dfrac{196.5}{1.50}$
$= 196.5\,\text{k} > 116\,\text{k}$ **OK**	$= 131\,\text{k} > 80\,\text{k}$ **OK**

Check web crippling.

$$\frac{l_b}{d} = \frac{8}{18.5} = 0.432 > 0.2 \quad \therefore \text{ Must use AISC Equation (J10-5b)}$$

$$R_n = 0.40t_w^2 \left[1 + \left(\frac{4l_b}{d} - 0.2\right)\left(\frac{t_w}{t_f}\right)^{1.5}\right]\sqrt{\frac{EF_{yw}t_f}{t_w}}$$

$$= (0.40)(0.495)^2\left[1 + \left(\frac{4 \times 8}{18.5} - 0.2\right)\left(\frac{0.495}{0.810}\right)^{1.5}\right]\sqrt{\frac{(29 \times 10^3)(36)(0.810)}{0.495}}$$

$$= 221.7\,\text{k}$$

LRFD $\phi = 0.75$	ASD $\Omega = 2.00$
$R_u = \phi R_n = (0.75)(221.7)$	$R_a = \dfrac{R_n}{\Omega} = \dfrac{221.7}{2.00} = 111\,\text{k}$
$= 166\,\text{k} > 116\,\text{k}$ **OK**	$>80\,\text{k}$ **OK**

Determine plate thickness.

$$n = \frac{10}{2} - 1.21 = 3.79\,\text{in}$$

LRFD $\phi_b = 0.90$	ASD $\Omega_b = 1.67$
$t = \sqrt{\dfrac{2R_u n^2}{\phi_b A_1 F_y}} = \sqrt{\dfrac{(2)(116)(3.79)^2}{(0.9)(80)(36)}}$	$t = \sqrt{\dfrac{2R_a n^2 \Omega_b}{A_1 F_y}} = \sqrt{\dfrac{(2)(80)(3.79)^2(1.67)}{(80)(36)}}$
$= 1.13\,\text{in}$	$= 1.15\,\text{in}$
Use PL $1\frac{1}{4} \times 8 \times 10$ (A36).	Use PL $1\frac{1}{4} \times 8 \times 10$ (A36).

On some occasions, the beam flanges alone probably provide sufficient bearing area, but bearing plates are nevertheless recommended, as they are useful in erection and ensure an even bearing surface for the beam. They can be placed separately from the beams and carefully leveled to the proper elevations. When the ends of steel beams are enclosed by the concrete or masonry walls, it is considered desirable to use some type of wall anchor to prevent the beam from moving longitudinally with respect to the wall. The usual anchor consists of a bent steel bar called a government anchor passing through the web of the beam and running parallel to the wall. Occasionally, clip angles attached to the web are used instead of government anchors. Should longitudinal loads of considerable size be anticipated, regular vertical anchor bolts may be used at the beam ends.

If we were to check to see if the flange thickness alone is sufficient, we would have $\left(\text{with } n = \dfrac{b_f}{2} - k \right) = \dfrac{7.64}{2} - 1.21 = 2.61 \text{ in.}$

LRFD $\phi_b = 0.90$	ASD $\Omega_b = 1.67$
$t = \sqrt{\dfrac{(2)(116)(2.61)^2}{(0.9)(8 \times 7.64)(36)}}$	$t = \sqrt{\dfrac{(2)(80)(2.61)^2(1.67)}{(8)(7.64)(36)}}$
$= 0.893 \text{ in} > t_f = 0.810 \text{ in for W18} \times 71$ **N.G.**	$= 0.910 > t_f = 0.810 \text{ in for W18} \times 71$ **N.G.**

$\therefore$ Flange t_f is not sufficient alone for either LRFD or ASD designs.

10.9   LATERAL BRACING AT MEMBER ENDS SUPPORTED ON BASE PLATES

The ends of beams and girders (and trusses) supported on base plates must be restrained against rotation about their longitudinal axes. This is clearly stated in Section F1 (2) of the AISC Specification. Stability at these locations may be obtained in several ways. Regardless of the method selected, the beams and bearing plates must be anchored to the supports.

Beam flanges may be connected to the floor or roof system which themselves must be anchored to prevent translation; the beam ends may be built into solid masonry or concrete walls; or transverse bearing stiffeners or end plates may be used as illustrated in Fig. 10.20.

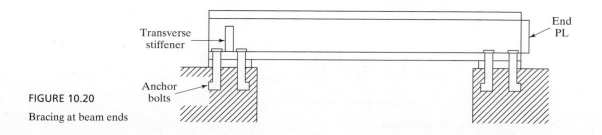

FIGURE 10.20

Bracing at beam ends

10.10 PROBLEMS FOR SOLUTION

Use LRFD for Probs. 10-1 to 10-9 except as indicated. Use both methods for all others.

10-1 to 10-5. *Considering moment only and assuming full lateral support for the compression flanges, select the lightest sections available, using 50 ksi steel and the LRFD method. The loads shown include the effect of the beam weights. Use elastic analysis, factored loads, and the 0.9 rule.*

10-1. (*Ans.* W24 × 62)

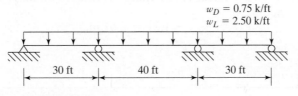

$$w_D = 0.75 \text{ k/ft}$$
$$w_L = 2.50 \text{ k/ft}$$

30 ft | 40 ft | 30 ft

FIGURE P10-1

10-2.

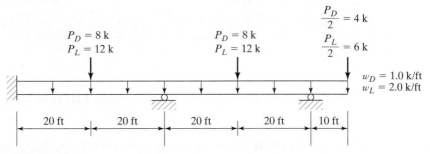

$$\frac{P_D}{2} = 4 \text{ k}$$
$$\frac{P_L}{2} = 6 \text{ k}$$

$$P_D = 8 \text{ k}$$
$$P_L = 12 \text{ k}$$

$$P_D = 8 \text{ k}$$
$$P_L = 12 \text{ k}$$

$$w_D = 1.0 \text{ k/ft}$$
$$w_L = 2.0 \text{ k/ft}$$

20 ft | 20 ft | 20 ft | 20 ft | 10 ft

FIGURE P10-2

10-3. (*Ans.* W21 × 50)

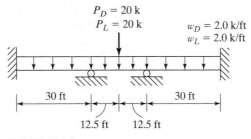

$$P_D = 20 \text{ k}$$
$$P_L = 20 \text{ k}$$

$$w_D = 2.0 \text{ k/ft}$$
$$w_L = 2.0 \text{ k/ft}$$

30 ft | 30 ft

12.5 ft | 12.5 ft

FIGURE P10-3

10-4.

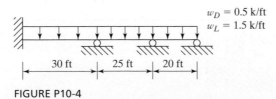

$$w_D = 0.5 \text{ k/ft}$$
$$w_L = 1.5 \text{ k/ft}$$

30 ft | 25 ft | 20 ft

FIGURE P10-4

10-5. (*Ans.* W21 × 44)

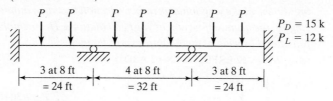

FIGURE P10-5

10-6. Repeat Prob. 10-1, using plastic analysis.

10-7. Repeat Prob. 10-3, using plastic analysis. (*Ans.* W21 × 48)

10-8. Repeat Prob. 10-5, using plastic analysis.

10-9. Three methods of supporting a roof are shown in Fig. P10-9. Using an elastic analysis with factored loads, F_y = 50 ksi, and assuming full lateral support in each case, select the lightest section if a dead uniform service load (including the beam self-weight) of 1.5 k/ft and a live uniform service load of 2.0 k/ft is to be supported. Consider moment only.

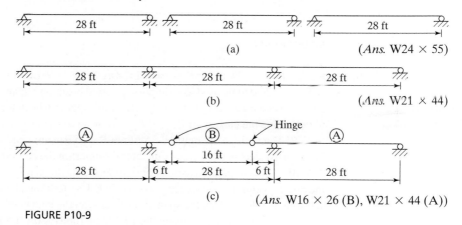

FIGURE P10-9

10-10. The welded plate girder section shown, made from 50 ksi steel, has full lateral support for its compression flange and is bent about its major axis. If C_b = 1.0, determine its design and allowable moments and shear strengths.

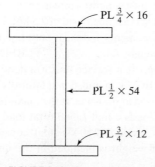

FIGURE P10-10

10-11 to 10-13. *Using F_y = 50 ksi, select the lightest available W-shape section for the span and load-ing shown. Make your initial member selection based on moment and check for shear. Neglect the beam self-weight in your calculations. The members are assumed to have full lateral bracing of their compression flanges.*

10-11. (*Ans.* W21 × 62 LRFD, W24 × 62 ASD)

FIGURE P10-11

10-12.

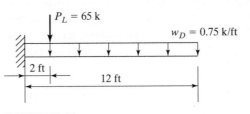

FIGURE P10-12

10-13. Repeat Prob. 10-11, using F_y = 36 ksi. (*Ans.* W24 × 76 LRFD, W24 × 84 ASD)

10-14. A W14 × 26 is to be used as a simply supported beam on a span of 12 ft with a single concentrated load located at 2 ft from the left support. Check for moment and shear to determine the maximum P_D and P_L permitted on the beam, using 50 ksi steel and the LRFD and the ASD methods. Assume P is 25 percent dead load and 75 percent live load. Neglect beam weight. Full lateral support of the compression flange exists.

10-15. A W10 × 17 is to be used as a simply supported beam on a span of 8 ft with a single moving concentrated service live load of 56 k. Use A992 steel and neglect the beam self-weight. Where along the span of the beam can the load be placed and not exceed the shear strength capacity using the LRFD method? (*Ans.* 1.5 ft ≤ x ≤ 6.5 ft)

10-16. A W12 × 40 consisting of 50 ksi steel is used as a simple beam for a span of 7.25 ft. If it has full lateral support, determine the maximum uniform loads w_u and w_a that it can support in addition to its own weight. Use LRFD and ASD methods and con-sider shear and moment only.

10-17. A 24-ft, simply supported beam must support a moving concentrated service live load of 50 k in addition to a uniform service dead load of 2.5 k/ft. Using 50 ksi steel, select the lightest section considering moments and shear only. Use LRFD and ASD meth-ods and neglect the beam self-weight. (*Ans.* W24 × 76 LRFD and ASD)

10-18. A 36-ft simple beam that supports a service uniform dead load of 1.0 k/ft and a service concentrated live load of 25 k at midspan is laterally unbraced except at its ends and at the concentrated load at midspan. If the maximum permissible center-line deflection under service loads equal $L/360$ total load and $L/1000$ live load, select the most economical W section of 50 ksi steel, considering moment, shear and deflection. The beam self-weight is included in the uniform dead load. Use C_b = 1.0.

10-19. Design a beam for a 30-ft simple span to support the working uniform loads of $w_D = 1.25$ k/ft (includes beam self-weight) and $w_L = 1.75$ k/ft. The maximum permissible total load deflection under working loads is 1/360 of the span. Use 50 ksi steel and consider moment, shear and deflection. The beam is to be braced laterally at its ends and the midspan only. (*Ans.* W24 × 76 LRFD and ASD)

10-20. Select the lightest W shape of A992 steel for uniform service dead load and the concentrated service live load shown in Fig. P10-20. The dead load includes the beam self-weight. The beam has lateral support for its compression flange and the ends and at the concentrated load. The maximum service live load deflection may not exceed 1/1000 of the span. Consider moment, shear and deflection.

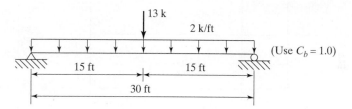

FIGURE P10-20

10-21. The cantilever beam shown in Fig. P10-21 is a W14 × 34 of A992 steel. There is no lateral support other than at the fixed end. Use an unbraced length equal to the span length. The uniform load is a service dead load that includes the beam weight, and the concentrated load is a service live load. Determine if the beam is adequate for moment, shear and an allowable total load deflection of 0.25 in. (*Ans.* OK for moment and deflection, N.G. for shear)

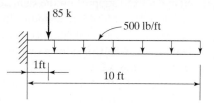

FIGURE P10-21

10-22. A W24 × 68 consisting of 50 ksi steel is used as a simply supported beam shown in Fig. P10-22 Include the beam weight and determine if the member has sufficient shear capacity using the equations from Chapter G of the AISC Specification (you may check your answer using the values from the tables), and determine whether or not the beam will meet the following deflection criteria: max LL = L/360 and max TL = L/240.

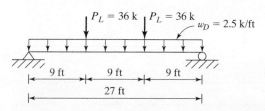

FIGURE P10-22

10-23. Select the lightest available W sections (F_y = 50 ksi) for the beams and girders shown in Fig. P10-23. The floor slab is 6 in reinforced concrete (weight = 145 lb/ft^3) and supports a 125 psf uniform live load. Assume that continuous lateral bracing of the compression flange is provided. The maximum permissible TL deflection is L/240. (*Ans.* Beam = W21 × 44 LRFD and ASD, Girder = W24 × 62 LRFD and ASD)

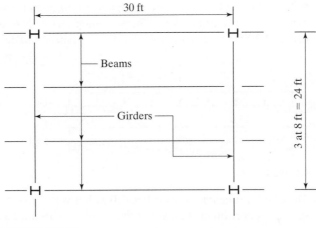

FIGURE P10-23

10-24. Repeat Prob. 10-23 if the live load is 250 psf.

10-25. Select the lightest available W section of 50 ksi steel for a beam that is simply supported on the left end and a fixed support on the right end of a 36-ft span. The member supports a service dead load of 2.4 k/ft, including its self-weight and a service live load of 3.0 k/ft. Assume full lateral support of the compression flange and maximum TL deflection of L/600. Consider moment, shear and deflection. (*Ans.* W30 × 108 LRFD, W30 × 116 ASD)

10-26. Repeat Prob. 10-25 if the nominal depth of the beam is limited to 27 in and lateral support of the compression flange is provided at the ends and the 1/3 points of the span. Use C_b = 1.0.

10-27. The beam shown in Fig. P10-27 is a W14 × 34 of A992 steel and has lateral support of the compression flange at the ends and at the points of the concentrated loads. The two concentrated loads are service live loads. Check the beam for shear and for Web Local Yielding and Web Crippling at the concentrated load if l_b = 6 in. Neglect the self-weight of the beam. (*Ans.* Shear and web crippling N.G., web local yielding OK)

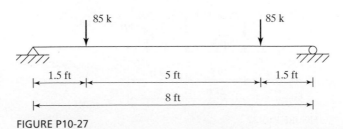

FIGURE P10-27

10-28. A 7-ft beam with full lateral support for its compression flange is supporting a moving concentrated live load of 58 k. Using 50 ksi steel, select the lightest W section. Assume the moving load can be placed anywhere in the middle 5 ft of the beam span. Choose a member based on moment then check if it is satisfactory for shear, and compute the minimum length of bearing required at the supports from the standpoint of web local yielding and web crippling. Neglect self-weight.

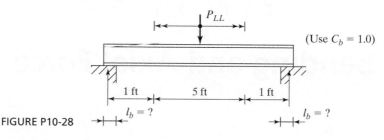

(Use $C_b = 1.0$)

FIGURE P10-28

10-29. A 12 ft beam with full lateral support of the compression flange is used on a simple span. A concentrated service live load of 64 k is applied 3 ft from the left support. Use A992 steel and neglect the member's self-weight. Select the lightest W section based on moment. Check the beam for Local Flange Bending at the concentrated load and for Web Local Yielding at the concentrated load if $l_b = 4$ in. Neglect beam self-weight. Resize beam if necessary. (*Ans.* W18 × 46 LRFD and ASD)

10-30. A W21 × 68 member is used as a simply supported beam with a span length of 12 ft. Determine C_b, since the lateral support of the compression flange is provided only at the ends. The member is uniformly loaded. The loads will produce factored moments of $M_{Dx} = 75$ ft-k, $M_{Lx} = 90$ ft-k and $M_{Dy} = 15$ ft-k, $M_{Ly} = 18$ ft-k. Is this member satisfactory for bending strength based on the interaction equation in Chapter H of the AISC Specification?

10-31. The 30-ft, simply supported beam shown in Fig. P10-31 has full support of its compression flange and is A992 steel. The beam supports a gravity service dead load of 132 lb/ft (includes beam weight) and gravity live load of 165 lb/ft. The loads are assumed to act through the c.g. of the section. Select the lightest available W10 section. (*Ans.* W10 × 22 LRFD, W10 × 26 ASD)

W10

3
12

FIGURE P10-31

10-32. Design a steel bearing plate from A572 (Grade 50) steel for a W18 × 35 beam, with end reactions of $R_D = 12$ k and $R_L = 16$ k. The beam will bear on a reinforced concrete wall with $f'_c = 3$ ksi. In the direction perpendicular to wall, the bearing plate maximum length of end bearing may not be longer than 6 in. W18 is A992 steel.

10-33. Design a steel bearing plate from A36 steel for a W24 × 55 beam supported by a reinforced concrete wall with $f'_c = 3$ ksi. The maximum beam reaction is $R_D = 30$ k and $R_L = 40$ k. The maximum length of end bearing perpendicular to the wall is the full wall thickness of 10 in. W24 is A992 steel. (*Ans.* Use PL11/8 × 9 3 0 ft to 8 in LRFD and ASD)

Bending and Axial Force

11.1   OCCURRENCE

Structural members that are subjected to a combination of bending and axial force are far more common than the student may realize. This section is devoted to listing some of the more obvious cases. Columns that are part of a steel building frame must nearly always resist sizable bending moments in addition to the usual compressive loads. It is almost impossible to erect and center loads exactly on columns, even in a testing lab; in an actual building, one can see that it is even more difficult. Even if building loads could be perfectly centered at one time, they would not stay in one place. Furthermore, columns may be initially crooked or have other flaws resulting in lateral bending. The beams framing into columns are commonly supported with framing angles or brackets on the sides of the columns. These eccentrically applied loads produce moments. Wind and other lateral loads cause columns to bend laterally, and the columns in rigid frame buildings are subjected to moments, even when the frame is supporting gravity loads alone. The members of bridge portals must resist combined forces, as do building columns. Among the causes of the combined forces are heavy lateral wind loads or seismic loads, vertical traffic loads—whether symmetrical or not—and the centrifugal effect of traffic on curved bridges.

The previous practice of the student has probably been to assume that truss members are only axially loaded. Purlins for roof trusses, however, are frequently placed between truss joints, causing the top chords to bend. Similarly, the bottom chords may be bent by the hanging of light fixtures, ductwork, and other items between the truss joints. All horizontal and inclined truss members have moments caused by their own weights, while all truss members—whether vertical or not—are subjected to secondary bending forces. Secondary forces are developed because the members are not connected with frictionless pins; as assumed in the usual analysis, the members' centers of gravity or those of their connectors do not exactly coincide at the joints, etc.

Moments in tension members are not as serious as those in compression members, because tension tends to reduce lateral deflections while compression increases them. Increased lateral deflections in turn result in larger moments, which cause larger

The Alcoa Building under construction in San Francisco, CA. (Courtesy of Bethlehem Steel Corporation.)

lateral deflections, and so on. It is hoped that members in such situations are stiff enough to prevent the additional lateral deflections from becoming excessive.

11.2   MEMBERS SUBJECT TO BENDING AND AXIAL TENSION

A few types of members subject to both bending and axial tension are shown in Fig. 11.1.

In Section H1 of the AISC Specification, the interaction equations that follow are given for symmetric shapes subjected simultaneously to bending and axial tensile forces. These equations are also applicable to members subjected to bending and compression forces, as will be described in Sections 11.3 to 11.9.

For $\dfrac{P_r}{P_c} \geq 0.2$,

$$\frac{P_r}{P_c} + \frac{8}{9}\left(\frac{M_{rx}}{M_{cx}} + \frac{M_{ry}}{M_{cy}}\right) \leq 1.0 \qquad \text{(AISC Equation H1-1a)}$$

and for $\dfrac{P_r}{P_c} < 0.2$,

$$\frac{P_r}{2P_c} + \left(\frac{M_{rx}}{M_{cx}} + \frac{M_{ry}}{M_{cy}}\right) \leq 1.0 \qquad \text{(AISC Equation H1-1b)}$$

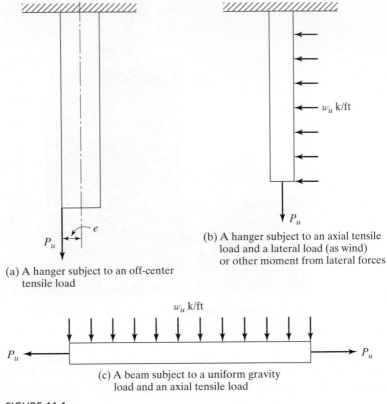

(a) A hanger subject to an off-center tensile load

(b) A hanger subject to an axial tensile load and a lateral load (as wind) or other moment from lateral forces

(c) A beam subject to a uniform gravity load and an axial tensile load

FIGURE 11.1

Some members subject to bending and axial tension.

in which

P_r = required axial tensile strength, P_u (LRFD) or P_a (ASD), kips

P_c = design axial tensile strength $(\phi_c P_n)$ or allowable axial tensile strength $\left(\dfrac{P_n}{\Omega_c}\right)$, kips

M_r = required flexural strength, M_u (LRFD) or M_a (ASD), ft-k

M_c = design flexural strength $(\phi_b M_n)$ or allowable flexural strength $\left(\dfrac{M_n}{\Omega_b}\right)$, ft-k

Usually, only a first-order analysis (that is, not including any secondary forces as described in the next section) is made for members subject to bending and axial tension. It is conservative to neglect the effects of tension forces acting with bending moments. However, the analyst may—in fact, is encouraged to—make second-order analyses for these members and use the results in his or her designs. Examples 11-1 and 11-2 illustrate the use of the interaction equations to review members subjected simultaneously to bending and axial tension.

Example 11-1

A 50 ksi W12 × 40 tension member with no holes is subjected to the axial loads $P_D = 25$ k and $P_L = 30$ k, as well as the bending moments $M_{Dy} = 10$ ft-k and $M_{Ly} = 25$ ft-k. Is the member satisfactory if $L_b < L_p$?

Solution

Using a W12 × 40 ($A = 11.7$ in^2)

LRFD	ASD
$P_r = P_u = (1.2)(25 \text{ k}) + (1.6)(30 \text{ k}) = 78$ k	$P_r = P_a = 25 \text{ k} + 30 \text{ k} = 55$ k
$M_{ry} = M_{uy} = (1.2)(10 \text{ ft-k}) + (1.6)(25 \text{ ft-k})$	$M_{ry} = M_{ay} = 10 \text{ ft-k} + 25 \text{ ft-k} = 35$ ft-k
$\qquad\qquad = 52$ ft-k	
$P_c = \phi P_n = \phi_t F_y A_g = (0.9)(50 \text{ksi})(11.7 \text{ in}^2)$	$P_c = \dfrac{P_n}{\Omega_c} = \dfrac{F_y A_g}{\Omega_c} = \dfrac{(50 \text{ ksi})(11.7 \text{ in}^2)}{1.67}$
$\qquad\qquad = 526.5$ k	$\qquad\qquad = 350.3$ k
$M_{cy} = \phi_b M_{py} = 63.0$ ft-k (AISC Table 3-4)	$M_{cy} = \dfrac{M_{cy}}{\Omega_b} = 41.9$ ft-k (AISC Table 3-4)
$\dfrac{P_r}{P_c} = \dfrac{78 \text{ k}}{526.5 \text{ k}} = 0.148 < 0.2$	$\dfrac{P_r}{P_c} = \dfrac{55 \text{ k}}{350.3 \text{ k}} = 0.157 < 0.2$
$\therefore$ **Must use AISC Eq. H1-1b**	$\therefore$ **Must use AISC Eq. H1-1b**
$\dfrac{P_r}{2P_c} + \left(\dfrac{M_{rx}}{M_{cx}} + \dfrac{M_{ry}}{M_{cy}} \right) \leq 1.0$	$\dfrac{P_r}{2P_c} + \left(\dfrac{M_{rx}}{M_{cx}} + \dfrac{M_{ry}}{M_{cy}} \right) \leq 1.0$
$\dfrac{78}{(2)(526.5)} + \left(0 + \dfrac{52}{63} \right)$	$\dfrac{55}{(2)(350.3)} + \left(0 + \dfrac{35}{41.9} \right)$
$\qquad = 0.899 < 1.0$ **OK**	$\qquad = 0.914 < 1.0$ **OK**

Example 11-2

A W10 × 30 tensile member with no holes, consisting of 50 ksi steel and with $L_b = 12.0$ ft, is subjected to the axial service loads $P_D = 30$ k and $P_L = 50$ k and to the service moments $M_{Dx} = 20$ ft-k and $M_{Lx} = 40$ ft-k. If $C_b = 1.0$, is the member satisfactory?

Solution

Using a W10 × 30 ($A = 8.84$ in^2, $L_p = 4.84$ ft and $L_r = 16.1$ ft, $\phi_b M_{px} = 137$ ft-k, BF for LRFD = 4.61, BF for ASD = 3.08 and $M_{px}/\Omega_b = 91.3$ ft-k from AISC Table 3-2)

LRFD	ASD
$P_r = P_u = (1.2)(30\,\text{k}) + (1.6)(50\,\text{k}) = 116\,\text{k}$	$P_r = P_a = 30\,\text{k} + 50\,\text{k} = 80\,\text{k}$
$M_{rx} = M_{ux} = (1.2)(20\,\text{ft-k}) + (1.6)(40\,\text{ft-k})$	$M_{rx} = M_{ax} = 20\,\text{ft-k} + 40\,\text{ft-k}$
$\qquad = 88\,\text{ft-k}$	$\qquad = 60\,\text{ft-k}$
$P_c = \phi P_n = \phi_t F_y A_g = (0.9)(50\,\text{ksi})(8.84\,\text{in}^2)$	$P_c = \dfrac{P_n}{\Omega_c} = \dfrac{F_y A_g}{\Omega_c} = \dfrac{(50\,\text{ksi})(8.84\,\text{in}^2)}{1.67}$
$\qquad = 397.8\,\text{k}$	$\qquad = 264.7\,\text{k}$
$M_{cx} = \phi_b M_{nx} = C_b[\phi_b M_{px} - BF(L_b - L_p)]$	$M_{cx} = \dfrac{M_{nx}}{\Omega_b} = C_b\left[\dfrac{M_{px}}{\Omega_b} - BF(L_b - L_p)\right]$
$\qquad = 1.0[137 - 4.61(12.0 - 4.84)]$	$\qquad = 1.0[91.3 - (3.08)(12 - 4.84)]$
$\qquad = 104.0\,\text{ft-k}$	$\qquad = 69.2\,\text{ft-k}$
$\dfrac{P_r}{P_c} = \dfrac{116}{397.8} = 0.292 > 0.2$	$\dfrac{P_r}{P_c} = \dfrac{80}{264.7} = 0.302 > 0.2$
$\therefore$ **Must use AISC Eq. H1-1a**	$\therefore$ **Must use AISC Eq. H1-1a**
$\dfrac{P_r}{P_c} + \dfrac{8}{9}\left(\dfrac{M_{rx}}{M_{cx}} + \dfrac{M_{ry}}{M_{cy}}\right) \le 1.0$	$\dfrac{P_r}{P_c} + \dfrac{8}{9}\left(\dfrac{M_{rx}}{M_{cx}} + \dfrac{M_{rx}}{M_{cy}}\right) \le 1.0$
$\dfrac{116}{397.8} + \dfrac{8}{9}\left(\dfrac{88}{104.0} + 0\right)$	$\dfrac{80}{264.7} + \dfrac{8}{9}\left(\dfrac{60}{69.2} + 0\right)$
$\qquad = 1.044 > 1.0$ **N.G.**	$\qquad = 1.073 > 1.0$ **N.G.**

11.3   FIRST-ORDER AND SECOND-ORDER MOMENTS FOR MEMBERS SUBJECT TO AXIAL COMPRESSION AND BENDING

When a beam column is subjected to moment along its unbraced length, it will be displaced laterally in the plane of bending. The result will be an increased or secondary moment equal to the axial compression load times the lateral displacement or eccentricity. In Fig. 11.2, we can see that the member moment is increased by an amount $P_{nt}\delta$, where P_{nt} is the axial compression force determined by a first-order analysis. This moment will cause additional lateral deflection, which will in turn cause a larger column moment, which will cause a larger lateral deflection, and so on until equilibrium is reached. M_r is the required moment strength of the member. M_{nt} is the first-order moment, assuming no lateral translation of the frame.

If a frame is subject to sidesway where the ends of the columns can move laterally with respect to each other, additional secondary moments will result. In Fig. 11.3, the secondary moment produced due to sidesway is equal to $P_{nt}\,\Delta$.

The moment M_r is assumed by the AISC Specification to equal M_{lt} (which is the moment due to the lateral loads) plus the moment due to $P_u\,\Delta$.

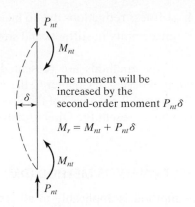

The moment will be increased by the second-order moment $P_{nt}\delta$

$$M_r = M_{nt} + P_{nt}\delta$$

FIGURE 11.2

Moment amplification of a column that is braced against sidesway.

The required total flexural strength of a member must at least equal the sum of the first-order and second-order moments. Several methods are available for determining this required strength, ranging from very simple approximations to very rigorous procedures.

A rigorous second-order inelastic analysis of the structure computes and takes into account the anticipated deformations in calculating the required maximum compressive strength, P_r, and the maximum required flexural strength, M_r. This method is usually more complex than is necessary for the structural design of typical structures. Should the designer make a second-order analysis, he or she should realize that it must account for the interaction of the various load effects. That is, one must consider combinations of loads acting at the same time. We cannot correctly make separate analyses and superimpose the results because it is inherently a non-linear problem.

The AISC Specification Chapter C.1 states that any rational method of design for stability that considers all of the effects list below is permitted.

1. flexural, shear, and axial member deformation, and all other deformations that contribute to displacement of the structure;
2. second-order effect (both P-Δ and P-δ effects);
3. geometric imperfections;

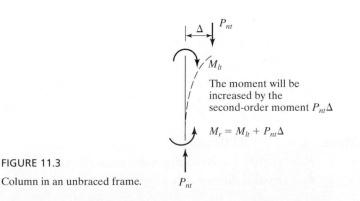

The moment will be increased by the second-order moment $P_{nt}\Delta$

$$M_r = M_{lt} + P_{nt}\Delta$$

FIGURE 11.3

Column in an unbraced frame.

4. stiffness reductions due to inelasticity;

5. uncertainty in stiffness and strength.

Three methods are presented in the AISC Specification. The *Direct Analysis Method* of design in Chapter C, the *Effective Length Method*, and the *First-Order Analysis Method* in Appendix 7 of the Specification are permitted alternatives. In this chapter, the authors present the Direct Analysis Method and the Effective Length Method.

11.4   DIRECT ANALYSIS METHOD (DM)

This method is applicable to all types of structures. It does not distinguish between building structural systems such as braced frames, moment frames, shear wall, or any combination of systems. It has the additional advantage of not having to calculate the effective length factor, K. This means that in determining the available axial compressive strength, P_c, $K = 1.0$ is used.

11.4.1   Second-Order Effects (C2.1 (2) – AISC Specification)

The required strength, P_r, can be determined using a rigorous second-order analysis that requires an iterative computer analysis of the model or by the approximate technique of utilizing an amplified first-order analysis using magnification factors, B_1 and B_2, that is specified in Appendix 8 and discussed in Section 11.5 of the text.

11.4.2   Stiffness Reduction (C2.3 – AISC Specification)

The Direct Analysis Method uses a reduced flexural and axial stiffness to account for the influence of inelasticity and uncertainty in strength and stiffness on second-order effects. In the analysis, reduced stiffness EI^*, is replaced with $0.8\tau_b EI$ and EA^*, is replaced with $0.8EA$.

The τ_b factor depends on the level of axial stress in the member which implies that the modulus of elasticity reduces as the material goes inelastic.

When: $\dfrac{\alpha P_r}{P_y} \leq 0.5$

$$\tau_b = 1.0 \qquad\qquad\qquad \text{(AISC Equation C2-2a)}$$

When: $\dfrac{\alpha P_r}{P_y} > 0.5$

$$\tau_b = 4\left(\dfrac{\alpha P_r}{P_y}\right)\left[1 - \dfrac{\alpha P_r}{P_y}\right] \qquad\qquad \text{(AISC Equation C2-2b)}$$

Where: $\alpha = 1.0$ (LRFD) and $\alpha = 1.6$ (ASD)

P_r = required axial compressive strength, kips

P_y = axial yield strength, kips

11.4.3 Notional Loads (C2.2b – AISC Specification)

To account for initial out-of-plumbness of columns (geometric imperfections), the Direct Analysis provisions require the application of notional loads. Notional loads are applied as lateral loads to a model of the structure that is based on its nominal geometry. The magnitude of the notional loads shall be:

$$N_i = 0.002\alpha Y_i \qquad\qquad \text{(AISC Equation C2-1)}$$

Where: $\alpha = 1.0$ (LRFD) and $\alpha = 1.6$ (ASD)
$\quad\quad N_i$ = notional load applied at level i, kips
$\quad\quad Y_i$ = gravity load applied at level i from load combinations, kips

Note: The 0.002 term represents an out-of-plumbness of 1/500 which is the maximum tolerance on column plumbness specified in the *AISC Code of Standard Practice for Steel Buildings and Bridges.*

The notional load shall be additive to other lateral loads and shall be applied in all load combinations, except as noted in AISC Specification section C2.2b(4). This section states that for structures where the ratio of maximum second-order drift to maximum first-order drift in all stories is equal to or less than 1.7,

$$B_2 = \frac{\Delta_{second\text{-}order}}{\Delta_{first\text{-}order}} \le 1.7$$

it is permissible to apply the notional load, N_i, only in gravity-only load combinations and not in combinations that include other lateral loads.

11.5 EFFECTIVE LENGTH METHOD (ELM)

This method, found in Appendix 7 of the AISC Specification, is applicable where the ratio of maximum second-order drift to maximum first-order drift in all stories is equal to or less than 1.5.

That is:
$$B_2 = \frac{\Delta_{second\text{-}order}}{\Delta_{first\text{-}order}} \le 1.5$$

The required strength, P_r, is calculated from an analysis conforming to the requirements of AISC Specification C2.1, except that the stiffness reduction indicated in C2.1 (2) need not be applied. The nominal stiffness of structural members is used.

Notional loads need only be applied in gravity-only load cases.

The K factor must be determined from a sidesway buckling analysis or the alignment charts as shown in Chapter 7 of this text. It is permitted to use a $K = 1.0$ in the design of all braced systems and in moment frames where the ratio of maximum second-order drift to maximum first-order drift in all stories is equal to or less than 1.1.

That is:
$$B_2 = \frac{\Delta_{second\text{-}order}}{\Delta_{first\text{-}order}} \le 1.1$$

TABLE 11.1 Comparison of Basic Stability Requirements with Specific Provisions

Basic Requirement in Section C1		Provision in Direct Analysis Method (DM)	Provision in Effective Length Method (ELM)
(1) Consider all deformations		C2.1(1). Consider all deformations	Same as DM (by reference to C2.1)
(2) Consider second-order effects (both P-Δ and P-δ)		C2.1(2). Consider second-order effects (P-Δ and P-δ)**	Same as DM (by reference to C2.1)
(3) Consider geometric imperfections	Effect of joint-position imperfections* on structure response	C2.2a. Direct modeling or C2.2b. Notional loads	Same as DM, second option only (by reference to C2.2b)
This includes joint-position imperfections(which affect structure response) and member imperfections (which affect structure response and member strength)*	Effect of member imperfections on structure response	Included in the stiffness reduction specified in C2.3	All these effects are considered by using KL from a sidesway buckling analysis in the member strength check. Note that the only difference between DM and ELM is that:
	Effect of member imperfections on member strength	Included in member strength formulas, with $KL = L$	
(4) Consider stiffness reduction due to inelasticity	Effect of stiffness reduction on structure response	Included in the stiffness reduction specified in C2.3	• DM uses reduced stiffness in the analysis; $KL = L$ in the member strength check
This affects structure response and member strength	Effect of stiffness reduction on member strength	Included in member strength formulas, with $KL = L$	
(5) Consider uncertainty in strength and stiffness	Effect of stiffness/strength uncertainty on structure response	Included in the stiffness reduction specified in C2.3	• ELM uses full stiffness in the analysis; KL from sideway buckling analysis in the member strength check for frame members
This affects structure response and member strength	Effect of stiffness/strength uncertainty on member strength	Included in member strength formulas, with $KL = L$	

*In typical building structures, the "joint-position imperfections" refers to column out-of-plumbness.
**Second-order effects may be considered either by rigorous second-order analysis or by the approximate technique (using B_1 and B_2) specified in Appendix 8.
Source: Commentary on the Specification, Section C2–Table C–C1.1, p. 16.1–273. June 23, 2010. "Copyright © American Institute of Steel Construction. Reprinted with permission. All rights reserved."

Table C-C1.1 of the commentary of Chapter C of the AISC Specification, which is reproduced in Table 11.1, presents a comparison of the basic stability requirements of the Direct Analysis Method (DM) and the Effective Length Method (ELM).

11.6 APPROXIMATE SECOND-ORDER ANALYSIS

In this chapter, the authors present the approximate second-order analysis given in Appendix 8 of the AISC Specification. We will make two first-order elastic analyses— one an analysis where the frame is assumed to be braced so that it cannot sway. We will call these moments M_{nt} and will multiply them by a magnification factor called B_1 to account for the P-δ effect (see Fig. 11.2). Then we will analyze the frame again, allowing it to sway. We will call these moments M_{lt} and will multiply them by a magnification factor called B_2 to account for the P-Δ effect (see Fig. 11.3). The final moment in a particular member will equal

$$M_r = B_1 M_{nt} + B_2 M_{lt}. \qquad \text{(AISC Equation A-8-1)}$$

The final axial strength P_r must equal

$$P_r = P_{nt} + B_2 P_{lt}. \qquad \text{(AISC Equation A-8-2)}$$

Instead of using the AISC empirical procedure described here, the designer may—and is encouraged to—use a theoretical second-order elastic analysis, provided that he or she meets the requirements of Chapter C of the Specification.

11.6.1   Magnification Factors

The magnification factors are B_1 and B_2. With B_1, the analyst attempts to estimate the $P_{nt}\delta$ effect for a column, whether the frame is braced or unbraced against sidesway. With B_2, he or she attempts to estimate the $P_{lt}\,\Delta$ effect in unbraced frames.

These factors are theoretically applicable when the connections are fully restrained or when they are completely unrestrained. The AISC Manual indicates that the determination of secondary moments in between those two classifications for partially restrained moment connections is beyond the scope of their specification. The terms *fully restrained* and *partially restrained* are discussed at length in Chapter 15 of this text.

The horizontal deflection of a multistory building due to wind or seismic load is called *drift*. It is represented by Δ in Fig. 11.3. Drift is measured with the so-called *drift index* Δ_H/L, where Δ_H is the first-order lateral inter-story deflection and L is the story height. For the comfort of the occupants of a building, the index usually is limited at working or service loads to a value between 0.0015 and 0.0030, and at factored loads to about 0.0040.

The expression to follow for B_1 was derived for a member braced against sidesway. It will be used only to magnify the M_{nt} moments (*those moments computed, assuming that there is no lateral translation of the frame*).

$$B_1 = \frac{C_m}{1 - \alpha\dfrac{P_r}{P_{e1}}} \geq 1.0 \qquad \text{(AISC Equation A-8-3)}$$

In this expression, C_m is a term that is defined in the next section of this chapter; α is a factor equal to 1.00 for LRFD and 1.60 for ASD; P_r is the required axial strength of the member; and P_{e1} is the member's Euler buckling strength calculated on the

$$P_{e1} = \frac{EI^*}{(K_1 L)^2} \qquad \text{(AISC Equation A-8-5)}$$

basis of zero sidesway. One is permitted to use the first-order estimate of P_r (that is, $P_r = P_{nt} + P_{lt}$) when calculating magnification factor, B_1. Also, K is the effective length factor in the plane of bending, determined based on the assumption of no lateral translation, set equal to 1.0, unless analysis justifies a smaller value. EI^* is $0.8 \, \tau_b EI$ when the direct analysis method is used, and EI when the effective length method is used.

In a similar fashion, P_{e2} is the elastic critical buckling resistance for the story in question, determined by a sidesway buckling analysis. For this analysis, K_2L is the effective length in the plane of bending, based on the sidesway buckling analysis. For this case, the sidesway buckling resistance may be calculated with the following expression, in which Σ is used to include all of the columns on that level or story:

$$P_{e \, story} = \Sigma \frac{\pi^2 EI}{(K_2L)^2}$$

Furthermore, the AISC permits the use of the following alternative expression for calculating $P_{e \, story}$

$$P_{e \, story} = R_M \frac{HL}{\Delta_H} \qquad \text{(AISC Equation A-8-7)}$$

Here, the factors are defined as follows:

$$R_M = 1 - 0.15 \left(\frac{P_{mf}}{P_{story}} \right) \qquad \text{(AISC Equation A-8-8)}$$

$H =$ story shear produced by the lateral loads used to compute Δ_H, kips

$L =$ story height, in

$\Delta_H =$ first-order interstory drift due to the lateral loads computed using stiffness required for the analysis method used, in

$P_{mf} =$ total vertical load in columns in the story that are part of moment frame, kips ($P_{mf} = 0$ for braced frame systems)

The values shown for P_{story} and $P_{e \, story}$ are for all of the columns on the floor in question. This is considered to be necessary because the B_2 term is used to magnify column moments for sidesway. For sidesway to occur in a particular column, it is necessary for all of the columns on the floor to sway simultaneously.

$$B_2 = \frac{1}{1 - \dfrac{P_{story}}{P_{e \, story}}} \qquad \text{(AISC Equation A-8-6)}$$

We must remember that the amplification factor B_2 is only applicable to moments caused by forces that cause sidesway and is to be computed for an entire story. (Of course, if you want to be conservative, you can multiply B_2 times the sum of the no-sway and the sway moments—that is, M_{nt} and M_{lt}—but that's probably overdoing it.) To use the B_2 value given by AISC Equation A-8-6, we must select initial member sizes (that is, so we can compute a value for $P_{e\ story}$ or Δ_H).

To calculate the values of P_{story} and $P_{e\ story}$ some designers will calculate the values for the columns in the one frame under consideration. This, however, is a rather bad practice, unless all the other frames on that level are exactly the same as the one under study.

11.6.2 Moment Modification or C_m Factors

In Section 11.6.1, the subject of moment magnification due to lateral deflections was introduced, and the factors B_1 and B_2 were presented with which the moment increases could be estimated. In the expression for B_1, a term C_m, called the *modification factor*, was included. The magnification factor B_1 was developed for the largest possible lateral displacement. On many occasions, the displacement is not that large, and B_1 overmagnifies the column moment. As a result, the moment may need to be reduced or modified with the C_m factor. In Fig. 11.4, we have a column bent in single curvature, with equal end moments such that the column bends laterally by an amount δ at mid-depth. The maximum total moment occurring in the column clearly will equal M plus the increased moment $P_{nt}\delta$. As a result, no modification is required and $C_m = 1.0$.

An entirely different situation is considered in Fig. 11.5, where the end moments tend to bend the member in reverse curvature. The initial maximum moment occurs at one of the ends, and we shouldn't increase it by a value $P_{nt}\delta$ that occurs some distance out in the column, because we will be overdoing the moment magnification. The purpose of the modification factor is to modify or reduce the magnified moment when the variation of the moments in the column is such that B_1 is made too large. If we didn't use a modification factor, we would end up with the same total moments in the columns of both Figs. 11.4 and 11.5, assuming the same dimensions and initial moments and load.

Modification factors are based on the rotational restraint at the member ends and on the moment gradients in the members. The AISC Specification (Appendix 8) includes two categories of C_m, as described in the next few paragraphs.

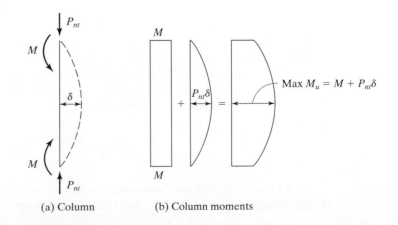

FIGURE 11.4

Moment magnification for column bent in single curvature.

(a) Column   (b) Column moments

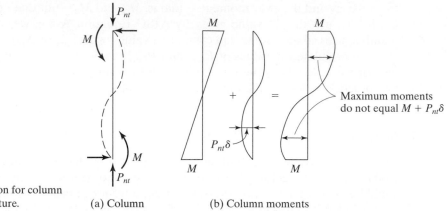

FIGURE 11.5

Moment magnification for column bent in double curvature.

(a) Column   (b) Column moments

In Category 1, the members are prevented from joint translation or sidesway, and they are not subject to transverse loading between supports. For such members, the modification factor is based on an elastic first-order analysis.

$$C_m = 0.6 - 0.4\frac{M_1}{M_2}$$ (AISC Equation A-8-4)

In this expression, M_1/M_2 is the ratio of the smaller moment to the larger moment at the ends of the unbraced length in the plane of bending under consideration. The ratio is negative if the moments cause the member to bend in single curvature, and positive if they bend the members in reverse or double curvature. As previously described, a member in single curvature has larger lateral deflections than a member bent in reverse curvature. With larger lateral deflections, the moments due to the axial loads will be larger.

Category 2 applies to members that are subjected to transverse loading between supports. The compression chord of a truss with a purlin load between its joints is a typical example of this category. The AISC Specification states that the value of C_m for this situation may be determined by rational analysis or by setting it conservatively equal to 1.0.

Instead of using these values for transversely loaded members, the values of C_m for Category 2 may be determined for various end conditions and loads by the values given in Table 11.2, which is a reproduction of Table C-A-8.1 of the Commentary on Appendix 8 of the AISC Specification. In the expressions given in the table, P_r is the required column axial load and P_{e1} is the elastic buckling load for a braced column for the axis about which bending is being considered.

$$P_{e1} = \frac{\pi^2 EI^*}{(K_1 L)^2}$$ (AISC Equation A-8-5)

In Table 11.2, note that some members have rotationally restrained ends and some do not. Sample values of C_m are calculated for four beam columns and shown in Fig. 11.6.

TABLE 11.2   Amplification Factors (ψ) and Modification Factors (C_m) for Beam Columns Subject to Transverse Loads between Joints.

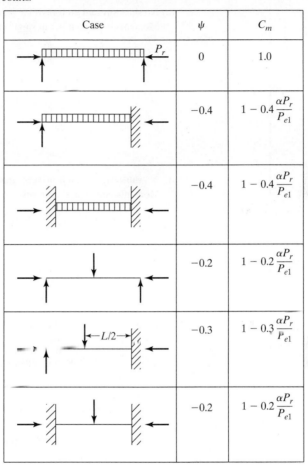

Case	ψ	C_m
	0	1.0
	−0.4	$1 - 0.4\dfrac{\alpha P_r}{P_{e1}}$
	−0.4	$1 - 0.4\dfrac{\alpha P_r}{P_{e1}}$
	−0.2	$1 - 0.2\dfrac{\alpha P_r}{P_{e1}}$
	−0.3	$1 - 0.3\dfrac{\alpha P_r}{P_{e1}}$
	−0.2	$1 - 0.2\dfrac{\alpha P_r}{P_{e1}}$

Source: Commentary on the Specification, Appendix 8–Table C–A–8.1, p16.1–525. June 22, 2010. "Copyright © American Institute of Steel Construction. Reprinted with permission. All rights reserved."

11.7   BEAM–COLUMNS IN BRACED FRAMES

The same interaction equations are used for members subject to axial compression and bending as were used for members subject to axial tension and bending. However, some of the terms involved in the equations are defined somewhat differently. For instance, P_a and P_u refer to compressive forces rather than tensile forces.

To analyze a particular beam column or a member subject to both bending and axial compression, we need to perform both a first-order and a second-order analysis to obtain the bending moments. The first-order moment is usually obtained by making an elastic analysis and consists of the moments M_{nt} (these are the moments in

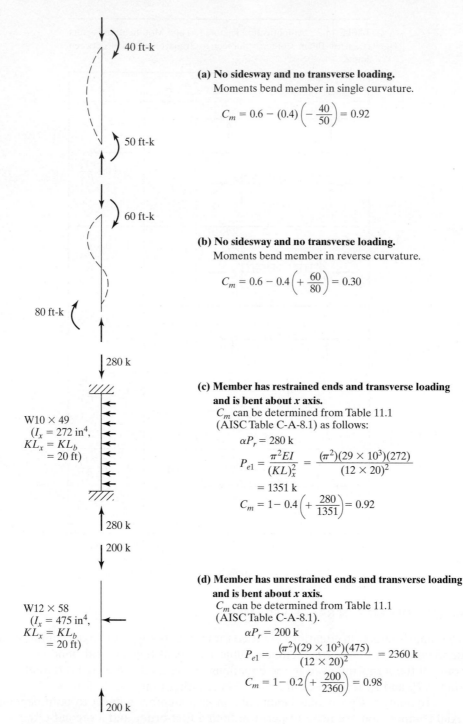

(a) No sidesway and no transverse loading.
Moments bend member in single curvature.

$$C_m = 0.6 - (0.4)\left(-\frac{40}{50}\right) = 0.92$$

(b) No sidesway and no transverse loading.
Moments bend member in reverse curvature.

$$C_m = 0.6 - 0.4\left(+\frac{60}{80}\right) = 0.30$$

(c) Member has restrained ends and transverse loading and is bent about x axis.
C_m can be determined from Table 11.1
(AISC Table C-A-8.1) as follows:

$$\alpha P_r = 280 \text{ k}$$

$$P_{e1} = \frac{\pi^2 EI}{(KL)_x^2} = \frac{(\pi^2)(29 \times 10^3)(272)}{(12 \times 20)^2}$$

$$= 1351 \text{ k}$$

$$C_m = 1 - 0.4\left(+\frac{280}{1351}\right) = 0.92$$

(d) Member has unrestrained ends and transverse loading and is bent about x axis.
C_m can be determined from Table 11.1
(AISC Table C-A-8.1).

$$\alpha P_r = 200 \text{ k}$$

$$P_{e1} = \frac{(\pi^2)(29 \times 10^3)(475)}{(12 \times 20)^2} = 2360 \text{ k}$$

$$C_m = 1 - 0.2\left(+\frac{200}{2360}\right) = 0.98$$

FIGURE 11.6

Example modification or C_m factors.

beam-columns caused by gravity loads) and the moments M_{lt} (these are the moments in beam-columns due to the lateral loads).

Theoretically, if both the loads and frame are symmetrical, M_{lt} will be zero. Similarly, if the frame is braced, M_{lt} will be zero. For practical purposes, you can have lateral deflections in taller buildings with symmetrical dimensions and loads.

Examples 11-3 to 11-5 illustrate the application of the interaction equations to beam-columns that are members of braced frames. In these examples, the approximate second-order analysis is used, only B_1 will be computed, as B_2 is not applicable. *It is to be remembered that C_m was developed for braced frames and thus must be used in these three examples for calculating B_1.* Also, the effective length method is used. This means that the values of axial loads and moments were determined in a first-order analysis using unreduced member stiffness and notional loads were added to the gravity-only load cases.

Example 11-3

A 12-ft W12 × 96 (50 ksi steel) is used as a beam-column in a braced frame. It is bent in single curvature with equal and opposite end moments and is not subjected to intermediate transverse loads. Is the section satisfactory if $P_D = 175$ k, $P_L = 300$ k, and first-order $M_{Dx} = 60$ ft-k and $M_{Lx} = 60$ ft-k?

Solution. Using a W12 × 96 ($A = 28.2$ in^2, $I_x = 833$ in^4, $\phi_b M_{px} = 551$ ft-k, $\dfrac{M_{px}}{\Omega_b} = 367$ ft-k, $L_p = 10.9$ ft, $L_r = 46.7$ ft, $BF = 5.78$ k for LRFD and 3.85 k for ASD)·

LRFD	ASD
$P_{nt} = P_u = (1.2)(175) + (1.6)(300) = 690$ k	$P_{nt} = P_a = 175 + 300 = 475$ k
$M_{ntx} = M_{ux} = (1.2)(60) + (1.6)(60) = 168$ ft-k	$M_{ntx} = M_{ax} = 60 + 60 = 120$ ft-k
For a braced frame, let $K = 1.0$	For a braced frame, let $K = 1.0$
$\therefore (KL)_x = (KL)_y = (1.0)(12) = 12$ ft	$\therefore (KL)_x = (KL)_y = (1.0)(12) = 12$ ft
$P_c = \phi_c P_n = 1080$ k (AISC Table 4-1)	$P_c = \dfrac{P_n}{\Omega_c} = 720$ k (AISC Table 4-1)
$P_r = P_{nt} + B_2\,P_{lt} = 690 + 0 = 690$ k	$P_r = P_{nt} + B_2\,P_{lt} = 475 + 0 = 475$ k
$\dfrac{P_r}{P_c} = \dfrac{690}{1080} = 0.639 > 0.2$	$\dfrac{P_r}{P_c} = \dfrac{475}{720} = 0.660 > 0.2$
$\therefore$ Must use AISC Eq. H1-1a	$\therefore$ Must use AISC Eq. H1-1a
$C_{mx} = 0.6 - 0.4\dfrac{M_1}{M_2}$	$C_{mx} = 0.6 - 0.4\dfrac{M_1}{M_2}$
$C_{mx} = 0.6 - 0.4\left(-\dfrac{168}{168}\right) = 1.0$	$C_{mx} = 0.6 - 0.4\left(-\dfrac{120}{120}\right) = 1.0$

(Continued)

LRFD	ASD
$P_{e1x} = \dfrac{\pi^2 E I_x}{(K_1 L_x)^2} = \dfrac{(\pi^2)(29{,}000)(833)}{(1.0 \times 12 \times 12)^2}$	$P_{e1x} = \dfrac{\pi^2 E I_x}{(K_1 L_x)^2} = \dfrac{(\pi^2)(29{,}000)(833)}{(1.0 \times 12 \times 12)^2}$
$= 11{,}498 \text{ k}$	$= 11{,}498 \text{ k}$
$B_{1x} = \dfrac{C_{mx}}{1 - \dfrac{\alpha P_r}{P_{e1x}}} = \dfrac{1.0}{1 - \dfrac{(1.0)(690)}{11{,}498}} = 1.064$	$B_{1x} = \dfrac{C_m}{1 - \dfrac{\alpha P_r}{P_{e1x}}} = \dfrac{1.0}{1 - \dfrac{(1.6)(475)}{11{,}498}} = 1.071$
$M_{rx} = B_{1x} M_{ntx} = (1.064)(168) = 178.8 \text{ ft-k}$	$M_{rx} = (1.071)(120) = 128.5 \text{ ft-k}$
Since $L_b = 12 \text{ ft} > L_p = 10.9 \text{ ft} < L_r = 46.6 \text{ ft}$	Since $L_b = 12 \text{ ft} > L_p = 10.9 \text{ ft} < L_r = 46.6 \text{ ft}$
$\therefore$ **Zone 2**	$\therefore$ **Zone 2**
$\phi_b M_{px} = 1.0[551 - (5.78)(12 - 10.9)] = 544.6 \text{ ft-k}$	$\dfrac{M_{px}}{\Omega_b} = 1.0[367 - 3.85(12 - 10.9)] = 362.7 \text{ ft-k}$
$\dfrac{P_r}{P_c} + \dfrac{8}{9}\left(\dfrac{M_{rx}}{M_{cx}} + \dfrac{M_{ry}}{M_{cy}}\right)$	$\dfrac{P_r}{P_c} + \dfrac{8}{9}\left(\dfrac{M_{rx}}{M_{cx}} + \dfrac{M_{ry}}{M_{cy}}\right) = \dfrac{475}{720} + \dfrac{8}{9}\left(\dfrac{128.5}{362.7} + 0\right)$
$= \dfrac{690}{1080} + \dfrac{8}{9}\left(\dfrac{178.8}{544.6} + 0\right) = 0.931 < 1.0 \text{ OK}$	$= 0.975 < 1.0 \text{ OK}$
$\therefore$ **Section is satisfactory.**	$\therefore$ **Section is satisfactory.**

In Part 6 of the AISC Manual, a somewhat simplified procedure for solving AISC Equations H1-1a and H1-1b is presented. The student, after struggling through Example 11-3, will surely be delighted to see these expressions, which are used for some of the remaining examples in this chapter.

Various parts of the equations are taken out, and numerical values substituted into them for W sections and recorded in Table 6-1 of the Manual. Each of these terms, such as p, b_x, and b_y, are shown on page 6-3 of the Manual. The revised forms of the equations, which follow, are given in the Manual on page 6-4:

$$pP_r + b_x M_{rx} + b_y M_{ry} \leq 1.0 \qquad \text{(Modified AISC Equation H1-1a)}$$

$$\frac{1}{2} pP_r + \frac{9}{8}(b_x M_{rx} + b_y M_{ry}) \leq 1.0 \quad \text{(Modified AISC Equation H1-1b)}$$

The value of p is based on the larger of $(KL)_y$ and equivalent $(KL)_y = (KL)_x/(r_x/r_y)$, and b_x is based on the unbraced length L_b. A single value of b_y applies for any W shape member because unbraced length is not a factor in weak-axis bending.

Example 11-3 is repeated as Example 11-4 with these simplified expressions. You must be sure to use the magnified values of M_{rx} and P_r in these equations.

Example 11-4

Repeat Example 11-3, using the AISC simplified method of Part 6 of the Manual and the values for K, L, P_r and M_{rx} determined in that earlier example.

Solution

LRFD	ASD
From Example 11-3 (LRFD)	From Example 11-3 (ASD)
$P_r = 690$ k	$P_r = 475$ k
$M_{rx} = 178.8$ ft-k	$M_{rx} = 128.5$ ft-k
From AISC Table 6-1 for a W12 × 96 with $KL = 12$ ft and $L_b = 12$ ft	From AISC Table 6-1 for a W12 × 96 with $KL = 12$ ft and $L_b = 12$ ft
$p = 0.924 \times 10^{-3}$	$p = 1.39 \times 10^{-3}$
$b_x = 1.63 \times 10^{-3}$	$b_x = 2.45 \times 10^{-3}$
$b_y = 3.51 \times 10^{-3}$ (from bottom of table)	$b_y = 5.28 \times 10^{-3}$ (from bottom of table)
Then with the modified equation $(0.924 \times 10^{-3})(690) + (1.63 \times 10^{-3})(178.8)$ $+ (3.51 \times 10^{-3})(0) = 0.929 < 1.0$	Then with the modified equation $(1.39 \times 10^{-3})(475) + (2.45 \times 10^{-3})(128.5)$ $+ (5.28 \times 10^{-3})(0) = 0.975 < 1.0$
Section is satisfactory.	**Section is satisfactory.**

In the examples of this chapter, the student may find that when it becomes necessary to compute a value for P_e, it is slightly easier to use a number given in AISC Table 4-1 for the appropriate steel section than to substitute into the equation $P_e = \dfrac{\pi^2 EI}{(K_1 L)^2}$. For instance, for a W12 × 96, we read at the bottom of the table a value given for P_{ex}, $(KL)^2/10^4 = 23{,}800$. If $KL = 12$ ft, P_{ex} will equal $\dfrac{(23{,}800)(10^4)}{(1.0 \times 12 \times 12)^2} = 11{,}478$ k.

Example 11-5

A 14-ft W14 × 120 (50 ksi steel) is used as a beam-column in a braced frame. It is bent in single curvature with equal and opposite moments. Its ends are rotationally restrained, and it is not subjected to intermediate transverse loads. Is the section satisfactory if $P_D = 70$ k, and $P_L = 100$ k and if it has the first-order moments $M_{Dx} = 60$ ft-k, $M_{Lx} = 80$ ft-k, $M_{Dy} = 40$ ft-k, and $M_{Ly} = 60$ ft-k?

Solution. Using a W14 $\times$ 120 ($A = 35.3$ in^2, $I_x = 1380$ in^4, $I_y = 495$ in^4, $Z_x = 212$ in^3, $Z_y = 102$ in^3, $L_p = 13.2$ ft, $L_r = 51.9$ ft, BF for LRFD = 7.65 k, and BF for ASD = 5.09 k).

LRFD	ASD
$P_{nt} = P_u = (1.2)(70) + (1.6)(100) = 244$ k	$P_{nt} = P_a = 70 + 100 = 170$ k
$M_{ntx} = M_{ux} = (1.2)(60) + (1.6)(80) = 200$ ft-k	$M_{ntx} = M_{ax} = 60 + 80 = 140$ ft-k
$M_{nty} = M_{uy} = (1.2)(40) + (1.6)(60) = 144$ ft-k	$M_{nty} = M_{ay} = 40 + 60 = 100$ ft-k
For a braced frame $K = 1.0$	For a braced frame $K = 1.0$
$KL = (1.0)(14) = 14$ ft	$KL = (1.0)(14) = 14$ ft
$P_c = \phi_c P_n = 1370$ k (AISC Table 4-1)	$P_c = \dfrac{P_n}{\Omega_c} = 912$ k (AISC Table 4-1)
$P_r = P_{nt} + \beta_2 P_{lt} = 244 + 0 = 244$ k	$P_r = P_{nt} + \beta_2 P_{lt} = 170 + 0 = 170$ k
$\dfrac{P_r}{P_c} = \dfrac{244}{1370} = 0.178 < 0.2$	$\dfrac{P_r}{P_c} = \dfrac{170}{912} = 0.186 < 0.2$
$\therefore$ Must use AISC Equation H1-1b	$\therefore$ Must use AISC Equation H1-1b
$C_{mx} = 0.6 - 0.4\left(-\dfrac{200}{200}\right) = 1.0$	$C_{mx} = 0.6 - 0.4\left(-\dfrac{140}{140}\right) = 1.0$
$P_{e1x} = \dfrac{(\pi^2)(29{,}000)(1380)}{(1.0 \times 12 \times 14)^2} = 13{,}995$ k	$P_{e1x} = \dfrac{(\pi^2)(29{,}000)(1380)}{(1.0 \times 12 \times 14)^2} = 13{,}995$ k
$B_{1x} = \dfrac{1.0}{1 - \dfrac{(1.0)(244)}{13{,}995}} = 1.018$	$B_{1x} = \dfrac{1.0}{1 - \dfrac{(1.6)(170)}{13{,}995}} = 1.020$
$M_{rx} = (1.018)(200) = 203.6$ ft-k	$M_{rx} = (1.020)(140) = 142.8$ ft-k
$C_{my} = 0.6 - 0.4\left(-\dfrac{144}{144}\right) = 1.0$	$C_{my} = 0.6 - 0.4\left(-\dfrac{100}{100}\right) = 1.0$
$P_{e1y} = \dfrac{(\pi^2)(29{,}000)(495)}{(1.0 \times 12 \times 14)^2} = 5020$ k	$P_{e1y} = \dfrac{(\pi^2)(29{,}000)(495)}{(1.0 \times 12 \times 14)^2} = 5020$ k
$B_{1y} = \dfrac{1.0}{1 - \dfrac{(1.0)(244)}{5020}} = 1.051$	$B_{1y} = \dfrac{1.0}{1 - \dfrac{(1.6)(170)}{5020}} = 1.057$
$M_{ry} = (1.051)(144) = 151.3$ ft-k	$M_{ry} = (1.057)(100) = 105.7$ ft-k
From AISC Table 6-1, for $KL = 14$ ft and $L_b = 14$ ft	From AISC Table 6-1, for $KL = 14$ ft and $L_b = 14$ ft
$p = 0.730 \times 10^{-3}, b_x = 1.13 \times 10^{-3},$ $b_y = 2.32 \times 10^{-3}$	$p = 1.10 \times 10^{-3}, b_x = 1.69 \times 10^{-3},$ $b_y = 3.49 \times 10^{-3}$

(Continued)

LRFD	ASD
$\dfrac{1}{2}p\,P_r + \dfrac{9}{8}(b_x M_{rx} + b_y M_{ry}) \leq 1.0$	$\dfrac{1}{2}p\,P_r + \dfrac{9}{8}(b_x M_{rx} + b_y M_{ry}) \leq 1.0$
$= \dfrac{1}{2}(0.730 \times 10^{-3})(244)$	$= \dfrac{1}{2}(1.10 \times 10^{-3})(170)$
$+ \dfrac{9}{8}(1.13 \times 10^{-3})(203.6)$	$+ \dfrac{9}{8}(1.69 \times 10^{-3})(142.8)$
$+ \dfrac{9}{8}(2.32 \times 10^{-3})(151.3)$	$+ \dfrac{9}{8}(3.49 \times 10^{-3})(105.7)$
$= 0.743 \leq 1.0$ OK	$= 0.780 \leq 1.0$ OK
Section is satisfactory but perhaps overdesigned.	**Section is satisfactory but perhaps overdesigned.**

Examples 11–3 and 11–5 are reworked using the Direct Analysis Method in Examples 11–6 and 11–7. The values of axial loads and moments given were determined in a first-order analysis using *reduced* member stiffness and application of the notional loads. The analysis yielded values that were essentially equal to the values from the Effective Length Method.

Example 11-6

A 12 ft long W12 × 96 (50 ksi steel) is used as a beam-column in a braced frame. It is bent in single curvature with equal and opposite end moments and is not subjected to intermediate transverse loads. Is the section satisfactory if $P_D = 175$ k, $P_L = 300$ k, and first-order $M_{Dx} = 60$ ft-k and $M_{Lx} = 60$ ft-k?

Solution. Using a W12 × 96 ($A = 28.2$ in², $I_x = 833$ in⁴, $\phi_b M_{px} = 551$ ft-k, $M_{px}/\Omega_b = 367$ ft-k, $L_p = 10.9$ ft, $L_r = 46.7$ ft, $BF = 5.78$ k for LRFD and 3.85 k for ASD).

LRFD	ASD
$P_{nt} = P_u = 1.2\,(175) + 1.6\,(300) = 690$ kips	$P_{nt} = P_a = 175 + 300 = 475$ kips
$M_{ntx} = M_{ux} = 1.2\,(60) + 1.6\,(60) = 168$ ft-k	$M_{ntx} = M_{ax} = 60 + 60 = 120$ ft-k
For Direct Analysis Method, $K = 1.0$	For Direct Analysis Method, $K = 1.0$
$\therefore\ (KL)_x = (KL)_y = 1.0(12) = 12$ ft	$\therefore\ (KL)_x = (KL)_y = 1.0(12) = 12$ ft
$P_c = \Phi_c P_n = 1080$ k (AISC Table 4-1)	$P_c = P_n / \Omega_c = 720$ k (AISC Table 4-1)
B_2 is not required, since braced frame, therefore	B_2 is not required, since braced frame, therefore
$P_r = P_{nt} + B_2 P_{lt} = 690 + 0 = 690$ k	$P_r = P_{nt} + B_2 P_{lt} = 475 + 0 = 475$ k
$\dfrac{P_r}{P_c} = \dfrac{690}{1080} = 0.639 > 0.2$	$\dfrac{P_r}{P_c} = \dfrac{475}{720} = 0.660 > 0.2$

(Continued)

LRFD	ASD
$\therefore$ Must use AISC Equation H1-1a	$\therefore$ Must use AISC Equation H1-1a
$C_m = 0.6 - 0.4 \dfrac{M_1}{M_2}$	$C_m = 0.6 - 0.4 \dfrac{M_1}{M_2}$
$= 0.6 - 0.4\left(-\dfrac{168}{168}\right) = 1.0$	$= 0.6 - 0.4\left(-\dfrac{120}{120}\right) = 1.0$
Determine τ_b:	Determine τ_b:
$\dfrac{\alpha P_r}{P_y} = \dfrac{1.0(690 \text{ k})}{(28.2 \text{ in}^2)(50 \text{ ksi})} = 0.49 < 0.5$	$\dfrac{\alpha P_r}{P_y} = \dfrac{1.6(475 \text{ k})}{(28.2 \text{ in}^2)(50 \text{ ksi})} = 0.539 > 0.5$
$\therefore \ \tau_b = 1.0$	$\therefore \ \tau_b = 4(\alpha P_r/P_y)[1-(\alpha P_r/P_y)]$
	$\tau_b = 4(0.539)[1-(0.539)] = 0.994$
$P_{elx} = \dfrac{\pi^2 0.8\tau_b EI^*}{(K_1 L_x)^2} = \dfrac{\pi^2 0.8(1.0)(29{,}000)(833)}{(1.0 \times 12 \times 12)^2} = 9198 \text{ k}$	$P_{elx} = \dfrac{\pi^2 0.8\tau_b EI^*}{(K_1 L_x)^2}$
	$= \dfrac{\pi^2 0.8(0.994)(29{,}000)(833)}{(1.0 \times 12 \times 12)^2} = 9143 \text{ k}$
$B_{1x} = \dfrac{C_{mx}}{1 - \dfrac{\alpha P_r}{P_{elx}}} = \dfrac{1.0}{1 - \dfrac{1.0(690)}{9198}} = 1.081$	$B_{1x} = \dfrac{C_{mx}}{1 - \dfrac{\alpha P_r}{P_{elx}}} = \dfrac{1.0}{1 - \dfrac{1.6(475)}{9143}} = 1.091$
$M_{rx} = B_{1x}M_{ntx} = 1.081(168) = 181.6 \text{ ft-k}$	$M_{rx} = B_{1x}M_{ntx} = 1.091(120) = 130.9 \text{ ft-k}$
Since $L_b = 12 \text{ ft} > L_p = 10.9 \text{ ft}, < L_r = 46.6 \text{ ft}$	Since $L_b = 12 \text{ ft} > L_p = 10.9 \text{ ft}, < L_r = 46.6 \text{ ft}$
$\therefore$ Zone 2	$\therefore$ Zone 2
$M_{cx} = \Phi_b M_{px} = 1.0[551-(5.78)(12-10.9)] = 544.6 \text{ ft-k}$	$M_c = M_{px}/\Omega_b = 1.0[367 - (3.85)(12 - 10.9)] = 362.7 \text{ ft-k}$
$\dfrac{P_r}{P_c} + \dfrac{8}{9}\left(\dfrac{M_{rx}}{M_{cx}} + \dfrac{M_{ry}}{M_{cy}}\right) \le 1.0$	$\dfrac{P_r}{P_c} + \dfrac{8}{9}\left(\dfrac{M_{rx}}{M_{cx}} + \dfrac{M_{ry}}{M_{cy}}\right) \le 1.0$
$\dfrac{690}{1080} + \dfrac{8}{9}\left(\dfrac{181.6}{544.6} + 0\right) = 0.935 \le 1.0$	$\dfrac{475}{720} + \dfrac{8}{9}\left(\dfrac{130.9}{362.7} + 0\right) = 0.981 \le 1.0$
$\therefore$ **Section is satisfactory.**	$\therefore$ **Section is satisfactory.**

Example 11-7

A 14 ft long W14 $\times$ 120 (50 ksi steel) is used as a beam-column in a braced frame. It is bent in single curvature with equal and opposite end moments. Its ends are rotationally restrained and it is not subjected to intermediate transverse loads. Is the section satisfactory if $P_D = 70$ k, $P_L = 100$ k, and if it has first-order moments $M_{Dx} = 60$ ft-k, $M_{Lx} = 80$ ft-k, $M_{Dy} = 40$ ft-k and $M_{Ly} = 60$ ft-k?

Solution. Using a W14 $\times$ 120 ($A = 35.3 \text{ in}^2$, $I_x = 1380 \text{ in}^4$, $I_y = 495 \text{ in}^4$, $Z_x = 212 \text{ in}^3$, $Z_y = 102 \text{ in}^3$, $L_p = 13.2 \text{ ft}$, $L_r = 51.9 \text{ ft}$, $BF = 7.65$ k for LRFD and 5.09 k for ASD).

LRFD	ASD
$P_{nt} = P_u = 1.2\,(70\text{ k}) + 1.6\,(100\text{ k}) = 244$ kips	$P_{nt} = P_a = 70\text{ k} + 100\text{ k} = 170$ kips
$M_{ntx} = M_{ux} = 1.2\,(60\text{ ft-k}) + 1.6\,(80\text{ ft-k}) = 200$ ft-k	$M_{ntx} = M_{ax} = 60\text{ ft-k} + 80\text{ ft-k} = 140$ ft-k
$M_{nty} = M_{uy} = 1.2\,(40\text{ ft-k}) + 1.6\,(60\text{ ft-k}) = 144$ ft-k	$M_{nty} = M_{ay} = 40\text{ ft-k} + 60\text{ ft-k} = 100$ ft-k

For Direct Analysis Method, $K = 1.0$ | For Direct Analysis Method, $K = 1.0$

$$\therefore \ (KL)_x = (KL)_y = 1.0(14) = 14 \text{ ft} \qquad\qquad \therefore \ (KL)_x = (KL)_y = 1.0(14) = 14 \text{ ft}$$

$P_c = \Phi_c P_n = 1370$ k (AISC Table 4-1) | $P_c = P_n/\Omega_c = 912$ k (AISC Table 4-1)

B_2 is not required, since braced frame, therefore | B_2 is not required, since braced frame, therefore

$P_r = P_{nt} + B_2 P_{lt} = 244 + 0 = 244$ k | $P_r = P_{nt} + B_2 P_{lt} = 170 + 0 = 170$ k

$$\frac{P_r}{P_c} = \frac{244}{1370} = 0.178 < 0.2 \qquad\qquad \frac{P_r}{P_c} = \frac{170}{912} = 0.186 < 0.2$$

$\therefore$ Must use AISC Equation H1-1b | $\therefore$ Must use AISC Equation H1-1b

$$C_{mx} = 0.6 - 0.4\left(-\frac{200}{200}\right) = 1.0 \qquad C_{mx} = 0.6 - 0.4\left(-\frac{140}{140}\right) = 1.0$$

Determine τ_b: | Determine τ_b:

$$\frac{\alpha P_r}{P_y} = \frac{1.0(244\text{ k})}{(35.3\text{ in}^2)(50\text{ ksi})} = 0.138 < 0.5 \qquad \frac{\alpha P_r}{P_y} = \frac{1.0(170\text{ k})}{(35.3\text{ in}^2)(50\text{ ksi})} = 0.154 > 0.5$$

$$\therefore \ \tau_b = 1.0 \qquad\qquad\qquad \therefore \ \tau_b = 1.0$$

$$P_{e1x} = \frac{\pi^2 0.8\tau_b EI^*}{(K_1 L_x)^2} = \frac{\pi^2 0.8(1.0)(29{,}000)(1380)}{(1.0 \times 12 \times 14)^2}$$
$$= 11{,}196 \text{ k}$$

$$P_{e1x} = \frac{\pi^2 0.8\tau_b EI^*}{(K_1 L_x)^2} = \frac{\pi^2 0.8(1.0)(29{,}000)(1380)}{(1.0 \times 12 \times 14)^2} = 11{,}196 \text{ k}$$

$$B_{1x} = \frac{C_{mx}}{1 - \dfrac{\alpha P_r}{P_{e1x}}} = \frac{1.0}{1 - \dfrac{1.0(244)}{11{,}196}} = 1.022$$

$$B_{1x} = \frac{C_{mx}}{1 - \dfrac{\alpha P_r}{P_{e1x}}} = \frac{1.0}{1 - \dfrac{1.6(170)}{11{,}196}} = 1.025$$

$M_{rx} = B_{1x} M_{ntx} = 1.022(200) = 204.5$ ft-k | $M_{rx} = B_{1x} M_{ntx} = 1.025(140) = 143.5$ ft-k

$$C_{my} = 0.6 - 0.4\left(-\frac{144}{144}\right) = 1.0 \qquad C_{my} = 0.6 - 0.4\left(-\frac{100}{100}\right) = 1.0$$

$$P_{e1y} = \frac{\pi^2 0.8\tau_b EI^*}{(K_1 L_y)^2} = \frac{\pi^2 0.8(1.0)(29{,}000)(495)}{(1.0 \times 12 \times 14)^2}$$
$$= 4016 \text{ k}$$

$$P_{e1y} = \frac{\pi^2 0.8\tau_b EI^*}{(K_1 L_y)^2} = \frac{\pi^2 0.8(1.0)(29{,}000)(495)}{(1.0 \times 12 \times 14)^2} = 4016 \text{ k}$$

$$B_{1y} = \frac{C_{my}}{1 - \dfrac{\alpha P_r}{P_{e1y}}} = \frac{1.0}{1 - \dfrac{1.0(244)}{4016}} = 1.065$$

$$B_{1y} = \frac{C_{my}}{1 - \dfrac{\alpha P_r}{P_{e1y}}} = \frac{1.0}{1 - \dfrac{1.6(170)}{4016}} = 1.073$$

$M_{ry} = B_{1y} M_{nty} = 1.065(144) = 153.3$ ft-k | $M_{ry} = B_{1y} M_{nty} = 1.073(100) = 107.3$ ft-k

(Continued)

LRFD	ASD
From AISC Table 6-1, for $KL = 14$ ft and $L_b = 14$ ft	From AISC Table 6-1, for $KL = 14$ ft and $L_b = 14$ ft
$p = 0.730 \times 10^{-3}, b_x = 1.13 \times 10^{-3}, b_y = 2.32 \times 10^{-3}$	$p = 1.10 \times 10^{-3}, b_x = 1.69 \times 10^{-3}, b_y = 3.49 \times 10^{-3}$
$\dfrac{1}{2} pP_r + \dfrac{9}{8}(b_x M_{rx} + b_y M_{ry}) \leq 1.0$	$\dfrac{1}{2} pP_r + \dfrac{9}{8}(b_x M_{rx} + b_y M_{ry}) \leq 1.0$
$\dfrac{1}{2}(0.730 \times 10^{-3})(244) + \dfrac{9}{8}(1.13 \times 10^{-3})(204.5)$	$\dfrac{1}{2}(1.10 \times 10^{-3})(170) + \dfrac{9}{8}(1.69 \times 10^{-3})(143.5)$
$+ \dfrac{9}{8}(2.32 \times 10^{-3})(153.3) = 0.749 \leq 1.0$ OK	$+ \dfrac{9}{8}(3.49 \times 10^{-3})(107.3) = 0.788 \leq 1.0$ OK
Section is satisfactory but perhaps overdesigned.	**Section is satisfactory but perhaps overdesigned.**

Example 11-8

For the truss shown in Fig. 11.7(a), a W8 × 35 is used as a continuous top chord member from joint L_0 to joint U_3. If the member consists of 50 ksi steel, does it have sufficient strength to resist the loads shown in parts (b) and (c) of the figure? The factored or LRFD loads are shown in part (b), while the service or ASD loads are shown in part (c). The 17.6 k and 12 k loads represent the reaction from a purlin. The compression flange of the W8 is braced only at the ends about the x-x axis, $L_x = 13$ ft, and at the ends and the concentrated load about the y-y axis, $L_y = 6.5$ ft and $L_b = 6.5$ ft.

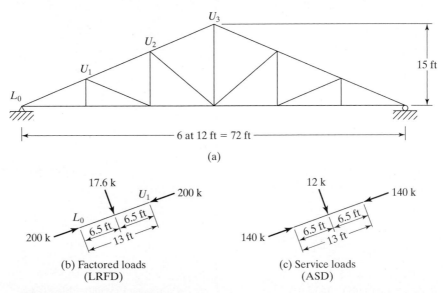

(a)

(b) Factored loads
(LRFD)

(c) Service loads
(ASD)

FIGURE 11.7

A truss whose top chord is subject to intermediate loads.

Solution. The Effective Length Method is used in this problem.

Using a W8 × 35 (A = 10.3 in^2, I_x = 127 in^4, r_x = 3.51 in, r_y = 2.03 in, L_P = 7.17 ft,

$\phi_b M_{Px}$ = 130 ft-k, $\dfrac{M_{Px}}{\Omega_b}$ = 86.6 ft-k, r_x/r_y = 1.73).

LRFD	ASD
$P_{nt} = P_u$ from figure = 200 k = P_r	$P_{nt} = P_a$ from figure = 140 k = P_r
Conservatively assume $K_x = K_y = 1.0$. In truth, the K-factor is somewhere between $K = 1.0$ (pinned–pinned end condition) and $K = 0.8$ (pinned–fixed end condition) for segment $L_o U_i$	Conservatively assume $K_x = K_y = 1.0$. In truth, the K-factor is somewhere between $K = 1.0$ (pinned–pinned end condition) and $K = 0.8$ (pinned–fixed end condition) for segment $L_o U_i$
$\left(\dfrac{KL}{r}\right)_x = \dfrac{(1.0)(12 \times 13)}{3.51} = 44.44 \leftarrow$	$\left(\dfrac{KL}{r}\right)_x = \dfrac{(1.0)(12 \times 13)}{3.51} = 44.44 \leftarrow$
$\left(\dfrac{KL}{r}\right)_y = \dfrac{(1.0)(12 \times 6.5)}{2.03} = 38.42$	$\left(\dfrac{KL}{r}\right)_y = \dfrac{(1.0)(12 \times 6.5)}{2.03} = 38.42$
From AISC Table 4-22, F_y = 50 ksi	From AISC Table 4-22, F_y = 50 ksi
$\phi_c F_{cr}$ = 38.97 ksi	$\dfrac{F_{cr}}{\Omega_c}$ = 25.91 ksi
$\phi_c P_n = (38.97)(10.3) = 401.4$ k = P_c	$\dfrac{P_n}{\Omega_c} = (25.91)(10.3) = 266.9$ k = P_r
$\dfrac{P_r}{P_c} = \dfrac{200}{401.4} = 0.498 > 0.2$	$\dfrac{P_r}{P_c} = \dfrac{140}{266.9} = 0.525 > 0.2$
$\therefore$ Must use AISC Eq. H1-1a	$\therefore$ Must use AISC Eq. H1-1a
Computing P_{e1x} and C_{mx}	Computing P_{e1x} and C_{mx}
$P_{e1x} = \dfrac{(\pi^2)(29{,}000)(127)}{(1.0 \times 12 \times 13)^2} = 1494$ k	$P_{e1x} = \dfrac{(\pi^2)(29{,}000)(127)}{(1.0 \times 12 \times 13)^2} = 1494$ k
	Computing C_m as in LRFD
From Table 11.1 For	From Table 11.1 For
$C_{mx} = 1 - 0.2\left(\dfrac{1.0\,(200)}{1494}\right) = 0.973$	$C_{mx} = 1 - 0.2\left(\dfrac{1.6\,(140)}{1494}\right) = 0.970$
For	For

(*Continued*)

$$C_{mx} = 1 - 0.3\left(\frac{1.0\,(200)}{1494}\right) = 0.960$$

Avg $C_{mx} = 0.967$

Computing M_{ux}

For   17.6 k

$$M_{ux} = \frac{PL}{4} = \frac{(17.6)(13)}{4} = 57.2 \text{ ft-k}$$

For   17.6 k

$$M_{ux} = \frac{3\,PL}{16} = \frac{(3)(17.6)(13)}{16} = 42.9 \text{ ft-k}$$

Avg $M_{ux} = 50.05$ ft-k $= M_{rx}$

$$B_{1x} = \frac{0.967}{1 - \dfrac{(1)(200)}{1494}} = 1.116$$

$M_r = (1.116)(50.05) = 55.86$ ft-k

Since $L_b = 6.5$ ft $< L_p = 7.17$ ft

$$\therefore \text{ Zone } ①$$

$\phi_b M_{nx} = 130$ ft-k $= M_{cx}$

Using Equation H1-1a

$$\frac{P_r}{P_c} + \frac{8}{9}\left(\frac{M_{rx}}{M_{cx}} + \frac{M_{ry}}{M_{cy}}\right) \leq 1.0$$

$$\frac{200}{401.4} + \frac{8}{9}\left(\frac{55.86}{130} + 0\right) \leq 1.0$$

$0.880 \leq 1.0$ Section OK

From AISC Table 6-1

$(KL)_y = 6.5$ ft

$$(KL)_{yEQUIV} = \frac{(KL)_x}{r_x/r_y} = \frac{13}{1.73} = 7.51 \text{ ft} \leftarrow$$

$P = 2.50 \times 10^{-3}$,   for $KL = 7.51$ ft

$b_x = 6.83 \times 10^{-3}$,   for $L_b = 6.5$ ft

$p\,P_r + b_x\,M_{rx} + b_y\,M_{ry} \leq 1.0$

$= (2.50 \times 10^{-3})\,(200) + (6.83 \times 10^{-3})\,(55.86) + 0$

$= 0.882 \leq 1.0$ Section OK

Section is Satisfactory.

$$C_{mx} = 1 - 0.3\left(\frac{1.6\,(140)}{1494}\right) = 0.955$$

Avg $C_{mx} = 0.963$

Computing M_{ax}

For   12 k

$$M_{ax} = \frac{(12)(13)}{4} = 39 \text{ ft-k}$$

For   12 k

$$M_{ax} = \frac{(3)(12)(13)}{16} = 29.25 \text{ ft-k}$$

Avg $M_{ax} = 34.13$ ft-k $= M_{rx}$

$$B_{1x} = \frac{0.967}{1 - \dfrac{(1.6)(140)}{1494}} = 1.138$$

$M_r = (1.138)(34.13) = 38.84$ ft-k

Since $L_b = 6.5$ ft $< L_P = 7.17$ ft

$$\therefore \text{ Zone } ①$$

$$\frac{M_{nx}}{\Omega_b} = 86.6 \text{ ft-k} = M_{cx}$$

Using Equation H1-1a

$$\frac{P_r}{P_c} + \frac{8}{9}\left(\frac{M_{rx}}{M_{cx}} + \frac{M_{ry}}{M_{cy}}\right) \leq 1.0$$

$$\frac{140}{266.9} + \frac{8}{9}\left(\frac{38.84}{88.6} + 0\right) \leq 1.0$$

$0.914 \leq 1.0$ Section OK

From AISC Table 6-1

$(KL)_y = 6.5$ ft

$$(KL)_{yEQUIV} = \frac{(KL)_x}{r_x/r_y} = \frac{13}{1.73} = 7.51 \text{ ft} \leftarrow$$

$P = 3.75 \times 10^{-3}$,   for $KL = 7.51$ ft

$b_x = 10.3 \times 10^{-3}$,   for $L_b = 6.5$ ft

$p\,P_r + b_x\,M_{rx} + b_y\,M_{ry} \leq 1.0$

$= (3.75 \times 10^{-3})\,(140) + (10.3 \times 10^{-3})\,(38.84) + 0$

$= 0.925 \leq 1.0$ Section OK

Section is Satisfactory.

11.8   BEAM-COLUMNS IN UNBRACED FRAMES

The maximum primary moments in unbraced frames almost always occur at the column ends. As you can see in Fig. 11.3, the maximum sidesway moments always occur at the member ends and the total moment for a particular column is determined by adding its primary end moment, M_{lt}, to its sidesway moment, $P_{nt} \Delta$. As described in Section 11.3, B_2 is the multiplier used in the Approximate Second-Order Analysis to account for $P\text{-}\Delta$ effect.

In Examples 11-9 and 11-10, beam-column sections will be analyzed using both the Direct Analysis Method and the Effective Length Method. In both examples, the approximate analysis method is used to account for the second-order effects. The magnification factors, B_1 and B_2, are determined for each beam-column in each direction of lateral translation. These second-order loads and moments are substituted into the appropriate interaction equation to determine if the section is satisfactory.

Example 11-9

Part (a) – Direct Analysis Method
A W10 × 39 of $F_y = 50$ ksi steel, is used for a 14 ft long beam-column in an unbraced frame about the x-x axis but is braced about the y-y axis. Based on a first-order analysis using the requirements of the Direct Analysis Method, the member supports the following factored loads: $P_{nt} = 130$ k, $P_{lt} = 25$ ft-k, $M_{ntx} = 45$ ft-k, and $M_{ltx} = 15$ ft-k. C_{mx} was determined to be 0.85. P_{story} is 1604 k and the ratio of P_{mf}/P_{story} is 0.333. H, the story shear, is equal to 33.4 k and the drift index (Δ_H/L) is 0.0025. Using the LRFD procedure, is the member satisfactory?

Solution

W10 × 39 $(A = 11.5 \text{ in}^2, I_x = 209 \text{ in}^4)$

$C_{mx} = 0.85$ (given) $\alpha = 1.0$ (LRFD)

$P_r = P_{nt} + P_{lt} = 130 \text{ k} + 25 \text{ k} = 155 \text{ k}$

Determine τ_b:

$$\frac{\alpha P_r}{P_y} = \frac{1.0(155 \text{ k})}{(11.5 \text{ in}^2)(50 \text{ ksi})} = 0.27 < 0.5$$

$\therefore \tau_b = 1.0$

$$P_{elx} = \frac{\pi^2 0.8 \tau_b E I^*}{(K_1 L_x)^2} = \frac{\pi^2 0.8 (1.0)(29{,}000)(209)}{(1.0 \times 12 \times 14)^2} = 1{,}696 \text{ k}$$

$$B_{1x} = \frac{C_{mx}}{1 - \dfrac{\alpha P_r}{P_{elx}}} = \frac{0.85}{1 - \dfrac{1.0(155)}{1{,}696}} = 0.94 < 1.0 \qquad \therefore B_{1x} = 1.0$$

$P_{story} = 1604 \text{ k}$ (given) $\Delta_H/L = 0.0025$ (given)

$H = 33.4 \text{ k}$ (given) $\alpha = 1.0$ (LRFD)

$$R_m = 1 - 0.15(P_{mf}/P_{story}) = 1 - 0.15(0.333) = 0.95$$

$$P_{e \, story \, x} = R_m \frac{(H)}{\dfrac{\Delta_H}{L}} = 0.95 \left(\frac{33.4}{0.0025}\right) = 12,692 \text{ k}$$

$$B_{2x} = \frac{1}{1 - \dfrac{\alpha(P_{story})}{P_{e \, story \, x}}} = \frac{1}{1 - \dfrac{1.0(1604)}{12,692}} = 1.15$$

$$\therefore \; P_r = P_{nt} + B_2 \, P_{lt} = 130 \text{ k} + 1.15(25 \text{ k}) = 158.8 \text{ k}$$

$$M_{rx} = B_{1x} \, M_{ntx} + B_{2x} \, M_{ltx} = 1.0(45 \text{ ft-k}) + 1.15(15 \text{ ft-k}) = 62.3 \text{ ft-k}$$

For Direct Analysis Method, $K = 1.0$

$$\therefore \; (KL)_x = (KL)_y = 1.0(14) = 14 \text{ ft}$$

$$P_c = \Phi_c P_n = 306 \text{ k} \qquad\qquad\qquad \text{(AISC Table 4-1)}$$

$$\frac{P_r}{P_c} = \frac{158.8}{306} = 0.52 > 0.2$$

$\therefore$ Must use AISC Equation H1-1a.

For W10 × 39, $\Phi M_{px} = 176$ ft-k, $L_p = 6.99$ ft, $L_r = 24.2$ ft

$BF = 3.78$ k, $L_b = 14$ ft, Zone 2, $C_b = 1.0$

$$\Phi M_{nx} = C_b \left[\Phi M_{px} - BF(L_b - L_p)\right] \Phi M_{px}$$

$$\Phi M_{nx} = 1.0[176 - 3.78(14 - 6.99)] = 149.5 \text{ ft-k}$$

Equation H1-1a:

$$\frac{158.8}{306} + \frac{8}{9}\left(\frac{62.3}{149.5} + 0\right) = 0.889 < 1.0 \quad \text{OK}$$

Additional check:

From Table 6-1 for $KL = 14$ ft and $L_b = 14$ ft

$$p = 3.27 \times 10^{-3}, b_x = 5.96 \times 10^{-3}$$

$$3.27 \times 10^{-3}(158.8) + 5.96 \times 10^{-3}(62.3) = 0.891 < 1.0 \quad \text{OK}$$

$\therefore$ **Section is satisfactory.**

Part (b) – Effective Length Method

Repeat using the same W10 × 39, 14 ft long section. Based on a first-order analysis using the requirement of the Effective Length Method, the member essentially has the same loads and moments. C_{mx} is still 0.85. P_{story} is 1604 k, the ratio of P_{mf}/P_{story} is 0.333 and H, the story shear, is equal to 33.4 k. The drift index (Δ_H/L) is reduced to 0.0020 due to the increased member stiffness in the analysis when compared the direct analysis method. K_x was determined to be 1.2 and the K_y is equal 1.0. Using the LRFD procedure, is the member satisfactory?

Solution

W10 × 39 ($A = 11.5$ in^2, $I_x = 209$ in^4)

$C_{mx} = 0.85$ (given) $\alpha = 1.0$ (LRFD)

$P_r = P_{nt} + P_{lt} = 130$ k $+ 25$ k $= 155$ k

$$P_{elx} = \frac{\pi^2 EI^*}{(K_1 L_x)^2} = \frac{\pi^2 (29,000)(209)}{(1.0 \times 12 \times 14)^2} = 2120 \text{ k}$$

$$B_{1x} = \frac{C_{mx}}{1 - \dfrac{\alpha P_r}{P_{elx}}} = \frac{0.85}{1 - \dfrac{1.0(155)}{2120}} = 0.92 < 1.0$$

$\therefore$ $B_{1x} = 1.0$

$P_{story} = 1604$ k (given) $\Delta_H/L = 0.0020$ (given)

$H = 33.4$ k (given) $\alpha = 1.0$ (LRFD)

$R_m = 1 - 0.15(P_{mf}/P_{story}) = 1 - 0.15(0.333) = 0.95$

$$P_{e\,story\,x} = R_m \left(\frac{H}{\dfrac{\Delta_H}{L}} \right) = 0.95 \left(\frac{33.4}{0.0020} \right) = 15,865 \text{ k}$$

$$B_{2x} = \frac{1}{1 - \dfrac{\alpha(P_{story})}{P_{e\,story\,x}}} = \frac{1}{1 - \dfrac{1.0(1604)}{15,865}} = 1.11$$

$\therefore$ $P_r = P_{nt} + B_2 P_{lt} = 130$ k $+ 1.11(25$ k$) = 157.8$ k

$M_{rx} = B_{1x} M_{ntx} + B_{2x} M_{ltx} = 1.0(45$ ft-k$) + 1.11(15$ ft-k$) = 61.7$ ft-k

Effective Length Method: $K_y = 1.0$ and $K_x = 1.2$

$\therefore$ $(KL)_y = 1.0(14) = 14$ ft $\leftarrow$ controls

Equivalent $(KL)_y = \dfrac{(KL)_x}{\dfrac{r_x}{r_y}} = \dfrac{(1.2)(14)}{2.16} = 7.78$ ft

$P_c = \Phi_c P_n = 306$ k   (AISC Table 4-1)

$\dfrac{P_r}{P_c} = \dfrac{157.8}{306} = 0.52 > 0.2$

∴ Must use AISC Equation H1-1a.

For W10 × 39, $\Phi M_{px} = 176$ ft-k, $L_p = 6.99$ ft, $L_r = 24.2$ ft

$BF = 3.78$ k, $L_b = 14$ ft, Zone 2, $C_b = 1.0$

$\Phi M_{nx} = C_b [\Phi M_{px} - BF(L_b - L_p)] \leq \Phi M_{px}$

$\Phi M_{nx} = 1.0[176 - 3.78(14-6.99)] = 149.5$ ft-k

Equation H1-1a:

$$\frac{157.8}{306} + \frac{8}{9}\left(\frac{61.7}{149.5} + 0\right) = 0.883 < 1.0 \quad \text{OK}$$

Additional check:

From Table 6-1 for $KL = 14$ ft and $L_b = 14$ ft

$p = 3.27 \times 10^{-3}, b_x = 5.96 \times 10^{-3}$

$3.27 \times 10^{-3} (157.8) + 5.96 \times 10^{-3} (61.7) = 0.884 < 1.0 \quad \text{OK}$

∴ **Section is satisfactory.**

Example 11-10

Part (a) – Direct Analysis Method

A 14 ft W10 × 45 of $F_y = 50$ ksi steel, is used in the same unbraced frame building as given in Example 11-9. The major difference is that this beam-column is bent about both the x-x axis and y-y axis. Based on a first-order analysis using the requirements of the Direct Analysis Method, the member supports the following factored loads: $P_{nt} = 65$ k, $P_{lt} = 30$ ft-k, $M_{ntx} = 50$ ft-k, $M_{ltx} = 20$ ft-k, $M_{nty} = 16$ ft-k, and $M_{lty} = 8$ ft-k. C_{mx} and C_{my} were determined to be 0.85. P_{story} is 1604 k and the ratio of P_{mf}/P_{story} is 0.333. H, the story shear, is equal to 33.4 k and the drift indices are $(\Delta_H/L)_x = 0.0025$ and $(\Delta_H/L)_y = 0.0043$. Using the LRFD procedure, is the member satisfactory?

Solution

W10 × 45 ($A = 13.3$ in², $I_x = 248$ in⁴, $I_y = 53.4$ in⁴)

$C_{mx} = C_{my} = 0.85$ (given) $\alpha = 1.0$ (LRFD)

$P_r = P_{nt} + P_{lt} = 65$ k $+ 30$ k $= 95$ k

Determine τ_b:

$$\frac{\alpha P_r}{P_y} = \frac{1.0(95 \text{ k})}{(13.3 \text{ in}^2)(50 \text{ ksi})} = 0.14 < 0.5$$

$\therefore \ \tau_b = 1.0$

$$P_{e1x} = \frac{\pi^2 0.8 \tau_b E I^*}{(K_1 L_x)^2} = \frac{\pi^2 0.8(1.0)(29{,}000)(248)}{(1.0 \times 12 \times 14)^2} = 2012 \text{ k}$$

$$B_{1x} = \frac{C_{mx}}{1 - \dfrac{\alpha P_r}{P_{e1x}}} = \frac{0.85}{1 - \dfrac{1.0(95)}{2012}} = 0.89 < 1.0 \qquad \therefore \ B_{1x} = 1.0$$

$$P_{e1y} = \frac{\pi^2 0.8 \tau_b E I^*}{(K_1 L_y)^2} = \frac{\pi^2 0.8(1.0)(29{,}000)(53.4)}{(1.0 \times 12 \times 14)^2} = 433 \text{ k}$$

$$B_{1y} = \frac{C_{my}}{1 - \dfrac{\alpha P_r}{P_{e1y}}} = \frac{0.85}{1 - \dfrac{1.0(95)}{433}} = 1.09 \qquad \therefore \ B_{1y} = 1.09$$

$P_{story} = 1604 \text{ k (given)} \qquad (\Delta_H/L)_x = 0.0025 \text{ and } (\Delta_H/L)_y = 0.0043 \text{ (given)}$

$H = 33.4 \text{ k (given)} \qquad \alpha = 1.0 \text{ (LRFD)}$

$R_m = 1 - 0.15(P_{mf}/P_{story}) = 1 - 0.15(0.333) = 0.95$

$$P_{e\,story\,x} = R_m\left(\frac{H}{\dfrac{\Delta_H}{L}}\right) = 0.95\left(\frac{33.4}{0.0025}\right) = 12{,}692 \text{ k}$$

$$B_{2x} = \frac{1}{1 - \dfrac{\alpha(P_{story})}{P_{e\,story\,x}}} = \frac{1}{1 - \dfrac{1.0(1604)}{12{,}692}} = 1.15$$

$$P_{e\,story\,y} = R_m\left(\frac{H}{\dfrac{\Delta_H}{L}}\right) = 0.95\left(\frac{33.4}{0.0043}\right) = 7379 \text{ k}$$

$$B_{2y} = \frac{1}{1 - \dfrac{\alpha(P_{story})}{P_{e\,story\,y}}} = \frac{1}{1 - \dfrac{1.0(1604)}{7379}} = 1.28$$

$\therefore \ P_r = P_{nt} + B_2 P_{lt} = 65 \text{ k} + 1.28(30 \text{ k}) = 103.4 \text{ k}$

$M_{rx} = B_{1x} M_{ntx} + B_{2x} M_{ltx} = 1.0(50 \text{ ft-k}) + 1.15(20 \text{ ft-k}) = 73.0 \text{ ft-k}$

$M_{ry} = B_{1y} M_{nty} + B_{2y} M_{lty} = 1.09(16 \text{ ft-k}) + 1.28(8 \text{ ft-k}) = 27.7 \text{ ft-k}$

For Direct Analysis Method, $K = 1.0$

$\therefore \ (KL)_x = (KL)_y = 1.0(14) = 14 \text{ ft}$

$P_c = \Phi_c P_n = 359 \text{ k}$ (AISC Table 4-1)

$$\frac{P_r}{P_c} = \frac{103.4}{359} = 0.29 > 0.2$$

∴ Must use AISC Equation H1-1a

For W10 × 45, $\Phi M_{px} = 206$ ft-k, $L_p = 7.10$ ft, $L_r = 26.9$ ft

$BF = 3.89$ k, $L_b = 14$ ft, Zone 2, $C_b = 1.0$

$\Phi M_{nx} = C_b [\Phi M_{px} - BF(L_b - L_p)] \leq \Phi M_{px}$

$\Phi M_{nx} = 1.0[206 - 3.89(14 - 7.10)] = 179.2$ ft-k

$\Phi M_{ny} = \Phi M_{py} = 76.1$ ft-k   (AISC Table 3-4)

Equation H1-1a:

$$\frac{103.4}{359} + \frac{8}{9}\left(\frac{73.0}{179.2} + \frac{27.7}{76.1}\right) = 0.974 < 1.0 \quad \text{OK}$$

Additional check:

From Table 6-1 for $KL = 14$ ft and $L_b = 14$ ft

$p = 2.78 \times 10^{-3}, b_x = 4.96 \times 10^{-3}, b_y = 11.7 \times 10^{-3}$

$2.78 \times 10^{-3} (103.4) + 4.96 \times 10^{-3} (73.0) + 11.7 \times 10^{-3}(27.7) = 0.974 < 1.0 \quad \text{OK}$

∴ **Section is satisfactory.**

Part (b) – Effective Length Method
Repeat using the same W10 × 45, 14 ft long section. Based on a first-order analysis using the requirement of the Effective Length Method, the member essentially has the same loads and moments. C_{mx} is still 0.85. P_{story} is 1604 k, the ratio of P_{mf}/P_{story} is 0.333 and H, the story shear, is equal to 33.4 k. The drift indices are reduced to $(\Delta_H/L)_x = 0.0020$ and $(\Delta_H/L)_y = 0.0034$ due to the increased member stiffness in the analysis when compared with the direct analysis method. K_x was determined to be 1.31 and the K_y equal to 1.25. Using the LRFD procedure, is the member satisfactory?

Solution

W10 × 45 ($A = 13.3$ in^2, $I_x = 248$ in^4, $I_y = 53.4$ in^4)

$C_{mx} = C_{my} = 0.85$ (given) $\alpha = 1.0$ (LRFD)

$P_r = P_{nt} + P_{lt} = 65$ k $+ 30$ k $= 95$ k

$$P_{e1x} = \frac{\pi^2 E I^*}{(K_1 L_x)^2} = \frac{\pi^2 (29,000)(248)}{(1.0 \times 12 \times 14)^2} = 2515 \text{ k}$$

$$B_{1x} = \frac{C_{mx}}{1 - \dfrac{\alpha P_r}{P_{e1x}}} = \frac{0.85}{1 - \dfrac{1.0(95)}{2515}} = 0.88 < 1.0 \qquad \therefore\ B_{1x} = 1.0$$

$$P_{e1y} = \frac{\pi^2 EI^*}{(K_1 L_x)^2} = \frac{\pi^2 (29{,}000)(53.4)}{(1.0 \times 12 \times 14)^2} = 542 \text{ k}$$

$$B_{1y} = \frac{C_{mx}}{1 - \dfrac{\alpha P_r}{P_{elx}}} = \frac{0.85}{1 - \dfrac{1.0(95)}{542}} = 1.03 \quad \therefore \quad B_{1y} = 1.03$$

$P_{story} = 1604 \text{ k (given)}$ $\qquad (\Delta_H/L)_x = 0.0020$ and $(\Delta_H/L)_y = 0.0034$ (given)

$H = 33.4 \text{ k (given)}$ $\qquad\qquad \alpha = 1.0 \text{ (LRFD)}$

$R_m = 1 - 0.15(P_{mf}/P_{story}) = 1 - 0.15(0.333) = 0.95$

$$P_{e \text{ story } x} = R_m \left(\frac{H}{\dfrac{\Delta_H}{L}} \right) = 0.95 \left(\frac{33.4}{0.0020} \right) = 15{,}865 \text{ k}$$

$$B_{2x} = \frac{1}{1 - \dfrac{\alpha(P_{story})}{P_{e \text{ story } x}}} = \frac{1}{1 - \dfrac{1.0(1604)}{15{,}865}} = 1.11$$

$$P_{e \text{ story } y} = R_m \left(\frac{H}{\dfrac{\Delta_H}{L}} \right) = 0.95 \left(\frac{33.4}{0.0034} \right) = 9332 \text{ k}$$

$$B_{2y} = \frac{1}{1 - \dfrac{\alpha(P_{story})}{P_{e \text{ story } y}}} = \frac{1}{1 - \dfrac{1.0(1604)}{9332}} = 1.21$$

$\therefore \; P_r = P_{nt} + B_2 P_{lt} = 65 \text{ k} + 1.21(30 \text{ k}) = 101.3 \text{ k}$

$M_{rx} = B_{1x}M_{ntx} + B_{2x}M_{ltx} = 1.0(50 \text{ ft-k}) + 1.11 (20 \text{ ft-k}) = 72.2 \text{ ft-k}$

$M_{ry} = B_{1y}M_{nty} + B_{2y}M_{lty} = 1.03(16 \text{ ft-k}) + 1.21(8 \text{ ft-k}) = 26.2 \text{ ft-k}$

Effective Length Method: $K_y = 1.25$ and $K_x = 1.31$

$\therefore \; (KL)_y = 1.25(14) = 17.5 \text{ ft} \leftarrow$ controls

Equivalent $(KL)_y = \dfrac{(KL)_x}{r_x/r_y} = \dfrac{(1.31)(14)}{2.15} = 8.53 \text{ ft}$

$P_c = \Phi_c P_n = 269.5 \text{ k}$ $\hfill$ (AISC Table 4-1)

$$\frac{P_r}{P_c} = \frac{101.3}{269.5} = 0.38 > 0.2$$

$\therefore \;$ Must use AISC Equation H1-1a

For W10 × 45, $\Phi M_{px} = 206 \text{ ft-k}$, $L_p = 7.10 \text{ ft}$, $L_r = 26.9 \text{ ft}$

$BF = 3.89$ k, $L_b = 14$ ft, Zone 2, $C_b = 1.0$

$\Phi M_{nx} = C_b [\Phi M_{px} - BF(L_b - L_p)] \le \Phi M_{px}$

$\Phi M_{nx} = 1.0[206 - 3.89(14 - 7.10)] = 179.2$ ft-k

$\Phi M_{ny} = \Phi M_{py} = 76.1$ ft-k   (AISC Table 3-4)

Equation H1-1a:

$$\frac{101.3}{268.5} + \frac{8}{9}\left(\frac{72.2}{179.2} + \frac{26.2}{76.1}\right) = 1.041 > 1.0 \quad \text{N.G.}$$

Additional check:

From Table 6-1 for $KL = 17.5$ ft and $L_b = 14$ ft

$p = 3.72 \times 10^{-3}, b_x = 4.96 \times 10^{-3}, b_y = 11.7 \times 10^{-3}$

$3.72 \times 10^{-3} (101.3) + 4.96 \times 10^{-3} (72.2) + 11.7 \times 10^{-3}(26.2) = 1.041 > 1.0$ N.G.

∴ **Section is NOT satisfactory.**

___

11.9   DESIGN OF BEAM–COLUMNS—BRACED OR UNBRACED

The design of beam–columns involves a trial-and-error procedure. A trial section is se-lected by some process and is then checked with the appropriate interaction equation. If the section does not satisfy the equation, or if it's too much on the safe side (that is, if it's overdesigned), another section is selected, and the interaction equation is applied again. Probably, the first thought of the reader is, "I sure hope that we can select a good section the first time and not have to go through all that rigamarole more than once or twice." We certainly can make a good estimate, and that's the topic of the remainder of this section.

A common method used for selecting sections to resist both moments and axial loads is the **equivalent axial load**, or **effective axial load, method**. With this method, the axial load (P_u or P_a) and the bending moment or bending moments M_{ux}, M_{uy}, or M_{ax} and M_{ay}) are replaced with a fictitious concentric load P_{ueq} or P_{aeq}, equivalent approx-imately to the actual axial load plus the moment effect.

It is assumed for this discussion that it is desired to select the most economical section to resist both a moment and an axial load. By a trial-and-error procedure, it is possible eventually to select the lightest section. Somewhere, however, there is a ficti-tious axial load that will require the same section as the one required for the actual mo-ment and the actual axial load. This fictitious load is called the equivalent axial load, or the effective axial load P_{ueq} or P_{aeq}.

Equations are used to convert the bending moment into an estimated equivalent axial load P'_u or P'_a, which is added to the design axial load P_u, or to P_a for ASD. The total of $P_u + P'_u$ or $P_a + P'_a$ is the equivalent or effective axial load P_{ueq} or P_{aeq}, and it is used to enter the concentric column tables of Part 4 of the AISC Manual for the choice of a trial section. In the approximate formula for P_{ueq} (or P_{aeq}) that follows, m is a factor given in Table 11.3 of this chapter. This table is taken from the second edition of the *Manual of Steel Construction Load and Resistance Factor Design* published in 1994 (where it was Table 3-2). The equivalent loads are estimated with the following expressions:

$$P_{ueq} = P_u + M_{ux}m + M_{uy}mu \qquad \text{(LRFD)}$$

or

$$P_{aeq} = P_a + M_{ax}m + M_{ay}mu \qquad \text{(ASD)}$$

To apply these expressions, a value of m is taken from the first approximation section of Table 11.3, and u is assumed equal to 2. *In applying the equation, the* moments M_{ux} and M_{uy} (or M_{ax} and M_{ay}) must be used in ft-k. The equations are solved for P_{ueq} or P_{aeq}. A column is selected from the concentrically loaded column tables for each load. Then the equation for P_{ueq} (or P_{aeq}) is solved again with a revised value of m from the subsequent approximations part of Table 11.3, and the value of u is kept equal to 2.0. (Actually, a more precise value of u for each column section was provided in the 1994 Manual.)

Limitations of P_{ueq} Formula

The application of the equivalent axial load formula and Table 11.3 results in econom-ical beam–column designs, unless the moment becomes quite large in comparison with the axial load. For such cases, the members selected will be capable of supporting the loads and moments, but may very well be rather uneconomical. The tables for

TABLE 11.3   Preliminary Beam–Column Design $F_y = 36$ ksi, $F_y = 50$ ksi

	Values of m												
F_y	36 ksi							50 ksi					
KL(ft)	10	12	14	16	18	20	22 and over	10	12	14	16	18	20	22 and over
1st Approximation													
All Shapes	2.0	1.9	1.8	1.7	1.6	1.5	1.3	1.9	1.8	1.7	1.6	1.4	1.3	1.2
Subsequent Approximation													
W4	3.1	2.3	1.7	1.4	1.1	1.0	0.8	2.4	1.8	1.4	1.1	1.0	0.9	0.8
W5	3.2	2.7	2.1	1.7	1.4	1.2	1.0	2.8	2.2	1.7	1.4	1.1	1.0	0.9
W6	2.8	2.5	2.1	1.8	1.5	1.3	1.1	2.5	2.2	1.8	1.5	1.3	1.2	1.1
W8	2.5	2.3	2.2	2.0	1.8	1.6	1.4	2.4	2.2	2.0	1.7	1.5	1.3	1.2
W10	2.1	2.0	1.9	1.8	1.7	1.6	1.4	2.0	1.9	1.8	1.7	1.5	1.4	1.3
W12	1.7	1.7	1.6	1.5	1.5	1.4	1.3	1.7	1.6	1.5	1.5	1.4	1.3	1.2
W14	1.5	1.5	1.4	1.4	1.3	1.3	1.2	1.5	1.4	1.4	1.3	1.3	1.2	1.2

Source: This table is from a paper in AISC *Engineering Journal* by Uang, Wattar, and Leet (1990).

concentrically loaded columns of Part 4 of the Manual are limited to the W14s, W12s, and shallower sections, but when the moment is large in proportion to the axial load, there will often be a much deeper and appreciably lighter section, such as a W27 or W30, that will satisfy the appropriate interaction equation.

Examples of the equivalent axial load method applied with the values given in Table 11.3 are presented in Examples 11-11 and 11-12. After a section is selected by the approximate P_{ueq} or P_{aeq} formula, it is necessary to check it with the appropriate interaction equation.

Example 11-11

Select a trial W section for both LRFD and ASD for the following data: $F_y = 50$ ksi, $(KL)_x = (KL)_y = 12$ ft, $P_{nt} = 690$ k and $M_{ntx} = 168$ ft-k for LRFD, and $P_{nt} = 475$ k and $M_{ntx} = 120$ ft-k for ASD. These were the values used in Example 11-3.

Solution

LRFD	ASD
Assume B_1 and $B_2 = 1.0$	Assume B_1 and $B_2 = 1.0$
$\therefore P_r = P_u = P_{nt} + B_2(P_{lt})$	$\therefore P_r = P_a = P_{nt} + B_2(P_{lt})$
$P_u = 690 + 0 = 690$ k	$P_a = 475 + 0 = 475$ k
and, $M_{rx} = M_{ux} = B_1(M_{ntx}) + B_2(M_{ltx})$	and, $M_{rx} = M_{ax} = B_1(M_{ntx}) + B_2(M_{ltx})$
$M_{ux} = 1.0(168) + 0 = 168$ ft-k	$M_{ax} = 1.0(120) + 0 = 120$ ft-k
$P_{ueq} = P_u + M_{ux}\, m + M_{uy}\, mu$	$P_{aeq} = P_a + M_{ax}m + M_{ay}mu$
From "1st Approximation" part of Table 11.3	From "1st Approximation" part of Table 11.3
$m = 1.8$ for $KL = 12$ ft, $F_y = 50$ ksi	$m = 1.8$ for $KL = 12$ ft, $F_y = 50$ ksi
$u = 2.0$ (assumed)	$u = 2.0$ (assumed)
$P_{ueq} = 690 + 168(1.8) + 0 = 992.4$ k	$P_{aeq} = 475 + 120(1.8) + 0 = 691.0$ k
1st trial section: W12 × 96 ($\Phi_c P_n = 1080$ k) from AISC Table 4-1	1st trial section: W12 × 96 ($P_n/\Omega_c = 720$ k) from AISC Table 4-1
From "Subsequent Approximation" part of Table 11.3, W12's	From "Subsequent Approximation" part of Table 11.3, W12's
$m = 1.6$	$m = 1.6$
$P_{ueq} = 690 + 168(1.6) + 0 = 958.8$ k	$P_{aeq} = 475 + 120(1.6) + 0 = 667.0$ k
Try W12 × 87, ($\Phi_c P_n = 981$ k > 958.8 k)	**Try W12 × 96**, ($P_n/\Omega_c = 720$ k > 667.0 k)

Note: These are trial sizes. B_1 and B_2, which were assumed, must be calculated and these W12 sections checked with the appropriate interaction equations.

Example 11-12

Select a trial W section for both LRFD and ASD for an unbraced frame and the following data: $F_y = 50$ ksi, $(KL)_x = (KL)_y = 10$ ft.

   For LRFD: $P_{nt} = 175$ k and $P_{lt} = 115$ k, $M_{ntx} = 102$ ft-k and $M_{ltx} = 68$ ft-k, $M_{nty} = 84$ ft-k and $M_{lty} = 56$ ft-k

   For ASD: $P_{nt} = 117$ k and $P_{lt} = 78$ k, $M_{ntx} = 72$ ft-k and $M_{ltx} = 48$ ft-k, $M_{nty} = 60$ ft-k and $M_{lty} = 40$ ft-k

Solution

LRFD	ASD
Assume B_{1x}, B_{1y}, B_{2x} and $B_{2y} = 1.0$	Assume B_{1x}, B_{1y}, B_{2x} and $B_{2y} = 1.0$
$\therefore\ P_r = P_u = P_{nt} + B_2(P_{lt})$	$\therefore\ P_r = P_a = P_{nt} + B_2(P_{lt})$
$P_u = 175 + 1.0(115) = 290$ k	$P_a = 117 + 1.0(78) = 195$ k
and, $M_{rx} = M_{ux} = B_{1x}(M_{ntx}) + B_{2x}(M_{ltx})$	and, $M_{rx} = M_{ax} = B_{1x}(M_{ntx}) + B_{2x}(M_{ltx})$
$M_{ux} = 1.0(102) + 1.0(68) = 170$ ft-k	$M_{ax} = 1.0(72) + 1.0(48) = 120$ ft-k
and, $M_{ry} = M_{uy} = B_{1y}(M_{nty}) + B_{2x}(M_{lty})$	and, $M_{ry} = M_{ay} = B_{1y}(M_{nty}) + B_{2x}(M_{lty})$
$M_{uy} = 1.0(84) + 1.0(56) = 140$ ft-k	$M_{ay} = 1.0(60) + 1.0(40) = 100$ ft-k
$P_{ueq} = P_u + M_{ux}m + M_{uy}mu$	$P_{aeq} = P_a + M_{ax}m + M_{ay}mu$
From "1st Approximation" part of Table 11.3	From "1st Approximation" part of Table 11.3
$m = 1.9$ for $KL = 10$ ft, $F_y = 50$ ksi	$m = 1.9$ for $KL = 10$ ft, $F_y = 50$ ksi
$u = 2.0$ (assumed)	$u = 2.0$ (assumed)
$P_{ueq} = 290 + 170(1.9) + 140(1.9)(2.0) = 1145$ k	$P_{aeq} = 195 + 120(1.9) + 100(1.9)(2.0) = 803$ k
1st trial section from Table 4.1:	1st trial section from Table 4.1:
W14 → W14 × 99 ($\Phi_c P_n = 1210$ k)	W14 → W14 × 99 ($P_n/\Omega_c = 807$ k)
W12 → W12 × 106 ($\Phi_c P_n = 1260$ k)	W12 → W12 × 106 ($P_n/\Omega_c = 838$ k)
W10 → W10 × 112 ($\Phi_c P_n = 1280$ k)	W10 → W10 × 112 ($P_n/\Omega_c = 851$ k)
Suppose we decide to use a W14 section:	Suppose we decide to use a W14 section:
From "Subsequent Approximation" part of Table 11.3, W14's	From "Subsequent Approximation" part of Table 11.3, W14's
$m = 1.5$	$m = 1.5$
$P_{ueq} = 290 + 170(1.5) + 140(1.5)(2.0) = 965$ k	$P_{aeq} = 195 + 120(1.5) + 100(1.5)(2.0) = 675$ k
Try W14 × 90, ($\Phi_c P_n = 1100$ k > 965 k)	**Try W14 × 90, ($P_n/\Omega_c = 735$ k > 675 k)**

Note: These are trial sizes. B_{1x}, B_{1y}, B_{2x} and B_{2y}, which were assumed, must be calculated and these W14 sections checked with the appropriate interaction equations.

Examples of complete beam-column design are presented in Examples 11-13 and 11-14. In the examples, a trial section is first determined using the equivalent load method as shown in Examples 11-11 and 11-12. The trial section is then checked with the appropriate interaction equation. Both examples use the Effective Length Method. Example 11-13 is a beam-column in a braced frame and Example 11-14 is a beam-column in an unbraced frame.

Example 11-13

Select the lightest W12 section for both LRFD and ASD for the following data: $F_y = 50$ ksi, $(KL)_x = (KL)_y = 12$ ft, $P_{nt} = 250$ k, $M_{ntx} = 180$ ft-k and $M_{nty} = 70$ ft-k for LRFD, and $P_{nt} = 175$ k, $M_{ntx} = 125$ ft-k and $M_{nty} = 45$ ft-k for ASD. $C_b = 1.0$, $C_{mx} = C_{my} = 0.85$.

Solution

LRFD	ASD
Assume $B_{1x} = B_{1y} = 1.0$, B_2 not required	Assume $B_{1x} = B_{1y} = 1.0$, B_2 not required
$\therefore\ P_r = P_u = P_{nt} + B_2(P_{lt})$	$\therefore\ P_r = P_a = P_{nt} + B_2(P_{lt})$
$P_u = 250 + 0 = 250$ k	$P_a = 175 + 0 = 175$ k
and, $M_{rx} = M_{ux} = B_1(M_{ntx}) + B_2(M_{ltx})$	and, $M_{rx} = M_{ax} = B_1(M_{ntx}) + B_2(M_{ltx})$
$M_{ux} = 1.0(180) + 0 = 180$ ft-k	$M_{ax} = 1.0(120) + 0 = 125$ ft-k
and, $M_{ry} = M_{uy} = B_1(M_{nty}) + B_2(M_{lty})$	and, $M_{ry} = M_{ay} = B_1(M_{nty}) + B_2(M_{lty})$
$M_{uy} = 1.0(70) + 0 = 70$ ft-k	$M_{ay} = 1.0(45) + 0 = 45$ ft-k
$P_{ueq} = P_u + M_{ux}m + M_{uy}mu$	$P_{aeq} = P_a + M_{ax}\,m + M_{ay}\,mu$
From "Subsequent Approximation" part of	From "Subsequent Approximation" part of
Table 11.3, W12's	Table 11.3, W12's
$m = 1.6$	$m = 1.6$
$u = 2.0$ (assumed)	$u = 2.0$ (assumed)
$P_{ueq} = 250 + 180(1.6) + 70(1.6)(2.0) = 762$ k	$P_{aeq} = 175 + 125(1.6) + 45(1.6)(2.0) = 519$ k
Try W12 × 72, $(\Phi_c P_n = 806$ k > 762 k) from Table 4.1	**Try W12 × 72,** $(P_n/\Omega_c = 536$ k > 519 k) from Table 4.1
From Table 6.1 for $KL = 12$ ft and $L_b = 12$ ft	From Table 6.1 for $KL = 12$ ft and $L_b = 12$ ft
$p = 1.24 \times 10^{-3}, b_x = 2.23 \times 10^{-3}, b_y = 4.82 \times 10^{-3}$	$p = 1.87 \times 10^{-3}, b_x = 3.36 \times 10^{-3}, b_y = 7.24 \times 10^{-3}$
$P_r/\Phi_c P_n = 250/806 = 0.310 > 0.2$ Use modified Equation H1-1a.	$P_r/P_n/\Omega_c = 175/536 = 0.326 > 0.2$ Use modified Equation H1-1a.
$1.24 \times 10^{-3}(250) + 2.23 \times 10^{-3}(180) + 4.82 \times 10^{-3}$ $(70) = 1.049 > 1.0$ N.G.	$1.87 \times 10^{-3}(175) + 3.36 \times 10^{-3}(125) + 7.24 \times 10^{-3}$ $(45) = 1.073 > 1.0$ N.G.

(Continued)

LRFD	ASD
Try W12 × 79, $(\Phi_c P_n = 887 \text{ k} > 762 \text{ k})$ from Table 4.1	Try W12 × 79, $(P_n/\Omega_c = 590 \text{ k} > 519 \text{ k})$ from Table 4.1
From Table 6.1 for $KL = 12$ ft and $L_b = 12$ ft	From Table 6.1 for $KL = 12$ ft and $L_b = 12$ ft
$p = 1.13 \times 10^{-3}, b_x = 2.02 \times 10^{-3}, b_y = 4.37 \times 10^{-3}$	$p = 1.69 \times 10^{-3}, b_x = 3.04 \times 10^{-3}, b_y = 6.56 \times 10^{-3}$
$1.13 \times 10^{-3} (250) + 2.02 \times 10^{-3} (180) + 4.37$ $\times 10^{-3}(70) = 0.952 < 1.0$ OK	$1.69 \times 10^{-3} (175) + 3.04 \times 10^{-3} (125) + 6.56 \times 10^{-3}$ $(45) = 0.971 < 1.0$ OK
Check $B_{1x} = B_{1y} = 1.0$	Check $B_{1x} = B_{1y} = 1.0$
$P_{e1x} = \dfrac{\pi^2 EI^*}{(K_1 L)^2} = \dfrac{\pi^2 (29{,}000)(662)}{(1.0 \times 12 \times 12)^2} = 9138 \text{ k}$	$P_{e1x} = \dfrac{\pi^2 EI^*}{(K_1 L)^2} = \dfrac{\pi^2 (29{,}000)(662)}{(1.0 \times 12 \times 12)^2} = 9138 \text{ k}$
$B_{1x} = \dfrac{C_{mx}}{1 - \dfrac{\alpha P_r}{P_{e1x}}} = \dfrac{0.85}{1 - \dfrac{1.0(250)}{9138}} = 0.87 < 1;$	$B_{1x} = \dfrac{C_{mx}}{1 - \dfrac{\alpha P_r}{P_{e1x}}} = \dfrac{0.85}{1 - \dfrac{1.6(175)}{9138}} = 0.88 < 1;$
$B_{1x} = 1.0$, OK	$B_{1x} = 1.0$, OK
$P_{e1y} = \dfrac{\pi^2 EI^*}{(K_1 L)^2} = \dfrac{\pi^2 (29{,}000)(216)}{(1.0 \times 12 \times 12)^2} = 2981 \text{ k}$	$P_{e1y} = \dfrac{\pi^2 EI^*}{(K_1 L)^2} = \dfrac{\pi^2 (29{,}000)(216)}{(1.0 \times 12 \times 12)^2} = 2981 \text{ k}$
$B_{1y} = \dfrac{C_{my}}{1 - \dfrac{\alpha P_r}{P_{e1y}}} = \dfrac{0.85}{1 - \dfrac{1.0(250)}{2981}} = 0.93 < 1;$	$B_{1y} = \dfrac{C_{my}}{1 - \dfrac{\alpha P_r}{P_{e1y}}} = \dfrac{0.85}{1 - \dfrac{1.6(175)}{2981}} = 0.94 < 1;$
$B_{1y} = 1.0$, OK	$B_{1y} = 1.0$, OK
With $B_{1x} = B_{1y} = 1.0$, section is sufficient based on previous check using modified Equation H1-1a.	With $B_{1x} = B_{1y} = 1.0$, section is sufficient based on previous check using modified Equation H1-1a.
Will perform additional check using Equation H1-1a:	Will perform additional check using Equation H1-1a:
For W12 × 79, $\Phi M_{px} = 446$ ft-k, $L_p = 10.8$ ft, $L_r = 39.9$ ft	For W12 × 79, $M_{px}/\Omega_b = 297$ ft-k, $L_p = 10.8$ ft, $L_r = 39.9$ ft
$BF = 5.67, L_b = 12$ ft, Zone 2, $C_b = 1.0, \Phi M_{py} = 204$ ft-k	$BF = 3.78, L_b = 12$ ft, Zone 2, $C_b = 1.0, M_{py}/\Omega_b = 135$ ft-k
$\Phi M_{nx} = C_b [\Phi M_{px} - BF(L_b - L_p)] \Phi M_{px}$	$M_{nx}/\Omega_b = C_b [M_{px}/\Omega_b - BF(L_b - L_p)] \le M_{px}/\Omega_b$
$\Phi M_{nx} = 1.0 [446 - 5.67(12 - 10.8)] = 439.2$ ft-k	$M_{nx}/\Omega_b = 1.0 [297 - 3.78(12 - 10.8)] = 292.4$ ft-k
$\Phi M_{ny} = \Phi M_{py} = 204$ ft-k	$M_{ny}/\Omega_b = M_{py}/\Omega_b = 135$ ft-k
Equation H1-1a:	Equation H1-1a:
$\dfrac{250}{887} + \dfrac{8}{9} \left(\dfrac{180}{439.2} + \dfrac{70}{204} \right) = 0.951 < 1.0$ OK	$\dfrac{175}{590} + \dfrac{8}{9} \left(\dfrac{125}{292.4} + \dfrac{45}{135} \right) = 0.973 < 1.0$ OK
Use W12 × 79, LRFD.	**Use W12 × 79, ASD.**

Example 11-14

Select a trial W12 section for both LRFD and ASD for an unbraced frame and the following data: $F_y = 50$ ksi, $L_x = L_y = 12$ ft, $K_x = 1.72$, $K_y = 1.0$, $P_{story} = 2400$ k (LRFD) = 1655k (ASD), $P_{e\ story} = 50,000$ k.

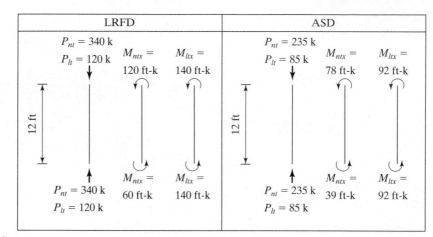

Solution

LRFD	ASD
Assume $B_{1x} = 1.0$	Assume $B_{1x} = 1.0$
$B_{2x} = \dfrac{1}{1 - \dfrac{\alpha P_{story}}{P_{e\ story}}} = \dfrac{1}{1 - \dfrac{1.0(2400)}{50,000}} = 1.05$	$B_{2x} = \dfrac{1}{1 - \dfrac{\alpha P_{story}}{P_{e\ story}}} = \dfrac{1}{1 - \dfrac{1.6(1655)}{50,000}} = 1.056$
Assume $(KL)_y = 1.0(12) = 12$ ft	Assume $(KL)_y = 1.0(12) = 12$ ft
$\therefore\ P_r = P_u = P_{nt} + B_2(P_{lt})$	$\therefore\ P_r = P_a = P_{nt} + B_2(P_{lt})$
$P_u = 340 + 1.05(120) = 466$ k	$P_a = 235 + 1.056(85) = 325$ k
and, $M_{rx} = M_{ux} = B_{1x}(M_{ntx}) + B_{2x}(M_{ltx})$	and, $M_{rx} = M_{ax} = B_{1x}(M_{ntx}) + B_{2x}(M_{ltx})$
$M_{ux} = 1.0(120) + 1.05(140) = 267$ ft-k	$M_{ax} = 1.0(78) + 1.056(92) = 176$ ft-k
$P_{ueq} = P_u + M_{ux}\,m + M_{uy}\,mu$	$P_{aeq} = P_a + M_{ax}m + M_{ay}mu$
From "Subsequent Approximation" part of Table 11.3, W12's	From "Subsequent Approximation" part of Table 11.3, W12's
$m = 1.6$	$m = 1.6$
$u = 2.0$ (assumed) not required	$u = 2.0$ (assumed) not required
$P_{ueq} = 466 + 267(1.6) + 0 = 894$ k	$P_{aeq} = 325 + 176(1.6) + 0 = 607$ k

(Continued)

LRFD	ASD
1st trial section from Table 4.1:	1st trial section from Table 4.1:
Try W12 × 87, ($\Phi_c P_n = 981$ k > 894 k)	Try W12x87, ($P_n/\Omega_c = 653$ k > 607 k)
Check $\Phi_c P_n$:	Check P_n/Ω_c:
$(KL)_y = 12$ ft	$(KL)_y = 12$ ft
Equivalent $(KL)_y = \dfrac{(KL)_x}{\dfrac{r_x}{r_y}} = \dfrac{(1.72)(12)}{1.75} = 11.79$ ft	Equivalent $(KL)_y = \dfrac{(KL)_x}{\dfrac{r_x}{r_y}} = \dfrac{(1.72)(12)}{1.75} = 11.79$ ft
$(KL)_y = 12$ ft controls so $\Phi_c P_n = 981$ k	$(KL)_y = 12$ ft controls so $P_n/\Omega_c = 653$ k
From Table 6-1 for $KL = 12$ ft and $L_b = 12$ ft	From Table 6-1 for $KL = 12$ ft and $L_b = 12$ ft
$p = 1.02 \times 10^{-3}, b_x = 1.82 \times 10^{-3}$	$p = 1.53 \times 10^{-3}, b_x = 2.74 \times 10^{-3}$
$P_r/\Phi_c P_n = 466/981 = 0.475 > 0.2$ Use modified H1-1a	$P_r/P_n/\Omega_c = 325/653 = 0.498 > 0.2$ Use modified H1-1a
$1.02 \times 10^{-3}(466) + 1.82 \times 10^{-3}(267) = 0.961 < 1.0$ OK	$1.53 \times 10^{-3}(325) + 2.74 \times 10^{-3}(176) = 0.979 < 1.0$ OK
Check $B_{1x} = 1.0$	Check $B_{1x} = 1.0$
$P_{e1x} = \dfrac{\pi^2 EI^*}{(K_1 L)^2} = \dfrac{\pi^2(29,000)(740)}{(1.0 \times 12 \times 12)^2} = 10{,}214$ k	$P_{e1x} = \dfrac{\pi^2 EI^*}{(K_1 L)^2} = \dfrac{\pi^2(29,000)(740)}{(1.0 \times 12 \times 12)^2} = 10{,}214$ k
Find C_{mx}: $\dfrac{M_1}{M_2} = \dfrac{60}{120} = +0.50$	Find C_{mx}: $\dfrac{M_1}{M_2} = \dfrac{39}{78} = +0.50$
$C_{mx} = 0.6 - 0.4(M_1/M_2) = 0.6 - 0.4(0.50) = 0.40$	$C_{mx} = 0.6 - 0.4(M_1/M_2) = 0.6 - 0.4(0.50) = 0.40$
$B_{1x} = \dfrac{C_{mx}}{1 - \dfrac{\alpha P_r}{P_{e1x}}} = \dfrac{0.40}{1 - \dfrac{1.0(466)}{10214}} = 0.42 < 1;$	$B_{1x} = \dfrac{C_{mx}}{1 - \dfrac{\alpha P_r}{P_{e1x}}} = \dfrac{0.40}{1 - \dfrac{1.6(325)}{10214}} = 0.42 < 1;$
$B_{1x} = 1.0,$ OK	$B_{1x} = 1.0,$ OK
With $B_{1x} = 1.0$, section is sufficient based on previous check using modified Equation H1-1a.	With $B_{1x} = 1.0$, section is sufficient based on previous check using modified Equation H1-1a.
Will perform additional check using Equation H1-1a:	Will perform additional check using Equation H1-1a:
For W12 × 87, $\Phi M_{px} = 495$ ft-k, $L_p = 10.8$ ft, $L_r = 43.1$ ft	For W12 × 87, $M_{px}/\Omega_b = 329$ ft-k, $L_p = 10.8$ ft, $L_r = 43.1$ ft
$BF = 5.73, L_b = 12$ ft, Zone 2, $C_b = 1.0$	$BF = 3.81, L_b = 12$ ft, Zone 2, $C_b = 1.0$
$\Phi M_{nx} = C_b (\Phi M_{px} - BF(L_b - L_p)) \le \Phi M_{px}$	$M_{nx}/\Omega_b = C_b[M_{px}/\Omega_b - BF(L_b - L_p)] M_{px}/\Omega_b$
$\Phi M_{nx} = 1.0[495 - 5.73(12-10.8)] = 488.1$ ft-k	$M_{nx}/\Omega_b = 1.0[329 - 3.81(12 - 10.8)] = 324.4$ ft-k
Equation H1-1a:	Equation H1-1a:
$\dfrac{466}{981} + \dfrac{8}{9}\left(\dfrac{267}{488.1} + 0\right) = 0.961 < 1.0$ OK	$\dfrac{325}{653} + \dfrac{8}{9}\left(\dfrac{176}{324.4} + 0\right) = 0.980 < 1.0$ OK
Use W12 × 87, LRFD.	**Use W12 × 87, ASD.**

11.10 PROBLEMS FOR SOLUTION

All problems are to be solved with both LRFD and ASD methods, unless noted otherwise.

Bending and Axial Tension

11-1 to 11-6. *Analysis Problems*

11-1. A W10 × 54 tension member with no holes and $F_y = 50$ ksi ($F_u = 65$ ksi) is subjected to service loads $P_D = 90$ k and $P_L = 120$ k and to service moments $M_{Dx} = 32$ ft-k and $M_{Lx} = 50$ ft-k. Is the member satisfactory if $L_b = 12$ ft and if $C_b = 1.0$? (*Ans.* OK 0.863– LRFD, OK 0.903 ASD)

11-2. A W8 × 35 tension member with no holes, consisting of $F_y = 50$ ksi steel ($F_u = 65$ ksi), is subjected to service loads $P_D = 50$ k and $P_L = 30$ k and to service moments $M_{Dx} = 35$ ft-k and $M_{Lx} = 25$ ft-k. Is the member satisfactory if $L_b = 12$ ft and if $C_b = 1.0$?

11-3. Repeat Prob. 11-2 if the member has 2 – 3/4 in diameter bolts in each flange and $U = 0.85$. (*Ans.* OK 0.920 LRFD, N.G. 1.015 ASD)

11-4. A W10 × 39 tension member with no holes and $F_y = 50$ ksi ($F_u = 65$ ksi) is subjected to service loads $P_D = 56$ k and $P_L = 73$ k that are placed with an eccentricity of 7 in with respect to the x axis. The member is to be 16 ft long and is braced laterally only at its supports. Is the member satisfactory if $C_b = 1.0$?

11-5. A W12 × 30 tension member with no holes is subjected to an axial load, P, which is 40 percent dead load and 60 percent live load and a uniform service wind load of 2.40 k/ft. The member is 14 ft long, laterally braced at its ends only and bending is about the x axis. Assume $C_b = 1.0$, $F_y = 50$ ksi and $F_u = 65$ ksi. What is the maximum value of P for this member to be satisfactory? (*Ans.* $P = 66.9$ k, LRFD; $P = 76.5$ k, ASD)

11-6. A W10 × 45 tension member with no holes and $F_y = 50$ ksi ($F_u = 65$ ksi) is subjected to service loads $P_D = 60$ k and $P_L = 40$ k and to service moments $M_{Dx} = 40$ ft-k, $M_{Lx} = 20$ ft-k, $M_{Dy} = 15$ ft-k and $M_{Ly} = 10$ ft-k. Is the member satisfactory if $L_b = 10.5$ ft and if $C_b = 1.0$?

11-7 to 11-9. *Design Problems*

11-7. Select the lightest available W12 section ($F_y = 50$ ksi, $F_u = 65$ ksi) to support service loads $P_D = 50$ k and $P_L = 90$ k and to service moments $M_{Dx} = 20$ ft-k and $M_{Lx} = 35$ ft-k. The member is 14 ft long and is laterally braced at its ends only. Assume $C_b = 1.0$. (*Ans.* W12 × 35, LRFD and ASD)

11-8. Select the lightest available W8 section ($F_y = 50$ ksi, $F_u = 65$ ksi) to support service loads $P_D = 55$ k and $P_L = 30$ k that are placed with an eccentricity of 2.5 in with respect to the y-axis. The member is 12 ft long and is braced laterally only at its supports. Assume $C_b = 1.0$.

11-9. Select the lightest available W10 section of A992 steel ($F_u = 65$ ksi) to support service loads $P_D = 30$ k and $P_L = 40$ k and to service moments $M_{Dx} = 40$ ft-k, $M_{Lx} = 55$ ft-k, $M_{Dy} = 8$ ft-k and $M_{Ly} = 14$ ft-k. The member is 12 ft long and is laterally braced at its ends only. Assume $C_b = 1.0$. (*Ans.* W10 × 54, LRFD and ASD)

Bending and Axial Compression

All problems (11-10 to 11-22) have loads and moments that were obtained from first-order analyses, and the approximate analysis method (Appendix 8 – AISC Specification) should be used to account for the second-order effects.

11-10 and 11-11. *Analysis problems in braced frames – using loads and moments obtained using the requirements of the Direct Analysis Method.*

11-10. A 14 ft long W12 × 65 beam-column is part of a braced frame and supports service loads of $P_D = 180$ k and $P_L = 110$ k. These loads are applied to the member at its upper end with an eccentricity of 3 in so as to cause bending about the major axis of the section. Check the adequacy of the member if it consists of A992 steel. Assume $C_b = 1.0$ and $C_{mx} = 1.0$.

11-11. A W10 × 60 beam-column member is in a braced frame and must support service loads of $P_D = 60$ k and $P_L = 120$ k and service moments $M_{Dx} = 30$ ft-k and $M_{Lx} = 60$ ft-k. These moments occur at one end while the opposite end is pinned. The member is 15 ft long, and $C_b = 1.0$. Is the member satisfactory if it consists of 50 ksi steel? (*Ans.* OK 0.931 LRFD, OK 0.954 ASD)

11-12 and 11-13. *Analysis problems in braced frames - using loads and moments obtained using the requirements of the Effective Length Method.*

11-12. A W 12 × 58 beam-column member in a braced frame must support the service loads in the Fig. P11-12. The member is A992 steel, C_b may be assumed to be 1.0, and bending is about the strong axis. The member is laterally braced at its ends only and $L_x = L_y = 16$ ft. Is the member adequate?

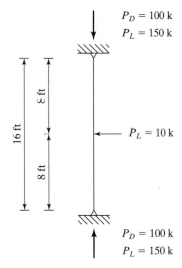

$P_D = 100$ k
$P_L = 150$ k

16 ft

3 ft

8 ft

$P_L = 10$ k

$P_D = 100$ k
$P_L = 150$ k

FIGURE P11-12

11-13. A W12 × 96 section consisting of 50 ksi steel is used for a 12 ft long beam-column in a braced frame. The member must support service loads $P_D = 85$ k and $P_L = 125$ k and service moments $M_{Dx} = 70$ ft-k, $M_{Lx} = 120$ ft-k, $M_{Dy} = 20$ ft-k and $M_{Ly} = 35$ ft-k. The moments are applied at both ends of the column so as to put it in single curvature about both axes. The column has $K_x = K_y = 1.0$ and $C_b = 1.0$. Is the member satisfactory? (*Ans.* N.G. 1.048 LRFD, N.G. 1.092 ASD)

11-14. *Analysis problem in an unbraced frame – using loads and moments obtained using the requirements of the Direct Analysis Method.*

A W12 × 72 section is used for a 16 ft long beam-column in an unbraced frame about the *x-x* axis but is braced about the *y-y* axis. Based on a first-order analysis, the member supports the following factored loads: P_{nt} load of 194 k and P_{lt} load of 152 k. The member also supports factored moments: M_{ntx} moment of 110 ft-k and M_{ltx} moment of 76 ft-k. The moments given are at the top of the member. The lower end has moments that are one-half of these values and bend the member in double curvature. There are no transverse loads between the ends. $P_{story} = 3000$ k, and $P_{e\ storyx} = 75,000$ k. Using the LRFD procedure, is the member satisfactory if $F_y = 50$ ksi and $C_b = 1.0$?

11-15. *Analysis problem in an unbraced frame – using loads and moments obtained using the requirements of the Effective Length Method.*

A W8 × 48 section consisting of 50 ksi steel is used for a 14 ft long beam-column in a one story building frame. It is unbraced in the plane of the frame (x-x axis) but is braced out of the plane of the frame (y-y axis) so that $K_y = 1.0$. K_x has been determined to equal 1.67. A first-order analysis has been completed and the results yielded a P_{nt} load of 52 k and a P_{lt} load of 16 k. The factored M_{ntx} moment was found to be 96 ft-k and a factored M_{ltx} moment of 50 ft-k. The member is pinned at the base, and is laterally braced only at the top and the bottom. There are no transverse loads between the ends and $C_b = 1.0$. The $P_{story} = 104$ k and the ratio of $P_{mf} / P_{story} = 1.0$. H, the story shear, is equal to 4.4 k, and the drift index (Δ_H/L) is 0.0025. Using the LRFD procedure, is the member adequate? (*Ans.* OK 0.985 LRFD)

11-16 and 11-17. *Design problems in braced frames – using loads and moments obtained using the Direct Analysis Method.*

11-16. Select the lightest W14, 15 ft long, beam-column that is not subjected to sidesway. The service loads are $P_D = 105$ k and $P_L = 120$ k. The service moments are $M_{Dx} = 85$ ft-k and $M_{Lx} = 95$ ft-k. The member consists of $F_y = 50$ ksi steel. Assume $C_b = 1.0$ and $C_{mx} = 0.85$.

11-17. Sidesway is prevented for the beam-column shown in the Fig. P11-17. If the first-order moments shown are about the x axis, select the lightest W8 if it consists of $F_y = 50$ steel. Assume $C_b = 1.0$. (*Ans.* W8 × 35 LRFD and ASD)

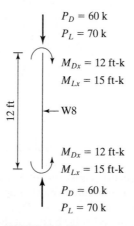

$P_D = 60$ k
$P_L = 70$ k

$M_{Dx} = 12$ ft-k
$M_{Lx} = 15$ ft-k

12 ft

← W8

$M_{Dx} = 12$ ft-k
$M_{Lx} = 15$ ft-k

$P_D = 60$ k
$P_L = 70$ k

FIGURE P11-17

11-18 and 11-19. *Design problems in braced frames – using loads and moments obtained using the Effective Length Method.*

11-18. Select the lightest W10 beam-column member in a braced frame that supports service loads of $P_D = 50$ k and $P_L = 75$ k. The member also supports the service moments

$M_{Dx} = 27.5$ ft-k, $M_{Lx} = 40$ ft-k, $M_{Dy} = 10$ ft-k and $M_{Ly} = 15$ ft-k. The member is 15 ft long and moments occur at one end while the other end is pinned. There are no transverse loads on the member and assume $C_b = 1.0$. Use 50 ksi steel.

11-19. Select the lightest W12, 15 ft long, $F_y = 50$ ksi steel beam-column for the first-order loads and moments shown in the Fig. P11-19. The column is part of a braced frame system. Assume $C_b = 1.0$. (*Ans.* W12 × 96 LRFD, W12 × 106 ASD)

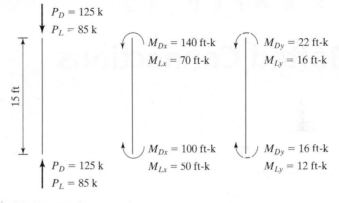

$P_D = 125$ k
$P_L = 85$ k

$M_{Dx} = 140$ ft-k $M_{Dy} = 22$ ft-k
$M_{Lx} = 70$ ft-k $M_{Ly} = 16$ ft-k

15 ft

$M_{Dx} = 100$ ft-k $M_{Dy} = 16$ ft-k
$P_D = 125$ k $M_{Lx} = 50$ ft-k $M_{Ly} = 12$ ft-k
$P_L = 85$ k

FIGURE P11-19

11-20. *Design problem in an unbraced frame – using loads and moments obtained using the requirements of the Direct Analysis Method.*

11-20. Using the LRFD procedure, select the lightest W12, 10 ft long beam-column in an unbraced frame. Based on a first-order analysis, the member supports the following factored loads: P_{nt} load of 140 k and P_{lt} load of 105 k. The member also supports factored moments: M_{ntx} moment of 60 ft-k, M_{ltx} moment of 90 ft-k, M_{nty} moment of 40 ft-k, and M_{lty} moment of 75 ft-k. $P_{story} = 5000$ k, $P_{e\ story\ x} = 40{,}000$ k, $P_{e\ story\ y} = 30{,}000$ k $C_{mx} = C_{my} = 1.0$, $C_b = 1.0$ and $F_y = 50$ ksi steel.

11-21 and 11-22. *Design problems in unbraced frames – using loads and moments obtained using the requirements of the Effective Length Method.*

11-21. Using the LRFD procedure, select the lightest W10, 16 ft long beam-column in an unbraced frame. Based on a first-order analysis, the member supports the following factored loads: P_{nt} load of 148 k and P_{lt} load of 106 k. The member also supports factored moments: M_{ntx} moment of 92 ft-k and M_{ltx} moment of 64 ft-k. The moments are equal at each end and the member is bent in single curvature. There are no transverse loads between the ends. $K_x = 1.5$, $K_y = 1.0$, $P_{story} = 2800$ k, $P_{e\ storyx} = 72{,}800$ k, $C_b = 1.0$ and $F_y = 50$ ksi. (*Ans.* W10 × 68 LRFD)

11-22. Using the LRFD procedure, select the lightest W12 section ($F_y = 50$ ksi steel) for a 14 ft long beam-column that is part of a portal unbraced frame. The given unfactored axial force and moment are due to wind. P_{lt} load is a 175 k wind load and M_{ltx} moment is 85 ft-k wind moment. $C_{mx} = 1.0$, $K_x = 1.875$ and $K_y = 1.0$. The $P_{story} = 1212.5$ k and the ratio of $P_{mf}/P_{story} = 0.20$. H, the story shear, is equal to 15.0 k, and the drift index (Δ_H/L) is 0.0020.

C H A P T E R   1 2

Bolted Connections

12.1 INTRODUCTION

For many years, riveting was the accepted method used for connecting the members of steel structures. For the last few decades, however, bolting and welding have been the methods used for making structural steel connections, and riveting is almost never used. This chapter and the next are almost entirely devoted to bolted connections, although some brief remarks are presented at the end of Chapter 13 concerning rivets.

Bolting of steel structures is a very rapid field erection process that requires less skilled labor than does riveting or welding. This gives bolting a distinct economic advantage over the other connection methods in the United States, where labor costs are so very high. Even though the purchase price of a high-strength bolt is several times that of a rivet, the overall cost of bolted construction is cheaper than that for riveted construction because of reduced labor and equipment costs and the smaller number of bolts required to resist the same loads.

Part 16.2 of the AISC Manual provides a copy of the "Specification for Structural Joints Using ASTM A325 or A490 Bolts," dated June 30, 2004, and published by the Research Council on Structural Connections (RCSC). There, the reader can find almost anything he or she would like to know about steel bolts. Included are types, sizes, steels, preparations needed for bolting, use of washers, tightening procedures, inspection, and so on.

12.2 TYPES OF BOLTS

There are several types of bolts that can be used for connecting steel members. They are described in the paragraphs that follow.

Unfinished bolts are also called *ordinary* or *common bolts*. They are classified by the ASTM as A307 bolts and are made from carbon steels with stress–strain characteristics very similar to those of A36 steel. They are available in diameters from 1/2 to 1 1/2 in in 1/8-in increments.

A307 bolts generally have square heads and nuts to reduce costs, but hexagonal heads are sometimes used because they have a slightly more attractive appearance, are easier to turn and to hold with the wrenches, and require less turning space. As they have relatively large tolerances in shank and thread dimensions, their design strengths are appreciably smaller than those for high-strength bolts. They are primarily used in light structures subjected to static loads and for secondary members (such as purlins, girts, bracing, platforms, small trusses, and so forth).

Designers often are guilty of specifying high-strength bolts for connections when common bolts would be satisfactory. *The strength and advantages of common bolts have usually been greatly underrated in the past.* The analysis and design of A307 bolted connections are handled exactly as are riveted connections in every way, except that the design stresses are slightly different.

High-strength bolts are made from medium carbon heat-treated steel and from alloy steel and have tensile strengths two or more times those of ordinary bolts. There are two basic types, the A325 bolts (made from a heat-treated medium carbon steel) and the higher strength A490 bolts (also heat-treated, but made from an alloy steel). High-strength bolts are used for all types of structures, from small buildings to skyscrapers and monumental bridges. These bolts were developed to overcome the weaknesses of rivets—primarily, insufficient tension in their shanks after cooling. The resulting rivet tensions may not be large enough to hold them in place during the application of severe impactive and vibrating loads. The result is that they may become loose and vibrate and may eventually have to be replaced. High-strength bolts may be tightened until they have very high tensile stresses so that the connected parts are clamped tightly together between the bolt and nut heads, permitting loads to be transferred primarily by friction.

Sometimes high strength bolts are needed with diameters and lengths larger than those available with A325 and A490 bolts. Should they be required with diameters exceeding 1 1/2 inches or lengths longer than 8 in, A449 bolts may be used, as well as A354 threaded rods. For anchor rods, ASTM F1554 threaded rods are preferred.

12.3   HISTORY OF HIGH-STRENGTH BOLTS

The joints obtained using high-strength bolts are superior to riveted joints in performance and economy, and they are the leading field method of fastening structural steel members. C. Batho and E. H. Bateman first claimed in 1934 that high-strength bolts could satisfactorily be used for the assembly of steel structures,[1] but it was not until 1947 that the Research Council on Riveted and Bolted Structural Joints of the Engineering Foundation was established. This group issued their first specifications in 1951, and high-strength bolts were adopted with amazing speed by building and bridge engineers for both static and dynamic loadings. They not only quickly became the leading method of making field connections, they also were found to have many applications for shop connections. The construction of the Mackinac Bridge in Michigan involved the use of more than one million high-strength bolts.

---

[1] C. Batho and E. H. Bateman, "Investigations on Bolts and Bolted Joints," H. M. Stationery Office (London, 1934). (In the United Kingdom, H. M. Stationery Office is rather like the U.S. Printing Office.)

High-strength bolt. (Courtesy of Bethlehem
Steel Corporation.)

Connections that were formerly made with ordinary bolts and nuts were not too
satisfactory when they were subjected to vibratory loads, because the nuts frequently
became loose. For many years, this problem was dealt with by using some type of
lock-nut, but the modern high-strength bolts furnish a far superior solution.

12.4   ADVANTAGES OF HIGH-STRENGTH BOLTS

Among the many advantages of high-strength bolts, which partly explain their great
success, are the following:

1. Smaller crews are involved, compared with riveting. Two two-person bolting crews
   can easily turn out over twice as many bolts in a day as the number of rivets driven
   by the standard four-person riveting crew. The result is quicker steel erection.
2. Compared with rivets, fewer bolts are needed to provide the same strength.
3. Good bolted joints can be made by people with a great deal less training and experience than is necessary to produce welded and riveted connections of equal quality.
   The proper installation of high-strength bolts can be learned in a matter of hours.
4. No erection bolts are required that may have to be later removed (depending on
   specifications), as in welded joints.
5. Though quite noisy, bolting is not nearly as loud as riveting.
6. Cheaper equipment is used to make bolted connections.
7. No fire hazard is present, nor danger from the tossing of hot rivets.
8. Tests on riveted joints and fully tensioned bolted joints under identical conditions
   show that bolted joints have a higher fatigue strength. Their fatigue strength is
   also equal to or greater than that obtained with equivalent welded joints.
9. Where structures are to be later altered or disassembled, changes in connections
   are quite simple because of the ease of bolt removal.

12.5   SNUG-TIGHT, PRETENSIONED, AND SLIP-CRITICAL BOLTS

High-strength bolted joints are said to be *snug-tight*, *pretensioned*, or *slip-critical*. These
terms are defined in the paragraphs to follow. The type of joint used is dependent on
the type of load that the fasteners will have to carry.

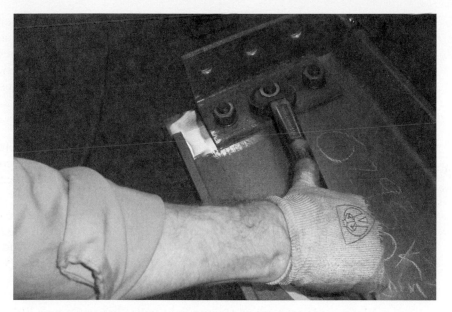

A spud wrench used by ironworkers for erecting structural steel and tightening bolts. One end of the wrench is sized for the hexagonal ends of bolts and nuts, while the other end is tapered to a rounded point and is used to align bolt holes between different connection pieces. (Courtesy of CMC South Carolina Steel.)

a. **Snug-tight bolts**

For most connections, bolts are tightened only to what is called a *snug-tight* condition. Snug-tight is the situation existing when all the plies of a connection are in firm contact with each other. It usually means the tightness produced by the full effort of a person using a spud wrench, or the tightness achieved after a few impacts of the pneumatic wrench. Obviously there is some variation in the degree of tightness achieved under these conditions. Snug-tight bolts must be clearly identified on both design and erection drawings.

Snug-tight bolts are permitted for all situations in which pretensioned or slip-critical bolts are not required. In this type of connection, the plies of steel being connected must be brought together so that they are solidly seated against each other, but they do not have to be in continuous contact. The installed bolts do not have to be inspected to determine their actual pretensioned stresses.

b. **Pretensioned joints**

The bolts in a pretensioned joint are brought to very high tensile stresses equal to approximately 70 percent of their minimum tensile stresses. To properly tighten them, it is necessary to first bring them to a snug-tight condition. Then they are further tightened by one of the four methods described in Section 12.6.

Pretensioned joints are required for connections subjected to appreciable load reversals where nearly full or full design loads are applied to them in one direction, after which the nearly full or full design loads are applied in the other direction. Such a condition is typical of seismic loadings, but not of wind loads. Pretensioned bolts are also required for joints subject to fatigue loads where there is no reversal of the load direction. In addition, they are used where the bolts are subjected to tensile fatigue stresses. A490 bolts should be pretensioned if they are subjected to tension or if they are subjected to combined shear and tension, whether or not there is fatigue. Pretensioned bolts are permitted when slip resistance is of no concern.

c. **Slip-critical joints**

The installation of slip-critical bolts is identical with that for pretensioned joints. The only difference between the two is in the treatment of the contact or faying surfaces. Their inspection is the same, except that the inspector needs to check the faying or contact surface for slip-critical joints.

Slip-critical joints are required only for situations involving shear or combined shear and tension. They are not required for situations involving only tension. In addition, they are to be used for joints with oversized holes and for joints with slotted holes where the load is applied approximately normal (within 80 to 100 degrees) to the long direction of the slot.

When loads are applied to snug-tight bolts, there may be a little slippage, as the holes are a little larger in diameter than the shanks of the bolts. As a result, the parts of the connection may bear against the bolts. You can see that if we have a fatigue situation with constantly changing loads, this is not a desirable situation.

For fatigue situations, and for connections subject to direct tension, it is desirable to use connections that will not slip. These are referred to as *slip-critical connections*. To achieve this situation, the bolts must be tightened until they reach a fully tensioned condition in which they are subject to extremely large tensile forces.

Fully tensioning bolts is an expensive process, as is the inspection necessary to see that they are fully tensioned. Thus, they should be used only where absolutely necessary, as where the working loads cause large numbers of stress changes resulting in fatigue problems. Section J of the AISC Commentary gives a detailed list of connections that must be made with fully tensioned bolts. Included in this list are connections for supports of running machinery or for live loads producing impact and stress reversal; column splices in all tier structures 200 ft or more in height; connections of all beams and girders to columns and other beams or girders on which the bracing of the columns is dependent for structures over 125 ft in height; and so on.

Snug-tight bolts have several advantages over fully tensioned ones. One worker can properly tighten bolts to a snug-tight condition with an ordinary spud wrench or with only a few impacts of an impact wrench. The installation is quick, and only a visual inspection of the work is needed. (Such is not the case for fully tensioned bolts.) Furthermore, snug-tight bolts may be installed with electric wrenches, thus eliminating the need for air compression on the site. As a result, the use of snug-tight bolts saves time and money and is safer than the procedure needed for fully tensioned bolts. *Therefore, for most situations, snug-tight bolts should be used*.

Tables 12.1 and 12.1M provide the minimum fastener tensions required for slip-resistant connections and for connections subject to direct tension. These are, respectively, reproductions of Tables J3.1 and J3.1M of the AISC Specification.

TABLE 12.1   Minimum Bolt Pretension, kips*

Bolt Size, in	Group A-A325 Bolts	Group B-A490 Bolts
$1/2$	12	15
$5/8$	19	24
$3/4$	28	35
$7/8$	39	49
1	51	64
$1\tfrac{1}{8}$	56	80
$1\tfrac{1}{4}$	71	102
$1\tfrac{3}{8}$	85	121
$1\tfrac{1}{2}$	103	148

* Equal to 0.70 of minimum tensile strength of bolts, rounded off to nearest kip, as specified in ASTM Specifications for A325 and A490M bolts with UNC threads.

TABLE 12.1M   Minimum Bolt Pretension, kN*

Bolt Size, mm	Group A-A325M Bolts	Group B-A490M Bolts
M16	91	114
M20	142	179
M22	176	221
M24	205	257
M27	267	334
M30	326	408
M36	475	595

* Equal to 0.70 of minimum tensile strength of bolts, rounded off to nearest kN, as specified in ASTM Specifications for A325M and A490M bolts with UNC threads.

Source: American Institute of Steel Construction, *Manual of Steel Construction* (Chicago: AISC, 2011), Table J3.1 and J3.1M, pp. 16.1–119. "Copyright © American Institute of Steel Construction. Reprinted with permission. All rights reserved."

The quality-control provisions specified in the manufacture of the A325 and A490 bolts are more stringent than those for the A449 bolts. As a result, despite the method of tightening, the A449 bolts may not be used in slip-resistant connections.

Although many engineers felt that there would be some slippage compared with rivets (because of the fact that the hot driven rivets more nearly filled the holes), it was found that there is less slippage in fully tensioned high-strength bolted joints than in riveted joints under similar conditions.

It is interesting that the nuts used for fully tensioned high-strength bolts almost never need special provisions for locking. Once these bolts are installed and sufficiently tightened to produce the tension required, there is almost no tendency for the nuts to come loose. There are, however, a few situations where they will work loose under heavy vibrating loads. What do we do then? Some steel erectors have replaced the offending bolts with longer ones with two fully tightened nuts. Others have welded the nuts onto the bolts. Apparently, the results have been somewhat successful.

12.6 METHODS FOR FULLY PRETENSIONING HIGH-STRENGTH BOLTS

We have already commented on the tightening required for snug-tight bolts. For fully tensioned bolts, several methods of tightening are available. These methods, including the turn-of-the-nut method, the calibrated wrench method, and the use of alternative design bolts and direct tension indicators, are permitted without preference by the Specification. For A325 and A490 bolts, the minimum pretension equals 70 percent of their specified minimum tensile strength.

12.6.1 Turn-of-the-Nut Method

The bolts are brought to a snug-tight condition and then, with an impact wrench, they are given from one-third to one full turn, depending on their length and the slope of the surfaces under their heads and nuts. Table 8-2, page 16.2-48 in the Manual, presents the amounts of turn to be applied. (The amount of turn given to a particular bolt can easily be controlled by marking the snug-tight position with paint or crayon.)

12.6.2 Calibrated Wrench Method

With this method, the bolts are tightened with an impact wrench that is adjusted to stall at that certain torque which is theoretically necessary to tension a bolt of that diameter and ASTM classification to the desired tension. Also, it is necessary that wrenches be calibrated daily and that hardened washers be used. Particular care needs to be given to protecting the bolts from dirt and moisture at the job site. The reader should refer to the "Specification for Structural Joints Using ASTM A325 or A490 Bolts" in Part 16.2 of the Manual for additional tightening requirements.

12.6.3 Direct Tension Indicator

The direct tension indicator (which was originally a British device) consists of a hardened washer that has protrusions on one face in the form of small arches. The arches will be flattened as a bolt is tightened. The amount of gap at any one time is a measure of the bolt tension.

12.6.4 Alternative Design Fasteners

In addition to the preceding methods, there are some alternative design fasteners that can be tensioned quite satisfactorily. Bolts with splined ends that extend beyond the threaded portion of the bolts, called **twist-off bolts**, are one example. Special wrench chucks are used to tighten the nuts until the splined ends shear off. This method of tightening bolts is quite satisfactory and will result in lower labor costs.

A maximum bolt tension is not specified in any of the preceding tightening methods. This means that the bolt can be tightened to the highest load that will not break it, and the bolt still will do the job. Should the bolt break, another one is put in, with no damage done. It might be noted that the nut is stronger than the bolt, and the bolt will break before the nut strips. (The bolt specification mentioned previously requires that a tension measuring device be available at the job site to ensure that specified tensions are achieved.)

An impact wrench used to tighten bolts to either a snug-tight or a fully tensioned condition. It may be electric, such as this one, or it may be pneumatic. (Courtesy of CMC South Carolina Steel.)

Torquing the nut for a high-strength bolt with an air-driven impact wrench.
(Courtesy of Bethlehem Steel Corporation.)

For fatigue situations where members are subjected to constantly fluctuating loads, the slip-resistant connection is very desirable. If, however, the force to be carried is less than the frictional resistance, and thus no forces are applied to the bolts, how

A "twist-off bolt," also called a "load-indicator bolt," or a "tension control bolt." Notice the spline at the end of the bolt shank. (Courtesy of CMC South Carolina Steel.)

could we ever have a fatigue failure of the bolts? Slip-resistant connections can be designed to prevent slipping either at the service load limit state or at the strength load limit state.

Other situations where slip-resistant connections are desirable include joints where bolts are used in oversized holes, joints where bolts are used in slotted holes and the loads are applied parallel or nearly so to the slots, joints that are subjected to significant force reversals, and joints where bolts and welds resist shear together on a common *faying surface*. (The faying surface is the contact, or shear area between the members.)

12.7 SLIP-RESISTANT CONNECTIONS AND BEARING-TYPE CONNECTIONS

When high-strength bolts are fully tensioned, they clamp the parts being connected tightly together. The result is a considerable resistance to slipping on the faying surface. This resistance is equal to the clamping force times the coefficient of friction.

If the shearing load is less than the permissible frictional resistance, the connection is referred to as **slip-resistant**. If the load exceeds the frictional resistance, the members will slip on each other and will tend to shear off the bolts; at the same time, the connected parts will push or bear against the bolts, as shown in Fig. 12.1 on page 401.

The surfaces of joints, including the area adjacent to washers, need to be free of loose scale, dirt, burrs, and other defects that might prevent the parts from solid seating. It is necessary for the surface of the parts to be connected to have slopes of not more than 1 to 20 with respect to the bolt heads and nuts, unless beveled washers are used. For slip-resistant joints, the faying surfaces must also be free from oil, paint, and lacquer. (Actually, paint may be used if it is proved to be satisfactory by test.)

If the faying surfaces are galvanized, the slip factor will be reduced to almost half of its value for clean mill scale surfaces. The slip factor, however, may be significantly improved if the surfaces are subjected to hand wire brushing or to "brush off" grit blasting. However, such treatments do not seem to provide increased slip resistance for sustained loadings where there seems to be a creeplike behavior.[2]

---

[2]J. W. Fisher and J. H. A. Struik, *Guide to Design Criteria for Bolted and Riveted Joints* (New York: John Wiley & Sons, 1974) pp. 205–206.

The AASHTO Specifications permit hot-dip galvanization if the coated surfaces are scored with wire brushes or sandblasted after galvanization and before steel erection.

The ASTM Specification permits the galvanization of the A325 bolts themselves, but not the A490 bolts. There is a danger of embrittlement of this higher-strength steel during galvanization due to the possibility that hydrogen may be introduced into the steel in the pickling operation of the galvanization process.

If special faying surface conditions (such as blast-cleaned surfaces or blast-cleaned surfaces with special slip-resistant coatings applied) are used to increase the slip resistance, the designer may increase the values used here to the ones given by the Research Council on Structural Joints in Part 16.2 of the AISC Manual.

12.8 MIXED JOINTS

Bolts may on occasion be used in combination with welds and on other occasions with rivets (as where they are added to old riveted connections to enable them to carry increased loads). The AISC Specification contains some specific rules for these situations.

12.8.1 Bolts in Combination with Welds

For new work, neither A307 common bolts nor high-strength bolts designed for bearing or snug-tight connections may be considered to share the load with welds. (Before the connection's ultimate strength is reached, the bolts will slip, with the result that the welds will carry a larger proportion of the load—the actual proportion being difficult to determine.) For such circumstances, welds will have to be proportioned to resist the entire loads.

If high-strength bolts are designed for slip-critical conditions, they may be allowed to share the load with welds. For such situations, the AISC Commentary J1.8 states that it is necessary to fully tighten the bolts before the welds are made. If the weld is made first, the heat from the weld may very well distort the connection so that we will not get the slip-critical resistance desired from the bolts. If the bolts are placed and fully tightened before the welds are made, the heat of the welding will not change the mechanical properties of the bolts. For such a situation, the loads may be shared if the bolts are installed in standard-size holes or in short slotted holes with the slots perpendicular to the load direction. However, the contribution of the bolts is limited to 50 percent of their available strength in a bearing-type connection.[3]

If we are making alterations for an existing structure that is connected with bearing or snug-tight bolts or with rivets, we can assume that any slipping that is going to occur has already taken place. Thus, if we are using welds in the alteration, we will design those welds neglecting the forces that would be produced by the existing dead load.

---

[3]Kulak and G. Y. Grondin, "Strength of Joints That Combine Bolts and Welds," *Engineering Journal* (Chicago: AISC, vol. 38 no. 2, 2nd Quarter, 2001) pp. 89–98.

12.8.2 High-Strength Bolts in Combination with Rivets

High-strength bolts may be considered to share loads with rivets for new work or for alterations of existing connections that were designed as slip-critical. (The ductility of the rivets allows the capacity of both sets of fasteners to act together.)

12.9   SIZES OF BOLT HOLES

In addition to the standard size bolt holes (STD), which are 1/16 in larger in diameter than the bolts, there are three types of enlarged holes: oversized, short-slotted, and long-slotted. Oversized holes will on occasion be very useful in speeding up steel erection. In addition, they give some latitude for adjustments in plumbing frames during erection. The use of nonstandard holes requires the approval of the designer and is subject to the requirements of Section J3 of the AISC Specification. Table 12.2 provides the nominal dimensions in inches for the various kinds of enlarged holes permitted by the AISC, while Table 12.2M provides the same information in millimeters. (These tables are, respectively, Tables J3.3 and J3.3M of the AISC Specification.)

The situations in which we may use the various types of enlarged holes are now described.

Oversized holes (OVS) may be used in all plies of connections as long as the applied load does not exceed the permissible slip resistance. They may not be used in bearing-type connections. It is necessary for hardened washers to be used over oversized holes that are located in outer plies. The use of oversized holes permits the use of larger construction tolerances.

Short-slotted holes (SSL) may be used regardless of the direction of the applied load for slip-critical connections. For bearing type connections, however, the slots must be perpendicular to the direction of loading. Should the load be applied in a direction approximately normal (between 80 and 100 degrees) to the slot, these holes may be used in any or all plies of connections for bearing-type connections. It is necessary to

TABLE 12.2   Nominal Hole Dimensions, Inches

		Hole Dimensions		
Bolt Diameter	Standard (Dia.)	Oversize (Dia.)	Short-slot (Width × Length)	Long-slot (Width × Length)
$\frac{1}{2}$	$\frac{9}{16}$	$\frac{5}{8}$	$\frac{9}{16} \times \frac{11}{16}$	$\frac{9}{16} \times 1\frac{1}{4}$
$\frac{5}{8}$	$\frac{11}{16}$	$\frac{13}{16}$	$\frac{11}{16} \times \frac{7}{8}$	$\frac{11}{16} \times 1\frac{9}{16}$
$\frac{3}{4}$	$\frac{13}{16}$	$\frac{15}{16}$	$\frac{13}{16} \times 1$	$\frac{13}{16} \times 1\frac{7}{8}$
$\frac{7}{8}$	$\frac{15}{16}$	$1\frac{1}{16}$	$\frac{15}{16} \times 1\frac{1}{8}$	$\frac{15}{16} \times 2\frac{3}{16}$
1	$1\frac{1}{16}$	$1\frac{1}{4}$	$1\frac{1}{16} \times 1\frac{5}{16}$	$1\frac{1}{16} \times 2\frac{1}{2}$
$\geq 1\frac{1}{8}$	$d + \frac{1}{16}$	$d + \frac{5}{16}$	$(d + \frac{1}{16}) \times (d + \frac{3}{8})$	$(d + \frac{1}{16}) \times (2.5 \times d)$

TABLE 12.2M   Nominal Hole Dimensions, mm

Bolt Diameter	Standard (Dia.)	Oversize (Dia.)	Short-slot (Width × Length)	Long-slot (Width × Length)
		Hole Dimensions		
M16	18	20	18 × 22	18 × 40
M20	22	24	22 × 26	22 × 50
M22	24	28	24 × 30	24 × 55
M24	27 [a]	30	27 × 32	27 × 60
M27	30	35	30 × 37	30 × 67
M30	33	38	33 × 40	33 × 75
≥M36	$d + 3$	$d + 8$	$(d + 3) \times (d + 10)$	$(d + 3) \times 2.5d$

[a] Clearance provided allows the use of a 1 in bolt if desirable.

Source: American Institute of Steel Construction, *Manual of Steel Construction* (Chicago: AISC, 2011), Table J3.3 and J3.3M, pp. 16.1–121. "Copyright © American Institute of Steel Construction. Reprinted with permission. All rights reserved."

use washers (hardened if high-strength bolts are being used) over short-slotted holes in an outer ply. The use of short-slotted holes provides for some mill and fabrication tolerances, but does not result in the necessity for slip-critical procedures.

Long-slotted holes (LSL) may be used in *only one* of the connected parts of slip-critical or bearing-type connections at any one faying surface. For slip-critical joints these holes may be used in any direction, but for bearing-type connections the loads must be normal (between 80 and 100 degrees) to the axes of the slotted holes. If long-slotted holes are used in an outer ply they will need to be covered with plate washers or a continuous bar with standard holes. For high-strength bolted connections, the washers or bar do not have to be hardened, but they must be made of structural-grade material and may not be less than 5/16 in thick. Long-slotted holes are usually used when connections are being made to existing structures where the exact positions of the members being connected are not known.

Generally, washers are used to prevent scoring or galling of members when bolts are tightened. Most persons think that they also serve the purpose of spreading out the clamping forces more uniformly to the connected members. Tests have shown, however, that standard size washers don't affect the pressure very much, except when oversized or short-slotted holes are used. Sections 2.5 and 2.6 of Part 16.2 of the Manual (page 16.2-13) provide detailed information concerning washers.

12.10   LOAD TRANSFER AND TYPES OF JOINTS

The following paragraphs present a few of the elementary types of bolted joints subjected to axial forces. (That is, the loads are assumed to pass through the centers of gravities of the groups of connectors.) For each of these joint types, some comments are made about the methods of load transfer. Eccentrically loaded connections are discussed in Chapter 13.

For this initial discussion, the reader is referred to part (a) of Fig. 12.1. It is assumed that the plates shown are connected with a group of snug-tight bolts. In other words, the bolts are not tightened sufficiently so as to significantly squeeze the plates together. If there is assumed to be little friction between the plates, they will slip a little due to the applied loads. As a result, the loads in the plates will tend to shear the connectors off on the plane between the plates and press or bear against the sides of the bolts, as shown in part (b) of the figure. These connectors are said to be in *single shear and bearing* (also called *unenclosed bearing*). They must have sufficient strength to satisfactorily resist these forces, and the members forming the joint must be sufficiently strong to prevent the connectors from tearing through.

When rivets were used instead of the snug-tight bolts, the situation was somewhat different because hot-driven rivets would cool and shrink and then squeeze or clamp the connected pieces together with sizable forces that greatly increased the friction

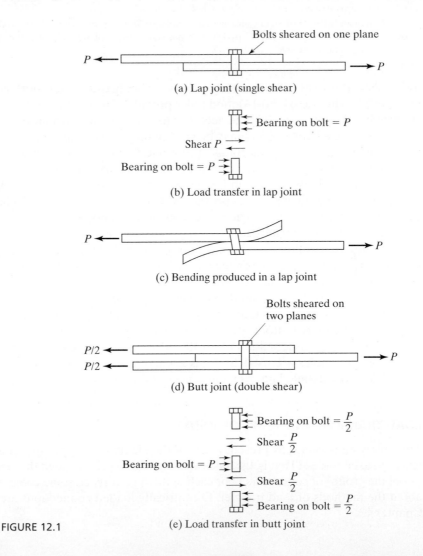

FIGURE 12.1

between the pieces. As a result, a large portion of the loads being transferred between the members was transferred by friction. The clamping forces produced in riveted joints, however, were generally not considered to be dependable, and for this reason specifications normally consider such connections to be snug-tight with no frictional resistance. The same assumption is made for A307 common bolts, as they are not tightened to large dependable tensions.

Fully tensioned high-strength bolts are in a different class altogether. By the tightening methods previously described, a very dependable tension is obtained in the bolts, resulting in large clamping forces and large dependable amounts of frictional resistance to slipping. Unless the loads to be transferred are larger than the frictional resistance, the entire forces are resisted by friction and the bolts are not really placed in shear or bearing. If the load exceeds the frictional resistance, there will be slippage, with the result that the bolts will be placed in shear and bearing.

12.10.1 The Lap Joint

The joint shown in part (a) of Fig. 12.1 is referred to as a *lap joint*. This type of joint has a disadvantage in that the center of gravity of the force in one member is not in line with the center of gravity of the force in the other member. A couple is present that causes an undesirable bending in the connection, as shown in part (c) of the figure. For this reason, the lap joint, which is desirably used only for minor connections, should be designed with at least two fasteners in each line parallel to the length of the member to minimize the possibility of a bending failure.

12.10.2 The Butt Joint

A *butt joint* is formed when three members are connected as shown in Fig. 12.1(d). If the slip resistance between the members is negligible, the members will slip a little and tend to shear off the bolts simultaneously on the two planes of contact between the members. Again, the members are bearing against the bolts, and the bolts are said to be in *double shear and bearing* (also called *enclosed bearing*). The butt joint is more desirable than the lap joint for two main reasons:

1. The members are arranged so that the total shearing force, P, is split into two parts, causing the force on each plane to be only about one-half of what it would be on a single plane if a lap joint were used. From a shear standpoint, therefore, the load-carrying ability of a group of bolts in double shear is theoretically twice as great as the same number of bolts in single shear.
2. A more symmetrical loading condition is provided. (In fact, the butt joint does provide a symmetrical situation if the outside members are the same thickness and resist the same forces. The result is a reduction or elimination of the bending described for a lap joint.)

12.10.3 Double-Plane Connections

The double-plane connection is one in which the bolts are subjected to single shear and bearing, but in which bending moment is prevented. This type of connection, which is shown for a hanger in Fig. 12.2(a), subjects the bolts to single shear.

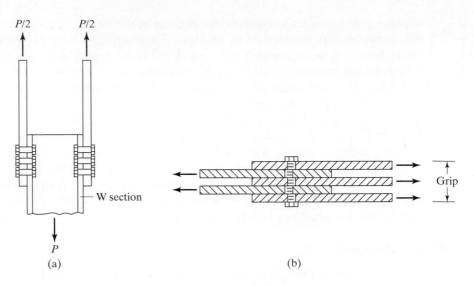

FIGURE 12.2 (a) Hanger connection. (b) Bolts in multiple shear.

12.10.4 Miscellaneous

Bolted connections generally consist of lap or butt joints or some combination of them, but there are other cases. For instance, there are occasionally joints in which more than three members are being connected and the bolts are in multiple shear, as shown in Fig. 12.2(b). In this figure, you can see how the loads are tending to shear this bolt on four separate planes (quadruple shear). Although the bolts in this connection are being sheared on more than two planes, the usual practice is to consider no more than double shear for strength calculations. It seems rather unlikely that shear failures can occur simultaneously on three or more planes. Several other types of bolted connections are discussed in this chapter and the next. These include bolts in tension, bolts in shear and tension, etc.

12.11 FAILURE OF BOLTED JOINTS

Figure 12.3 shows several ways in which failure of bolted joints can occur. To design bolted joints satisfactorily, it is necessary to understand these possibilities. These are described as follows:

1. The possibility of failure in a lap joint by shearing of the bolt on the plane between the members (single shear) is shown in part (a).
2. The possibility of a tension failure of one of the plates through a bolt hole is shown in part (b).
3. A possible failure of the bolts and/or plates by bearing between the two is given in part (c).
4. The possibility of failure due to the shearing out of part of the member is shown in part (d).
5. The possibility of a shear failure of the bolts along two planes (double shear) is shown in part (e).

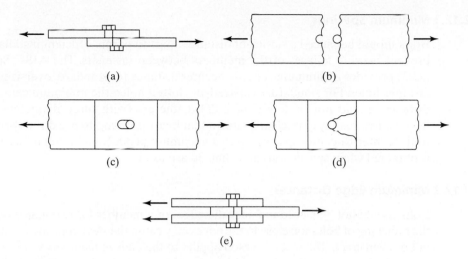

FIGURE 12.3

(a) Failure by single shearing of bolt. (b) Tension failure of plate. (c) Crushing failure of plate. (d) Shear failure of plate behind bolt. (e) Double shear failure of a butt joint.

12.12   SPACING AND EDGE DISTANCES OF BOLTS

Before minimum spacings and edge distances can be discussed, it is necessary for a few terms to be explained. The following definitions are given for a group of bolts in a connection and are shown in Fig. 12.4:

Pitch is the center-to-center distance of bolts in a direction parallel to the axis of the member.

Gage is the center-to-center distance of bolt lines perpendicular to the axis of the member.

The *edge distance* is the distance from the center of a bolt to the adjacent edge of a member.

The *distance between bolts* is the shortest distance between fasteners on the same or different gage lines.

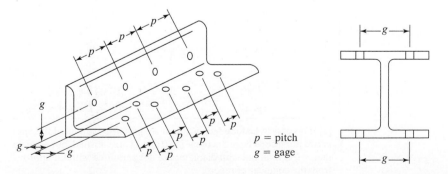

p = pitch
g = gage

FIGURE 12.4

12.12.1 Minimum Spacings

Bolts should be placed a sufficient distance apart to permit efficient installation and to prevent bearing failures of the members between fasteners. The AISC Specification (J3.3) provides a minimum center-to-center distance for standard, oversized, or slotted fastener holes. For standard, oversized, or slotted holes, the minimum center-to-center distance should not be less than 2 2/3 diameters (with three diameters being preferred). Test results have clearly shown that bearing strengths are directly proportional to the center-to-center spacing up to a maximum of 3d. No additional bearing strength is obtained when spacings greater than 3d are used.

12.12.2 Minimum Edge Distances

Bolts should not be placed too near the edges of a member for two major reasons. First, the punching of holes too close to the edges may cause the steel opposite the hole to bulge out or even crack. The second reason applies to the ends of members where there is danger of the fastener tearing through the metal. The usual practice is to place the fastener a minimum distance from the edge of the plates equal to about 1.5 to 2.0 times the fastener diameter so that the metal there will have a shearing strength at least equal to that of the fasteners. For more exact information, it is necessary to refer to the specification. The AISC Specification (J3.4) states that the distance from the center of a standard hole to the edge of a connected part may not be less than the applicable value given in Table 12.3 or 12.3M. (Tables J3.4 and J3.4M in the Manual.)

The minimum edge distance from the center of an oversized hole or a slotted hole to the edge of a connected part must equal the minimum distance required for a standard hole plus an increment C_2, values of which are provided in Table 12.4 or Table 12.4M. (These tables are, respectively, Tables J3.5 and J3.5M from the AISC

TABLE 12.3 Minimum Edge Distance[a] from Center of Standard Hole[b] to Edge of Connected Part, inches	
Bolt Diameter (in)	Minimum Edge Distance (in)
$1/2$	$3/4$
$5/8$	$7/8$
$3/4$	1
$7/8$	$1\frac{1}{8}$
1	$1\frac{1}{4}$
$1\frac{1}{8}$	$1\frac{1}{2}$
$1\frac{1}{4}$	$1\frac{5}{8}$
Over $1\frac{1}{4}$	$1\frac{1}{4} \times$ Diameter

[a] If necessary, lesser edge distances are permitted provided the appropriate provisions from Sections J3.10 and J4 are satisfied, but edge distances less than one bolt diameter are not permitted without approval from the engineer of record.
[b] For oversized or slotted holes, see Table J3.5.

TABLE 12.3M   Minimum Edge Distance[a] from Center of Standard Hole[b] to Edge of Connected Part, mm

Bolt Diameter (mm)	Minimum Edge Distance (mm)
16	22
20	26
22	28
24	30
27	34
30	38
36	46
Over 36	$1.25d$

[a] If necessary, lesser edge distances are permitted provided the appropriate provisions from Sections J3.10 and J4 are satisfied, but edge distances less than one bolt diameter are not permitted without approval from the engineer of record.

[b] For oversized or slotted holes, see Table J3.5M.

Source: American Institute of Steel Construction, *Manual of Steel Construction Load & Resistance Factor Design*, 14th ed. (Chicago: AISC, 2011), Table J3.4 and J3.4M, p. 16.1–123. "Copyright © American Institute of Steel Construction. Reprinted with permission. All rights reserved."

TABLE 12.4   Values of Edge Distance Increment C_2, Inches

Normal Diameter of Fastener (in)	Oversized Holes	Slotted Holes		Long Axis Parallel to Edge
		Long Axis Perpendicular to Edge		
		Short Slots	Long Slots [a]	
$\leq \frac{7}{8}$	$\frac{1}{16}$	$\frac{1}{8}$	$\frac{3}{4}d$	0
1	$\frac{1}{8}$	$\frac{1}{8}$		
$\geq 1\frac{1}{8}$	$\frac{1}{8}$	$\frac{3}{16}$		

[a] When length of slot is less than maximum allowable (see Table 12.2 herein), C_2 is permitted to be reduced by one-half the difference between the maximum and actual slot lengths.

TABLE 12.4M   Values of Edge Distance Increment C_2, mm

Nominal Diameter of Fastener (mm)	Oversized Holes	Slotted Holes		Long Axis Parallel to Edge
		Long Axis Perpendicular to Edge		
		Short Slots	Long Slots [a]	
≤ 22	2	3	$0.75d$	0
24	3	3		
≥ 27	3	5		

[a] When length of slot is less than maximum allowable (see Table 12.2M in text), C_2 is permitted to be reduced by one-half the difference between the maximum and actual slot lengths.

Source: American Institute of Steel Construction, *Manual of Steel Construction*, 14th ed. (Chicago: AISC, 2011), Table J3.5 and J3.5M, p. 16.1–124. "Copyright © American Institute of Steel Construction. Reprinted with permission. All rights reserved."

Specification.) As will be seen in the pages to follow, the computed bearing strengths of connections will have to be reduced if these requirements are not met.

12.12.3 Maximum Spacing and Edge Distances

Structural steel specifications provide maximum edge distances for bolted connections. The purpose of such requirements is to reduce the chances of moisture getting between the parts. When fasteners are too far from the edges of parts being connected, the edges may sometimes separate, thus permitting the entrance of moisture. When this happens and there is a failure of the paint, corrosion will develop and accumulate, causing increased separations between the parts. The AISC maximum permissible edge distance (J3.5) is 12 times the thickness of the connected part, but not more than 6 in (150 mm).

The maximum edge distances and spacings of bolts used for weathering steel are smaller than they are for regular painted steel subject to corrosion, or for regular unpainted steel not subject to corrosion. One of the requirements for using weathering steel is that it must not be allowed to be constantly in contact with water. For this reason, the AISC Specification tries to insure that the parts of a built-up weathering steel member are connected tightly together at frequent intervals to prevent forming of pockets that might catch and hold water. The AISC Specification (J3.5) states that the maximum spacing of bolts center-to-center for painted members, or for unpainted members not subject to corrosion, is 24 times the thickness of the thinner plate, not to exceed 12 in (305 mm). For unpainted members consisting of weathering steel subject to atmospheric corrosion, the maximum is 14 times the thickness of the thinner plate, not to exceed 7 in (180 mm).

Holes cannot be punched very close to the web-flange junction of a beam or the junction of the legs of an angle. They can be drilled, but this rather expensive practice should not be followed unless there is an unusual situation. Even if the holes are drilled in these locations, there may be considerable difficulty in placing and tightening the bolts in the limited available space.

12.13   BEARING-TYPE CONNECTIONS—LOADS PASSING THROUGH CENTER OF GRAVITY OF CONNECTIONS

12.13.1 Shearing Strength

In bearing-type connections, it is assumed that the loads to be transferred are larger than the frictional resistance caused by tightening the bolts, with the result that the members slip a little on each other, putting the bolts in shear and bearing. The design or LRFD strength of a bolt in single shear equals ϕ times the nominal shearing strength of the bolt in ksi times its cross-sectional area. The allowable ASD strength equals its nominal shearing strength divided by Ω times its cross-sectional area. The LRFD ϕ value is 0.75 for high-strength bolts, while for ASD Ω it is 2.00.

The nominal shear strengths of bolts and rivets are given in Table 12.5 (Table J3.2 in the AISC Specification). For A325 bolts, the values are 54 ksi if threads are not excluded from shear planes and 68 ksi if threads are excluded. (The values are 68 ksi and 84 ksi,

TABLE 12.5   Nominal Strength of Fasteners and Threaded Parts, ksi (MPa)

Description of Fasteners	Nominal Tensile Strength, F_{nt}, ksi (MPa)[a]	Nominal Shear Strength in Bearing-Type Connections, F_{nv}, ksi (MPa)[b]
A307 bolts	45 (310)	27 (188)[c][d]
Group A (A325 type) bolts, when threads are not excluded from shear planes	90 (620)	54 (372)
Group A (A325 type) bolts, when threads are excluded from shear planes	90 (620)	68 (457)
Group B (A490 type) bolts, when threads are not excluded from shear planes	113 (780)	68 (457)
Group B (A490 type) bolts, when threads are excluded from shear planes	113 (780)	84 (579)
Threaded parts meeting the requirements of Section A3.4 of the Manual, when threads are not excluded from shear planes	$0.75F_u$	$0.450F_u$
Threaded parts meeting the requirements of Section A3.4 of the Manual, when threads are excluded from shear planes	$0.75F_u$	$0.563F_u$

[a] For high-strength bolts subjected to tensile fatigue loading, see Appendix 3.
[b] For end loaded connections with a fastener pattern length greater than 38 in (965 mm), F_{nv} shall be reduced to 83.3 percent of the tabulated values. Fastener pattern length is the maximum distance parallel to the line of force between the centerline of the bolts connecting two parts with one faying surface.
[c] For A307 bolts, the tabulated values shall be reduced by 1 percent for each $1/16$ in (2 mm) over 5 diameters of length in the grip.
[d] Threads permitted in shear planes.

Source: American Institute of Steel Construction, *Manual of Steel Construction*, 14th ed. (Chicago: AISC, 2011), Table J3.2, p. 16.1–120. "Copyright © American Institute of Steel Construction. Reprinted with permission. All rights reserved."

respectively, for A490 bolts.) Should a bolt be in double shear, its shearing strength is considered to be twice its single shear value.

The student may very well wonder what is done in design practice concerning threads excluded or not excluded from the shear planes. If normal bolt and member sizes are used, the threads will almost always be excluded from the shear plane. It is true, however, that some extremely conservative individuals always assume that the threads are not excluded from the shear plane.

Sometimes the designer needs to use high-strength bolts with diameters larger than those of available A325 and A490 bolts. One example is the use of very large bolts for fastening machine bases. For such situations, AISC Specification A3.3 permits the use of the quenched and tempered A449 bolts. (**Quenching** is the heating of steel to approximately 1650°F, followed by its quick cooling in water, oil, brine, or molten lead. This process produces very strong and hard steels, but, at the same time, steels that are more susceptible to residual stresses. For this reason, after quenching is done, the steel is tempered. **Tempering** is the reheating of steel to a temperature of perhaps 1100° or 1150°F, after which the steel is allowed to air cool. The internal stresses are reduced and the steel is made tougher and more ductile.)

12.13.2 Bearing Strength

The bearing strength of a bolted connection is not, as you might expect, determined from the strength of the bolts themselves; rather, it is based upon the strength of the parts being connected and the arrangement of the bolts. In detail, its computed strength is dependent upon the spacing of the bolts and their edge distances, the specified tensile strength F_u of the connected parts, and the thickness of the connected parts.

Expressions for the nominal bearing strengths (R_n values) at bolt holes are provided in Section J3.10 of the AISC Specification. To determine ϕR_n and $\dfrac{R_n}{\Omega}$, ϕ is 0.75 and Ω is 2.00. The various expressions listed there include nominal bolt diameters (d), the thicknesses of members bearing against the bolts (t), and the clear distances (l_c) between the edges of holes and the edges of the adjacent holes or edges of the material in the direction of the force. Finally, F_u is the specified minimum tensile strength of the connected material.

To be consistent throughout this text, the author has conservatively assumed that the diameter of a bolt hole equals the bolt diameter, plus 1/8 in. This dimension is used in computing the value of L_c for substituting into the expressions for R_n.

The expressions to follow are used to compute the nominal bearing strengths of bolts used in connections that have standard, oversized, or short-slotted holes, regardless of the direction of loading. They also are applicable to connections with long-slotted holes if the slots are parallel to the direction of the bearing forces.

a. If deformation around bolt holes is a design consideration (that is, if we want deformations to be ≤ 0.25 in), then

$$R_n = 1.2\,l_c t F_u \leq 2.4\,dt F_u \qquad \text{(AISC Equation J3-6a)}$$

For the problems considered in this text, we will normally assume that deformations around the bolt holes are important. Thus, unless specifically stated otherwise, Equation J3-6a will be used for bearing calculations.

If deformation around bolt holes is not a design consideration (that is, if deformations > 0.25 in are acceptable), then

$$R_n = 1.5\,l_c t F_u \leq 3.0\,dt F_u \qquad \text{(AISC Equation J3-6b)}$$

b. For bolts used in connections with long-slotted holes, the slots being perpendicular to the forces,

$$R_n = 1.0\,l_c t F_u \le 2.0\,dt F_u \qquad \text{(AISC Equation J3-6c)}$$

As described in Section 12.9 of this chapter, oversized holes cannot be used in bearing connections, but short-slotted holes can be used in bearing-type connections—if the loads are perpendicular to the long directions of the slots.

Tests of bolted joints have shown that neither the bolts nor the metal in contact with the bolts actually fail in bearing. However, these tests also have shown that the efficiency of the connected parts in tension and compression is affected by the magnitude of the bearing stress. Therefore, the nominal bearing strengths given by the AISC Specification are values above which they feel the strength of the connected parts is impaired. In other words, these apparently very high design bearing stresses are not really bearing stresses at all, but, rather, indexes of the efficiencies of the connected parts. If bearing stresses larger than the values given are permitted, the holes seem to elongate more than about 1/4 in and impair the strength of the connections.

From the preceding, we can see that the bearing strengths given are not specified to protect fasteners from bearing failures, because they do not need such protection. Thus, the same bearing values will be used for a particular joint, regardless of the grades of bolts used and regardless of the presence or absence of bolt threads in the bearing area.

12.13.3 Minimum Connection Strength

Example 12-1 illustrates the calculations involved in determining the strength of the bearing-type connection shown in Fig. 12.5. By a similar procedure, the number of bolts required for a certain loading condition is calculated in Example 12-2. In each case, the bearing thickness to be used equals the smaller total thickness on one side or the other, since the steel grade for all plates is the same and since the edge distances are identical for all plates. For instance, in Fig. 12.6, the bearing thickness equals the smaller of $2 \times 1/2$ in on the left or 3/4 in on the right.

In the connection tables of the AISC Manual and in various bolt literature, we constantly see abbreviations used when referring to various types of bolts. For instance, we may see A325-SC, A325-N, A325-X, A490-SC, and so on. These are used to represent the following:

A325-SC—slip-critical or fully tensioned A325 bolts

A325-N—snug-tight or bearing A325 bolts with threads *included* in the shear planes

A325-X—snug-tight or bearing A325 bolts with threads *excluded* from the shear planes

---

Example 12-1

Determine the design strength $\phi_c P_n$ and the allowable strength $\dfrac{P_n}{\Omega}$ for the bearing-type connection shown in Fig. 12.5. The steel is A36($F_y = 36$ ksi and $F_u = 58$ ksi), the bolts

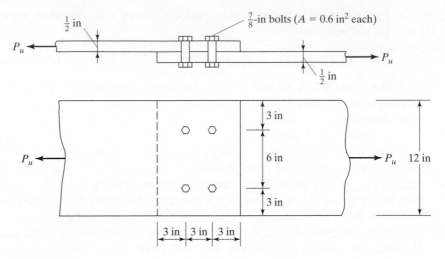

FIGURE 12.5

are 7/8-in A325, the holes are standard sizes, and the threads are excluded from the shear plane. Assume that deformations at bolt holes are a design consideration.

Solution

(a) Gross section yielding of plates

$$P_n = F_y A_g = (36 \text{ ksi})\left(\frac{1}{2} \text{ in} \times 12 \text{ in}\right) = 216 \text{ k}$$

LRFD $\phi_t = 0.9$	ASD $\Omega_t = 1.67$
$\phi_{tn} = (0.9)(216) = 194.4 \text{ k}$	$\dfrac{P_n}{\Omega} = \dfrac{216}{1.67} = 129.3 \text{ k}$

(b) Tensile rupture strength of plates

$$A_n = 6.00 \text{ in}^2 - (2)\left(\frac{7}{8} \text{ in} + \frac{1}{8} \text{ in}\right)\left(\frac{1}{2} \text{ in}\right) = 5.00 \text{ in}^2$$

$U = 1.0$ as all parts connected

$A_e = U A_n = (1.00)(5.00) = 5.00 \text{ in}^2 < 0.85 A_g$

$\qquad = (0.85)(6.00) = 5.10 \text{ in}^2$ as per AISC Spec. J4.1

$P_n = F_u A_e = (58 \text{ ksi})(5.00 \text{ in}^2) = 290 \text{ k}$

LRFD $\phi_t = 0.75$	ASD $\Omega_t = 2.00$
$\phi_t P_n = (0.75)(290) = 217.5 \text{ k}$	$\dfrac{P_n}{\Omega_t} = \dfrac{290}{2.00} = 145 \text{ k}$

(c) Bearing strength of bolts

$$l_c = \text{lesser of } 3 - \frac{1}{2} \text{ or } 3 - 1 = 2.00 \text{ in}$$

$$R_n = 1.2 l_c t F_u (\text{No. of bolts}) \leq 2.4 \, dt \, F_u (\text{No. of bolts})$$

$$= (1.2)(2.00 \text{ in})\left(\frac{1}{2} \text{ in}\right)(58 \text{ ksi})(4)$$

$$= 278.4 \text{ k} > (2.4)\left(\frac{7}{8} \text{ in}\right)\left(\frac{1}{2} \text{ in}\right)(58 \text{ ksi})(4) = 243.6 \text{ k}$$

LRFD $\phi = 0.75$	ASD $\Omega = 2.00$
$\phi R_n = (0.75)(243.6) = 182.7$ k	$\dfrac{R_n}{\Omega} = \dfrac{243.6}{2.00} = 121.8$ k

(d) Shearing strength of bolts

$$R_n = F_{nv} A_b (\text{No. of bolts}) = (68 \text{ ksi})(0.6 \text{ in}^2)(4) = 163.2 \text{ k}$$

LRFD $\phi = 0.75$	ASD $\Omega = 2.00$
$\phi R_n = (0.75)(163.2) = 122.4$ k	$\dfrac{R_n}{\Omega} = \dfrac{163.2 \text{ k}}{2.00} = 81.6$ k
LRFD = 122.4 k (controls)	**ASD = 81.6 k (controls)**

Example 12-2

How many 3/4-in A325 bolts in standard-size holes with threads excluded from the shear plane are required for the bearing-type connection shown in Fig. 12.6? Use $F_u = 58$ ksi and assume edge distances to be 2 in and the distance center-to-center of holes to be 3 in. Assume that deformation at bolt holes is a design consideration. $P_u = 345$ k (LRFD). $P_a = 230$ k (ASD).

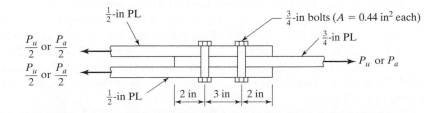

FIGURE 12.6

Solution. Bolts in double shear and bearing on 3/4 in

Bearing strength of 1 bolt

$$L_c = \text{lesser of } 2 - \frac{\frac{3}{4} + \frac{1}{8}}{2} = 1.56 \text{ in} \quad \text{or} \quad 3 - (2)\left(\frac{\frac{3}{4} + \frac{1}{8}}{2}\right) = 2.125 \text{ in}$$

$$R_n = 1.2 l_c t F_u \leq 2.4 dt F_u$$

$$= (1.2)(1.56 \text{ in})\left(\frac{3}{4} \text{ in}\right)(58 \text{ ksi}) = 81.4 \text{ k} > (2.4)\left(\frac{3}{4} \text{ in}\right)\left(\frac{3}{4} \text{ in}\right)(58 \text{ ksi}) = 78.3 \text{ k}$$

Shearing strength of 1 bolt

$$R_n = (2 \times 0.44 \text{ in}^2)(68 \text{ ksi}) = 59.8 \text{ k} \leftarrow \text{controls}$$

LRFD $\phi = 0.75$	ASD $\Omega = 2.00$
$\phi R_n = (0.75)(59.8) = 44.8 \text{ k}$	$\dfrac{R_n}{\Omega} = \dfrac{59.8}{2.00} = 29.9 \text{ k}$
No. of bolts reqd. $= \dfrac{P_u}{\phi R_n}$	No. of bolts reqd. $= \dfrac{P_a}{R_n/\Omega}$
$= \dfrac{345}{44.8} = 7.70$	$= \dfrac{230}{29.9} = 7.69$
Use eight $\frac{3}{4}$-in bearing type A325 bolts.	**Use eight $\frac{3}{4}$-in bearing type A325 bolts.**

Where cover plates are bolted to the flanges of W sections, the bolts must carry the longitudinal shear on the plane between the plates and the flanges. With reference to the cover-plated beam of Fig. 12.7, the unit longitudinal shearing stress to be resisted between a cover plate and the W flange can be determined with the expression $f_v = VQ/Ib$. The total shear force across the flange for a 1-in length of the beam equals $(b)(1.0)(VQ/Ib) = VQ/I$.

The AISC Specification (E6.2) provides a maximum permissible spacing for bolts used in the outside plates of built-up members. It equals the thinner outside plate thickness times $0.75\sqrt{E/F_y}$ and may not be larger than 12 in.

The spacing of pairs of bolts at a particular section in Fig. 12.7 can be determined by dividing the LRFD design shear strength of two bolts by the factored shear per in or by dividing the ASD allowable shear design strength of 2 bolts by the service shear per inch. Theoretically, the spacings will vary as the external shear varies along the span. Example 12-3 illustrates the calculations involved in determining bolt spacing for a cover-plated beam.

The reader should note that AISC Specification F13.3 states that the total cross-sectional area of the cover plates of a bolted girder may not be greater than 70 percent of the total flange area.

Example 12-3

At a certain section in the cover-plated beam of Fig. 12.7, the external factored shears are $V_u = 275$ k and $V_a = 190$ k. Determine the spacing required for 7/8-in A325 bolts used in a bearing-type connection. Assume that the bolt threads are excluded from the shear plane, the edge distance is 3.5 in, $F_y = 50$ ksi, and $F_u = 65$ ksi. Deformation at bolt holes is a design consideration.

Solution

Checking AISC Specification F13.3

$$A \text{ of 1 cover plate} = \left(\frac{3}{4}\right)(16) = 12.00 \text{ in}^2$$

$$A \text{ of 1 flange} = 12.00 + (12.5)(1.15) = 26.38 \text{ in}^2$$

$$\text{Plate area} \div \text{flange area} = \frac{12.00}{26.38} < 0.70 \qquad \text{(OK)}$$

Computing shearing force to be taken

$$I_g = 3630 + (2)\left(\frac{3}{4} \times 16\right)\left(\frac{22.1}{2} + \frac{0.75}{2}\right)^2 = 6760 \text{ in}^4$$

LRFD	ASD
Factored shear per in $= \dfrac{V_u Q}{I}$ for LRFD	Service load shear per in $= \dfrac{V_a Q}{I}$ for ASD
$= \dfrac{(275)\left(\dfrac{3}{4} \times 16 \times 11.425\right)}{6760} = 5.578$ k/in	$= \dfrac{(190)\left(\dfrac{3}{4} \times 16 \times 11.425\right)}{6760} = 3.853$ k/in

Bolts in single shear and bearing on 0.75 in

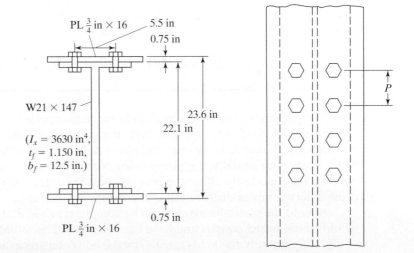

PL $\frac{3}{4}$ in × 16   5.5 in
0.75 in

W21 × 147

($I_x = 3630$ in^4,
$t_f = 1.150$ in,
$b_f = 12.5$ in.)

23.6 in
22.1 in

0.75 in

PL $\frac{3}{4}$ in × 16

P

FIGURE 12.7

Bearing strength of 2 bolts

$$L_c = 3.5 - \frac{\frac{7}{8} + \frac{1}{8}}{2} = 3.0 \text{ in}$$

$$R_n = 1.2 L_c t F_u \leq 2.4 dt F_u$$

$$= (2)(1.2)(3.0 \text{ in})\left(\frac{3}{4} \text{ in}\right)(65 \text{ ksi})$$

$$= 351 \text{ k} > (2)(2.4)\left(\frac{7}{8} \text{ in}\right)\left(\frac{3}{4} \text{ in}\right)(65 \text{ ksi}) = 204.8 \text{ k}$$

Shearing strength of 2 bolts

$$A = 0.60 \text{ in}^2 \text{ each bolt}$$

LRFD $\phi = 0.75$	ASD $\Omega = 2.00$
$\phi R_n = (0.75)(81.6) = 61.2 \text{ k}$	$\dfrac{R_n}{\Omega} = \dfrac{81.6}{2.00} = 40.8 \text{ k}$
Spacing of bolts $= \dfrac{61.2}{5.578}$	Spacing reqd. for bolts
$= 10.97 \text{ in}$	$= \dfrac{40.8}{3.853} = 10.59 \text{ in}$

$$\text{Max spacing by AISC (E6.2)} = (t)\left(0.75\sqrt{\frac{E}{F_y}}\right)$$

$$= \left(\frac{3}{4}\right)(0.75)\sqrt{\frac{29 \times 10^3}{50}} = 13.55 \text{ in} \leq 12 \text{ in}$$

Now that we have the calculated spacing of the pairs of bolts, we can see that L_c in the direction of the force is $>L_c$ to the edge of the member. ∴ there will be no change in the nominal bearing strength of the bolts.

Use $\frac{7}{8}$-in A325 bolts 10 in on center for both LRFD and ASD

The assumption has been made that the loads applied to a bearing-type connection are equally divided between the bolts if edge distances and spacings are satisfactory. For this distribution to be correct, the plates must be perfectly rigid and the bolts perfectly elastic, but actually, the plates being connected are elastic, too, and have deformations that decidedly affect the bolt stresses. The effect of these deformations is to cause a very complex distribution of load in the elastic range.

Should the plates be assumed to be completely rigid and nondeforming, all bolts would be deformed equally and have equal stresses. This situation is shown in part (a) of Fig. 12.8. Actually, the loads resisted by the bolts of a group are probably never equal

Bridge over Allegheny River at Kittaning, PA. (Courtesy of the American Bridge Company.)

(in the elastic range) when there are more than two bolts in a line. Should the plates be deformable, the plate stresses, and thus the deformations, will decrease from the ends of the connection to the middle, as shown in part (b) of Fig. 12.8. The result is that the highest stressed elements of the top plate will be over the lowest stressed elements of the lower plate, and vice versa. The slip will be greatest at the end bolts and smallest at the middle bolts. The bolts at the ends will then have stresses much greater than those in the inside bolts.

The greater the spacing is of bolts in a connection, the greater will be the variation in bolt stresses due to plate deformation; therefore, the use of compact joints is very desirable, as they will tend to reduce the variation in bolt stresses. It might be interesting to consider a theoretical (although not practical) method of roughly equalizing bolt stresses. The theory would involve the reduction of the thickness of the plate toward its end, in proportion to the reduced stresses, by stepping. This procedure, which is shown in Fig. 12.8(c), would tend to equalize the deformations of the plate and thus of the bolt stresses. A similar procedure would be to scarf the overlapping plates.

The calculation of the theoretically correct elastic stresses in a bolted group based on plate deformations is a tedious problem and is rarely if ever handled in the design office. On the other hand, the analysis of a bolted joint based on the plastic theory is a very simple problem. In this theory, the end bolts are assumed to be stressed to their yield point. Should the total load on the connection be increased, the end bolts will deform without resisting additional load, the next bolts in the line will

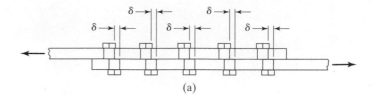

(a)

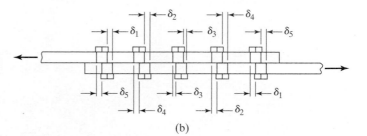

(b)

FIGURE 12.8

(a) Assuming nondeforming plates.
(b) Assuming deformable plates.
(c) Stepped joint (impractical).

(c)

Three high-strength bolted structures in Constitution Plaza Complex, Hartford, CT, using approximately 195,000 bolts. (Courtesy of Bethlehem Steel Corporation.)

have their stresses increased until they too are at the yield point, and so forth. Plastic analysis seems to justify to a certain extent the assumption of rigid plates and equal bolt stresses that is usually made in design practice. This assumption is used in the example problems of this chapter.

When there are only a few bolts in a line, the plastic theory of equal stresses seems to be borne out very well, but when there are a large number of bolts in a line, the situation changes. Tests have clearly shown that the end bolts will fail before the full redistribution takes place.[4]

For load-carrying bolted joints, it is common for specifications to require a minimum of two or three fasteners. The feeling is that a single connector may fail to live up to its specified strength because of improper installation, material weakness, etc., but if several fasteners are used, the effects of one bad fastener in the group will be overcome.

12.14 SLIP-CRITICAL CONNECTIONS—LOADS PASSING THROUGH CENTER OF GRAVITY OF CONNECTIONS

Almost all bolted connections with standard size holes are designed as bearing-type connections. On some occasions, however, particularly in bridges, it is felt that slipping should be prevented. High-strength bolted connections may be designed such that slipping is prevented either at the service load limit state or at the strength limit state. These are referred to as slip-critical connections.

Slip-critical connections should be used only when the engineer feels that slipping will adversely affect the serviceability of a structure. For such a structure, slipping may cause excessive distortion of the structure or a reduction in strength or stability, even if the strength of the connection is adequate. For one example, it is felt necessary to use slip-critical connections when oversized holes are used or when slots parallel to the load direction of force are planned. Other situations where slip-critical bolts are desirable are described in Section J3.8 of the Commentary to the AISC Specification.

If bolts are tightened to their required tensions for slip-critical connections (see Tables 12.1 and 12.1M), there is very little chance of their bearing against the plates that they are connecting. In fact, tests show that there is very little chance of slip occurring unless there is a calculated shear of at least 50 percent of the total bolt tension. As we have said all along, this means that slip-critical bolts are not stressed in shear; however, AISC Specification J3.8 provides *design shear strengths* (they are really design friction values on the faying surfaces) so that the designer can handle the connections in just about the same manner he or she handles bearing-type connections.

Although there is little or no bearing on the bolts used in slip-critical connections, the AISC in its Section J3.10 states that bearing strength is to be checked for both bearing-type and slip-critical connections because a possibility still exists that slippage could occur; therefore, the connection must have sufficient strength as a bearing-type connection.

AISC Specification J3.8 states that the nominal slip resistance of a connection (R_n) shall be determined with the expression

$$R_n = \mu D_u h_f T_b n_s \qquad \text{(AISC Equation J3-4)}$$

in which

μ = mean slip coefficient = 0.30 for Class A faying surfaces and 0.5 for Class B faying surfaces. Section 3 of Part 16.2 of the AISC Manual provides detailed information concerning these two surfaces. Briefly, Class A denotes unpainted clean, mill scale surfaces or surfaces with Class A coatings on blast-cleaned steel surfaces. Class B surfaces are unpainted blast-cleaned steel surfaces or surfaces with Class B coatings

D_u = 1.13. This is a multiplier that gives the ratio of the mean installed pretension to the specified minimum pretension given in Table 12.1 of this text (Table J3.1 in the AISC Specification)

h_f = factor for fillers, determined as follows:

(1) Where bolts have been added to distribute loads in the filler, h_f = 1.0
(2) Where bolts have not been added to distribute the load in the filler,
   (i) For one filler between connected parts, h_f = 1.0
   (ii) For two or more fillers between connected parts, h_f = 0.85

T_b = minimum fastener tension, as given in Table 12.1 of this text
n_s = number of slip planes

For standard size and short-slotted holes perpendicular to the direction of the load

$$\phi = 1.00 \text{ (LRFD)} \qquad\qquad \Omega = 1.50 \text{ (ASD)}$$

For oversized and short-slotted holes parallel to the direction of the load

$$\phi = 0.85 \text{ (LRFD)} \qquad\qquad \Omega = 1.76 \text{ (ASD)}$$

For long-slotted holes

$$\phi = 0.70 \text{ (LRFD)} \qquad\qquad \Omega = 2.14 \text{ (ASD)}$$

It is permissible to introduce finger shims up to 1/4-in thick into slip-critical connections with standard holes without the necessity of reducing the bolt design strength values to those specified for slotted holes (AISC Specification Section J3.2).

The preceding discussion concerning slip-critical joints does not present the whole story, because during erection the joints may be assembled with bolts, and as the members are erected their weights will often push the bolts against the side of the holes before they are tightened and put them in some bearing and shear.

The majority of bolted connections made with standard-size holes can be designed as bearing-type connections without the need to worry about serviceability. Furthermore, if connections are made with three or more bolts in standard-size holes or are used with slots perpendicular to the force direction, slip probably cannot occur, because at least one and perhaps more of the bolts will be in bearing before the external loads are applied.

Sometimes, bolts are used in situations where deformations can cause increasing loads that may be larger than the strength limit states. These situations may occur in connections that make use of oversize holes or slotted holes which are parallel to the load.

When a slip-critical connection is being used with standard or short-slotted holes perpendicular to the direction of the load, the slip will not result in an increased load and a $\phi = 1.0$ or $\Omega = 1.5$ is used. For oversized and short-slotted holes parallel to the direction of load which could result in an increased load situation, a value of $\phi = 0.85$ or $\Omega = 1.76$ is used. Similarly, for long-slotted holes a value of $\phi = 0.70$ or $\Omega = 2.14$ is applicable.

Example 12-4 that follows illustrates the design of slip-critical bolts for a lap joint. The example presents the determination of the number of bolts required for the limit state of slip.

Example 12-4

For the lap joint shown in Fig. 12.9, the axial service loads are $P_D = 27.5$ k and $P_L = 40$ k. Determine the number of 1-in A325 slip-critical bolts in standard-size holes needed for the limit state of slip if the faying surface is Class A. The edge distance is 1.75 in, and the c. to c. spacing of the bolts is 3 in. $F_y = 50$ ksi. $F_u = 65$ ksi.

Solution

Loads to be resisted

LRFD	ASD
$P_u = (1.2)(27.5) + (1.6)(40) = 97$ k	$P_a = 27.5 + 40 = 67.5$ k

Nominal strength of 1 bolt

$$R_n = \mu D_u h_f T_b n_s \qquad \text{(AISC Equation J3-4)}$$

$\mu = 0.30$ for Class A surface

$D_u = 1.13$ multiplier

$h_f = 1.00$ factor for filler

$T_b = 51$ k minimum bolt pretension   (AISC Table J3.1)

$n_s = 1.0 = $ number of slip planes

$R_n = (0.30)(1.13)(1.00)(51 \text{ k})(1.00) = 17.29$ k/bolt

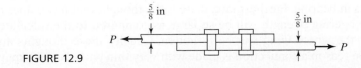

FIGURE 12.9

(a) Slip-critical design to prevent slip

LRFD $\phi = 1.00$	ASD $\Omega = 1.50$
$\phi R_n = (1.00)(17.29) = 17.29$ k	$\dfrac{R_n}{\Omega} = \dfrac{17.29}{1.50} = 11.53$ k
No. of bolts reqd.	No. of bolts reqd.
$= \dfrac{97}{17.29} = 5.61$	$= \dfrac{67.5}{11.53} = 5.86$
Use 6 bolts.	**Use 6 bolts.**

Bearing strength of 6 bolts

$$L_c = \text{lesser of } 3 - \left(1 + \frac{1}{8}\right) = 1.875 \text{ in or } 1.75 - \frac{1 + \dfrac{1}{8}}{2} = 1.187 \text{ in}$$

$$\text{Total } R_n = (6)(1.5 l_c t F_u) \le (6)(2.4 dt F_u)$$

$$= (6)\left(1.5 \times 1.187 \text{ in} \times \frac{5}{8} \text{ in} \times 65 \text{ ksi}\right)$$

$$= 434 \text{ k} < (6)\left(2.4 \times 1.00 \text{ in} \times \frac{5}{8} \text{ in} \times 65 \text{ ksi}\right) = 585 \text{ k}$$

LRFD $\phi = 0.75$	ASD $\Omega = 2.00$
$\phi R_n = (0.75)(434) = 326$ k > 97 k OK	$\dfrac{R_n}{\Omega} = \dfrac{434}{2.00} = 217$ k > 67.5 k OK

Shearing strength of 6 bolts (single shear)

$$\text{Total } R_n = 6 F_{nv} A_b = (6)(68 \text{ ksi})(0.785 \text{ in}^2) = 320.3 \text{ k}$$

LRFD $\phi = 0.75$	ASD $\Omega = 2.00$
$\phi R_n = (0.75)(320.3) = 240.2$ k > 97 k OK	$\dfrac{R_n}{\Omega} = \dfrac{320.3}{2.00} = 160.2$ k > 67.5 k OK
Use 6 bolts.	**Use 6 bolts.**

The reader may think that bearing strength checks for slip-critical connections are a waste of time. He or she may feel the connections are not going to slip and put the bolts in bearing. Furthermore, there is the thought that if slip does occur, the calculated bolt bearing strength will be so large as compared to the calculated shearing strength that the whole thing can be forgotten. Usually, these thoughts are correct, but if for some reason a connection is made with very thin parts, bearing may very well control.

Example 12-5

Repeat Example 12-4 if the plates have long-slotted holes in the direction of the load. Assume that deformations of the connections will cause an increase in the critical load. Therefore, design the connection to prevent slip at the limit state of slip.

Solution

$P_u = 97$ k and $P_a = 67.5$ k from Example 12-4 solution.

Nominal strength of 1 bolt

$$R_n = \mu D_u h_f T_b n_s$$
$$\mu = 0.30 \text{ for Class A surface}$$
$$D_u = 1.13 \text{ multiplier}$$
$$h_f = 1.00 \text{ factor for filler}$$
$$T_b = 51 \text{ k minimum bolt pretension}$$
$$n_s = 1.0 = \text{number of slip planes}$$
$$R_n = (0.30)(1.13)(1.0)(51)(1.0) = 17.29 \text{ k/bolt}$$

Number of bolts required for long-slotted holes

LRFD $\phi = 0.70$	ASD $\Omega = 2.14$
$\phi R_n = (0.70)(17.29) = 12.10$ k	$\dfrac{R_n}{\Omega} = \dfrac{17.29}{2.14} = 8.08$ k
No. reqd. $= \dfrac{97}{12.10} = 8.02$ bolts	No. reqd. $= \dfrac{67.5}{8.08} = 8.35$ bolts

Note: Shear and bearing were checked in Example 12-4 and are obviously ok here as they are higher than they were before.

Ans.   **Use 9 bolts.**   **Use 9 bolts.**

12.15 PROBLEMS FOR SOLUTION

For each of the problems listed, the following information is to be used, unless otherwise indicated (a) AISC Specification; (b) standard-size holes; (c) members have clean mill-scale surfaces (Class A); (d) $F_y = 36$ ksi and $F_u = 58$ ksi unless otherwise noted, (e) deformation at service loads is a design consideration. Do not consider block shear, unless specifically requested.

12-1 to 12-5. *Determine the LRFD design tensile strength and the ASD allowable tensile strength for the member shown, assuming a bearing-type connection.*

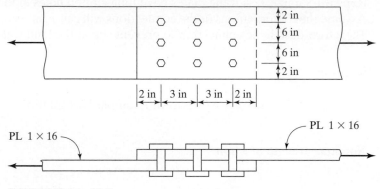

FIGURE P12-1 to 12-5

12-1. A325 $\frac{3}{4}$-in bolts, threads excluded from shear plane. (*Ans.* 202.0 k, 134.7 k)

12-2. A325 1-in bolts, threads excluded from shear plane.

12-3. A490 1-in bolts, threads not excluded from shear plane. (*Ans.* 281.6 k, 190.7 k)

12-4. $\frac{7}{8}$-in A325 bolts, threads excluded from shear plane.

12-5. $\frac{3}{4}$-in A490 bolts, threads not excluded from shear plane. (*Ans.* 202.0 k, 134.7 k)

12-6 to 12-10. *Determine the LRFD design tensile strength and the ASD allowable tensile strength for the member and the bearing-type connections.*

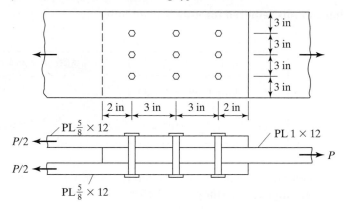

FIGURE P12-6 to 12-10

12-6. A325 $\frac{3}{4}$-in bolts, threads excluded from shear planes.

12-7. A490 $\frac{7}{8}$-in bolts, threads not excluded from shear planes. (*Ans.* 388.8 k, 258.7 k)

12-8. A490 $\frac{3}{4}$-in bolts, threads excluded from shear plane.

12-9. A steel with $F_y = 50$ ksi, $F_u = 70$ ksi, $\frac{7}{8}$-in A490 bolts, threads excluded from shear plane. (*Ans.* 472.5 k, 315 k)

12-10. A steel with $F_y = 50$ ksi, $F_u = 70$ ksi, 1-in A490 bolts, threads excluded from shear planes.

12-11 to 12-13. *How many bolts are required for LRFD and ASD for the bearing-type connection shown, if $P_D = 50\ k$ and $P_L = 100\ k$?*

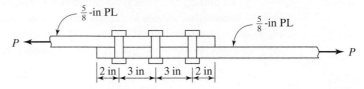

FIGURE P12-11 to 12-13

12-11. A325 $\frac{3}{4}$-in bolts, threads excluded from shear plane. (*Ans.* 10 both LRFD and ASD)

12-12. $F_y = 50$ ksi, $F_u = 70$ ksi, $\frac{3}{4}$-in A325 bolts, threads excluded from shear plane.

12-13. A490 1-in bolts, threads not excluded from shear plane. (*Ans.* 6 both LRFD and ASD)

12-14 to 12-16. *How many bolts are required (LRFD and ASD) for the bearing-type connection shown if $P_D = 120\ k$ and $P_L = 150\ k$?*

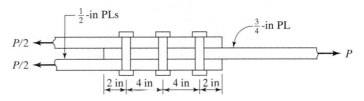

FIGURE P12-14 to 12-16

12-14. A325 $\frac{7}{8}$-in bolts, threads excluded from shear planes.

12-15. A490 $\frac{3}{4}$-in bolts, threads not excluded from shear planes. (*Ans.* 9 or 10 both LRFD and ASD)

12-16. A325 1-in bolts, threads not excluded from shear planes.

12-17. The truss member shown in the accompanying illustration consists of two C12 × 25s (A36 steel) connected to a 1-in gusset plate. How many $\frac{7}{8}$-in A325 bolts (threads excluded from shear plane) are required to develop the full design tensile capacity of the member if it is used as a bearing-type connection? Assume $U = 0.85$. Use both LRFD and ASD methods. (*Ans.* 8 both LRFD and ASD)

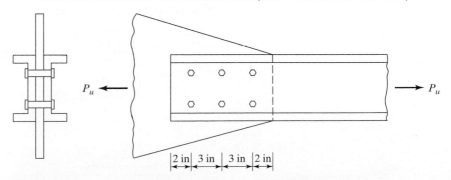

FIGURE P12-17

12-18. Repeat Prob. 12-17, using $\frac{3}{4}$-in A490 bolts (threads excluded).

12-19. Rework Prob. 12-17, if $\frac{7}{8}$-in A490 bolts are used (threads not excluded). $F_y = 50$ ksi and $F_u = 65$ ksi. (*Ans.* 9 both LRFD and ASD)

12-20. For the connection shown in the accompanying illustration, $P_u = 360$ k and $P_a = 260$ k. Determine by LRFD and ASD the number of 1-in A325 bolts required for a bearing-type connection, using A36 steel. Threads are excluded from shear plane.

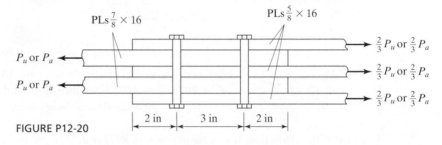

FIGURE P12-20

12-21. Rework Prob. 12-20, using $\frac{7}{8}$-in A490 bolts (threads not excluded from shear plane). (*Ans.* 12 LRFD and 13 or 14 ASD)

12-22. How many $\frac{3}{4}$-in A490 bolts (threads excluded from shear planes) in a bearing-type connection are required to develop the design tensile strength of the member shown? Assume that A36 steel is used and that there are two lines of bolts in each flange (at least three in a line 4 in o.c.). LRFD and ASD. Do not consider block shear.

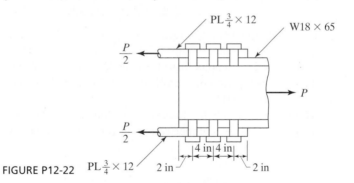

FIGURE P12-22

12-23. For the beam shown in the accompanying illustration, what is the required spacing of $\frac{3}{4}$-in A490 bolts (threads not excluded from shear plane) in a bearing-type

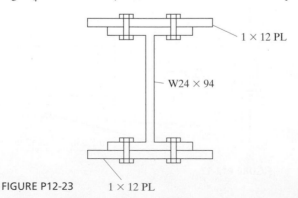

FIGURE P12-23   1 × 12 PL

connection at a section where the external shear $V_D = 80$ k and $V_L = 160$ k? LRFD and ASD. Assume $L_c = 1.50$ in. (*Ans.* 8 in both LRFD and ASD)

12-24. The cover-plated section shown in the accompanying illustration is used to support a uniform load $w_D = 10$ k/ft (includes beam weight effect and $w_L = 12.5$ k/ft for an 24-ft simple span). If $\frac{7}{8}$-in A325 bolts (threads excluded) are used in a bearing-type connection, work out a spacing diagram for the entire span, for LRFD only.

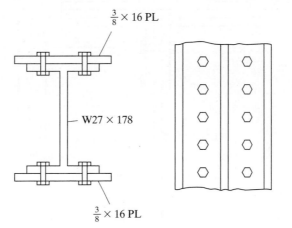

$\frac{3}{8} \times 16$ PL

W27 × 178

$\frac{3}{8} \times 16$ PL

FIGURE P12-24

12-25. For the section shown in the accompanying illustration, determine, for ASD only, the required spacing of $\frac{7}{8}$-in A490 bolts (threads excluded) for a bearing-type connection if the member consists of A572 grade 60 steel ($F_u = 75$ ksi). $V_D = 100$ k and $V_L = 140$ k. Assume Class A surfaces and $L_c = 1.0$ in (*Ans.* 6 in both)

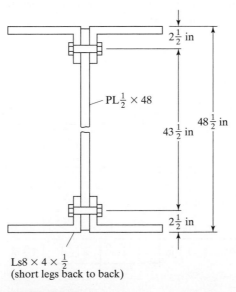

$2\frac{1}{2}$ in

PL$\frac{1}{2} \times 48$

$48\frac{1}{2}$ in

$43\frac{1}{2}$ in

$2\frac{1}{2}$ in

Ls8 × 4 × $\frac{1}{2}$
(short legs back to back)

FIGURE P12-25

12-26.   For an external shear V_u of 600 k, determine by LRFD the spacing required for 1-in A325 web bolts (threads excluded) in a bearing-type connection for the built-up section shown in the accompanying illustration. Assume that $l_c = 1.5$ in and A36 steel.

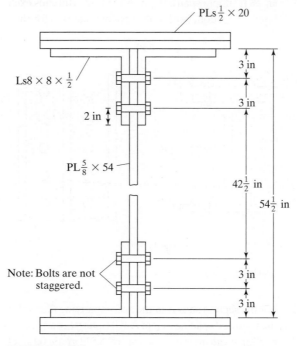

FIGURE P12-26

12-27.   Determine the design strength P_u and the allowable strength P_a for the connection shown if $\frac{7}{8}$-in A325 bolts (threads excluded) are used in a slip-critical connection with a factor for fillers, $h_f = 1.0$. Assume A36 steel and Class B faying surface and standard size holes. (*Ans.* 132.2 k, 88.1 k)

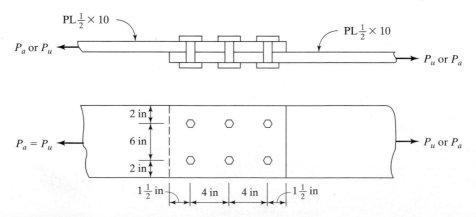

FIGURE P12-27

12-28 to 12-33.   *Repeat these problems, using the loads given and determine the number of bolts required for a slip-critical connection. Assume Class A surfaces, standard-size holes, $h_f = 1.00$, and l_c values of 1.50 in, UNO.*

12-28.   Prob. 12-6. $P_D = 100$ k, $P_L = 150$ k

12-29.   Prob. 12-11. $P_D = 50$ k, $P_L = 100$ k (*Ans.* 24 both LRFD, ASD)

12-30.   Prob. 12-13. $P_D = 75$ k, $P_L = 160$ k

12-31.   Prob. 12-14. $P_D = 120$ k, $P_L = 150$ k (*Ans.* 16 both LRFD, ASD)

12-32.   Prob. 12-16. $P_D = 40$ k, $P_L = 100$ k

12-33.   Prob. 12-20. (*Ans.* 11 LRFD, 12 ASD)

12-34 and 12-35.   *Using the bearing-type connection from each problem given, determine the number of 1-in A490 bolts required, by LRFD and ASD, for a slip-critical connection. Assume long-slotted holes in the direction of the load, Class A faying surfaces, $h_f = 1.00$, and $l_c = 1.25$ in.*

12-34.   Prob. 12-12.

12-35.   Prob. 12-15. (*Ans.* 11 or 12 LRFD, 12 ASD)

12-36.   Determine the design tensile strength P_u and the allowable tensile strength P_a of the connection shown if eight $\frac{7}{8}$-in A325 bearing-type bolts (threads excluded from shear plane) are used in each flange. Include block shear in your calculations. A36 steel is used.

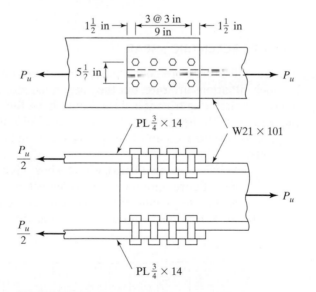

FIGURE P12-36

12-37.   Repeat Prob. 12-36, using $\frac{7}{8}$-in A490 bearing-type bolts. $F_y = 50$ ksi and $F_u = 65$ ksi (*Ans.* 604.8 k, 403.2 k)

CHAPTER 13

Eccentrically Loaded Bolted Connections and Historical Notes on Rivets

13.1 BOLTS SUBJECTED TO ECCENTRIC SHEAR

Eccentrically loaded bolt groups are subjected to shears and bending moments. You might think that such situations are rare, but they are much more common than most people suspect. For instance, in a truss it is desirable to have the center of gravity of a member lined up exactly with the center of gravity of the bolts at its end connections. This feat is not quite as easy to accomplish as it may seem, and connections are often subjected to moments.

Eccentricity is quite obvious in Fig. 13.1(a), where a beam is connected to a column with a plate. In part (b) of the figure, another beam is connected to a column with a pair of web angles. It is obvious that this connection must resist some moment, because the center of gravity of the load from the beam does not coincide with the reaction from the column.

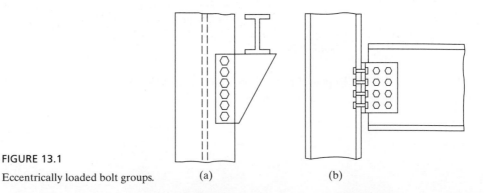

FIGURE 13.1

Eccentrically loaded bolt groups.   (a)   (b)

430

Bridge over New River Gorge near Charleston in Fayette County, W.V. (Courtesy of the American Bridge Company.)

In general, specifications for bolts and welds clearly state that the center of gravity of the connection should coincide with the center of gravity of the member, unless the eccentricity is accounted for in the calculations. However, Section J1.7 of the AISC Specification provides some exceptions to this rule. It states that the rule is not applicable to the end connections of statically loaded single angles, double angles, and similar members. In other words, the eccentricities between the centers of gravity of these members and the centers of gravity of the connections may be ignored unless fatigue loadings are involved. *Furthermore, the eccentricity between the gravity axes and the gage lines of bolted members may be neglected for statically loaded members.*

The AISC Specification presents values for computing the design strengths of individual bolts, but does not specify a method for computing the forces on these fasteners when they are eccentrically loaded. As a result, the method of analysis to be used is left up to the designer.

Three general approaches for the analysis of eccentrically loaded connections have been developed through the years. The first of the methods is the very conservative *elastic method* in which friction or slip resistance between the connected parts is neglected. In addition, these connected parts are assumed to be perfectly rigid. This type of analysis has been commonly used since at least 1870.[1,2]

Tests have shown that the elastic method usually provides very conservative results. As a consequence, various *reduced* or *effective eccentricity methods* have been proposed.[3] The analysis is handled just as it is in the elastic method, except that smaller eccentricities, and thus smaller moments, are used in the calculations.

The third method, called the *instantaneous center of rotation method*, provides the most realistic values compared with test results, but is extremely tedious to apply, at least with handheld calculators. Tables 7-7 to 7-14 in Part 7 of the Manual for eccentrically loaded bolted connections are based on the ultimate strength method and enable us to solve most of these types of problems quite easily, as long as the bolt patterns are symmetrical. The remainder of this section is devoted to these three analysis methods.

13.1.1 Elastic Analysis

For this discussion, the bolts of Fig. 13.2(a) are assumed to be subjected to a load P that has an eccentricity of e from the c.g. (center of gravity) of the bolt group. To consider the force situation in the bolts, an upward and downward force—each equal to P—is assumed to act at the c.g. of the bolt group. This situation, shown in part (b) of the figure, in no way changes the bolt forces. The force in a particular bolt will, therefore, equal P divided by the number of bolts in the group, as seen in part (c), plus the force due to the moment caused by the couple, shown in part (d) of the figure.

The magnitude of the forces in the bolts due to the moment Pe will now be considered. The distances of each bolt from the c.g. of the group are represented by the values d_1, d_2, etc., in Fig. 13.3. The moment produced by the couple is assumed to cause the plate to rotate about the c.g. of the bolt connection, with the amount of rotation or

---

[1] W. McGuire, *Steel Structures* (Englewood Cliffs, NJ: Prentice-Hall, 1968), p. 813.
[2] C. Reilly, "Studies of Iron Girder Bridges," *Proc. Inst. Civil Engrs.* 29 (London, 1870).
[3] T.R. Higgins, "New Formulas for Fasteners Loaded Off Center," *Engr. News Record* (May 21, 1964).

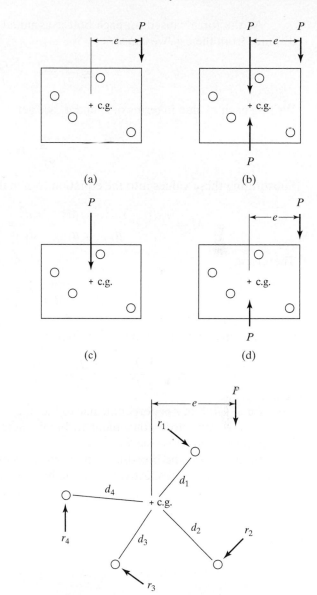

FIGURE 13.2

(a)   (b)   (c)   (d)

FIGURE 13.3

strain at a particular bolt being proportional to its distance from the c.g. (For this derivation, the gusset plates are again assumed to be perfectly rigid and the bolts are assumed to be perfectly elastic.) Stress is greatest at the bolt that is the greatest distance from the c.g., because stress is proportional to strain in the elastic range.

The rotation is assumed to produce forces of r_1, r_2, r_3, and r_4, respectively, from the bolts in the figure. The moment transferred to the bolts must be balanced by resisting moments of the bolts as shown in Equation (1)

$$M_{c.g.} = Pe = r_1d_1 + r_2d_2 + r_3d_3 + r_4d_4 \qquad (1)$$

As the force caused on each bolt is assumed to be directly proportional to the distance from the c.g., we can write

$$\frac{r_1}{d_1} = \frac{r_2}{d_2} = \frac{r_3}{d_3} = \frac{r_4}{d_4}$$

Writing each r value in terms of r_1 and d_1, we get

$$r_1 = \frac{r_1 d_1}{d_1} \quad r_2 = \frac{r_1 d_2}{d_1} \quad r_3 = \frac{r_1 d_3}{d_1} \quad r_4 = \frac{r_1 d_4}{d_1}$$

Substituting these values into the equation (original) and simplifying yields

$$M = \frac{r_1 d_1^2}{d_1} + \frac{r_1 d_2^2}{d_1} + \frac{r_1 d_3^2}{d_1} + \frac{r_1 d_4^2}{d_1} = \frac{r_1}{d_1}(d_1^2 + d_2^2 + d_3^2 + d_4^2)$$

Therefore,

$$M = \frac{r_1 \Sigma d^2}{d_1}$$

The force on each bolt can now be written as

$$r_1 = \frac{M d_1}{\Sigma d^2} \quad r_2 = \frac{d_2}{d_1} r_1 = \frac{M d_2}{\Sigma d^2} \quad r_3 = \frac{M d_3}{\Sigma d^2} \quad r_4 = \frac{M d_4}{\Sigma d^2}$$

Each value of r is perpendicular to the line drawn from the c.g. to the particular bolt. It is usually more convenient to break these reactions down into vertical and horizontal components. See Fig. 13.4.

In this figure, the horizontal and vertical components of the distance d_1 are represented by h and v, respectively, and the horizontal and vertical components of force

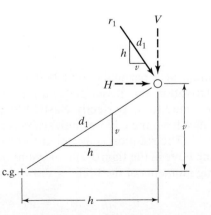

FIGURE 13.4

r_1 are represented by H and V, respectively. It is now possible to write the following ratio from which H can be obtained:

$$\frac{r_1}{d_1} = \frac{H}{v}$$

$$H = \frac{r_1 v}{d_1} = \left(\frac{M d_1}{\Sigma d^2}\right)\left(\frac{v}{d_1}\right)$$

Therefore,

$$H = \frac{Mv}{\Sigma d^2}$$

By a similar procedure,

$$V = \frac{Mh}{\Sigma d^2}$$

Example 13-1

Determine the force in the most stressed bolt of the group shown in Fig. 13.5, using the elastic analysis method.

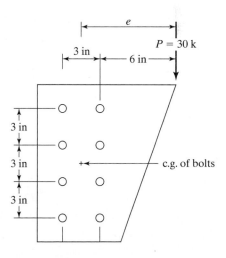

FIGURE 13.5

Solution. A sketch of each bolt and the forces applied to it by the direct load and the clockwise moment are shown in Fig. 13.6. From this sketch, the student can see that the upper right-hand bolt and the lower right-hand bolt are the most stressed and that their respective stresses are equal:

$$e = 6 + 1.5 = 7.5 \text{ in}$$

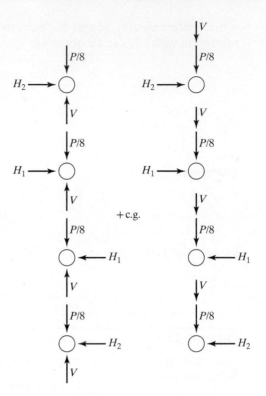

FIGURE 13.6

$$M = Pe = (30\text{ k})(7.5\text{ in}) = 225\text{ in-k}$$

$$\Sigma d^2 = \Sigma h^2 + \Sigma v^2$$

$$\Sigma d^2 = (8)(1.5)^2 + (4)(1.5^2 + 4.5^2) = 108\text{ in}^2$$

For lower right-hand bolt

$$H = \frac{Mv}{\Sigma d^2} = \frac{(225\text{ in-k})(4.5\text{ in})}{108\text{ in}^2} = 9.38\text{ k} \leftarrow$$

$$V = \frac{Mh}{\Sigma d^2} = \frac{(225\text{ in-k})(1.5\text{ in})}{108\text{ in}^2} = 3.13\text{ k} \downarrow$$

$$\frac{P}{8} = \frac{30\text{ k}}{8} = 3.75\text{ k} \downarrow$$

These components for the lower right-hand bolt are sketched as follows:

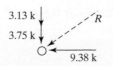

The resultant force applied to this bolt is

$$R = \sqrt{(3.13 + 3.75)^2 + (9.38)^2} = 11.63 \text{ k}$$

If the eccentric load is inclined, it can be broken down into vertical and horizontal components, and the moment of each about the c.g. of the bolt group can be determined. Various design formulas can be developed that will enable the engineer to directly design eccentric connections, but the process of assuming a certain number and arrangement of bolts, checking stresses, and redesigning probably is just as satisfactory.

The trouble with this inaccurate, but very conservative, method of analysis is that, in effect, we are assuming that there is a linear relation between loads and deformations in the fasteners; further, we assume that their yield stress is not exceeded when the ultimate load on the connection is reached. Various experiments have shown that these assumptions are incorrect.

Summing up this discussion, we can say that the elastic method is easier to apply than the instantaneous center of rotation method to be described in Section 13.1.3. However, it is probably too conservative, as it neglects the ductility of the bolts and the advantage of load redistribution.

13.1.2 Reduced Eccentricity Method

The elastic analysis method just described appreciably overestimates the moment forces applied to the connectors. As a result, quite a few proposals have been made through the years that make use of an effective eccentricity, in effect taking into account the slip resistance on the faying or contact surfaces. One set of reduced eccentricity values that were fairly common at one time follow:

1. With one gage line of fasteners and where n is the number of fasteners in the line,

$$e_{\text{effective}} = e_{\text{actual}} - \frac{1 + 2n}{4}$$

2. With two or more gage lines of fasteners symmetrically placed and where n is the number of fasteners in each line,

$$e_{\text{effective}} = e_{\text{actual}} - \frac{1 + n}{2}$$

The reduced eccentricity values for two fastener arrangements are shown in Fig. 13.7.

To analyze a particular connection with the reduced eccentricity method, the value of $e_{\text{effective}}$ is computed as described before and is used to compute the eccentric moment. Then the elastic procedure is used for the remainder of the calculations.

13.1.3 Instantaneous Center of Rotation Method

Both the elastic and reduced eccentricity methods for analyzing eccentrically loaded fastener groups are based on the assumption that the behavior of the fasteners is elastic. A much more realistic method of analysis is the instantaneous center of

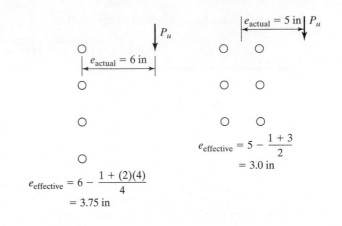

$$e_{\text{effective}} = 6 - \frac{1 + (2)(4)}{4}$$
$$= 3.75 \text{ in}$$

$$e_{\text{effective}} = 5 - \frac{1 + 3}{2}$$
$$= 3.0 \text{ in}$$

FIGURE 13.7

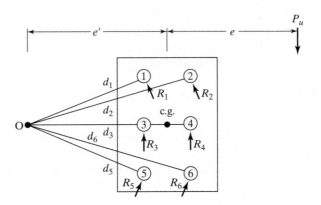

FIGURE 13.8

rotation method, which is described in the next few paragraphs. The values given in the AISC Manual for eccentrically loaded fastener groups were computed by this method.

If one of the outermost bolts in an eccentrically loaded connection begins to slip or yield, the connection will not fail. Instead, the magnitude of the eccentric load may be increased, the inner bolts will resist more load, and failure will not occur until all of the bolts slip or yield.

The eccentric load tends to cause both a relative rotation and translation of the connected material. In effect, this is equivalent to pure rotation of the connection about a single point called the *instantaneous center of rotation*. An eccentrically loaded bolted connection is shown in Fig. 13.8, and the instantaneous center is represented by point 0. It is located a distance e' from the center of gravity of the bolt group.

The deformations of these bolts are assumed to vary in proportion to their distances from the instantaneous center. The ultimate shear force that one of them can resist is not equal to the pure shear force that a bolt can resist. Rather, it is dependent upon the load-deformation relationship in the bolt. Studies by Crawford and Kulak[4] have shown that this force may be closely estimated with the following expression:

$$R = R_{ult}(1 - e^{-10\Delta})^{0.55}$$

In this formula, R_{ult} is the ultimate shear load for a single fastener equaling 74 k for a 3/4-in diameter A325 bolt, e is the base of the natural logarithm (2.718), and Δ is the total deformation of a bolt. Its maximum value experimentally determined is 0.34 in. The Δ values for the other bolts are assumed to be in proportion to R as their d distances are to d for the bolt with the largest d. The coefficients 10.0 and 0.55 also were experimentally obtained. Figure 13.9 illustrates this load-deformation relationship.

This expression clearly shows that the ultimate shear load taken by a particular bolt in an eccentrically loaded connection is affected by its deformation. Thus, the load applied to a particular bolt is dependent upon its position in the connection with respect to the instantaneous center of rotation.

The resisting forces of the bolts of the connection of Fig. 13.8 are represented with the letters R_1, R_2, R_3, and so on. Each of these forces is assumed to act in a direction perpendicular to a line drawn from point 0 to the center of the bolt in question. For this symmetrical connection, the instantaneous center of rotation will fall somewhere on a horizontal line through the center of gravity of the bolt group. This is the case because the sum of the horizontal components of the R forces must be zero, as also must be the sum of the moments of the horizontal components about point 0. The position of point 0 on the horizontal line may be found by a tedious trial-and-error procedure to be described here.

With reference to Fig. 13.8, the moment of the eccentric load about point 0 must be equal to the summation of the moments of the R resisting forces about the same point. If we knew the location of the instantaneous center, we could compute R values for the bolts with the Crawford-Kulak formula and determine P_u from the expression to follow, in which e and e' are distances shown in Figs. 13.8 and 13.11.

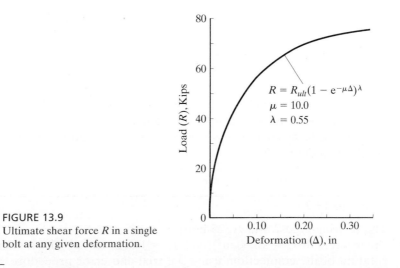

FIGURE 13.9
Ultimate shear force R in a single bolt at any given deformation.

[4]S. F. Crawford and G. L. Kulak, "Eccentrically Loaded Bolt Connections," *Journal of Structural Division*, ASCE 97, ST3 (March 1971), pp. 765–783.

$$P_u(e' + e) = \Sigma Rd$$

$$P_u = \frac{\Sigma Rd}{e' + e}$$

To determine the design strength of such a connection according to the AISC Specification, we can replace R_{ult} in the Crawford-Kulak formula with the design shearing strength of one bolt in a connection where the load is not eccentric. For instance, if we have 7/8-in A325 bolts (threads excluded from shear plane) in single shear bearing on a sufficient thickness so that bearing does not control, R_{ult} will equal, for the LRFD method,

$$R_{ult} = \phi F_n A_b = (0.75)(68 \text{ ksi})(0.60 \text{ in}^2) = 30.6 \text{ k}$$

The location of the instantaneous center is not known, however. Its position is estimated, the R values determined, and P_u calculated as described. It will be noted that P_u must be equal to the summation of the vertical components of the R resisting forces (ΣR_v). If the value is computed and equals the P_u computed by the preceding formula, we have the correct location for the instantaneous center. If not, we try another location, and so on.

In Example 13-2, the author demonstrates the very tedious trial-and-error calculations necessary to locate the instantaneous center of rotation for a symmetrical connection consisting of four bolts. In addition, the LRFD design strength of the connection ϕR_n and the allowable strength R_n/Ω are determined.

To solve such a problem, it is very convenient to set the calculations up in a table similar to the one used in the solution to follow. In the table shown, the h and v values given are the horizontal and vertical components of the d distances from point 0 to the centers of gravity of the individual bolts. The bolt that is located at the greatest distance from point 0 is assumed to have a Δ value of 0.34 in. The Δ values for the other bolts are assumed to be proportional to their distances from point 0. The Δ values so determined are used in the R formula.

A set of tables entitled "Coefficients C for Eccentrically Loaded Bolt Groups" is presented in Tables 7-7 to 7-14 of the AISC Manual. The values in these tables were determined by the procedure described here. A large percentage of the practical cases that the designer will encounter are included in the tables. Should some other situation not covered be faced, the designer may very well decide to use the more conservative elastic procedure previously described.

Example 13-2

The bearing-type 7/8-in A325 bolts of the connection of Fig. 13.10 have a nominal shear strength $r_n = (0.60 \text{ in}^2)(68 \text{ ksi}) = 40.8 \text{ k}$. Locate the instantaneous center of rotation of the connection, using the trial-and-error procedure, and determine the value of P_u.

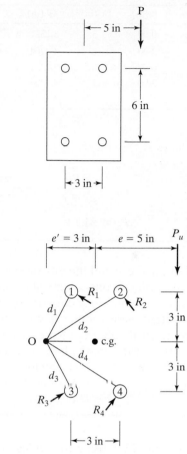

FIGURE 13.10

FIGURE 13.11

Solution. *By trial and error:* Try a value of $e' = 3$ in, reference being made to Fig. 13.11. In the accompanying table, Δ for bolt 1 equals $(3.3541/5.4083)\,(0.34) = 0.211$ in and R for the same bolt equals $30.6(1 - e^{-(10)(0.211)})^{0.55}$.

Bolt No.	h (in)	v (in)	d (in)	Δ (in)	R(kips)	R_v(kips)	Rd (k-in)
1	1.5	3	3.3541	0.211	28.50	12.74	95.58
2	4.5	3	5.4083	0.34	30.03	24.99	162.43
3	1.5	3	3.3541	0.211	28.50	12.74	95.58
4	4.5	3	5.4083	0.34	30.03	24.99	162.43
						$\Sigma = 75.46$	$\Sigma = 516.03$

$$P_u = \frac{\Sigma Rd}{e' + e} = \frac{516.03}{3 + 5} = 64.50 \text{ k not} = 75.46 \text{ k} \qquad \text{N.G.}$$

After several trials, assume that $e' = 2.40$ in.

Bolt No.	h (in)	v (in)	d (in)	Δ (in)	R (kips)	R_v (kips)	Rd (k-in)
1	0.90	3	3.1321	0.216	28.61	8.22	89.62
2	3.90	3	4.9204	0.34	30.03	23.81	147.78
3	0.90	3	3.1321	0.216	28.61	8.22	89.62
4	3.90	3	4.9204	0.34	30.03	23.81	147.78
						$\Sigma = 64.06$	$\Sigma = 474.80$

Then, we have

$$P_u = \frac{\Sigma Rd}{e' + e} = \frac{474.80}{2.4 + 5} = 64.16 \text{ k almost} = 64.06 \text{ k} \qquad \text{OK}$$

$$P_u = 64.1 \text{ k}$$

Although the development of this method of analysis was actually based on bearing-type connections where slip may occur, both theory and load tests have shown that the method may conservatively be applied to slip-critical connections.[5]

The instantaneous center of rotation may be expanded to include inclined loads and unsymmetrical bolt arrangements, but the trial-and-error calculations with a hand calculator are extraordinarily long for such situations.

Examples 13-3 and 13-4 provide illustrations of the use of the ultimate strength tables in Part 7 of the AISC Manual, for both analysis and design.

Example 13-3

Repeat Example 13-2, using the tables in Part 7 of the Manual. These tables are entitled "Coefficients C for Eccentrically Loaded Bolt Groups." Determine both LRFD design strength and ASD allowable strength of connection.

Solution. Enter Manual Table 7-8 with angle $= 0°$, $s = 6$ in, $e_x = 5$ in, and $n = 2$ vertical rows.

$$C = 2.24$$

$$r_n = F_{nv} A_g = (68 \text{ ksi})(0.6 \text{ in}^2) = 40.8 \text{ k}$$

(From statement in Example 13-2, shear controls are not checked, and thus bearing is not checked.)

$$R_n = Cr_n = (2.24)(40.8) = 91.4 \text{ k}$$

---

[5]G. L. Kulak, "Eccentrically Loaded Slip-Resistant Connections," *Engineering Journal*, AISC, vol. 12, no. 2 (2nd Quarter, 1975), pp. 52–55.

LRFD $\phi = 0.75$	ASD $\Omega = 2.00$
$\phi R_n = (0.75)(91.4) = \mathbf{68.6\ k}$ Generally agrees with trial-and-error solution in preceding example.	$\dfrac{R_n}{\Omega} = \dfrac{91.4}{2.00} = \mathbf{45.7\ k}$

Example 13-4

Using both LRFD and ASD, determine the number of 7/8-in A325 bolts in standard-size holes required for the connection shown in Fig. 13.12. Use A36 steel and assume that the connection is to be a bearing type with threads excluded from the shear plane. Further assume that the bolts are in single shear and bearing on 1/2 in. Use the instantaneous center of rotation method as presented in the tables of Part 7 of the AISC Manual. Assume that $L_c = 1.0$ in and deformation at bolt holes at service loads is not a design consideration.

Solution

$$e_x = e = 5\tfrac{1}{2} \text{ in} = 5.5 \text{ in}$$

Bolts in single shear and bearing on 1/2 in:

$$r_n = \text{nominal shear strength per fastener}$$

$$= F_{nv}A_b = (68 \text{ ksi})(0.6 \text{ in}^2) = 40.8 \text{ k} \leftarrow$$

$$r_n = \text{nominal bearing strength per fastener}$$

$$= 1.5 l_c t F_u = (1.5)(1.0 \text{ in})\left(\tfrac{1}{2} \text{ in}\right)(58 \text{ ksi}) = 43.5 \text{ k}$$

$$< 3.0 \, dt F_u = (3.0)\left(\tfrac{7}{8}\right)\left(\tfrac{1}{2}\right)(58) = 76.1 \text{ k}$$

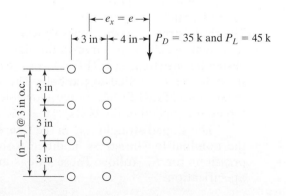

FIGURE 13.12

With reference to Table 7-8 in the Manual, the value of C_{min} required to provide a sufficient number of bolts can be determined as follows:

LRFD $\phi = 0.75$	ASD $\Omega = 2.00$
$P_u = (1.2)(35) + (1.6)(45) = 114 \text{ k}$	$P_a = 35 + 45 = 80 \text{ k}$
$C_{min} = \dfrac{P_u}{\phi r_n} = \dfrac{114}{(0.75)(40.8)} = 3.73$	$C_{min} = \dfrac{\Omega P_a}{r_n} = \dfrac{(2.00)80}{40.8} = 3.92$
* Use four $\frac{7}{8}$ A325 bolts in each row, as described next.	

* With $e_x = 5 \ 1/2$ in and a vertical spacing s of 3 in, we move horizontally in the table until we find the number of bolts in each vertical row so as to provide a C of 3.73 or more. With $e_x = 5$ in and $n = 4$, we find $C = 4.51$. Then, with $e_x = 6$ in and $n = 4$, we find $C = 4.03$. Interpolating for $e_x = 5 \ 1/2$ in, we find $C = 4.27 > 3.73$,   OK for LRFD.

Since $C = 4.27 > 3.92$,   OK for ASD

Use four 7/8 in A325 $\times$ bolts in each row (for both LRFD and ASD).

Note: If the situation faced by the designer does not fit the eccentrically loaded bolt group tables given in Part 7 of the AISC Manual, it is recommended that the conservative elastic procedure be used to handle the problem, whether analysis or design.

13.2 BOLTS SUBJECTED TO SHEAR AND TENSION (BEARING-TYPE CONNECTIONS)

The bolts used for a large number of structural steel connections are subjected to a combination of shear and tension. One quite obvious case is shown in Fig. 13.13, where a diagonal brace is attached to a column. The vertical component of force in the figure, V, is trying to shear the bolts off at the face of the column, while the horizontal component of force, H, is trying to fracture them in tension.

Tests on bearing-type bolts subject to combined shear and tension show that their strengths can be represented with an elliptical interaction curve, as shown in Fig. 13.14. The three straight dashed lines shown in the figure can be used quite accurately to represent the elliptical curve. In this figure, the horizontal dashed line represents the design tensile stress of LRFD ϕF_{nt} or the allowable tensile stress of ASD F_{nt}/Ω if no shear forces is applied to the bolts. The vertical dashed line represents the design shear stress of LRFD ϕF_{nv} or the allowable shear stress of the ASD F_{nv}/Ω if no tensile forces are applied to the bolts.

The sloped straight line in the figure is represented by the expression for F'_{nt}, the nominal tensile stress modified to include the effects of shearing force. Expressions for F'_{nt} follow. These values are provided in Section J3.7 of the AISC Specification.

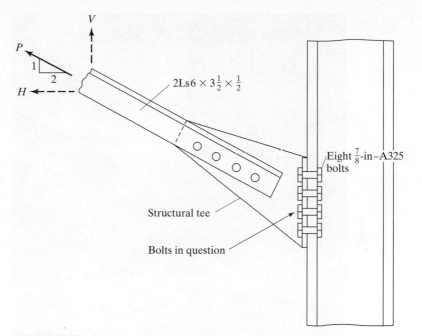

FIGURE 13.13

Combined shear and tension connection.

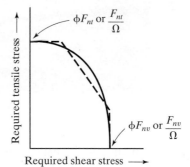

FIGURE 13.14

Bolts in a bearing-type connection subject to combined shear and tension.

For LRFD ($\phi = 0.75$)

$$F'_{nt} = 1.3F_{nt} - \frac{F_{nt}}{\phi F_{nv}} f_{rv} \le F_{nt} \qquad \text{(AISC Equation J3-3a)}$$

For ASD ($\Omega = 2.00$)

$$F'_{nt} = 1.3F_{nt} - \frac{\Omega F_{nt}}{F_{nv}} f_{rv} \le F_{nt} \qquad \text{(AISC Equation J3-3b)}$$

APD Building, Dublin, GA. (Courtesy Britt, Peters and Associates.)

in which

F_{nt} is the nominal tensile stress from Table 12-5 (AISC Table J3.2), ksi.

F_{nv} is the nominal shear stress from Table 12-5 (AISC Table J3.2), ksi.

f_{rv} is the required shear stress using LRFD or ASD load combinations, ksi.

The available shear stress of the fastener shall equal or exceed the required shear stress, f_{rv}.

The AISC Specification (J3.7) says that if the required stress, f, in either shear or tension, is equal to or less than 30 percent of the corresponding available stress, it is not necessary to investigate the effect of combined stress.

Example 13-5

The tension member previously shown in Fig. 13.13 has eight 7/8-in A325 high-strength bolts in a bearing-type connection. Is this a sufficient number of bolts to resist the applied loads $P_D = 80$ k and $P_L = 100$ k, using the LRFD and ASD specifications, if the bolt threads are excluded from the shear planes?

Solution

LRFD $\phi = 0.75$	ASD $\Omega = 2.00$
$P_u = (1.2)(80) + (1.6)(100) = 256$ k	$P_a = 80 + 100 = 180$ k
$V = \dfrac{1}{\sqrt{5}}(256) = 114.5$ k	$V = \dfrac{1}{\sqrt{5}}(180) = 80.5$ k
$H = \dfrac{2}{\sqrt{5}}(256) = 229$ k	$H = \dfrac{2}{\sqrt{5}}(180) = 161$ k
$F_{nt} = 90$ ksi	$F_{nt} = 90$ ksi
$F_{nv} = 68$ ksi	$F_{nv} = 68$ ksi
$f_{rv} = \dfrac{114.5 \text{ k}}{(8)(0.6 \text{ in}^2)} = 23.85$ ksi	$f_v = \dfrac{80.5 \text{ k}}{(8)(0.6 \text{ in}^2)} = 16.77$ ksi
$f_{rt} = \dfrac{229 \text{ k}}{(8)(0.6 \text{ in}^2)} = 47.7$ ksi	$f_t = \dfrac{161 \text{ k}}{(8)(0.6 \text{ in}^2)} = 33.54$ ksi
$F'_{nt} = 1.3F_{nt} - \dfrac{F_{nt}}{\phi F_{nv}} f_{rv} \le F_{nt}$	$F'_{nt} = 1.3F_{nt} - \dfrac{\Omega F_{nt}}{F_{nv}} f_{rv} \le F_{nt}$
$= (1.3)(90) - \dfrac{90}{(0.75)(68)}(23.85)$	$= (1.3)(90) - \dfrac{(2.00)(90)}{68}(16.77)$
$= 74.9$ ksi < 90 ksi	$= 72.6$ ksi < 90 ksi
$\phi F'_{nt} = (0.75)(74.9) = 56.2$ ksi > 47.7 ksi	$\dfrac{F'_{nt}}{\Omega} = \dfrac{72.6}{2.00} = 36.3$ ksi > 33.54 ksi
Connection is OK.	**Connection is OK.**

13.3   BOLTS SUBJECTED TO SHEAR AND TENSION (SLIP-CRITICAL CONNECTIONS)

When an axial tension force is applied to a slip-critical connection, the clamping force will be reduced, and the design shear strength must be decreased in some proportion to the loss in clamping or prestress. This is accomplished in the AISC Specification (Section J3.9) by multiplying the available slip resistance of the bolts (as determined in AISC Section J.8) by a factor k_{sc}.

For LRFD

$$k_{sc} = 1 - \frac{T_u}{D_u T_b n_b} \qquad \text{(AISC Equation J3-5a)}$$

For ASD

$$k_{sc} = 1 - \frac{1.5T_a}{D_u T_b n_b} \qquad \text{(AISC Equation J3-5b)}$$

Here, the factors are defined as follows:

T_u = the tension force due to the LRFD load combination (that is, $\dfrac{P_u}{n_b}$)

D_u = a multiplier = 1.13, previously defined in Section 12.14 (AISC Section J3.8)

T_b = minimum fastener tension, as given in Table 12.1 (Table J3.1, AISC)

n_b = the number of bolts carrying the applied tension

T_a = the tension force due to the ASD load combination (that is, $\dfrac{P_a}{n_b}$)

Example 13-6

A group of twelve 7/8-in A325 high-strength bolts with standard holes is used in a lap joint for a slip-critical joint designed to prevent slip. The connection is to resist the service shear loads $V_D = 40$ k and $V_L = 50$ k, as well as the tensile service loads $T_D = 50$ k and $T_L = 50$ k. Is the connection satisfactory if the faying surface is Class B and the factor for fillers, h_f, is 1.00?

Solution

R_n for 1 bolt in an ordinary slip-critical connection

$$R_n = \mu D_u h_f T_b n_s = (0.50)(1.13)(1.00)(39)(1) = 22.03 \text{ k/bolt}$$

LRFD ϕ = 1.00	ASD Ω = 1.50
$V_u = (1.2)(40) + (1.6)(50) = 128$ k	$V_a = 40 + 50 = 90$ k
$T_u = (1.2)(50) + (1.6)(50) = 140$ k	$T_a = 50 + 50 = 100$ k
$\phi R_n = (1.0)(22.03) = 22.03$ k/bolt	$\dfrac{R_n}{\Omega} = \dfrac{22.03}{1.50} = 14.69$ k/bolt
Reduction due to tensile load	Reduction due to tensile load
$k_{sc} = 1 - \dfrac{T_u}{D_u T_b n_b}$	$k_{sc} = 1 - \dfrac{1.5 T_a}{D_u T_b n_b}$
$= 1 - \dfrac{140}{(1.13)(39)(12)} = 0.735$	$= 1 - \dfrac{(1.5)(100)}{(1.13)(39)(12)} = 0.716$
Reduced ϕR_n/bolt	Reduced $\dfrac{R_n}{\Omega}$/bolt
$= (0.735)(22.03 \text{ k}) = 16.20$ k/bolt	$= (0.716)(14.69) = 10.52$ k/bolt
Design slip resistance for 12 bolts = $(12)(16.20) = 194.4$ k	Allowable slip resistance for 12 bolts = $(12)(10.52) = 126.2$ k
>128 k **OK**	>90 k **OK**
Connection is satisfactory.	**Connection is satisfactory.**

13.4   TENSION LOADS ON BOLTED JOINTS

Bolted and riveted connections subjected to pure tensile loads have been avoided as much as possible in the past by designers. The use of tensile connections was probably used more often for wind-bracing systems in tall buildings than for any other situation.

Other locations exist, however, where they have been used, such as hanger connections for bridges, flange connection for piping systems, etc. Figure 13.15 shows a hanger-type connection with an applied tensile load.

Hot-driven rivets and fully tensioned high-strength bolts are not free to shorten, with the result that large tensile forces are produced in them during their installation. These initial tensions are actually close to their yield points. There has always been considerable reluctance among designers to apply tensile loads to connectors of this type for fear that the external loads might easily increase their already present tensile stresses and cause them to fail. The truth of the matter, however, is that when external tensile loads are applied to connections of this type, the connectors probably will experience little, if any, change in stress.

Fully tensioned high-strength bolts actually prestress the joints in which they are used against tensile loads. (Think of a prestressed concrete beam that has external compressive loads applied at each end.) The tensile stresses in the connectors squeeze together the members being connected. If a tensile load is applied to this connection at the contact surface, it cannot exert any additional load on the bolts until the members are pulled apart and additional strains put on the bolts. The members cannot be pulled apart until a load is applied that is larger than the total tension in the connectors of the connection. This statement means that the joint is prestressed against tensile forces by the amount of stress initially put in the shanks of the connectors.

Another way of saying this is that if a tensile load P is applied at the contact surface, it tends to reduce the thickness of the plates somewhat, but, at the same time, the contact pressure between the plates will be correspondingly reduced, and the plates will tend to expand by the same amount. The theoretical result, then, is no change in plate thickness and no change in connector tension. This situation continues until P equals the connector tension. At this time an increase in P will result in separation of the plates, and thereafter the tension in the connector will equal P.

Should the load be applied to the outer surfaces, there will be some immediate strain increase in the connector. This increase will be accompanied by an expansion of the plates, even though the load does not exceed the prestress, but the increase will be very slight because the load will go to the plate and connectors roughly in proportion

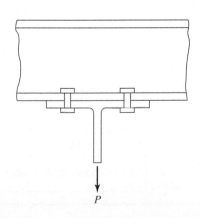

FIGURE 13.15
Hanger connection.

P

to their stiffness. As the plate is stiffer, it will receive most of the load. An expression can be developed for the elongation of the bolt, on the basis of the bolt area and the assumed contact area between the plates. Depending on the contact area assumed, it will be found that, unless P is greater than the bolt tension, its stress increase will be in the range of 10 percent. Should the load exceed the prestress, the bolt stress will rise appreciably.

The preceding rather lengthy discussion is approximate, but should explain why an ordinary tensile load applied to a bolted joint will not change the stress situation very much.

The AISC nominal tensile strength of bolted or threaded parts is given by the expression to follow, which is independent of any initial tightening force:

$$R_n = F_n A_b, \text{ with } F_n = F_{nt} \text{ for tension or } F_{nv} \text{ for shear.} \quad \text{(AISC Equation J3-1)}$$

When fasteners are loaded in tension, there is usually some bending due to the deformation of the connected parts. As a result, the value of ϕ for LRFD is a rather small 0.75, and Ω for ASD is a rather large 2.00. Table 12.5 of this text (AISC Table J3.2) gives values of F_{nt}, the nominal tensile strength (ksi) for the different kinds of connectors, with the values of threaded parts being quite conservative.

In this expression, A_b is the nominal body area of the unthreaded portion of a bolt, or its threaded part not including upset rods. An upset rod has its ends made larger than the regular rod, and the threads are placed in the enlarged section so that the area at the root of the thread is larger than that of the regular rod. An upset rod was shown in Fig. 4.3. The use of upset rods is not usually economical and should be avoided unless a large order is being made.

If an upset rod is used, the nominal tensile strength of the threaded portion is set equal to $0.75F_u$ times the cross-sectional area at its major thread diameter. This value must be larger than F_y times the nominal body area of the rod before upsetting.

Example 13-7 illustrates the determination of the strength of a tension connection.

Example 13-7

Determine the design tensile strength (LRFD) and the allowable tensile strength (ASD) of the bolts for the hanger connection of Fig. 13.15 if eight 7/8-in A490 high-strength bolts with threads excluded from the shear plane are used. Neglect prying action.

Solution

$$R_n \text{ for 8 bolts} = 8F_{nt}A_b = (8)(113 \text{ ksi})(0.6 \text{ in}^2) = 542.4 \text{ k}$$

LRFD $\phi = 0.75$	ASD $\Omega = 2.00$
$\phi R_n = (0.75)(542.4) = \textbf{406.8 k}$	$\dfrac{R_n}{\Omega} = \dfrac{542.4}{2.00} = \textbf{271.2 k}$

13.5   PRYING ACTION

A further consideration that should be given to tensile connections is the possibility of prying action. A tensile connection is shown in Fig. 13.16(a) that is subjected to prying action as illustrated in part (b) of the same figure. Should the flanges of the connection be quite thick and stiff or have stiffener plates like those in Fig. 13.16(c), the prying action will probably be negligible, but this would not be the case if the flanges are thin and flexible and have no stiffeners.

It is usually desirable to limit the number of rows of bolts in a tensile connection, because a large percentage of the load is carried by the inner rows of multi-row connections, even at ultimate load. The tensile connection shown in Fig. 13.17 illustrates this point, as the prying action will throw a large part of the load to the inner connectors, particularly if the plates are thin and flexible. For connections subjected to pure tensile loads, estimates should be made of possible prying action and its magnitude.

The additional force in the bolts resulting from prying action should be added to the tensile force resulting directly from the applied forces. The actual determination

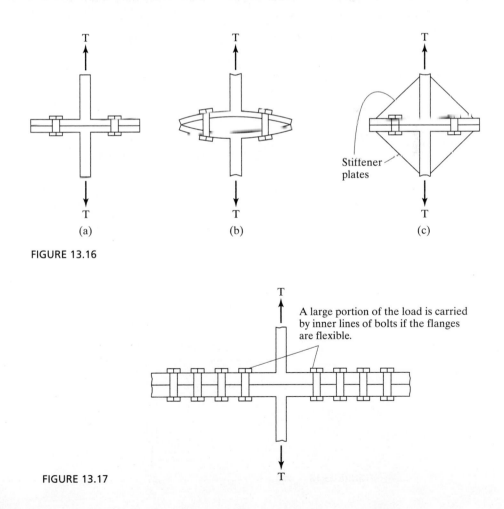

FIGURE 13.16

FIGURE 13.17

of prying forces is quite complex, and research on the subject is still being conducted. Several empirical formulas have been developed that approximate test results. Among these are the AISC expressions included in this section.

Hanger and other tension connections should be so designed as to prevent significant deformations. The most important item in such designs is the use of rigid flanges. Rigidity is more important than bending resistance. To achieve this goal, the distance b shown in Fig. 13.18 should be made as small as possible, with a minimum value equal to the space required to use a wrench for tightening the bolts. Information concerning wrench clearance dimensions is presented in a table entitled "Entering and Tightening Clearance" in Tables 7-16 and 7-17 of Part 7 of the AISC Manual.

Prying action, which is present only in bolted connections, is caused by the deformation of the connecting elements when tensile forces are applied. The results are increased forces in some of the bolts above the forces caused directly by the tensile forces. Should the thicknesses of the connected parts be as large as or larger than the values given by the AISC formulas to follow, which are given on pages 9–10 in the Manual, prying action is considered to be negligible. Reference is here made to Fig. 13.18 for the terms involved in the formulas.

For LRFD

$$t_{min} = \sqrt{\frac{4.44Tb'}{pF_u}}$$

For ASD

$$t_{min} = \sqrt{\frac{6.66Tb'}{pF_u}}$$

The following terms are defined for these formulas:

$$T = \text{required strength of each bolt} = r_{ut}$$

$$\text{or} \quad r_{at} = \frac{T_u \text{ or } T_a}{\text{no. of bolts}}, \text{ kips}$$

$$b' = \left(b - \frac{d_b}{2}\right), \text{ in}$$

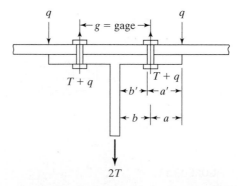

FIGURE 13.18

b = distance from ⊄ of bolt to face of tec (for an angle b is measured to ⊄ of angle leg) in

d_b = bolt diameter

p = tributary length per pair of bolts (⊥ to plane of paper) preferably not >g, in

F_u = specified minimum tensile strength of the connecting element, ksi

Example 13-8, which follows, presents the calculation of the minimum thickness needed for a structural tee flange so that prying action does not have to be considered for the bolts.

Example 13-8

A 10-in long WT8 × 22.5 (t_f = 0.565 in, t_w = 0.345 in, and b_f = 7.04 in) is connected to a W36 × 150 as shown in Fig. 13.19, with six 7/8-in A325 high-strength bolts spaced 3 in o.c. If A36 steel is used, F_u = 58 ksi, is the flange sufficiently thick if prying action is considered? P_D = 30 k and P_L = 40 k.

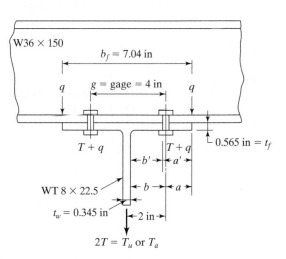

FIGURE 13.19

Solution

LRFD	ASD
$T_u = (1.2)(30) + (1.6)(40) = 100 \text{ k}$	$T_a = 30 + 40 = 70 \text{ k}$
$T = r_{ut} = \dfrac{100}{6} = 16.67 \text{ k each bolt}$	$T = r_{at} = \dfrac{70}{6} = 11.67 \text{ k each bolt}$

$$b' = \left(2 - \frac{0.345}{2}\right) - \frac{0.875}{2} = 1.39 \text{ in}$$

$$d_b = 0.875 \text{ in}$$

$$p = 3 \text{ in}$$

LRFD	ASD
$t_{min} = \sqrt{\dfrac{(4.44)(16.67)(1.39)}{(3)(58)}}$	$t_{min} = \sqrt{\dfrac{(6.66)(11.67)(1.39)}{(3)(58)}}$
$= 0.769 \text{ in} > t_f = 0.565 \text{ in}$	$= 0.788 \text{ in} > t_f = 0.565 \text{ in}$
$\therefore$ Prying action must be considered.	$\therefore$ Prying action must be considered.

Note: Though not presented here in, pages 9-10 through 9-13 in the AISC Manual provide equations for computing the extra tensile force (q) caused by prying action.

13.6   HISTORICAL NOTES ON RIVETS

Rivets were the accepted method for connecting the members of steel structures for many years. Today, however, they no longer provide the most economical connections and are obsolete. It is doubtful that you could find a steel fabricator who can do riveting. It is, however, desirable for the designer to be familiar with rivets, even though he or she will probably never design riveted structures. He or she may have to analyze an existing riveted structure for new loads or for an expansion of the structure. The purpose of these sections is to present only a very brief introduction to the analysis and design of rivets. One advantage of studying these obsolete connectors is that, while doing so, you automatically learn how to analyze A307 common bolts. These bolts are handled exactly as are rivets, except that the design stresses are slightly different. Example 13-11 illustrates the design of a connection with A307 bolts.

The rivets used in construction work were usually made of a soft grade of steel that would not become brittle when heated and hammered with a riveting gun to form the head. The typical rivet consisted of a cylindrical shank of steel with a rounded head on one end. It was heated in the field to a cherry-red color (approximately 1800°F), inserted in the hole, and a head formed on the far end, probably with a portable rivet gun powered by compressed air. The rivet gun, which had a depression in its head to give the rivet head the desired shape, applied a rapid succession of blows to the rivet.

For riveting done in the shop, the rivets were probably heated to a light cherry-red color and driven with a pressure-type riveter. This type of riveter, usually called a "bull" riveter, squeezed the rivet with a pressure of perhaps as high as 50 to 80 tons (445 to 712 kN) and drove the rivet with one stroke. Because of this great pressure, the rivet in its soft state was forced to fill the hole very satisfactorily. This type of riveting

The U.S. Customs Court, Federal Office Building under construction in New York City. (Courtesy of Bethlehem Steel Corporation.)

was much to be preferred over that done with the pneumatic hammer, but no greater nominal strengths were allowed by riveting specifications. The bull riveters were built for much faster operation than were the portable hand riveters, but the latter riveters were needed for places that were not easily accessible (i.e., field erection).

As the rivet cooled, it shrank, or contracted, and squeezed together the parts being connected. The squeezing effect actually caused considerable transfer of stress between the parts being connected to take place by friction. The amount of friction was not dependable, however, and the specifications did not permit its inclusion in the strength of a connection. Rivets shrink diametrically as well as lengthwise and actually become somewhat smaller than the holes that they are assumed to fill. (Permissible strengths for rivets were given in terms of the nominal cross-sectional areas of the rivets before driving.)

Some shop rivets were driven cold with tremendous pressures. Obviously, the cold-driving process worked better for the smaller-size rivets (probably 3/4 in in diameter or less), although larger ones were successfully used. Cold-driven rivets fill the holes better, eliminate the cost of heating, and are stronger because the steel is cold worked. There is, however, a reduction of clamping force, since the rivets do not shrink after having been driven.

13.7 TYPES OF RIVETS

The sizes of rivets used in ordinary construction work were 3/4 in and 7/8 in in diameter, but they could be obtained in standard sizes from 1/2 in to 1 1/2 in in 1/8-in increments. (The smaller sizes were used for small roof trusses, signs, small towers, etc., while the larger sizes were used for very large bridges or towers and very tall buildings.) The use of more than one or two sizes of rivets or bolts on a single job is usually undesirable, because it is expensive and inconvenient to punch different-size holes in a member in

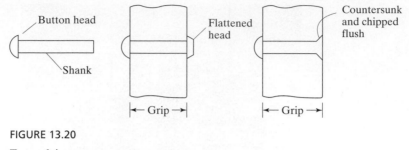

FIGURE 13.20

Types of rivets.

the shop, and the installation of different-size rivets or bolts in the field may be confusing. Some cases arise where it is absolutely necessary to have different sizes, as where smaller rivets or bolts are needed for keeping the proper edge distance in certain sections, but these situations should be avoided if possible.

Rivet heads, usually round in shape, were called *button heads*; but if clearance requirements dictated, the head was flattened or even countersunk and chipped flush. These situations are shown in Fig. 13.20.

The countersunk and chipped-flush rivets did not have sufficient bearing areas to develop full strength, and the designer usually discounted their computed strengths by 50 percent. A rivet with a flattened head was preferred over a countersunk rivet, but if a smooth surface was required, the countersunk and chipped-flush rivet was necessary. This latter type of rivet was appreciably more expensive than the button head type, in addition to being weaker; and it was not used unless absolutely necessary.

There were three ASTM classifications for rivets for structural steel applications, as described in the paragraphs that follow.

13.7.1   ASTM Specification A502, Grade 1

These rivets were used for most structural work. They had a low carbon content of about 0.80 percent, were weaker than the ordinary structural carbon steel, and had a higher ductility. The fact that these rivets were easier to drive than the higher-strength rivets was the main reason that, when rivets were used, they probably were A502, Grade 1, regardless of the strength of the steel used in the structural members.

13.7.2   ASTM Specification A502, Grade 2

These carbon-manganese rivets had higher strengths than the Grade 1 rivets and were developed for the higher-strength steels. Their higher strength permitted the designer to use fewer rivets in a connection and thus smaller gusset plates.

13.7.3   ASTM Specification A502, Grade 3

These rivets had the same nominal strengths as the Grade 2 rivets, but they had much higher resistance to atmospheric corrosion, equal to approximately four times that of carbon steel without copper.

13.8   STRENGTH OF RIVETED CONNECTIONS—RIVETS IN SHEAR AND BEARING

The factors determining the strength of a rivet are its grade, its diameter, and the thickness and arrangement of the pieces being connected. The actual distribution of stress around a rivet hole is difficult to determine, if it can be determined at all; and to simplify the calculations, it is assumed to vary uniformly over a rectangular area equal to the diameter of the rivet times the thickness of the plate.

The strength of a rivet in single shear is the nominal shearing strength times the cross-sectional area of the shank of the rivet. Should a rivet be in double shear, its shearing strength is considered to be twice its single-shear value.

AISC Appendix 5.2.6 indicates that, in checking older structures with rivets, the designer is to assume that the rivets are ASTM A502, grade 1, unless a higher grade is determined by documentation or testing. The nominal shearing strength of A502, grade 1 rivets was 25 ksi, and ϕ was 0.75.

Examples 13-9 and 13-10 illustrate the calculations necessary either to determine the LRFD design and the ASD allowable strengths of existing connections or to design riveted connections. Little comment is made here concerning A307 bolts. The reason is that all the calculations for these fasteners are made exactly as they are for rivets, except that the shearing strengths given by the AISC Specification are different. Only one brief example with common bolts (Example 13-11) is included.

The AISC Specification does not today include rivets, and thus ϕ and Ω values are not included therein. For the example problems to follow, the author uses the rivet ϕ values which were given in the third edition of the LRFD Specification. The Ω values were then determined by the author with the expression $\Omega - 1.50/\phi$, as they were throughout the present specification.

Example 13-9

Determine the LRFD design strength ϕP_n and the ASD allowable strength P_n/Ω of the bearing-type connection shown in Fig. 13.21. A36 steel and A502, Grade 1 rivets

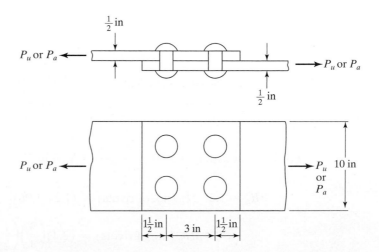

FIGURE 13.21

are used in the connection, and it is assumed that standard-size holes are used and that edge distances and center-to-center distances are = 1.5 in and 3 in, respectively. Neglect block shear. The rivets are 3/4 in in diameter and their F_{nv} is 25 ksi.

Solution. Design tensile force applied to plates

$$A_g = \left(\frac{1}{2} \text{ in}\right)(10 \text{ in}) = 5.00 \text{ in}^2$$

$$A_n = \left[\left(\frac{1}{2} \text{ in}\right)(10 \text{ in}) - (2)\left(\frac{7}{8} \text{ in}\right)\left(\frac{1}{2} \text{ in}\right)\right] = 4.125 \text{ in}^2$$

For tensile yielding

$$P_n = F_y A_g = (36 \text{ ksi})(5.00 \text{ in}^2) = 180 \text{ k}$$

LRFD $\phi_t = 0.90$	ASD $\Omega_t = 1.67$
$\phi_t P_n = (0.90)(180) = 162 \text{ k}$	$\dfrac{P_n}{\Omega_t} = \dfrac{180}{1.67} = 107.8 \text{ k}$

For tensile rupture

$$A_e = UA_n = 1.0 \times 4.125 \text{ in}^2 = 4.125 \text{ in}^2$$
$$P_n = F_u A_e = (58 \text{ ksi})(4.125 \text{ in}^2) = 239.25 \text{ k}$$

LRFD $\phi_t = 0.75$	ASD $\Omega_t = 2.00$
$\phi_t P_n = (0.75)(239.25) = 179.4 \text{ k}$	$\dfrac{P_n}{\Omega_t} = \dfrac{239.25}{2.00} = 119.6 \text{ k}$

Rivets in single shear and bearing on 1/2 in

$$R_n = (A_{rivet})(F_{nv})(\text{no. of rivets}) = (0.44 \text{ in}^2)(25 \text{ ksi})(4) = 44.0 \text{ k}$$

LRFD $\phi = 0.75$	ASD $\Omega = 2.00$	
$\phi R_n = (0.75)(44.0) = 33.0 \text{ k}$	$\dfrac{R_n}{\Omega} = \dfrac{44.0}{2.00} = 22.0 \text{ k}$	← Controls

For bearing with $l_c = 1.50 - \dfrac{\frac{3}{4} + \frac{1}{8}}{2} = 1.06 \text{ in}$

$$R_n = 1.2 l_c t F_u \text{ (no. of rivets)} = (1.2)(1.06)\left(\frac{1}{2}\right)(58)(4) = 147.55 \text{ k}$$

$$< 2.4 \, dt F_u \text{ (no. of rivets)} = (2.4)\left(\frac{3}{4}\right)\left(\frac{1}{2}\right)(58)(4) = 208.8 \text{ k}$$

LRFD ϕ = 0.75	ASD Ω = 2.00
$\phi R_n = (0.75)(147.55) = 110.7$ k	$\dfrac{R_n}{\Omega} = \dfrac{147.55}{2.0} = 73.8$ k

$$\textbf{\textit{Ans. }} \boldsymbol{P_u = 33.0 \text{ k}} \qquad\qquad \textbf{\textit{Ans. }} \boldsymbol{P_a = 22.0 \text{ k}}$$

Example 13-10

How many 7/8-in A502, Grade 1 rivets are required for the connection shown in Fig. 13.22 if the plates are A36, standard-size holes are used, and the edge distances and center-to-center distances are 1.5 in and 3 in, respectively? Solve by both the LRFD and ASD methods. For ASD, use the Ω values assumed for Example 13-9. $P_u = 170$ k and $P_a = 120$ k.

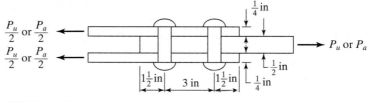

FIGURE 13.22

Solution

Rivets in double shear and bearing on 1/2 in

Nominal double shear strength of 1 rivet

$$r_n = (2A_{rivet})(F_{nv}) = (2 \times 0.6 \text{ in}^2)(25.0 \text{ ksi}) = 30 \text{ k}$$

LRFD ϕ = 0.75	ASD Ω = 2.00	
$\phi r_n = (0.75)(30.0) = 22.5$ k	$\dfrac{r_n}{\Omega} = \dfrac{30.00}{2.00} = 15.0$ k	$\leftarrow$ Controls

Nominal bearing strength of 1 rivet

$$l_c = 1.50 - \dfrac{\dfrac{7}{8} + \dfrac{1}{8}}{2} = 1.00 \text{ in}$$

$$r_n = 1.2 l_c\, t F_u = (1.2)(1.0)\left(\dfrac{1}{2}\right)(58) = 34.8 \text{ k}$$

$$< 2.4 \, dt F_u = (2.4)\left(\dfrac{7}{8}\right)\left(\dfrac{1}{2}\right)(58) = 60.9 \text{ k}$$

LRFD $\phi = 0.75$	ASD $\Omega = 2.00$
$\phi r_n = (0.75)(34.8) = 26.1$ k/rivet	$\dfrac{r_n}{\Omega} = \dfrac{34.8}{2.00} = 17.4$ k/rivet

LRFD	ASD
No. of rivets reqd	No of rivets reqd
$= \dfrac{170}{22.5} = 7.56$	$= \dfrac{120}{15.0} = 8$
Use eight $\frac{7}{8}$-in	Use eight $\frac{7}{8}$-in
A502, grade 1 rivets.	A502, grade 1 rivets.

Example 13-11

Repeat Example 13-10, using 7/8-in A307 bolts, for which $F_{nv} = 27$ ksi.

Solution. Bolts in double shear and bearing on .5 in:

Nominal shear strength of 1 bolt

$$r_n = (A_{bolt})(F_{nv}) = (2 \times 0.6 \text{ in}^2)(27 \text{ ksi}) = 32.4 \text{ k}$$

LRFD $\phi = 0.75$	ASD $\Omega = 2.00$	
$\phi r_n = (0.75)(32.4) = 24.3$ k	$\dfrac{r_n}{\Omega} = \dfrac{32.4}{2.00} = 16.2$ k	$\leftarrow$ Controls

Nominal bearing strength of 1 bolt

$$r_n = 1.2\, l_c\, t F_u = (1.2)(1.0)\left(\frac{1}{2}\right)(58) = 34.8 \text{ k}$$

LRFD $\phi = 0.75$	ASD $\Omega = 2.00$
$\phi r_n = (0.75)(34.8) = 26.1$ k	$\dfrac{r_n}{\Omega} = \dfrac{34.8}{2.00} = 17.4$ k

LRFD	ASD
No. of bolts reqd	No. of bolts reqd
$= \dfrac{170}{24.3} = 7.00$	$= \dfrac{120}{16.2} = 7.41$
Use seven $\frac{7}{8}$-in A307 bolts.	Use eight $\frac{7}{8}$-in A307 bolts.

13.9 PROBLEMS FOR SOLUTION

For each of the problems listed, the following information is to be used, unless otherwise indicated: (a) A36 steel; (b) standard-size holes; (c) threads of bolts excluded from shear plane.

13-1 to 13-7. *Determine the resultant load on the most stressed bolt in the eccentrically loaded connections shown, using the elastic method.*

13-1. (*Ans.* 23.26 k)

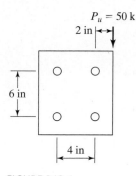

FIGURE P13-1

13-2.

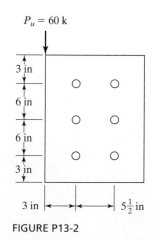

FIGURE P13-2

13-3. (*Ans.* 16.49 k)

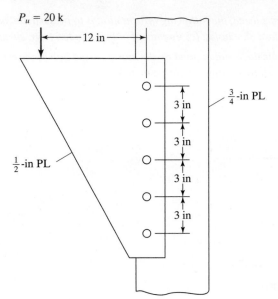

FIGURE P13-3

13-4.

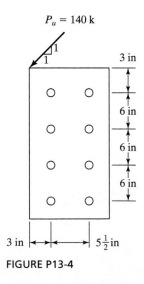

FIGURE P13-4

13-5. (*Ans.* 21.87 k)

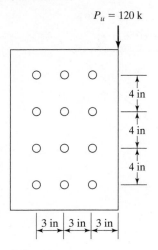

$P_u = 120$ k

4 in

4 in

4 in

3 in   3 in   3 in

FIGURE P13-5

13-6.

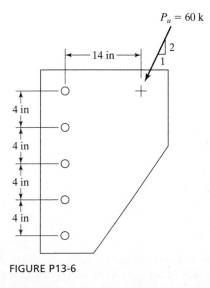

$P_u = 60$ k

14 in

2

1

4 in

4 in

4 in

4 in

FIGURE P13-6

13-7. (*Ans.* 33.75 k)

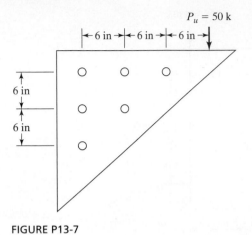

FIGURE P13-7

13-8. Repeat Prob. 13-2, using the reduced eccentricity method given in Section 13.1.2

13-9. Using the elastic method, determine the LRFD design strength and the ASD allowable strength of the bearing-type connection shown. The bolts are 3/4 in A325 and are in single shear and bearing on 5/8 in. The holes are standard sizes and the bolt threads are excluded from the shear plane.(*Ans.* 58.0 k LRFD, 38.7 k ASD)

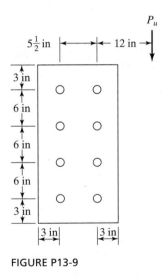

FIGURE P13-9

13-10. Using the elastic method, determine the ASD allowable strength P_n/Ω and the LRFD design strength, ϕP_n for the slip-critical connection shown. The 7/8-in A325 bolts are in "double shear." All plates are 1/2-in thick. Surfaces are Class A. Holes are standard sizes and $h_f = 1.0$.

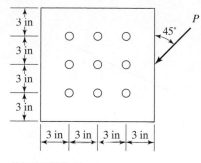

FIGURE P13-10

13-11. Repeat Prob. 13-9, using the ultimate strength tables entitled "Coefficients C for Eccentrically Loaded Bolt Groups" in Part 7 of the AISC Manual. (*Ans.* 73.6 k)

13-12. Repeat Prob. 13-10, using the ultimate strength tables entitled "Coefficients C for Eccentrically Loaded Bolt Groups" in Part 7 of the AISC Manual.

13-13. Is the bearing-type connection shown in the accompanying illustration sufficient to resist the 200 k load that passes through the center of gravity of the bolt group, according to the LRFD and ASD specifications? (*Ans.* $\phi F'_{nt} = 54.7$ ksi, $F'_{nt}/\Omega = 37.8$ ksi. Therefore connection is satisfactory.)

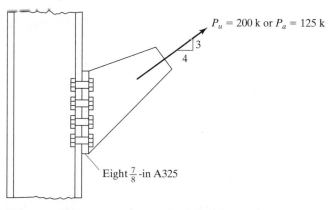

FIGURE P13-13

13-14. Repeat Prob. 13-13 if slip-critical bolts for the required strength level are used and if surfaces are Class A, $h_f = 1.00$ and standard size holes are used.

13-15. If the load shown in the accompanying bearing-type illustration passes through the center of gravity of the bolt group, how large can it be, according to both the LRFD and ASD specifications? Bolt threads are excluded from the shear planes. (*Ans.* 148.7 k LRFD, 99.2 k ASD)

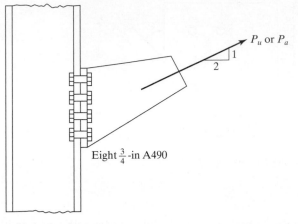

FIGURE P13-15

13-16. Repeat Prob. 13-15 if bolts are A325.

13-17 to 13-24. *Solve these problems by both the LRFD and ASD methods.*

13-17. Determine the number of 3/4-in A325 bolts required in the angles and in the flange of the W shape shown in the accompanying illustration if a bearing-type (snug-tight) connection is used. Use 50 ksi steel, $F_u = 65$ ksi, $L_c = 1.0$ in. Deformation around the bolt holes is a design consideration. (*Ans.* 4 in angle, both LRFD & ASD; 8 in W section, both LRFD & ASD)

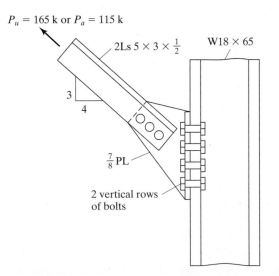

FIGURE P13-17

13-18. Determine the design strength ϕP_n and the allowable strength P_n/Ω of the connection shown if 3/4-in A502, Grade 1 rivets and A36 steel are used. Assume $F_v = 25$ ksi and threads excluded from shear planes.

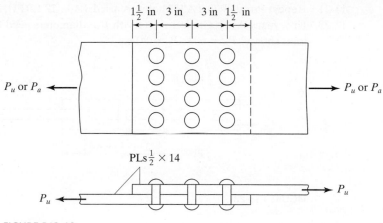

FIGURE P13-18

13-19. The truss tension member shown consists of a single-angle 5 × 3 × 5/16 and is connected to a 1/2-in gusset plate with five 7/8-in A502, Grade 1 rivets. Determine ϕP_n and P_n/Ω if U is assumed to equal 0.9. Neglect block shear. Steel is A36. $F_v = 25$ ksi. (*Ans.* 56.2 k, 37.5 k)

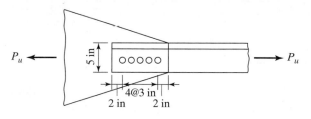

FIGURE P13-19

13-20. How many 7/8-in A502, Grade 1 rivets are needed to carry the load shown in the accompanying illustration if $F_v = 25$ ksi?

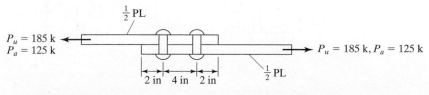

FIGURE P13-20

13-21. Repeat Prob. 13-20 if A307 bolts are used. (*Ans.* 22 LRFD; 22 ASD)

13-22. How many A502, Grade 1 rivets with 1-in diameters need to be used for the butt joint shown? $P_D = 60$ k, $P_L = 80$ k.

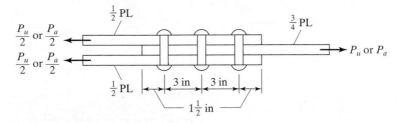

FIGURE P13-22

13-23. For the connection shown in the accompanying illustration, $P_u = 475$ k and $P_a = 320$ k, determine the number of 7/8-in A502, Grade 2 rivets required. $F_v = 25$ ksi (*Ans.* 22 LRFD; 22 ASD)

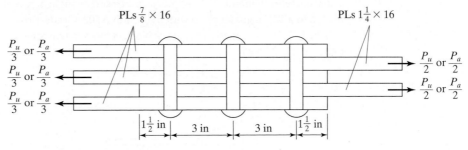

FIGURE P13-23

13-24. For the A36 beam shown in the accompanying illustration, what is the required spacing of 7/8-in A307 bolts if $V_u = 140$ k and $V_a = 100$ k? Assume that $F_v = 25$ ksi, $L_c = 1.50$ in.

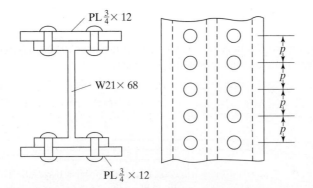

FIGURE P13-24

CHAPTER 14

Welded Connections

14.1 GENERAL

Welding is a process by which metallic parts are connected by heating their surfaces to a plastic or fluid state and allowing the parts to flow together and join (with or without the addition of other molten metal). It is impossible to determine when welding originated, but it was at least several thousand years ago. Metal-working, including welding, was quite an art in ancient Greece three thousand years ago, but welding had undoubtedly been performed for many centuries before that. Ancient welding probably was a forging process in which the metals were heated to a certain temperature (not to the melting stage) and hammered together.

Although modern welding has been available for many years, it has come into its own only in the last few decades for the building and bridge phases of structural engineering. The adoption of structural welding was quite slow for several decades, because many engineers thought that welding had two major disadvantages: (1) Welds had reduced fatigue strength, compared with riveted and bolted connections, and (2) it was impossible to ensure a high quality of welding without unreasonably extensive and costly inspection.

These attitudes persisted for many years, although tests began to indicate that neither reason was valid. Regardless of their validity, these views were widely held and undoubtedly slowed down the use of welding—particularly for highway bridges and, to an even greater extent, railroad bridges. Today, most engineers agree that welded joints have considerable fatigue strength. They will also admit that the rules governing the qualification of welders, the better techniques applied, and the excellent workmanship requirments of the AWS (American Welding Society) specifications make the inspection of welding a much less difficult problem. Furthermore, the chemistry of steels manufactured today is especially formulated to improve their weldability. Consequently, welding is now permitted for almost all structural work.

On the subject of welding, it is interesting to consider welded ships. Ships are subjected to severe impactive loadings that are difficult to predict, yet naval architects use

Roof framing for Cherokee Central Schools, Cherokee, NC. (Courtesy of CMC South Carolina Steel.)

all-welded ships with great success. A similar discussion can be made for airplanes and aeronautical engineers. The slowest adoption of structural welding was for railroad bridges. These bridges are undoubtedly subjected to heavier live loads, larger vibrations, and more stress reversals than highway bridges; but are their stress situations as serious and as difficult to predict as those for ships and planes?

14.2   ADVANTAGES OF WELDING

Today, it is possible to make use of the many advantages that welding offers, since the fatigue and inspection fears have been largely eliminated. Following are several of the many advantages that welding offers:

1. To most designers, the first advantage is economic, because the use of welding permits large savings in pounds of steel used. Welded structures allow the elimination of a large percentage of the gusset and splice plates necessary for bolted structures, as well as the elimination of bolt heads. In some bridge trusses, it may be possible to save up to 15 percent or more of the steel weight by welding.

2. Welding has a much wider range of application than bolting. Consider a steel pipe column and the difficulties of connecting it to other steel members by bolting. A bolted connection may be virtually impossible, but a welded connection presents few difficulties. Many similar situations can be imagined in which welding has a decided advantage.

3. Welded structures are more rigid, because the members often are welded directly to each other. Frequently, the connections for bolted structures are made through

intermediate connection angles or plates that deform due to load transfer, making the entire structure more flexible. On the other hand, greater rigidity can be a disadvantage where simple end connections with little moment resistance are desired. In such cases, designers must be careful as to the type of joints they specify.

4. The process of fusing pieces together creates the most truly continuous structures. Fusing results in one-piece construction, and because welded joints are as strong as or stronger than the base metal, no restrictions have to be placed on the joints. This continuity advantage has permitted the erection of countless slender and graceful statically indeterminate steel frames throughout the world. Some of the more outspoken proponents of welding have referred to bolted structures, with their heavy plates and abundance of bolts, as looking like tanks or armored cars compared with the clean, smooth lines of welded structures. For a graphic illustration of this advantage, compare the moment-resisting connections of Fig. 15.5.

5. It is easier to make changes in design and to correct errors during erection (and less expensive) if welding is used. A closely related advantage has certainly been illustrated in military engagements during the past few wars by the quick welding repairs made to military equipment under battle conditions.

6. Another item that is often important is the relative silence of welding. Imagine the importance of this fact when working near hospitals or schools or when making additions to existing buildings. Anyone with close-to-normal hearing who has attempted to work in an office within several hundred feet of a bolted job can attest to this advantage.

7. Fewer pieces are used, and as a result, time is saved in detailing, fabrication, and field erection.

14.3 AMERICAN WELDING SOCIETY

The American Welding Society's *Structural Welding Code*[1] is the generally recognized standard for welding in the United States. The AISC Specification clearly states that the provisions of the AWS Code apply under the AISC Specification, with only a few minor exceptions, and these are listed in AISC Specification J2. Both the AWS and the AASHTO Specifications cover dynamically loaded structures: Generally, the AWS specification is used for designing the welds for buildings subject to dynamic loads.

14.4 TYPES OF WELDING

Although both gas and arc welding are available, almost all structural welding is arc welding. Sir Humphry Davy discovered in 1801 how to create an electric arc by bringing close together two terminals of an electric circuit of relatively high voltage. Although he is generally given credit for the development of modern welding, a good many years elapsed after his discovery before welding was actually performed with the electric arc. (His work was of the greatest importance to the modern structural world, but it is

---

[1]American Welding Society, *Structural Welding Code-Steel*, AWS D.1.1-00 (Miami: AWS, 2006).

interesting to note that many people say his greatest discovery was not the electric arc, but rather a laboratory assistant whose name was Michael Faraday.) Several Europeans formed welds of one type or another in the 1880s with the electric arc, while in the United States the first patent for arc welding was given to Charles Coffin of Detroit in 1889.[2]

The figures shown in this chapter illustrate the necessity of supplying additional metal to the joints being welded to give satisfactory connections. In electric-arc welding, the metallic rod, which is used as the electrode, melts off into the joint as it is being made. When gas welding is used, it is necessary to introduce a metal rod known as a *filler* or *welding rod*.

In gas welding, a mixture of oxygen and some suitable type of gas is burned at the tip of a torch or blowpipe held in the welder's hand or by machine. The gas used in structural welding usually is acetylene, and the process is called *oxyacetylene welding*. The flame produced can be used for flame cutting of metals, as well as for welding. Gas welding is fairly easy to learn, and the equipment used is rather inexpensive. It is a slow process, however, compared with other means of welding, and normally it is used for repair and maintenance work and not for the fabrication and erection of large steel structures.

In arc welding, an electric arc is formed between the pieces being welded and an electrode held in the operator's hand with some type of holder, or by an automatic machine. The arc is a continuous spark that, upon contact, brings the electrode and the pieces being welded to the melting point. The resistance of the air or gas between the electrode and the pieces being welded changes the electrical energy into heat. A temperature of 6000 to 10,000°F is produced in the arc. As the end of the electrode melts, small droplets, or globules, of the molten metal are formed and actually are forced by the arc across to the pieces being connected, which penetrate the molten metal to become a part of the weld. The amount of penetration can be controlled by the amount of current consumed. Since the molten droplets of the electrodes actually are propelled into the weld, arc welding can be successfully used for overhead work.

A pool of molten steel can hold a fairly large amount of gases in solution and, if not protected from the surrounding air, will chemically combine with oxygen and nitrogen. After cooling, the welds will be somewhat porous due to the little pockets formed by the gases. Such welds are relatively brittle and have much less resistance to corrosion. A welded joint can be shielded by using an electrode coated with certain mineral compounds. The electric arc causes the coating to melt and creates an inert gas or vapor around the area being welded. The vapor acts as a shield around the molten metal and keeps it from coming freely in contact with the surrounding air. It also deposits a slag in the molten metal, which has less density than the base metal and comes to the surface to protect the weld from the air while the weld cools. After cooling, the slag can easily be removed by peening and wire brushing (such removal being absolutely necessary before painting or application of another weld layer). The elements of the shielded arc welding process are shown in Fig. 14.1. This figure is taken from the *Procedure Handbook of Arc Welding Design and Practice* published by the Lincoln Electric Company. *Shielded metal arc welding* is abbreviated here with the letters SMAW.

---

[2]Lincoln Electric Company, *Procedure Handbook of Arc Welding Design and Practice*, 11th ed. Part I (Cleveland, OH, 1957).

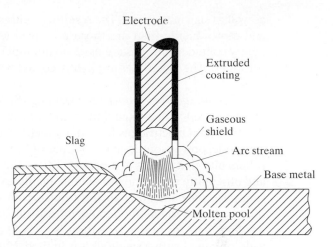

FIGURE 14.1

Elements of the shielded metal arc welding process (SMAW).

The type of welding electrode used is very important because it decidedly affects the weld properties such as strength, ductility, and corrosion resistance. Quite a number of different kinds of electrodes are manufactured, the type to be used for a certain job being dependent upon the metal to be welded, the amount of material that needs to be added, the position of the work, etc. The electrodes fall into two general classes—the *lightly coated electrodes* and the *heavily coated electrodes*.

The heavily coated electrodes are normally used in structural welding because the melting of their coatings produces very satisfactory vapor shields around the work,

Shielded metal arc welding (SMAW) and electrode just before starting an arc to fillet weld the clip angle to the beam web. (Courtesy of CMC South Carolina Steel.)

as well as slag in the weld. The resulting welds are stronger, more resistant to corrosion, and more ductile than are those produced with lightly coated electrodes. When the lightly coated electrodes are used, no attempt is made to prevent oxidation, and no slag is formed. The electrodes are lightly coated with some arc-stabilizing chemical such as lime.

Submerged (or *hidden*) *arc welding* (SAW) is an automatic process in which the arc is covered with a mound of granular fusible material and is thus hidden from view. A bare metal electrode is fed from a reel, melted, and deposited as filler material. The electrode, power source, and a hopper of flux are attached to a frame that is placed on rollers and that moves at a certain rate as the weld is formed. SAW welds are quickly and efficiently made and are of high quality, exhibiting high impact strength and corrosion resistance and good ductility. Furthermore, they provide deeper penetration, with the result that the area effective in resisting loads is larger. A large percentage of the welding done for bridge structures is SAW. If a single electrode is used, the size of the weld obtained with a single pass is limited. Multiple electrodes may be used, however, permitting much larger welds.

Welds made by the SAW process (automatic or semiautomatic) are consistently of high quality and are very suitable for long welds. One disadvantage is that the work must be positioned for near-flat or horizontal welding.

Another type of welding is *flux-cored arc welding* (FCAW). In this process, a flux-filled steel tube electrode is continuously fed from a reel. Gas shielding and slag are formed from the flux. The AWS Specification (4.14) provides limiting sizes for welding electrode diameters and weld sizes, as well as other requirements pertaining to welding procedures.

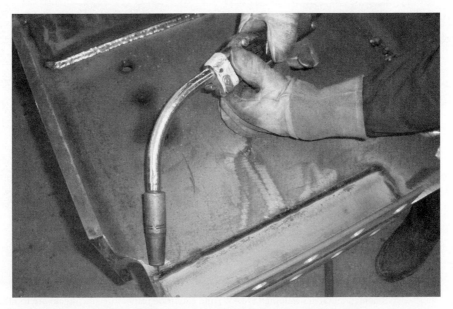

Flux-cored arc welding (FCAW). (Courtesy of CMC South Carolina Steel.)

14.5   PREQUALIFIED WELDING

The AWS accepts four welding processes as being prequalified. In this context, the word *prequalified* means that processes are acceptable without the necessity of further proof of their suitability by procedure qualification tests. What we are saying is that, based on many years of experience, sound weld metal with the desired properties can be deposited if the work is performed in accordance with the requirements of the Structural Welding Code of the AWS. The processes that are listed in AWS Specification 1.3.1 are (1) shielded metal arc welding (SMAW), (2) submerged arc welding (SAW), (3) gas metal arc welding (GMAW), and (4) flux-cored arc welding (FCAW). The SMAW process is the usual process applied for hand welding, while the other three are typically automatic or semiautomatic.

14.6   WELDING INSPECTION

Three steps must be taken to ensure good welding for a particular job: (1) establishment of good welding procedures, (2) use of prequalified welders, and (3) employment of competent inspectors in both the shop and the field.

When the procedures established by the AWS and AISC for good welding are followed, and when welders are used who have previously been required to prove their ability, good results usually are obtained. To make absolutely sure, however, well-qualified inspectors are needed.

Good welding procedure involves the selection of proper electrodes, current, and voltage; the properties of base metal and filler; and the position of welding—to name only a few factors. The usual practice for large jobs is to employ welders who have certificates showing their qualifications. In addition, it is not a bad practice to have each person make an identifying mark on each weld so that those frequently doing poor work can be identified. This practice tends to improve the general quality of the work performed.

14.6.1   Visual Inspection

Another factor that will cause welders to perform better work is simply the presence of an inspector who they feel knows good welding when he or she sees it. A good inspector should have done welding and spent much time observing the work of good welders. From this experience, he or she should be able to know if a welder is obtaining satisfactory fusion and penetration. He or she also should be able to recognize good welds in regard to shape, size, and general appearance. For instance, the metal in a good weld should approximate its original color after it has cooled. If it has been overheated, it may have a rusty and reddish-looking color. An inspector can use various scales and gages to check the sizes and shapes of welds.

Visual inspection by a competent person usually gives a good indication of the quality of welds, but is not a perfect source of information, especially regarding the subsurface condition of the weld. It surely is the most economical inspection method and is particularly useful for single-pass welds. This method, however, is good only for picking up surface imperfections. There are several methods for determining the internal

Lincoln ML-3 Squirtwelder mounted on a self-propelled
trackless trailer deposits this 1/4-in web-to-flange weld at
28 in/min. (Courtesy of the Lincoln Electric Company.)

soundness of a weld, including the use of penetrating dyes and magnetic particles,
ultrasonic testing, and radiographic procedures. These methods can be used to detect
internal defects such as porosity, weld penetration, and the presence of slag.

14.6.2 Liquid Penetrants

Various types of dyes can be spread over weld surfaces. These dyes will penetrate into
the surface cracks of the weld. After the dye has penetrated into the crack, the excess
surface material is wiped off and a powdery developer is used to draw the dye out of
the cracks. The outlines of the cracks can then be seen with the eye. Several variations
of this method are used to improve the visibility of the defects, including the use of flu-
orescent dyes. After the dye is drawn from the cracks, they stand out brightly under a
black light.[3] Like visual inspection, this method enables us to detect cracks that are
open to the surface.

---

[3]James Hughes, "It's Superinspector." *Steelways*, 25. no. 4 (New York: American Iron and Steel Institute.
September/October, 1969), pp. 19–21.

14.6.3   Magnetic Particles

In this method, the weld being inspected is magnetized electrically. Cracks that are at or near the surface of the weld cause north and south poles to form on each side of the cracks. Dry iron powdered filings or a liquid suspension of particles is placed on the weld. These particles form patterns when many of them cling to the cracks, showing the locations of cracks and indicating their size and shape. Only cracks, seams, inclusions, etc., within about 1/10 in of the surface can be located by this method. A disadvantage is that if multilayer welds are used, the method has to be applied to each layer.

14.6.4   Ultrasonic Testing

In recent years, the steel industry has applied ultrasonics to the manufacture of steel. Although the equipment is expensive, the method is quite useful in welding inspections as well. Sound waves are sent through the material being tested and are reflected from the opposite side of the material. These reflections are shown on a cathode ray tube. Defects in the weld will affect the time of the sound transmission. The operator can read the picture on the tube and then locate flaws and learn how severe they are. Ultrasonic testing can sucessfully be used to locate discontinuities in carbon and low-alloy steels, but it doesn't work too well for some stainless steels or for extremely coarse-grained steels.

14.6.5   Radiographic Procedures

The more expensive radiographic methods can be used to check occasional welds in important structures. From these tests, it is possible to make good estimates of the percentage of bad welds in a structure. Portable x-ray machines (where access is not a problem) and radium or radioactive cobalt for making pictures are excellent, but expensive methods of testing welds. These methods are satisfactory for butt welds (such as for the welding of important stainless steel piping at chemical and nuclear projects), but they are not satisfactory for fillet welds, because the pictures are difficult to interpret. A further disadvantage of such methods is the radioactive danger. Careful procedures have to be used to protect the technicians as well as nearby workers. On a construction job, this danger generally requires night inspection of welds, when only a few workers are near the inspection area. (Normally, a very large job would be required before the use of the extremely expensive radioactive materials could be justified.)

A properly welded connection can always be made much stronger—perhaps as much as two times stronger—than the plates being connected. As a result, the actual strength is much greater than is required by the specifications. The reasons for this extra strength are as follows: The electrode wire is made from premium steel, the metal is melted electrically (as is done in the manufacture of high-quality steels), and the cooling rate is quite rapid. As a result, it is rare for a welder to make a weld of less strength than required by the design.

14.7   CLASSIFICATION OF WELDS

Three separate classifications of welds are described in this section. These classifications are based on the types of welds made, the positions of the welds, and the types of joints used.

14.7.1   Type of Weld

The two main types of welds are the *fillet welds* and the *groove welds*. In addition, there are plug and slot welds, which are not as common in structural work. These four types of welds are shown in Fig. 14.2.

Fillet welds are those made where parts lap over each other, as shown in Fig. 14.2(a). They may also be used in tee joints, (as illustrated in Fig. 14.4). Fillet welds are the most economical welds to use, as little preparation of the parts to be connected is necessary. In addition, these welds can be made very well by welders of somewhat lesser skills than those required for good work with other types of welds.

The fillet welds will be shown to be weaker than groove welds, although most structural connections (about 80 percent) are made with fillet welds. Any person who has experience in steel structures will understand why fillet welds are more common than groove welds. Groove welds, as illustrated in Fig. 14.2(b) and (c) (which are welds made in grooves between the members to be joined) are used when the members to be connected are lined up in the same plane. To use them, the members have to fit almost

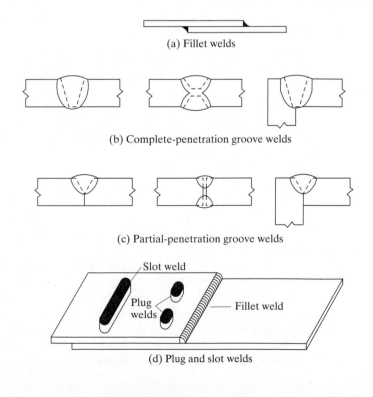

(a) Fillet welds

(b) Complete-penetration groove welds

(c) Partial-penetration groove welds

Slot weld

Plug welds

Fillet weld

(d) Plug and slot welds

FIGURE 14.2

Four types of structural welds.

perfectly, and unfortunately, the average steel structure does not fit together that way. Have you ever seen steel workers pulling and ramming steel members to get them into position? When members are allowed to lap over each other, larger tolerances are allowable in erection, and fillet welds are used. Nevertheless, groove welds are quite common for many connections, such as column splices, butting of beam flanges to columns, etc., and they make up about 15 percent of structural welding. Groove welds can be either *complete-penetration* welds, which extend for the full thickness of the part being connected, or *partial penetration* welds, which extend for only part of the member thickness.

Groove welds are generally more expensive than fillet welds because of the costs of preparation. In fact, groove welds can cost up to 50 to 100 percent more than fillet welds.

A plug weld is a circular weld that passes through one member into another, thus joining the two together. A slot weld is a weld formed in a slot, or elongated hole, that joins one member to the other member through the slot. The slot may be partly or fully filled with weld material. These welds are shown in Fig. 14.2(d). These two expensive types of welds may occasionally be used when members lap over each other and the desired length of fillet welds cannot be obtained. They may also be used to stitch together parts of a member, such as the fastening of cover plates to a built-up member.

A plug or slot weld is not generally considered suitable for transferring tensile forces perpendicular to the faying surface, because there is not usually much penetration of the weld into the member behind the plug or slot and the fact is that resistance to tension is provided primarily by penetration.

Structural designers accept plug and slot welds as being satisfactory for stitching the different parts of a member together, but many designers are not happy using these welds for the transmission of shear forces. The penetration of the welds from the slots or plugs into the other members is questionable; in addition, there can be critical voids in the welds that cannot be detected with the usual inspection procedures.

14.7.2 Position

Welds are referred to as *flat, horizontal, vertical,* or *overhead*—listed in order of their economy, with the flat welds being the most economical and the overhead welds being the most expensive. A moderately skilled welder can do a very satisfactory job with a flat weld, but it takes the very best to do a good job with an overhead weld. Although the flat welds often are done with an automatic machine, most structural welding is done by hand. We indicated previously that the assistance of gravity is not necessary for the forming of good welds, but it does speed up the process. The globules of the molten electrodes can be forced into the overhead welds against gravity, and good welds will result; however, they are slow and expensive to make, so it is desirable to avoid them whenever possible. These types of welds are shown in Fig. 14.3.

14.7.3 Type of Joint

Welds can be further classified according to the type of joint used: *butt, lap, tee, edge, corner*, etc. See Fig. 14.4.

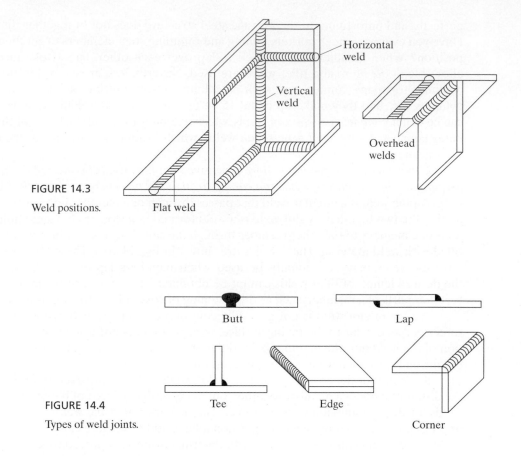

FIGURE 14.3

Weld positions.

FIGURE 14.4

Types of weld joints.

14.8   WELDING SYMBOLS

Figure 14.5 presents the various welding symbols developed by the American Welding Society. With this excellent shorthand system, a great deal of information can be presented in a small space on engineering plans and drawings. These symbols require only a few lines and numbers and remove the necessity of drawing in the welds and making long descriptive notes. It is certainly desirable for steel designers and draftsmen to use this standardized system. If most of the welds on a drawing are the same size, a note to that effect can be given and the symbols omitted, except for the off-size welds.

The purpose of this section is to give a general idea of the appearance of welding symbols and the information they can convey. (For more detailed information, refer to the AISC Handbook and to other materials published by the AWS.) The information presented in Fig. 14.5 may be quite confusing; for this reason, a few very common symbols for fillet welds are presented in Fig. 14.6, together with an explanation of each.

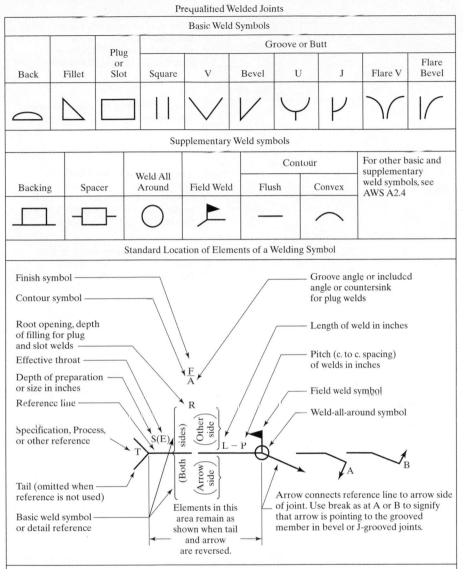

Prequalified Welded Joints

Basic Weld Symbols

Back	Fillet	Plug or Slot	Groove or Butt						
			Square	V	Bevel	U	J	Flare V	Flare Bevel

Supplementary Weld symbols

Backing	Spacer	Weld All Around	Field Weld	Contour		For other basic and supplementary weld symbols, see AWS A2.4
				Flush	Convex	

Standard Location of Elements of a Welding Symbol

Finish symbol

Contour symbol

Root opening, depth of filling for plug and slot welds

Effective throat

Depth of preparation or size in inches

Reference line

Specification, Process, or other reference

Tail (omitted when reference is not used)

Basic weld symbol or detail reference

Groove angle or included angle or countersink for plug welds

Length of weld in inches

Pitch (c. to c. spacing) of welds in inches

Field weld symbol

Weld-all-around symbol

F
A

R

S(E)

T

(Both sides) (Arrow side) (Other side)

L – P

Arrow connects reference line to arrow side of joint. Use break as at A or B to signify that arrow is pointing to the grooved member in bevel or J-grooved joints.

Elements in this area remain as shown when tail and arrow are reversed.

A

B

Note:
Size, weld symbol, length of weld, and spacing must read in that order, from left to right, along the reference line. Neither orientation of reference nor location of the arrow alters this rule.
The perpendicular leg of ⊾, V, ⊬, ⊮, weld symbols must be at left.
Dimensions of fillet welds must be shown on both the arrow side and the other side.
Symbols apply between abrupt changes in direction of welding unless governed by the "all around" symbol or otherwise dimensioned.
These symbols do not explicitly provide for the case that frequently occurs in structural work, where duplicate material (such as stiffeners) occurs on the far side of a web or gusset plate. The fabricating industry has adopted this convention: that when the billing of the detail material discloses the existence of a member on the far side as well as on the near side, the welding shown for the near side shall be duplicated on the far side.

FIGURE 14.5

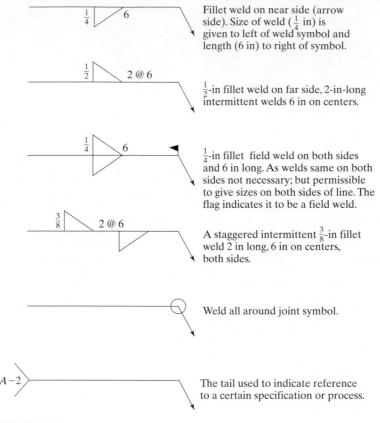

Fillet weld on near side (arrow side). Size of weld ($\frac{1}{4}$ in) is given to left of weld symbol and length (6 in) to right of symbol.

$\frac{1}{2}$-in fillet weld on far side, 2-in-long intermittent welds 6 in on centers.

$\frac{1}{4}$-in fillet field weld on both sides and 6 in long. As welds same on both sides not necessary; but permissible to give sizes on both sides of line. The flag indicates it to be a field weld.

A staggered intermittent $\frac{3}{8}$-in fillet weld 2 in long, 6 in on centers, both sides.

Weld all around joint symbol.

The tail used to indicate reference to a certain specification or process.

FIGURE 14.6

Sample weld symbols.

14.9 GROOVE WELDS

When complete penetration groove welds are subjected to axial tension or axial compression, the weld stress is assumed to equal the load divided by the net area of the weld. Three types of groove welds are shown in Fig. 14.7. The square groove joint,

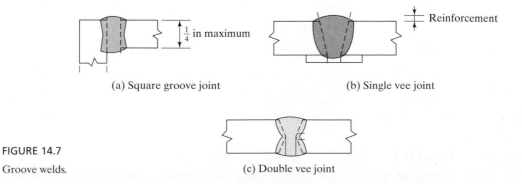

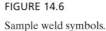

$\frac{1}{4}$ in maximum

Reinforcement

(a) Square groove joint

(b) Single vee joint

FIGURE 14.7

Groove welds.

(c) Double vee joint

shown in part (a) of the figure, is used to connect relatively thin material up to a maximum of 1/4 in thickness. As the material becomes thicker, it is necessary to use the single-vee groove welds and the double-vee groove welds illustrated in parts (b) and (c), respectively, of Fig. 14.7. For these two welds, the members are bevelled before welding to permit full penetration of the weld.

The groove welds shown in Fig. 14.7 are said to have *reinforcement*. Reinforcement is added weld metal that causes the throat dimension to be greater than the thickness of the welded material. Because of reinforcement, groove welds may be referred to as 125 percent, 150 percent, etc., according to the amount of extra thickness at the weld. There are two major reasons for having reinforcement: (1) reinforcement gives a little extra strength, because the extra metal takes care of pits and other irregularities; and (2) the welder can easily make the weld a little thicker than the welded material. It would be a difficult, if not impossible, task to make a perfectly smooth weld with no places that were thinner or thicker than the material welded.

Reinforcement undoubtedly makes groove welds stronger and better when they are subjected to static loads. When the connection is to be subjected to vibrating loads repeatedly, however, reinforcement is not as satisfactory, because stress concentrations develop in the reinforcement and contribute to earlier failure. For these kinds of cases, a common practice is to provide reinforcement and grind it off flush with the material being connected (AASHTO Section 10.34.2.1).

Figure 14.8 shows some of the edge preparations that may be necessary for groove welds. In part (a), a bevel with a feathered edge is shown. When feathered edges are used, there is a problem with burn-through. This may be lessened if a *land* is used, such as the one shown in part (b) of the figure, or a backup strip or backing bar, as shown in part (c). The backup strip is often a 1/4-in copper plate. Weld metal does not stick to copper, and copper has a very high conductivity that is useful in carrying away excess heat and reducing distortion. Sometimes, steel backup strips are used, but they will become a part of the weld and are thus left in place. A land should not be used together with a backup strip, because there is a high possibility that a gas pocket might be formed, preventing full penetration. When double bevels are used, as shown in part (d) of the figure, spacers are sometimes provided to prevent burn-through. The spacers are removed after one side is welded.

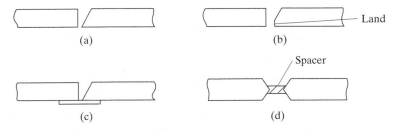

FIGURE 14.8

Edge preparation for groove welds. (a) Bevel with feathered edge. (b) Bevel with a land. (c) Bevel with a backup plate. (d) Double bevel with a spacer.

From the standpoints of strength, resistance to impact stress repetition, and amount of filler metal required, groove welds are preferable to fillet welds. From other standpoints, however, they are not so attractive, and the vast majority of structural welding is fillet welding. Groove welds have higher residual stresses, and the preparations (such as scarfing and veeing) of the edges of members for groove welds are expensive, but the major disadvantages more likely lie in the problems involved with getting the pieces to fit together in the field. (The advantages of fillet welds in this respect were described in Section 14.7.) For these reasons, field groove joints are not used often, except on small jobs and where members may be fabricated a little long and cut in the field to the lengths necessary for precise fitting.

14.10 FILLET WELDS

Tests have shown that fillet welds are stronger in tension and compression than they are in shear, so the controlling fillet weld stresses given by the various specifications are shearing stresses. When practical, it is desirable to try to arrange welded connections so that they will be subjected to shearing stresses only, and not to a combination of shear and tension or shear and compression.

When fillet welds are tested to failure with loads parallel to the weld axes, they seem to fail by shear at angles of about 45° through the throat. Their strength is therefore assumed to equal the design shearing stress or allowable shear stress times the theoretical throat area of the weld. The theoretical throats of several fillet welds are shown in Fig. 14.9. The throat area equals the theoretical throat distance times the length of the weld. In this figure, the root of the weld is the point at which the faces of the original metal pieces intersect, and the theoretical throat of the weld is the shortest distance from the root of the weld to its diagrammatic face.

For the 45° or equal leg fillet, the throat dimension is 0.707 times the leg of the weld, but it has a different value for fillet welds with unequal legs. The desirable fillet weld has a flat or slightly convex surface, although the convexity of the weld does not

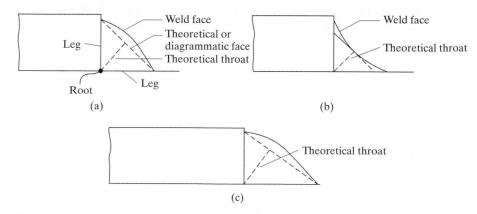

FIGURE 14.9

(a) Convex surface. (b) Concave surface. (c) Unequal leg fillet weld.

add to its calculated strength. At first glance, the concave surface would appear to give the ideal fillet weld shape, because stresses could apparently flow smoothly and evenly around the corner, with little stress concentration. Years of experience, however, have shown that single-pass fillet welds of a concave shape have a greater tendency to crack upon cooling, and this factor has proved to be of greater importance than the smoother stress distribution of convex types.

When a concave weld shrinks, the surface is placed in tension, which tends to cause cracks.

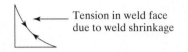

Tension in weld face
due to weld shrinkage

When the surface of a convex weld shrinks, it does not place the outer surface in tension; rather, as the face shortens, it is placed in compression.

Compression in weld face
due to weld shrinkage

Another item of importance pertaining to the shape of fillet welds is the angle of the weld with respect to the pieces being welded. The desirable value of this angle is in the vicinity of 45°. For 45° fillet welds, the leg sizes are equal, and such welds are referred to by the leg sizes (e.g., a 1/4-in fillet weld). Should the leg sizes be different (not a 45° weld), both leg sizes are given in describing the weld (e.g., a 3/8-by-1/2-in fillet weld).

The effective throat thickness may be increased in the calculations if it can be shown that consistent penetration beyond the root of the diagrammatic weld is obtained by the weld process being used (AISC specification J2.2a). Submerged arc welding is one area in which consistent extra penetration has, in the past, been assured to occur.

14.11 STRENGTH OF WELDS

In the discussion that follows, reference is made to Fig. 14.10. The stress in a fillet weld is usually said to equal the load divided by the effective throat area of the weld, with no consideration given to the direction of the load. Tests have shown, however, that transversely loaded fillet welds are appreciably stronger than ones loaded parallel to the weld's axis.

Transverse fillet welds are stronger for two reasons: First, they are more uniformly stressed over their entire lengths, while longitudinal fillet welds are stressed unevenly, due to varying deformations along their lengths; second, tests show that failure occurs at angles other than 45°, giving them larger effective throat areas.

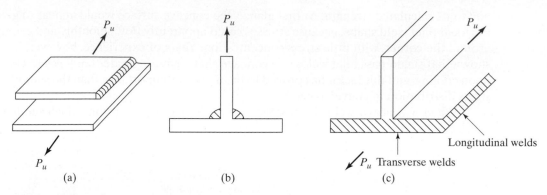

FIGURE 14.10

(a) Longitudinal fillet weld. (b) Transverse fillet weld. (c) Transverse and longitudinal welds.

The method of determining the strength of fillet welds along their longitudinal axes, regardless of the load directions, is usually used to simplify computations. It is rather common for designers to determine the strength of all fillet welds by assuming that the loads are applied in the longitudinal direction.

14.12   AISC REQUIREMENTS

When welds are made, the electrode material should have properties of the base metal. If the properties are comparable, the weld metal is referred to as the *matching base metal*. (That is, their nominal strengths are similar.)

Table 14.1 (which is Table J2.5 of the AISC Specification) provides nominal strengths for various types of welds, including fillet welds, plug and slot welds, and complete-penetration and partial-penetration groove welds.

The design strength of a particular weld (ϕR_n) and the allowable strength R_n/Ω of welded joints shall be the lower value of the base material strength determined according to the limit states of tensile rupture and shear rupture, and the weld metal strength determined according to the limit state of rupture by the expressions to follow:

For the base metal, the nominal strength is

$$R_n = F_{nBM}A_{BM}$$ (AISC Equation J2-2)

For the weld metal, the nominal strength is

$$R_n = F_{nw}A_{we}$$ (AISC Equation J2-3)

TABLE 14.1 Available Strength of Welded Joints, ksi (MPa)

Load Type and Direction Relative to Weld Axis	Pertinent Metal	ϕ and Ω	Nominal Strength (F_{nBM} or F_{nw}) ksi (MPa)	Effective Area (A_{BM} or A_{we}) in² (mm²)	Required Filler Metal Strength Level [a][b]
COMPLETE-JOINT-PENETRATION GROOVE WELDS					

Load Type and Direction Relative to Weld Axis	Pertinent Metal	ϕ and Ω	Nominal Strength (F_{nBM} or F_{nw}) ksi (MPa)	Effective Area (A_{BM} or A_{we}) in² (mm²)	Required Filler Metal Strength Level [a][b]
COMPLETE-JOINT-PENETRATION GROOVE WELDS					
Tension Normal to weld axis	Strength of the joint is controlled by the base metal.				Matching filler metal shall be used. For T and corner joints with backing left in place, notch tough filler metal is required. See Section J2.6.
Compression Normal to weld axis	Strength of the joint is controlled by the base metal.				Filler metal with a strength level equal to or one strength level less than matching filler metal is permitted.
Tension or Compression Parallel to weld axis	Tension or compression in parts joined parallel to a weld need not be considered in design of welds joining the parts.				Filler metal with a strength level equal to or less than matching filler metal is permitted.
Shear	Strength of the joint is controlled by the base metal.				Matching filler metal shall be used. [c]
PARTIAL-JOINT-PENETRATION GROOVE WELDS INCLUDING FLARE VEE GROOVE AND FLARE BEVEL GROOVE WELDS					
Tension Normal to weld axis	Base	$\phi = 0.75$ $\Omega = 2.00$	F_u	Effective Area	
	Weld	$\phi = 0.80$ $\Omega = 1.88$	$0.60F_{EXX}$	See J2.1a	
Compression Column to Base Plate and column splices designed per J1.4(a)	Compressive stress need not be considered in design of welds joining the parts.				Filler metal with a strength level equal to or less than matching filler metal is permitted.
Compression Connections of members designed to bear other than columns as described in J1.4(b)	Base	$\phi = 0.90$ $\Omega = 1.67$	F_y	See J4	
	Weld	$\phi = 0.80$ $\Omega = 1.88$	$0.60F_{EXX}$	See J2.1a	
Compression Connections not finished-to-bear	Base	$\phi = 0.90$ $\Omega = 1.67$	F_y	See J4	
	Weld	$\phi = 0.80$ $\Omega = 1.88$	$0.90F_{EXX}$	See J2.1a	
Tension or Compression Parallel to weld axis	Tension or compression in parts joined parallel to a weld need not be considered in design of welds joining the parts.				
Shear	Base	Governed by J4			
	Weld	$\phi = 0.75$ $\Omega = 2.00$	$0.60F_{EXX}$	See J2.1a	

(Continued)

TABLE 14.1 Continued

Load Type and Direction Relative to Weld Axis	Pertinent Metal	ϕ and Ω	Nominal Strength (F_{nBM} or F_{nw}) ksi (MPa)	Effective Area (A_{BM} or A_{we}) in² (mm²)	Required Filler Metal Strength Level[a][b]
FILLET WELDS INCLUDING FILLETS IN HOLES AND SLOTS AND SKEWED T-JOINTS					
Shear	Base	Governed by J4			Filler metal with a strength level equal to or less than matching filler metal is permitted.
	Weld	$\phi = 0.75$ $\Omega = 2.00$	$0.60F_{EXX}^{[d]}$	See J2.2a	
Tension or Compression Parallel to weld axis	Tension or compression in parts joined parallel to a weld need not be considered in design of welds joining the parts.				
PLUG AND SLOT WELDS					
Shear Parallel to faying surface on the effective area	Base	Governed by J4			Filler metal with a strength level equal to or less than matching filler metal is permitted.
	Weld	$\phi = 0.75$ $\Omega = 2.00$	$0.60F_{EXX}$	J2.3a	

[a] For matching weld metal see AWS D1.1, Section 3.3.
[b] Filler metal with a strength level one strength level greater than matching is permitted.
[c] Filler metals with a strength level less than matching may be used for groove welds between the webs and flanges of built-up sections transferring shear loads, or in applications where high restraint is a concern. In these applications, the weld joint shall be detailed and the weld shall be designed using the thickness of the material as the effective throat, $\phi = 0.80$, $\Omega = 1.88$ and $0.60F_{EXX}$ as the nominal strength.
[d] Alternatively, the provisions of J2.4(a) are permitted, provided the deformation compatibility of the various weld elements is considered. Alternatively, Sections J2.4(b) and (c) are special applications of J2.4(a) that provide for deformation compatibility.

Source: AISC Specification, Table J2.5, p. 16.1–114 and 16.1–115, June 22, 2010. "Copyright © American Institute of Steel Construction. Reprinted with permission. All rights reserved."

In the preceding equations,

F_{nBM} = the nominal stress of the base metal, ksi

F_{nw} = the nominal stress of the weld metal, ksi

A_{BM} = effective area of the base metal, in²

A_{we} = effective area of the weld, in²

Table 14.1 (Table J2.5 in AISC Specification) provides the weld values needed to use these equations: ϕ, Ω, F_{BM}, and F_w. Limitations on these values are also given in this table.

The filler metal electrodes for shielded arc welding are listed as E60XX, E70XX, etc. In this classification, the letter E represents an electrode, while the first set of digits (60, 70, 80, 90, 100, or 110) indicates the minimum tensile strength of the weld, in ksi.

The remaining digits may be used to specify the type of coating. Because strength is the most important factor to the structural designer, we usually specify electrodes as E70XX, E80XX, or simply E70, E80, and so on. For the usual situation, E70 electrodes are used for steels with F_y values from 36 to 60 ksi, while E80 is used when F_y is 65 ksi.

In addition to the nominal stresses given in Table 14.1, there are several other provisions applying to welding given in Section J2.2b of the LRFD Specification. Among the more important are the following:

1. The minimum length of a fillet weld may not be less than four times the nominal leg size of the weld. Should its length actually be less than this value, the weld size considered effective must be reduced to one-quarter of the weld length.

2. The maximum size of a fillet weld along edges of material less than 1/4 in thick equals the material thickness. For thicker material, it may not be larger than the material thickness less 1/16 in, unless the weld is specially built out to give a full-throat thickness. For a plate with a thickness of 1/4 in or more, it is desirable to keep the weld back at least 1/16 in from the edge so that the inspector can clearly see the edge of the plate and thus accurately determine the dimensions of the weld throat.

As a general statement, the weldability of a material improves as the thickness to be welded decreases. The problem with thicker material is that thick plates take heat from welds more rapidly than thin plates, even if the same weld sizes are used. (The problem can be alleviated somewhat by preheating the metal to be welded to a few hundred degrees Fahrenheit and holding it there during the welding operation.)

3. The minimum permissible size fillet welds of the AISC Specification are given in Table 14.2 (Table J2.4 of the AISC Specification). They vary from 1/8 in for 1/4 in or thinner material up to 5/16 in for material over 3/4 in in thickness. The smallest practical weld size is about 1/8 in, and the most economical size is probably about 1/4 or 5/16 in. The 5/16-in weld is about the largest size that can be made in one

TABLE 14.2 Minimum Size of Fillet Welds

Material Thickness of Thinner Part Joined, in (mm)	Minimum Size of Fillet Weld,[a] in (mm)
To $\frac{1}{4}$ (6) inclusive	$\frac{1}{8}$ (3)
Over $\frac{1}{4}$ (6) to $\frac{1}{2}$ (13)	$\frac{3}{16}$ (5)
Over $\frac{1}{2}$ (13) to $\frac{3}{4}$ (19)	$\frac{1}{4}$ (6)
Over $\frac{3}{4}$ (19)	$\frac{5}{16}$ (8)

[a] Leg dimension of fillet welds. Single pass welds must be used. See Section J2.2b of the LRFD Specification for maximum size of fillet welds.

Source: AISC Specification, Table J2.4, p. 16.1–111, June 22, 2010.

pass with the shielded metal arc welded process (SMAW); with the submerged arc process (SAW), 1/2 in is the largest size.

These minimum sizes were not developed on the basis of strength considerations, but rather because thick materials have a quenching or rapid cooling effect on small welds. If this happens, the result is often a loss in weld ductility. In addition, the thicker material tends to restrain the weld material from shrinking as it cools, with the result that weld cracking can result and present problems.

Note that the minimum sizes given in Table 14.2 are dependent on the thinner of the two parts being joined. It may be larger, however, if so required by the calculated strength.

4. Sometimes, end returns or boxing is used at the end of fillet welds, as shown in Fig. 14.11. In the past, such practices were recommended to provide better fatigue resistance and to make sure that weld thicknesses were maintained over their full lengths. Recent research has shown that such returns are not necessary for developing the capacity of such connections. End returns are also used to increase the plastic deformation capability of such connections (AISC Commentary J2.2b).

5. When longitudinal fillet welds are used for the connection of plates or bars, their length may not be less than the perpendicular distance between them, because of shear lag (discussed in Chapter 3).

6. For lap joints, the minimum amount of lap permitted is equal to five times the thickness of the thinner part joined, but may not be less than 1 in (AISC J2.2b). The purpose of this minimum lap is to keep the joint from rotating excessively.

7. Should the actual length (l) of an end-loaded fillet weld be greater than 100 times its leg size (w), the AISC Specification (J2.2b) states that, due to stress variations along the weld, it is necessary to determine a smaller or effective length for strength determination. This is done by multiplying l by the term β, as given in the following equation in which w is the weld leg size:

$$\beta = 1.2 - 0.002 \, (l/w) \leq 1.0 \qquad \text{(AISC Equation J2-1)}$$

If the actual weld length is greater than 300 w, the effective length shall be taken as 180 w.

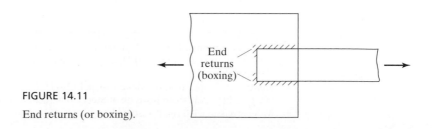

FIGURE 14.11

End returns (or boxing).

The all-welded 56-story Toronto
Dominion Bank Tower. (Courtesy of the
Lincoln Electric Company.)

14.13 DESIGN OF SIMPLE FILLET WELDS

Examples 14-1 and 14-2 demonstrate the calculations used to determine the strength
of various fillet welded connections; Example 14-3 presents the design of such a
connection. In these and other problems, weld lengths are selected no closer than the
nearest 1/4 in, because closer work cannot be expected in shop or field.

Example 14-1

a. Determine the design strength of a 1-in length of a 1/4-in fillet weld formed by
the shielded metal arc process (SMAW) and E70 electrodes with a minimum

tensile strength $F_{EXX} = 70$ ksi. Assume that load is to be applied parallel to the weld length.

b. Repeat part (a) if the weld is 20 in long.

c. Repeat part (a) if the weld is 30 in long.

Solution

a. $R_n = F_{nw}A_{we}$

$\quad = $ (nominal strength of base metal $0.60\ F_{EXX}$)(throat t)(weld length)

$\quad = (0.60 \times 70 \text{ ksi})\left(\dfrac{1}{4} \text{ in} \times 0.707 \times 1.0\right) = 7.42 \text{ k/in}$

LRFD $\phi = 0.75$	ASD $\Omega = 2.00$
$\phi R_n = (0.75)(7.42) = \mathbf{5.56 \text{ k/in}}$	$\dfrac{R_n}{\Omega} = \dfrac{7.42}{2.00} = \mathbf{3.71 \text{ k/in}}$

b. Length, $l = 20$ in

LRFD	ASD
$\dfrac{l}{w} = \dfrac{20}{\frac{1}{4}} = 80 < 100$	$\dfrac{L}{w} = \dfrac{20}{\frac{1}{4}} = 80 < 100$
$\therefore \beta = 1.0$	$\therefore \beta = 1.0$
$\phi R_n L = (5.56)(20) = \mathbf{111.2 \text{ k}}$	$\dfrac{R_n}{\Omega} L = (3.71)(20) = \mathbf{74.2 \text{ k}}$

c. Length, $l = 30$ in

LRFD	ASD
$\dfrac{l}{w} = \dfrac{30}{\frac{1}{4}} = 120 > 100$	$\dfrac{L}{w} = \dfrac{30}{\frac{1}{4}} = 120 > 100$
$\therefore \beta = 1.2 - (0.002)(120) = 0.96$	$\therefore \beta = 1.2 - (0.002)(120) = 0.96$
$\phi R_n \beta L = (5.56)(0.96)(30) = \mathbf{160.1 \text{ k}}$	$\dfrac{R_n}{\Omega}\beta L = (3.71)(0.96)(30) = \mathbf{106.8 \text{ k}}$

Fillet welds may not be designed with a stress that is greater than the design stress on the adjacent members being connected. If the external force applied to the member (tensile or compressive) is parallel to the axis of the weld metal, the design strength may not exceed the axial design strength of the member.

Example 14-2 illustrates the calculations necessary to determine the design strength of plates connected with longitudinal fillet welds. In this example, the shearing strength of the weld per inch controls, and it is multiplied by the total length of the welds to give the total capacity of the connections.

Example 14-2

What is the design strength of the connection shown in Fig. 14.12 if the plates consist of A572 Grade 50 steel ($F_u = 65$ ksi)? E70 electrodes were used, and the 7/16-in fillet welds were made by the SMAW process.

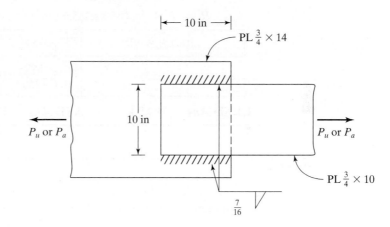

FIGURE 14.12

Solution

$$\text{Weld strength} = F_{we}A_{we} = (0.60 \times 70 \text{ ksi})\left(\frac{7}{16} \text{ in} \times 0.707 \times 20 \text{ in}\right) = 259.8 \text{ k}$$

$$\text{Checking the length to weld size ratio } \frac{L}{w} = \frac{10 \text{ in}}{{}^{7}\!/_{16} \text{ in}} = 22.86 < 100$$

∴ No reduction in weld strength is required as $\beta = 1.0$.

LRFD ϕ = 0.75	ASD Ω = 2.00	
$\phi R_n = (0.75)(259.8) = 194.9$ k	$\dfrac{R_n}{\Omega} = \dfrac{259.8}{2.00} = 129.9$ k	← controls

Check tensile yielding for $\dfrac{3}{4} \times 10$ *PL*

$$R_n = F_y A_g = (50 \text{ ksi})\left(\frac{3}{4} \text{ in} \times 10 \text{ in}\right) = 375 \text{ k}$$

LRFD ϕ_t = 0.90	ASD Ω_t = 1.67
$\phi_t R_n = (0.90)(375) = 337.5$ k	$\dfrac{R_n}{\Omega_t} = \dfrac{375}{1.67} = 224.6$ k

Check tensile rupture strength for $\frac{3}{4} \times 10$ PL

$$A_e = A_g U$$

since the weld length, $l = 10$ in, is equal to the distance between the welds, $U = 0.75$ (see Case 4, AISC Table D3.1)

$$A_e = \frac{3}{4} \text{ in} \times 10 \text{ in} \times 0.75 = 5.62 \text{ in}^2$$

$$R_n = F_u A_e = (65 \text{ ksi})(5.62 \text{ in}^2) = 365.3 \text{ k}$$

LRFD $\phi_t = 0.75$	ASD $\Omega_t = 2.00$
$\phi R_n = (0.75)(365.3) = 274.0$ k	$\dfrac{R_n}{\Omega_t} = \dfrac{365.3}{2.00} = 182.7$ k

LRFD Ans = 194.9 k   **ASD Ans = 129.9 k**

Example 14-3

Using 50 ksi steel and E70 electrodes, design SMAW fillet welds to resist a full-capacity load on the 3/8 × 6-in member shown in Fig. 14.13.

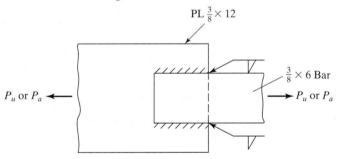

PL $\frac{3}{8} \times 12$

$\frac{3}{8} \times 6$ Bar

P_u or P_a

P_u or P_a

FIGURE 14.13

Solution

Tensile yield strength of gross section of $\frac{3}{8} \times 6$ bar

$$R_n = F_y A_g = (50 \text{ ksi})\left(\frac{3}{8} \text{ in} \times 6 \text{ in}\right) = 112.5 \text{ k}$$

LRFD $\phi_t = 0.90$	ASD $\Omega_t = 1.67$	
$\phi_t R_n = (0.90)(112.5) = 101.2$ k	$\dfrac{R_n}{\Omega_t} = \dfrac{112.5}{1.67} = 67.4$ k	← controls

Tensile rupture strength of $\frac{3}{8} \times 6$ bar, assume $U = 1.0$ (conservative)

$$A_e = \frac{3}{8} \text{ in} \times 6 \text{ in} \times 1.0 = 2.25 \text{ in}^2$$

$$R_n = F_u A_e = (65 \text{ ksi})(2.25 \text{ in}^2) = 146.2 \text{ k}$$

LRFD $\phi_t = 0.75$	$\Omega_t = 2.00$
$\phi_t R_n = (0.75)(146.2) = 109.6 \text{ k}$	$\dfrac{R_n}{\Omega_t} = \dfrac{146.2}{2.00} = 73.1 \text{ k}$

∴ Tensile capacity of bar is controlled by yielding.

Design of weld

$$\text{Maximum weld size} = \frac{3}{8} - \frac{1}{16} = \frac{5}{16} \text{ in}$$

$$\text{Minimum weld size} = \frac{3}{16} \text{ in (Table 14.2)}$$

Use $\dfrac{5}{16}$ weld (maximum size with one pass)

$$R_n \text{ of weld per in} = F_w A_{we} = (0.60 \times 70 \text{ ksi})\left(\frac{5}{16} \text{ in} \times 0.707\right)$$

$$= 9.28 \text{ k/in}$$

LRFD $\phi = 0.75$	ASD $\Omega = 2.00$
$\phi R_n = (0.75)(9.28) = 6.96 \text{ k/in}$	$\dfrac{R_n}{\Omega} = \dfrac{9.28}{2.00} = 4.64 \text{ k/in}$
Weld length reqd $= \dfrac{101.2}{6.96}$	Weld length reqd $= \dfrac{67.4}{4.64}$
$= 14.54 \text{ in or } 7\frac{1}{2} \text{ in each side}$	$= 14.53 \text{ in or } 7\frac{1}{2} \text{ in each side}$
$\dfrac{L}{w} = \dfrac{7.5}{\frac{5}{16}} = 24 < 100 \text{ OK } \beta = 1.00$	$\dfrac{L}{w} = \dfrac{7.5}{\frac{5}{16}} = 24 < 100 \text{ OK } \beta = 1.00$

Use $7\frac{1}{2}$-in welds each side.   **Use $7\frac{1}{2}$-in welds each side.**

Author's Note: With a weld length of only 7 1/2 in and a distance between the welds of 6 in, the *U* factor will be 0.75 and the tensile capacity will be reduced (controlled by rupture). One might consider using a smaller weld size, possibly the minimum value of 3/16 in, to increase the weld length.

AISC Section J2.4 states that the strength of fillet welds loaded transversely in a plane through their centers of gravity may be determined with the following equation in which $\phi = 0.75$, $\Omega = 2.00$, and θ is the angle between the line of action of the load and the longitudinal axis of the weld:

$$F_{nw} = (0.6F_{EXX})(1.0 + 0.50 \sin^{1.5}\theta) \quad \text{(AISC Equation J2-5)}$$

As the angle θ is increased, the strength of the weld increases. Should the load be perpendicular to the longitudinal axis of the weld, the result will be a 50 percent increase in the computed weld strength. Example 14-4 illustrates the application of this Appendix expression.

Example 14-4

a. Determine the LRFD design and the ASD allowable strengths of the $\frac{1}{4}$-in SMAW fillet welds formed with E70 electrodes, which are shown in part (a) of Fig. 14.14. The load is applied parallel to the longitudinal axis of the welds.

b. Repeat part (a) if the load is applied at a 45° angle, with the longitudinal axis of the welds as shown in part (b) of the figure.

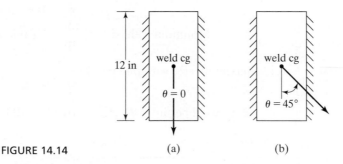

FIGURE 14.14   (a)   (b)

Solution

a. Load paralled to longitudinal axis of welds

Effective throat $t = (0.707)\left(\dfrac{1}{4}\right) = 0.177$ in

$$R_n = F_{nw}A_{we} = (0.60 \times 70 \text{ ksi})(0.177 \text{ in} \times 24 \text{ in}) = 178.4 \text{ k}$$

LRFD $\phi = 0.75$	ASD $\Omega = 2.00$
$\phi R_n = (0.75)(178.4) = $ **133.8 k**	$\dfrac{R_n}{\Omega} = \dfrac{178.4}{2.00} = $ **89.2 k**

b. Load applied at a 45° angle, with longitudinal axes of welds

$$R_n = 0.60F_{EXX}(1.0 + 0.50 \sin^{1.5}\theta)(0.177 \text{ in} \times 24 \text{ in})$$

$$= (0.60 \times 70 \text{ ksi})(1 + 0.50 \times \sin^{1.5}45°)(0.177 \text{ in} \times 24 \text{ in}) = 231.4 \text{ k}$$

LFRD $\phi = 0.75$	ASD $\Omega = 2.00$
$\phi R_n = (0.75)(231.4) = $ **173.6 k**	$\dfrac{R_n}{\Omega} = \dfrac{231.4}{2.00} = $ **115.7 k**

The values for ϕR_n and R_n/Ω are increased by almost 30 percent above the values in part (a), where the weld was longitudinally loaded.

14.14   DESIGN OF CONNECTIONS FOR MEMBERS WITH BOTH LONGITUDINAL AND TRANSVERSE FILLET WELDS

For this discussion, the fillet welds of Fig. 14.15 are considered. To determine the total nominal strength of the welds, it seems logical that we should add the nominal strength of the side welds, calculated with $R_n = F_w A_w$ with $F_w = 0.60 F_{EXX}$, to the nominal strength of the end or transverse welds, calculated with $R_n = 0.60 F_{EXX}(1.0 + 0.50 \sin^{1.5}\theta) A_w$. This procedure, however, is not correct, because the less ductile transverse welds will reach their ultimate deformation capacities before the side or longitudinal welds reach their maximum strengths. As a result of this fact, the AISC in its Section J2.4c states that the total nominal strength of a connection with side and transverse welds is to equal the larger of the values obtained with the following two equations:

$$R_n = R_{nwl} + R_{nwt} \qquad \text{(AISC Equation J2-10a)}$$

$$R_n = 0.85 R_{nwl} + 1.5 R_{nwt} \qquad \text{(AISC Equation J2-10b)}$$

In these expressions, R_{nwl} is the total nominal strength of the longitudinal or side fillet welds, calculated with $R_{wl} = F_{nw} A_{we}$. The total nominal strength of the transversely loaded fillet welds, R_{wt}, is also calculated with $F_{nw} A_{we}$ and not with $0.60 F_w(1 + 0.50 \sin^{1.5}\theta) A_w$. Example 14.5 illustrates the calculation of the LRFD design strength ϕR_n and the ASD allowable strength R_n/Ω for the connection of Fig. 14.15, with its side and transverse welds.

Example 14-5

Determine the total LRFD design strength and the total ASD allowable strength of the 5/16-in E70 fillet welds shown in Fig. 14.15.

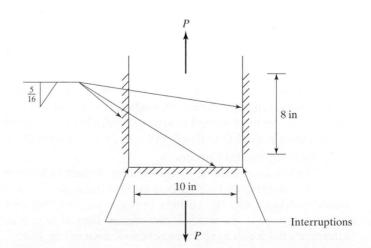

FIGURE 14.15

Solution

$$\text{Effective throat } t = (0.707)\left(\frac{5}{16} \text{ in}\right) = 0.221 \text{ in}$$

$$R_{wl} = R_n \text{ for side welds} = F_{nw}A_{we} = (0.60 \times 70 \text{ ksi})(2 \times 8 \text{ in} \times 0.221 \text{ in})$$
$$= 148.5 \text{ k}$$

$$R_{wt} = R_n \text{ for transverse end weld} = F_{nw}A_{we}$$
$$= (0.60 \times 70 \text{ ksi})(10 \text{ in} \times 0.221 \text{ in}) = 92.8 \text{ k}$$

Applying AISC Equations J2-10a and J2-10b

$$R_n = R_{nwl} + R_{nwt} = 148.5 \text{ k} + 92.8 \text{ k} = 241.3 \text{ k}$$

$$R_n = 0.85R_{nwl} + 1.5R_{nwt} = (0.85)(148.5 \text{ k}) + (1.5)(92.8 \text{ k}) = 265.4 \text{ k} \leftarrow \text{controls}$$

LRFD $\phi = 0.75$	ASD $\Omega = 2.00$
$\phi R_n = (0.75)(265.4) = $ **199 k**	$\dfrac{R_n}{\Omega} = \dfrac{265.4}{2.00} = $ **132.7 k**

14.15   SOME MISCELLANEOUS COMMENTS

1. Weld Terminations

Generally speaking, the termination points of fillet welds do not have much effect on the strength and serviceability of connections. For certain situations, though, this is not altogether correct. For instance, notches may not only adversely affect the static strength of a connection, but also may decidedly reduce the resistance of the weld to the development of cracks when cyclic loads of sufficient size and frequency are applied. For such situations, it is wise to terminate the welds before the end of a joint is reached.

Should welds be terminated one or two weld sizes (or leg sizes) from the end of joints, their strengths will be only negligibly reduced. In fact, such length reductions are usually not even considered in strength calculations. A detailed discussion of this topic is presented in the Manual in Section J2.2b of the AISC Commentary.

2. Welding around Corners

The reader should be aware of an important fact concerning the use of longitudinal and transverse end welds. It is quite difficult for the welder to deposit continuous welds evenly around the corners between longitudinal and transverse welds without causing a gouge in the welds at the corner. Therefore, it is generally a good practice to interrupt fillet welds at such corners, as shown in Fig. 14.15.

Generally speaking, the location of the ends of fillet welds do not affect their strength or serviceability. However, should cyclic loads of sufficient size and frequency be applied in situations where the welds are uneven or have notches at the corners, there can be a considerable loss of strength. For such situations, the welds are often specified to be terminated before the corners are reached.

3. Strength of 1/16-in Welds for Calculation Purposes

It is rather convenient for design purposes to know the strength of a 1/16-in fillet weld 1 in long. Though this size is below the minimum permissible size given in Table 14.2, the calculated strength of such a weld is useful for determining weld sizes for calculated forces. For a 1-in-long SMAW weld with the load parallel to weld axis, we have the following:

For LRFD

$$\phi F_w A_w = (0.75)(0.60F_{EXX})\left(0.707 \times \frac{1}{16}\right)(1.0) = 0.0199F_{EXX}$$

For ASD

$$\frac{F_w A_w}{\Omega} = \frac{(0.60F_{EXX})\left(0.707 \times \frac{1}{16}\right)(1.00)}{2.00} = 0.0133F_{EXX}$$

For LRFD with E70 electrodes, $R_n = (0.0199)(70) = 1.39$ k/in. If we are designing a fillet weld to resist a factored force of 6.5 k/in, the required LRFD weld size is $6.5/1.39 = 4.68$ sixteenths of an inch, say, 5/16 in.

For ASD with E70 electrodes, $R_n/\Omega = (0.0133)(70) = 0.931$ k/in. If we are designing a fillet weld to resist a service load force of 4.4 k/in, the required ASD weld size is $4.4/0.931 = 4.73$ sixteenths of an inch, say, 5/16 in.

14.16   DESIGN OF FILLET WELDS FOR TRUSS MEMBERS

Should the members of a welded truss consist of single angles, double angles, or similar shapes and be subjected to static axial loads only, the AISC Specification (J1.7) permits the connections to be designed by the procedures described in the preceding section. The designers can select the weld size, calculate the total length of the weld required, and place the welds around the member ends as they see fit. (It would not make sense, of course, to place the weld all on one side of a member, such as for the angle of Fig. 14.16, because of the rotation possibility.)

It should be noted that the centroid of the welds and the centroid of the statically loaded angle do not coincide in the connection shown in this figure. If a welded connection is subjected to varying stresses (such as those occurring in a bridge member), it is essential to place the welds so that their centroid will coincide with the centroid of

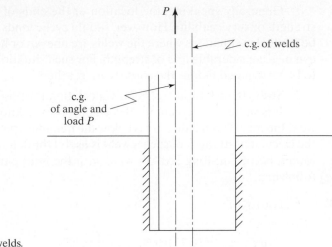

FIGURE 14.16

Eccentrically loaded welds.

the member (or the resulting torsion must be accounted for in design). If the member being connected is symmetrical, the welds will be placed symmetrically; if the member is not symmetrical, the welds will not be symmetrical.

The force in an angle, such as the one shown in Fig. 14.17, is assumed to act along its center of gravity. If the center of gravity of weld resistance is to coincide with the angle force, the welds must be asymmetrically placed, or in this figure L_1 must be longer than L_2. (When angles are connected by bolts, there is usually an appreciable amount of eccentricity, but in a welded joint, eccentricity can be fairly well eliminated.) The information necessary to handle this type of weld design can be easily expressed in equation form, but only the theory behind the equations is presented here.

For the angle shown in Fig. 14.17, the force acting along line L_2 (designated here as P_2) can be determined by taking moments about point A. The member force and the

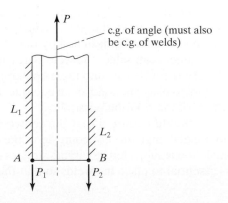

FIGURE 14.17

weld resistance should coincide, and the moments of the two about any point must be zero. If moments are taken about point A, the force P_1 (which acts along line L_1) will be eliminated from the equation, and P_2 can be determined. In a similar manner, P_1 can be determined by taking moments about point B or by $\Sigma V = 0$. Example 14-6 illustrates the design of fillet welds of this type.

There are other possible solutions for the design of the welds for the angle considered in Fig. 14.17. Although the 7/16-in weld is the largest one permitted at the edges of the 1/2-in angle, a larger weld could be used on the other side next to the outstanding angle leg. From a practical point of view, however, the welds should be the same size, because different-size welds slow the welder down due to the need to change electrodes to make different sizes.

Should cyclic or fatigue-type loads be involved, and should the centers of gravity of the loads and welds not coincide, the life of the connection may be severely reduced. Appendix 3 of the AISC Specification addresses the topic of fatigue design.

Example 14-6

Use $F_y = 50$ ksi and $F_u = 65$ ksi, E70 electrodes, and the SMAW process to design side fillet welds for the full capacity of the $5 \times 3 \times 1/2$-in angle tension member shown in Fig. 14.18. Assume that the member is subjected to repeated stress variations, making any connection eccentricity undesirable. Check block shear strength of the member. Assume that the WT chord member has adequate strength to develop the weld strengths and that the thickness of its web is 1/2 in. Assume that $U = 0.8 l$.

Solution

Tensile yielding on gross section

$$P_n = F_y A_g = (50 \text{ ksi})(3.75 \text{ in}^2) = 187.5 \text{ k}$$

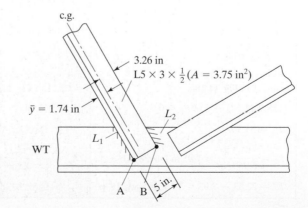

FIGURE 14.18

Tensile rupture on net section

$$A_e = UA_g = (0.87)(3.75 \text{ in}^2) = 3.26 \text{ in}^2$$

$$P_n = F_u A_e = (65 \text{ ksi})(3.26 \text{ in}^2) = 211.9 \text{ k}$$

LRFD	ASD
For tensile yielding ($\phi_t = 0.90$)	For tensile yielding ($\Omega_t = 1.67$)
$\phi_t P_n = (0.9)(187.5) = 168.7 \text{ k}$	$\dfrac{P_n}{\Omega_t} = \dfrac{187.5}{1.67} = 112.3 \text{ k}$
For tensile rupture ($\phi_t = 0.75$)	For tensile rupture ($\Omega_t = 2.00$)
$\phi_t P_n = (0.75)(211.9) = 158.9 \text{ k} \leftarrow$	$\dfrac{P_n}{\Omega_t} = \dfrac{211.9}{2.00} = 105.9 \text{ k} \leftarrow$

Maximum weld size $= \dfrac{1}{2} - \dfrac{1}{16} = \dfrac{7}{16}$ in

Use $\frac{5}{16}$-in weld (largest that can be made in single pass)

Effective throat t of weld $= (0.707)\left(\dfrac{5}{16} \text{ in}\right) = 0.221$ in

LRFD	ASD
Design strength/in of $\frac{5}{16}$-in welds ($\phi = 0.75$)	Allowable strength/in of $\frac{5}{16}$-in welds ($\Omega = 2.00$)
$= (0.75)(0.60 \times 70)(0.221)(1)$	$= \dfrac{(0.60 \times 70)(0.221)(1)}{2.00}$
$= 6.96 \text{ k/in}$	$= 4.64 \text{ k/in}$
Weld length reqd $= \dfrac{158.9}{6.96}$	Weld length reqd $= \dfrac{105.9}{4.64}$
$= 22.83 \text{ in}$	$= 22.82 \text{ in}$
Taking moments about point A in Fig. 14.18	Taking moments about point A in Fig. 14.18
$(158.9)(1.74) - 5.00P_2 = 0$	$(105.9)(1.74) - 5.00P_2 = 0$
$P_2 = 55.3 \text{ k}$	$P_2 = 36.85 \text{ k}$
$L_2 = \dfrac{55.3 \text{ k}}{6.96 \text{ k/in}} = 7.95 \text{ in (say, 8 in)}$	$L_2 = \dfrac{36.85 \text{ k}}{4.64 \text{ k/in}} = 7.94 \text{ in (say, 8 in)}$
$L_1 = 22.83 - 7.95 = 14.88 \text{ in (say, 15 in)}$	$L_1 = 22.82 - 7.94 = 14.88 \text{ in (say, 15 in)}$

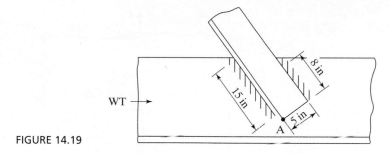

FIGURE 14.19

Welds arranged as shown in Fig. 14.19.

Checking block shearing strength, assuming dimensions previously described

$$R_n = 0.6F_u A_{nv} + U_{bs}F_u A_{nt} \leq 0.6F_y A_{gv} + U_{bs}F_u A_{nt}$$

$$= (0.6)(65)(15 + 8)\left(\frac{1}{2}\right) + (1.00)(65)\left(5 \times \frac{1}{2}\right)$$

$$\leq (0.6)(50)(15 + 8)\left(\frac{1}{2}\right) + (1.00)(65)\left(5 \times \frac{1}{2}\right)$$

$$= 611 \text{ k} > 507.5 \text{ k}$$

$$\therefore R_n = 507.5 \text{ k}$$

LRFD $\phi = 0.75$	ASD $\Omega = 2.00$
$\phi R_n = (0.75)(507.5) = 380.6 \text{ k} > 158.9 \text{ k}$ **OK**	$\dfrac{R_n}{\Omega} = \dfrac{507.5}{2.00} = 253.8 \text{ k} > 105.9 \text{ } k$ **OK**

14.17   PLUG AND SLOT WELDS

On some occasions, the spaces available for fillet welds may not be sufficient to support the applied loads. The plate shown in Fig. 14.20 falls into this class. Even using the maximum fillet weld size (7/16 in here) will not support the load. It is assumed that space is not available to use transverse welds at the plate ends in this case.

One possibility for solving this problem involves the use of a slot weld, as shown in the figure. There are several AISC requirements pertaining to slot welds that need to be mentioned here. AISC Specification J2.3 states that the width of a slot may not be less than the member thickness plus 5/16 in (rounded off to the next greater odd 1/16 in, since structural punches are made in these diameters), nor may it be greater than 2.25 times the weld thickness. For members up to 5/8-in thickness, the weld thickness

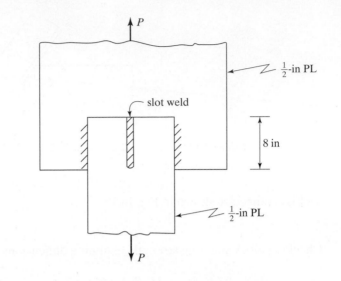

FIGURE 14.20

must equal the plate thickness; for members greater than 5/8-in thickness, the weld thickness may not be less than one-half the member thickness (or 5/8 in). The maximum length permitted for slot welds is ten times the weld thickness. The limitations given in specifications for the maximum sizes of plug or slot welds are caused by the detrimental shrinkage that occurs around these types of welds when they exceed certain sizes. Should holes or slots larger than those specified be used, it is desirable to use fillet welds around the borders of the holes or slots rather than using a slot or plug weld. Slot and plug welds are normally used in conjunction with fillet welds in lap joints. Sometimes, plug welds are used to fill in the holes temporarily used for erection bolts for beam and column connections. They may or may not be included in the calculated strength of these joints.

The LRFD design strength of a plug or slot weld is equal to its design stress ϕF_w times its area in the shearing plane. The ASD allowable strength equals F_w/Ω times the same shearing area. The shearing area is the area of contact at the base of the plug or slot. The length required for a slot weld can be determined from the expression to follow:

$$L = \frac{\text{load}}{(\text{width})(\text{design stress})}$$

Example 14-7

Design SMAW fillet welds and a slot weld to connect the plates shown in Fig. 14.20 if $P_D = 110$ k, $P_L = 120$ k, $F_y = 50$ ksi, $F_u = 65$ ksi, and E70 electrodes are used. As shown in the figure, the plates may lap over each other by only 8 in due to space limitations.

Solution

$$\text{Maximum weld size} = \frac{1}{2} - \frac{1}{16} = \frac{7}{16} \text{ in}$$

$$\text{Effective throat thickness} = (0.707)\left(\frac{7}{16} \text{ in}\right) = 0.309 \text{ in}$$

$$R_n = \text{nominal capacity of fillet welds}$$

$$= (0.60 \times 70 \text{ ksi})(2 \times 8 \text{ in} \times 0.309 \text{ in}) = 207.6 \text{ k}$$

LRFD $\phi = 0.75$	ASD $\Omega = 2.00$
$P_u = (1.2)(110) + (1.6)(120) = 324 \text{ k}$	$P_a = 110 + 120 = 230 \text{ k}$
$\phi R_n = (0.75)(207.6) = 155.7 \text{ k}$	$\dfrac{R_n}{\Omega} = \dfrac{207.6}{2.00} = 103.8 \text{ k}$
$< 324 \text{ k}$ $\therefore$ **Try slot weld.**	$< 230 \text{ k}$ $\therefore$ **Try slot weld.**

$$\text{Minimum width of slot} = t \text{ of PL} + \frac{5}{16} = \frac{1}{2} + \frac{5}{16} = \frac{13}{16} \text{ in}$$

$$\text{Maximum width of slot} = \left(2\frac{1}{4}\right)\left(\frac{1}{2} \text{ in}\right) = 1\frac{1}{8} \text{ in}$$

$$\text{Rounding slot width to next larger odd } \frac{1}{16} \text{ in} = \frac{15}{16} \text{ in}$$

$$\text{Try } \frac{15}{16}\text{-in wide slot.}$$

LRFD $\phi = 0.75$	ASD $\Omega = 2.00$
ϕR_n for all welds must $= 324 \text{ k}$	$\dfrac{R_n}{\Omega}$ for all welds must $= 230 \text{ k}$
$155.7 + \phi(R_n \text{ of slot weld}) = 324$	$103.8 + \dfrac{R_n}{\Omega}$ for slot weld $= 230$
$155.7 + (0.75)\left(\dfrac{15}{16} \times L \times 0.6 \times 70\right) = 324$	$103.8 + \dfrac{^{15}/_{16} \times L \times 0.6 \times 70}{2.00} = 230$
L reqd $= 5.70 \text{ in}$	L reqd $= 6.41 \text{ in}$
Use 6-in slot weld.	**Use $6\frac{1}{2}$-in slot weld.**

14.18   SHEAR AND TORSION

Fillet welds are frequently loaded with eccentrically applied loads, with the result that the welds are subjected to either shear and torsion or to shear and bending. Figure 14.21 is presented to show the difference between the two situations. Shear and torsion, shown in part (a) of the figure, are the subject of this section, while shear and bending, shown in part (b) of the figure, are discussed in Section 14.19.

As is the case for eccentrically loaded bolt groups (Section 13.1), the AISC Specification provides the design strength of welds, but does not specify a method of analysis for eccentrically loaded welds. It's left to the designer to decide which method to use.

14.18.1 Elastic Method

Initially, the very conservative elastic method is presented. In this method, friction or slip resistance between the connected parts is neglected, the connected parts are assumed to be perfectly rigid, and the welds are assumed to be perfectly elastic.

For this discussion, the welded bracket of part (a) of Fig. 14.21 is considered. The pieces being connected are assumed to be completely rigid, as they were in bolted connections. The effect of this assumption is that all deformation occurs in the weld. The

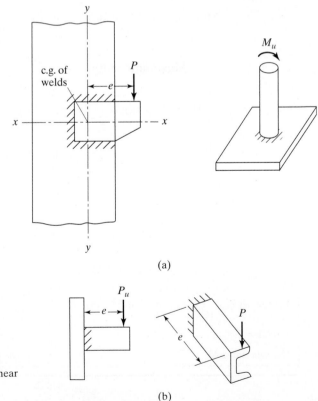

FIGURE 14.21

(a) Welds subjected to shear and torsion. (b) Welds subjected to shear and bending.

weld is subjected to a combination of shear and torsion, as was the eccentrically loaded bolt group considered in Section 13.1. The force caused by torsion can be computed from the following familiar expression:

$$f = \frac{Td}{J}$$

In this expression, T is the torsion, d is the distance from the center of gravity of the weld to the point being considered, and J is the polar moment of inertia of the weld. It is usually more convenient to break the force down into its vertical and horizontal components. In the following expressions, f_h and f_v are the horizontal and vertical components, respectively, of the force f:

$$f_h = \frac{Tv}{J} \quad f_v = \frac{Th}{J}$$

Notice that the formulas are almost identical to those used for determining stresses in bolt groups subject to torsion. These components are combined with the usual direct shearing stress, which is assumed to equal the reaction divided by the total length of the welds. For design of a weld subject to shear and torsion, it is convenient to assume a 1-in weld and to compute the stresses on a weld of that size. Should the assumed weld be overstressed, a larger weld is required; if it is understressed, a smaller one is desirable.

Although the calculations probably will show the weld to be overstressed or understressed, the calculation does not have to be repeated, because a ratio can be set up to give the weld size for which the load would produce a computed stress exactly equal to the design stress. Note that the use of a 1-in weld simplifies the units, because 1 in of length of weld is 1 in² of weld, and the computed stresses are said to be either kips per square inch or kips per inch of length. Should the calculations be based on some size other than a 1-in weld, the designer must be very careful to keep the units straight, particularly in obtaining the final weld size. To further simplify the calculations, the welds are assumed to be located at the edges where the fillet welds are placed, rather than at the centers of their effective throats. As the throat dimensions are rather small, this assumption changes the results very little. Example 14-8 illustrates the calculations involved in determining the weld size required for a connection subjected to a combination of shear and torsion.

Example 14-8

For the A36 bracket shown in Fig. 14.22(a), determine the fillet weld size required if E70 electrodes, the AISC Specification, and the SMAW process are used.

Solution. Assuming a 1-in weld as shown in part (b) of Fig. 14.22

$$A = 2(4 \text{ in}^2) + 10 \text{ in}^2 = 18 \text{ in}^2$$
$$\bar{x} = \frac{(4 \text{ in}^2)(2 \text{ in})(2)}{18 \text{ in}^2} = 0.89 \text{ in}$$

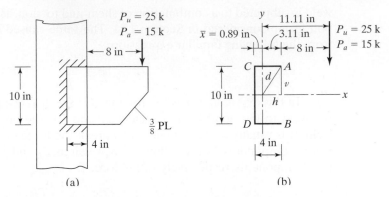

FIGURE 14.22   (a)   (b)

$$I_x = \left(\frac{1}{12}\right)(1)(10)^3 + (2)(4)(5)^2 = 283.3 \text{ in}^4$$

$$I_y = 2\left(\frac{1}{12}\right)(1)(4)^3 + 2(4)(2 - 0.89)^2 + (10)(0.89)^2 = 28.4 \text{ in}^4$$

$$J = 283.3 + 28.4 = 311.7 \text{ in}^4$$

According to our previous work, the welds perpendicular to the direction of the loads are appreciably stronger than the welds parallel to the loads. However, to simplify the calculations, the author conservatively assumes that all the welds have design strengths or allowable strengths per inch equal to the values for the welds parallel to the loads.

$$R_n \text{ for a 1 in weld} = 0.707 \times 1 \times 0.6 \times 70 = 29.69 \text{ ksi}$$

LRFD $\phi = 0.75$	ASD $\Omega = 2.00$
$\phi R_n = (0.75)(29.69) = 22.27 \text{ ksi}$	$\dfrac{R_n}{\Omega} = \dfrac{29.69}{2.00} = 14.84 \text{ ksi}$
Forces @ points C & D	**Forces @ points C & D**
$f_h = \dfrac{(25 \times 11.11)(5)}{311.7} = 4.46 \text{ k/in}$	$f_h = \dfrac{(15)(11.11)(5)}{311.7} = 2.67 \text{ k/in}$
$f_v = \dfrac{(25 \times 11.11)(0.89)}{311.7} = 0.79 \text{ k/in}$	$f_v = \dfrac{(15)(11.11)(0.89)}{311.7} = 0.48 \text{ k/in}$
$f_s = \dfrac{25}{18} = 1.39 \text{ k/in}$	$f_s = \dfrac{15}{18} = 0.83 \text{ k/in}$
$f_r = \sqrt{(0.79 + 1.39)^2 + (4.46)^2}$	$f_r = \sqrt{(0.48 + 0.83)^2 + (2.67)^2}$
$= 4.96 \text{ k/in}$	$= 2.97 \text{ k/in}$
Size $= \dfrac{4.96 \text{ k/in}}{22.27 \text{ k/in}^2} = 0.223 \text{ in, say } \frac{1}{4} \text{ in}$	Size $= \dfrac{2.97 \text{ k/in}}{14.84 \text{ k/in}^2} = 0.200 \text{ in, say } \frac{1}{4} \text{ in}$
Forces @ points A & B	**Forces @ points A & B**
$f_h = \dfrac{(25 \times 11.11)(5)}{311.7} = 4.46 \text{ k/in}$	$f_h = \dfrac{(15 \times 11.11)(5)}{311.7} = 2.67 \text{ k/in}$

(Continued)

LRFD $\phi = 0.75$	ASD $\Omega = 2.00$
$f_v = \dfrac{(25 \times 11.11)(3.11)}{311.7} = 2.77 \text{ k/in}$	$f_v = \dfrac{(15 \times 11.11)(3.11)}{311.7} = 1.66 \text{ k/in}$
$f_s = \dfrac{25}{18} = 1.39 \text{ k/in}$	$f_s = \dfrac{15}{18} = 0.83 \text{ k/in}$
$f_r = \sqrt{(2.77 + 1.39)^2 + (4.46)^2}$	$f_r = \sqrt{(1.66 + 0.83)^2 + (2.67)^2}$
$\quad = 6.10 \text{ k/in}$	$\quad = 3.65 \text{ k/in}$
$\text{Size} = \dfrac{6.10}{22.27} = 0.274 \text{ in, say } \dfrac{5}{16} \text{ in}$	$\text{Size} = \dfrac{3.65}{14.84} = 0.246 \text{ in, say } \dfrac{1}{4} \text{ in}$
Use $\frac{5}{16}$-in fillet welds, E70, SMAW.	Use $\frac{1}{4}$-in fillet weld, E70, SMAW.

14.18.2 Ultimate Strength Method

An ultimate strength analysis of eccentrically loaded welded connections is more realistic than the more conservative elastic procedure just described. For the discussion that follows, the eccentrically loaded fillet weld of Fig. 14.23 is considered. As for eccentrically loaded bolted connections, the load tends to cause a relative rotation and translation between the parts connected by the weld.

Even if the eccentric load is of such a magnitude that it causes the most stressed part of the weld to yield, the entire connection will not yield. The load may be increased, the less stressed fibers will begin to resist more of the load, and failure will not occur until all the weld fibers yield. The weld will tend to rotate about its instantaneous center of rotation. The location of this point (which is indicated by the letter O in the figure) is dependent upon the location of the eccentric load, the geometry of the weld, and the deformations of the different elements of the weld.

If the eccentric load P_u or P_a is vertical and if the weld is symmetrical about a horizontal axis through its center of gravity, the instantaneous center will fall somewhere on the horizontal x axis. Each differential element of the weld will provide a resisting force R. As shown in Fig. 14.23, each of these resisting forces is assumed to act perpendicular to a ray

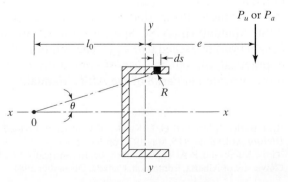

FIGURE 14.23

drawn from the instantaneous center to the center of gravity of the weld element in question.

Studies have been made to determine the maximum shear forces that eccentrically loaded weld elements can withstand.[4, 5] The results, which depend on the load deformation relationship of the weld elements, may be represented either with curves or in formula fashion. The ductility of the entire weld is governed by the maximum deformation of the weld element that first reaches its limit. (The element that is located at the greatest distance from the weld's instantaneous center probably reaches its limit first.)

Just as in eccentrically loaded bolted connections, the location of the instantaneous center of rotation is determined by trial and error. Unlike eccentrically loaded bolt groups, however, the strength and deformation of welds are dependent on the angle θ that the force in each element makes with the axis of that element. The deformation of each element is proportional to its distance from the instantaneous center. At maximum stress, the weld deformation is Δ_{max} which is determined with the following expression, in which w is the weld leg size:

$$\Delta_{max} = 1.087w(\theta + 6)^{-0.65} \leq 0.17w$$

The deformation in a particular element is assumed to vary directly in proportion to its distance from the instantaneous center:

$$\Delta_r = \frac{l_r}{l_m}\Delta_m$$

The nominal shear strength of a weld segment at a deformation Δ is

$$R_n = 0.6 \, F_{EXX} A_{we} \, (1.0 + 0.50 \sin^{1.5} \theta)[p(1.9 - 0.9p)]^{0.3}$$

where θ is the angle of loading measured from the weld longitudinal axis, and p is the ratio of the deformation of an element to its deformation at ultimate stress.

We can assume a location of the instantaneous center, determine R_n values for the different elements of the weld, and compute ΣR_x and ΣR_y. The three equations of equilibrium ($\Sigma M = 0$, $\Sigma R_x = 0$, and $\Sigma R_y = 0$) will be satisfied if we have the correct location of the instantaneous center. If they are not satisfied, we will try another location, and so on. Finally, when the equations are satisfied, the value of P_u can be computed as $\sqrt{(\Sigma R_x)^2 + (\Sigma R_y)^2}$.

Detailed information on the use of these expressions is given in Part 8 of the AISC Manual. The AISC Specification permits weld strengths in eccentrically loaded connections to exceed $0.6F_{EXX}$. The solution of these kinds of problems is completely impractical without the use of computers or of computer-generated tables such as those provided in Part 8 of the AISC Manual.

---

[4]L. J. Butler, S. Pal, and G. L. Kulak, "Eccentrically Loaded Weld Connections," *Journal of the Structural Division*, vol. 98, no. ST5, May 1972, pp. 989–1005.

[5]G. L. Kulak and P. A. Timler, "Tests on Eccentrically Loaded Fillet Welds," Dept of Civil Engineering, University of Alberta, Edmonton, Canada, December 1984.

TABLE 14.3   Electrode Strength Coefficient, C_1

Electrode	F_{EXX} (ksi)	C_1
E60	60	0.857
E70	70	1.00
E80	80	1.03
E90	90	1.16
E100	100	1.21
E110	110	1.34

The values given in the tables of Part 8 of the AISC Manual were developed by the ultimate strength method. Using the tables, the nominal strength R_n of a particular connection can be determined from the following expression in which C is a tabular coefficient, C_1 is a coefficient depending on the electrode number and given in Table 14.3 (Table 8-3 in AISC manual), D is the weld size in sixteenths of an inch, and l is the length of the vertical weld:

$$R_n = CC_1Dl$$

$$\phi = 0.75 \text{ and } \Omega = 2.00$$

The Manual includes tables for both vertical and inclined loads (at angles from the vertical of 0° to 75°). The user is warned not to interpolate for angles in between these values because the results may not be conservative. Therefore, the user is advised to use the value given for the next-lower angle. If the connection arrangement being considered is not covered by the tables, the conservative elastic procedure previously described may be used. Examples 14-9 and 14-10 illustrate the use of these ultimate strength tables.

Example 14-9

Repeat Example 14-8, using the AISC tables that are based on an ultimate strength analysis. The connection is redrawn in Fig. 14.24.

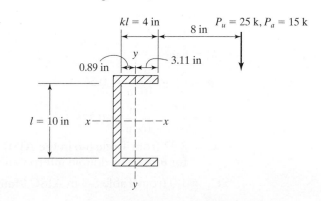

FIGURE 14.24

Solution

$$e_x = 11.11 \text{ in}$$

$$l = 10 \text{ in}$$

$$a = \frac{e_x}{l} = \frac{11.11 \text{ in}}{10 \text{ in}} = 1.11$$

$$k = \frac{kl}{l} = \frac{4 \text{ in}}{10 \text{ in}} = 0.40$$

$$C = 1.31 \text{ from Table 8-8, in the AISC Manual for } \theta$$
$$= 0° \text{ by straight-line interpolation}$$

$$C_1 = 1.0 \text{ from Table 8-3 in AISC Manual (E70 electrodes)}$$

LRFD $\phi = 0.75$	ASD $\Omega = 2.00$
D_{min} = weld size reqd $= \dfrac{P_u}{\phi C C_1 l}$	D_{min} = weld size reqd $= \dfrac{\Omega P_a}{C C_1 l}$
$= \dfrac{25}{(0.75)(1.31)(1.0)(10)}$	$= \dfrac{(2.00)(15)}{(1.31)(1.0)(10)}$
	$= 2.29 \text{ sixteenths}$
$= 2.54 \text{ sixteenths} = 0.159 \text{ in}$	$= 0.143 \text{ in}$
(compared with 0.273 in obtained with elastic method)	(compared with 0.246 in obtained with the elastic method)
Use $\frac{3}{16}$-in fillet weld, E70, SMAW.	Use $\frac{3}{16}$-in fillet weld, E70, SMAW.

Example 14-10

Determine the weld size required for the situation shown in Fig. 14.25, using the AISC tables that are based on an ultimate strength analysis (A36 steel, E70 electrodes).

Solution

$$e_x = al = 9 \text{ in}$$

$$l = 16 \text{ in}$$

$$kl = 6 \text{ in}$$

$$a = \frac{al}{l} = \frac{9 \text{ in}}{16 \text{ in}} = 0.562$$

$$k = \frac{kl}{l} = \frac{6 \text{ in}}{16 \text{ in}} = 0.375$$

$$C = 3.32 \text{ from Table 8-6 in the AISC Manual}$$
$$\text{for } \theta = 0° \text{ by double interpolation}$$

$$C_1 = 1.0 \text{ from Table 8-3 in AISC Manual (E70 electrodes)}$$

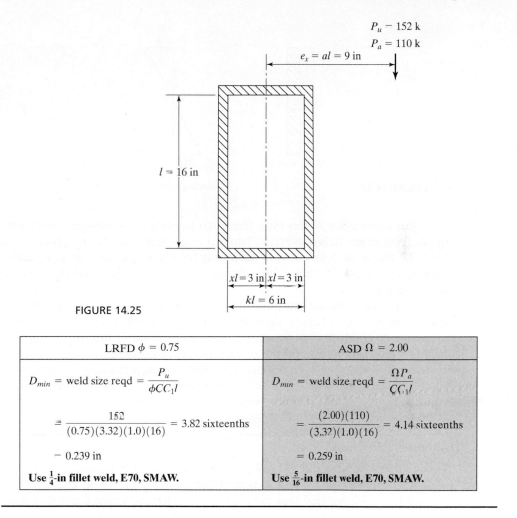

$$P_u = 152 \text{ k}$$
$$P_a = 110 \text{ k}$$
$$e_x = al = 9 \text{ in}$$

$l = 16$ in

$xl = 3$ in | $xl = 3$ in

$kl = 6$ in

FIGURE 14.25

LRFD $\phi = 0.75$	ASD $\Omega = 2.00$
D_{min} = weld size reqd = $\dfrac{P_u}{\phi C C_1 l}$	D_{min} = weld size reqd = $\dfrac{\Omega P_a}{C C_1 l}$
$= \dfrac{152}{(0.75)(3.32)(1.0)(16)} = 3.82$ sixteenths	$= \dfrac{(2.00)(110)}{(3.32)(1.0)(16)} = 4.14$ sixteenths
$= 0.239$ in	$= 0.259$ in
Use $\frac{1}{4}$-in fillet weld, E70, SMAW.	Use $\frac{5}{16}$-in fillet weld, E70, SMAW.

14.19   SHEAR AND BENDING

The welds shown in Fig. 14.21(b) and in Fig. 14.26 are subjected to a combination of shear and bending.

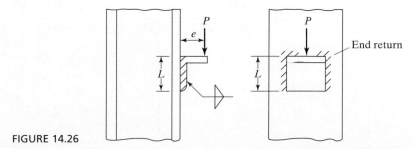

e

P

P

End return

L

L

FIGURE 14.26

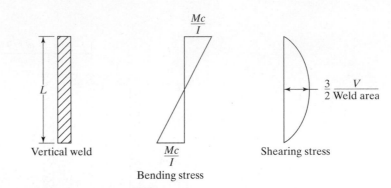

FIGURE 14.27

Vertical weld

Bending stress

Shearing stress

For short welds of this type, the usual practice is to consider a uniform variation of shearing stress. If, however, the bending stress is assumed to be given by the flexure formula, the shear does not vary uniformly for vertical welds, but as a parabola with a maximum value 1 1/2 times the average value. These stress and shear variations are shown in Fig. 14.27.

The reader should carefully note that the maximum shearing stresses and the maximum bending stresses occur at different locations. Therefore, it probably is not necessary to combine the two stresses at any one point. If the weld is capable of withstanding the worst shear and the worst moment individually, it is probably satisfactory. In Example 14-11, however, a welded connection subjected to shear and bending is designed by the usual practice of assuming a uniform shear distribution in the weld and combining the value vectorially with the maximum bending stress.

Example 14-11

Using E70 electrodes, the SMAW process, and the LRFD Specification, determine the weld size required for the connection of Fig. 14.26 if $P_D = 10$ k, $P_L = 20$ k, $e = 2$ 1/2 in, and $L = 8$ in. Assume that the member thicknesses do not control weld size.

Solution

Initially, assume fillet welds with 1-in leg sizes.

LRFD $\phi = 0.75$	ASD $\Omega_t = 2.00$
$P_u = (1.2)(10) + (1.6)(20) = 44$ k	$P_a = 10 + 20 = 30$ k
$f_v = \dfrac{P_u}{A} = \dfrac{44}{(2)(8)} = 2.75$ k/in	$f_v = \dfrac{P_a}{A} = \dfrac{30}{(2)(8)} = 1.88$ k/in
$f_b = \dfrac{Mc}{I} = \dfrac{(44 \times 2.5)(4)}{2\left(\dfrac{1}{12}\right)(1)(8)^3} = 5.16$ k/in	$f_b = \dfrac{Mc}{I} = \dfrac{(30 \times 2.5)(4)}{2\left(\dfrac{1}{12}\right)(1)(8)^3} = 3.52$ k/in

(Continued)

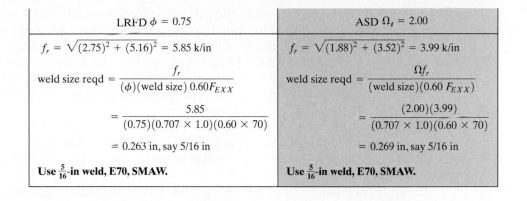

LRFD $\phi = 0.75$	ASD $\Omega_t = 2.00$
$f_r = \sqrt{(2.75)^2 + (5.16)^2} = 5.85 \text{ k/in}$	$f_r = \sqrt{(1.88)^2 + (3.52)^2} = 3.99 \text{ k/in}$
weld size reqd $= \dfrac{f_r}{(\phi)(\text{weld size})\,0.60 F_{EXX}}$	weld size reqd $= \dfrac{\Omega f_r}{(\text{weld size})(0.60\,F_{EXX})}$
$= \dfrac{5.85}{(0.75)(0.707 \times 1.0)(0.60 \times 70)}$	$= \dfrac{(2.00)(3.99)}{(0.707 \times 1.0)(0.60 \times 70)}$
$= 0.263 \text{ in, say 5/16 in}$	$= 0.269 \text{ in, say 5/16 in}$
Use $\frac{5}{16}$-in weld, E70, SMAW.	Use $\frac{5}{16}$-in weld, E70, SMAW.

The subject of shear and bending for welds is a very practical one, as it is the situation commonly faced in moment resisting connections. This topic is considered at some length in Chapter 15.

14.20   FULL-PENETRATION AND PARTIAL-PENETRATION GROOVE WELDS

14.20.1 Full-Penetration Groove Welds

When plates with different thicknesses are joined, the strength of a full-penetration groove weld is based on the strength of the thinner plate. Similarly, if plates of different strengths are joined, the strength of a full-penetration weld is based on the strength of the weaker plate. Notice that no allowances are made for the presence of reinforcement—that is, for any extra weld thickness.

Full-penetration groove welds are the best type of weld for resisting fatigue failures. In fact, in some specifications they are the only groove welds permitted if fatigue is possible. Furthermore, a study of some specifications shows that allowable stresses for fatigue situations are increased if the crowns or reinforcement of the groove welds have been ground flush.

14.20.2 Partial-Penetration Groove Welds

When we have groove welds that do not extend completely through the full thickness of the parts being joined, they are referred to as *partial-penetration groove welds*. Such welds can be made from one or both sides, with or without preparation of the edges (such as bevels). Partial-penetration welds are shown in Fig. 14.28.

FIGURE 14.28

Partial-penetration groove welds.

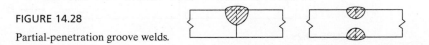

Partial-penetration groove welds often are economical in cases in which the welds are not required to develop large forces in the connected materials, such as for column splices and for the connecting together of the various parts of built-up members.

In Table 14.1, we can see that the design stresses are the same as for full-penetration welds when we have compression or tension parallel to the axis of the welds. When we have tension transverse to the weld axis, there is a substantial strength reduction because of the possiblity of high stress concentrations.

Examples 14-12 and 14-13 illustrate the calculations needed to determine the strength of full-penetration and partial-penetration groove welds. The design strengths of both full-penetration and partial-penetration groove welds are given in Table 14.1 of this text (AISC Table J2.5).

In part (b) of Example 14-13, it is assumed that two W sections are spliced together with a partial-penetration groove weld, and the design shear strength of the member is determined. To do this, it is necessary to compute the following: (1) the shear rupture strength of the base material as per AISC Section J2.4, (2) the shear yielding strength of the connecting elements as per AISC Section J4.3, and (3) the shear strength of the weld as per AISC Section J2.2 and AISC Table J2.5. The design shear strength of the member is the least of these three values, which are described as follows:

1. Shear fracture of base material $= F_n A_{ns}$ with $\phi = 0.75$, $\Omega = 2.00$, $F_n = 0.6F_u$, and $A_{ns} =$ net area subject to shear.
2. Shear yielding of connecting elements $= \phi R_n =$ with $(0.60A_{vg})\,F_y$, with $\phi = 0.75$, $\Omega = 2.00$, and $A_{vg} =$ gross area subjected to shear.
3. Shear yielding of the weld $= F_w = (0.60F_{EXX})A_w$, with $\phi = 0.75$, $\Omega = 2.00$, and $A_w = A_{eff} =$ area of weld.

Example 14-12

a. Determine the LRFD design strength and the ASD allowable strength of a SMAW full-penetration groove weld for the plates shown in Fig. 14.29. Use $F_y = 50$ ksi and E70 electrodes.

b. Repeat part (a) if a partial-joint-penetration groove weld (45° bevel) is used with a depth of $\frac{1}{2}$ in.

FIGURE 14.29

Solution

a. **Full-penetration groove weld**

Tension yielding strength

$$R_n = F_y A_g$$

$$= (50 \text{ ksi})\left(\frac{3}{4} \text{ in} \times 6 \text{ in}\right) = 225 \text{ k}$$

LRFD ϕ = 0.90	ASD Ω = 1.67
$\phi R_n = (0.9)(225) = $ **202.5 k**	$\dfrac{R_n}{\Omega} = \dfrac{225}{1.67} = $ **134.7 k**

← controls

Tension rupture strength

$$R_n = F_u A_e \quad \text{where } A_e = A_g U$$

$$\text{and } U = 1.0$$

$$= (65 \text{ ksi})\left(\frac{3}{4} \text{ in}\right)(6 \text{ in})(1.0) = 292.5 \text{ k}$$

LRFD ϕ = 0.75	ASD Ω = 2.00
$\phi R_n = (0.75)(292.5) = $ **219.4 k**	$\dfrac{R_n}{\Omega} = \dfrac{292.5}{2.00} = $ **146.2 k**

b. **Partial-joint-penetration groove weld**

Weld values

Effective throat of weld

$$= \frac{1}{2} - \frac{1}{8} = \frac{3}{8} \text{ in as reqd in AISC Table J2.1}$$

$$R_n = (0.60 \times 70 \text{ ksi})\left(\frac{3}{8} \text{ in} \times 6 \text{ in}\right) = 94.5 \text{ k}$$

LRFD ϕ = 0.80	ASD Ω = 1.88
$\phi R_n = (0.80)(94.5) = $ **75.6 k**	$\dfrac{R_n}{\Omega} = \dfrac{94.5}{1.88} = $ **50.3 k**

← controls

Base metal values

$$R_n = \text{strength of base metal} = F_u A_e$$

$$= (65 \text{ ksi})\left(\frac{3}{8} \text{ in} \times 6 \text{ in}\right) = 146.3 \text{ k}$$

LRFD $\phi = 0.75$	ASD $\Omega = 2.00$
$\phi R_n = (0.75)(146.3) = \mathbf{109.7\ k}$	$\dfrac{R_n}{\Omega} = \dfrac{146.3}{2.00} = \mathbf{73.1\ k}$

Example 14-13

a. A full-penetration groove weld made with E70 electrodes is use to splice together the two halves of a 50 ksi ($F_u = 65$ ksi) W21 × 166. Determine the shear strength of the splice.

b. Repeat part (a) if two vertical partial-penetration welds (60° V-groove) (E70 electrodes) with throat thicknesses of 1/4 in are used.

Solution

Using a W21 × 166 ($d = 22.5$ in, $t_w = 0.750$ in)

a. **Strength of the joint is controlled by the base metal (J4.2)**

Shear yielding strength

$$R_n = 0.60\ F_y A_{gv}$$
$$= 0.60\ (50\ \text{ksi})\ (0.75\ \text{in})\ (22.5\ \text{in}) = 506.2\ \text{k}$$

LRFD $\phi = 1.00$	ASD $\Omega = 1.50$
$\phi R_n = (1.0)(506.2) = \mathbf{506.2\ k}$	$\dfrac{R_n}{\Omega} = \dfrac{506.2}{1.50} = \mathbf{337.5\ k}$

Shear rupture strength

$$R_n = 0.60\ F_u A_{nv}$$
$$= 0.60\ (65\ \text{ksi})\ (0.75\ \text{in})\ (22.5\ \text{in}) = 658.1\ \text{k}$$

LRFD $\phi = 0.75$	ASD $\Omega = 2.00$	
$\phi R_n = (0.75)(658.1) = \mathbf{493.6\ k}$	$\dfrac{R_n}{\Omega} = \dfrac{658.1}{2.00} = \mathbf{329.0\ k}$	← controls

b.   **Base metal values (J4.2)**

Shear yielding strength

$$R_n = 0.60\ F_y A_{gv}$$

$$= 0.60\ (50\ \text{ksi})\ (0.75\ \text{in})\ (22.5\ \text{in}) = 506.2\ \text{k}$$

LRFD ϕ = 1.00	ASD Ω = 1.50
$\phi R_n = (1.00)(506.2) = $ **506.2 k**	$\dfrac{R_n}{\Omega} = \dfrac{506.2}{1.50} = $ **337.5 k**

Shear rupture strength

$$R_n = 0.60\ F_u A_{nv}$$

$$= 0.60(65\,\text{ksi})\left(2 \times \frac{1}{4}\,\text{in}\right)(22.5\,\text{in}) = 438.7\ \text{k}$$

LRFD ϕ = 0.75	ASD Ω = 2.00	
$\phi R_n = (0.75)(438.7) = $ **329.0 k**	$\dfrac{R_n}{\Omega} = \dfrac{438.7}{2.00} = $ **219.4 k**	← controls

Weld values (60° V-groove weld)

$$R_n = 0.6\ F_{EXX} A_{we}$$

$$= 0.6(70\,\text{ksi})\left(2 \times \frac{1}{4}\,\text{in}\right)(22.5\,\text{in}) = 472.5\ \text{k}$$

LRFD ϕ = 0.75	ASD Ω = 2.00
$\phi R_n = (0.75)(472.5) = $ **354.4 k**	$\dfrac{R_n}{\Omega} = \dfrac{472.5}{2.00} = $ **236.3 k**

14.21   PROBLEMS FOR SOLUTION

Unless otherwise noted, A36 steel is to be used for all problems.

14-1. A 1/4-in fillet weld, SMAW process, is used to connect the members shown in the accompanying illustration. Determine the LRFD design load and the ASD allowable load that can be applied to this connection, including the plates, using the AISC Specification and E70 electrodes. (*Ans.* 97.2 k, 64.7 k)

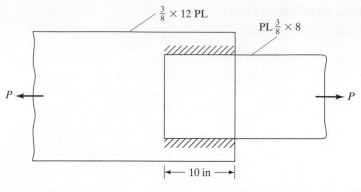

$\frac{3}{8} \times 12$ PL

PL $\frac{3}{8} \times 8$

P

P

$\longleftarrow$ 10 in $\longrightarrow$

FIGURE P14-1

14-2. Repeat Prob. 14-1 if the weld lengths are 24 in.

14-3. Rework Prob. 14-1 if A572 grade 65 steel and E80 electrodes are used.
(*Ans.* 127.3 k, 84.8 k)

14-4. Determine the LRFD design strength and the ASD allowable strength of the 5/16-in fillet welds shown, if E70 electrodes are used.

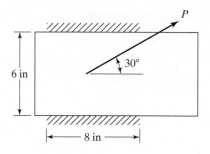

P

6 in

$30°$

8 in

FIGURE P14-4

14-5. (a) Repeat Prob. 14-4 if 1/4-in welds are used and if $\theta = 45°$. (*Ans.* 115.6 k, LRFD; 92.3 k, ASD)

(b) Repeat part (a) if $\theta = 15°$. (*Ans.* 95.0 k, LRFD; 63.3 k, ASD)

14-6. Using both the LRFD and ASD methods, design maximum-size SMAW fillet welds for the plates shown, if $P_D = 40$ k, $P_L = 60$ k and E70 electrodes are used.

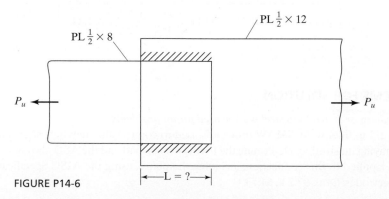

PL $\frac{1}{2} \times 12$

PL $\frac{1}{2} \times 8$

P_u

P_u

$\longleftarrow L = ? \longrightarrow$

FIGURE P14-6

14-7. Repeat Prob. 14-6 if $\frac{5}{16}$-in welds are used. (*Ans.* 10.5 in, 11 in)

14-8. Calculate ϕR_n for Prob. 14-6, using $\frac{5}{16}$ in side welds 8 in long and a vertical end weld at the end of the $\frac{1}{2} \times 8$ plate. Also use A572 grade 65 steel and E80 electrodes.

14-9. Rework Prob. 14-8, using $\frac{1}{4}$ in side welds 10 in long and welds at the end of the $\frac{1}{2} \times 8$ PL and E70 electrodes. (*Ans.* 161.5 k, 107.6 k)

14-10. The 5/8 × 8-in PL shown in the accompanying illustration is to be connected to a gusset plate with 1/4-in SMAW fillet welds. Determine ϕR_n and $\dfrac{R_n}{\Omega}$ of the bar if E70 electrodes are used.

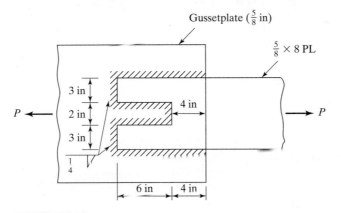

FIGURE P14-10

14-11. Design by LRFD and ASD maximum size side SMAW fillet welds required to develop the loads $P_D = 70$ k and $P_L = 60$ k for an L6 × 4 × 1/2, using E70 electrodes and 50 ksi steel. The member is connected on the sides of the 6-in leg and is subject to alternating loads. (*Ans.* $L_1 = 12.5$ in, $L_2 = 6.5$ in (LRFD); $L_1 = 13.5$ in, $L_2 = 7.0$ in (ASD))

14-12. Rework Prob. 14-11, using side welds and a weld at the end of the angle.

14-13. Rework Prob. 14-11, using E80 electrodes. (*Ans.* $L_1 = 11$ in, $L_2 = 5.5$ in (LRFD); $L_1 = 12$ in, $L_2 = 6$ in (ASD))

14-14. One leg of an 8 × 8 × 3/4 angle is to be connected with side welds and a weld at the end of the angle to a plate behind, to develop the loads $P_D = 170$ k and $P_L = 200$ k. Balance the fillet welds around the center of gravity of the angle. Using LRFD and ASD methods, determine weld lengths if E70 electrodes and maximum weld size is used.

14-15. It is desired to design 5/16-in SMAW fillet welds necessary to connect a C10 × 30 made from A36 steel to a 3/8-in gusset plate. End, side, and slot welds may be used to develop the loads $P_D = 80$ k and $P_L = 120$ k. Use both ASD and LRFD procedures. No welding is permitted on the back of the channel. Use E70 electrodes. It is assumed that, due to space limitations, the channel can lap over the gusset plate by a maximum of 8 in. (*Ans.* $\frac{15}{16} \times 4$ in slot LRFD, $\frac{15}{16} \times 4\frac{1}{4}$ in slot ASD).

14-16. Rework Prob. 14-15, using A572 grade 60 steel, E80 electrodes, and 5/16-in fillet welds.

14-17. Using the elastic method, determine the maximum force per inch to be resisted by the fillet weld shown in the accompanying illustration. (*Ans.* 11.77 k/in)

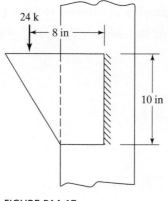

FIGURE P14-17

14-18. Using the elastic method, determine the maximum force to be resisted per inch by the fillet weld shown in the accompanying illustration.

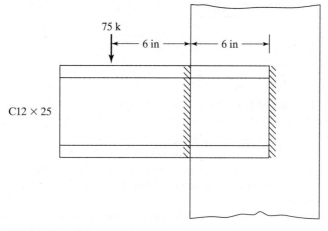

FIGURE P14-18

14-19. Using the elastic method, rework Prob. 14-18 if welds are used on the top and bottom of the channel in addition to those shown in the figure. (*Ans.* 5.88 k/in)

14-20. Using the elastic method, determine the maximum force per inch to be resisted by the fillet welds shown in the accompanying illustration.

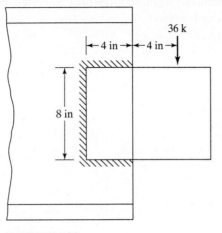

FIGURE P14-20

14-21. Determine the maximum eccentric loads ϕP_n that can be applied to the connection shown in the accompanying illustration if 1/4-in SMAW fillet welds are used. Assume plate thickness is 1/2 in and use E70 electrodes. (a) Use elastic method. (b) Use AISC tables and the ultimate strength method. (*Ans.* (a) 27.0 k, (b) 62.4 k)

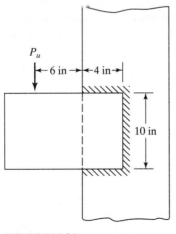

FIGURE P14-21

14-22. Rework Prob. 14-21 if 5/16-in fillet welds are used and the vertical weld is 8 in high.

14-23. Using the LRFD method and E70 electrodes, determine the fillet weld size required for the connection of Prob. 14-17 if $P_D = 10$ k, $P_L = 10$ k, and the height of the weld is 12 in. (a) Use elastic method. (b) Use AISC tables and the ultimate strength method. (*Ans.* (a) 7/16 in, (b) 1/4 in)

14-24. Repeat part (a) of Prob. 14-23, using the ASD method.

14-25. Using E70 electrodes and the SMAW process, determine the LRFD fillet weld size required for the bracket shown in the accompanying illustration. (a) Use elastic method. (b) Use AISC tables and the ultimate strength method.
(*Ans.* (a) 3/8 in, (b) 1/4 in)

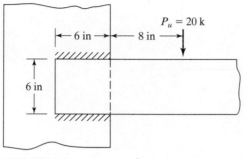

FIGURE P14-25

14-26. Rework Prob. 14-25 if the load is increased from 20 to 25 k and the horizontal weld lengths are increased from 6 to 8 in.

14-27. Rework Prob. 14-25, using ASD with $P_a = 11$ k. (*Ans.* (a) 5/16 in, (b) 3/16 in)

14-28. Using LRFD only, determine the fillet weld size required for the connection shown in the accompanying illustration. E70. (a) Use elastic method. (b) Use AISC tables and the ultimate strength method.

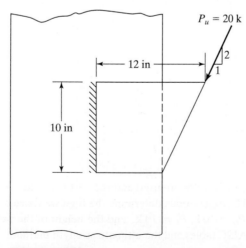

FIGURE P14-28

14-29. Using the ASD method only, determine the SMAW fillet weld size required for the connection shown in the accompanying illustration. E70. What angle thickness should be used? (a) Use elastic method. (b) Use AISC tables and the ultimate strength method. (*Ans.* (a) 3/16 in, (b) 1/8 in)

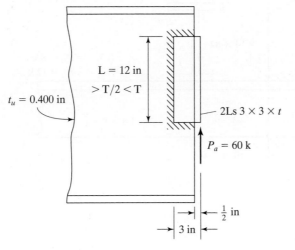

FIGURE P14-29

14-30. Assuming that the LRFD method is to be used, determine the fillet weld size required for the connection shown in the accompanying illustration. Use E70 electrodes and the elastic method.

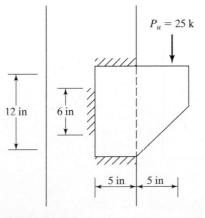

FIGURE P14-30

14-31. Determine the fillet weld size required by the ASD method for the connection shown in the accompanying illustration. E70. The SMAW process is to be used. (a) Use the elastic method. (b) Use ASD tables and the ultimate strength method. (*Ans.* (a) 3/8 in, (b) 3/16 in)

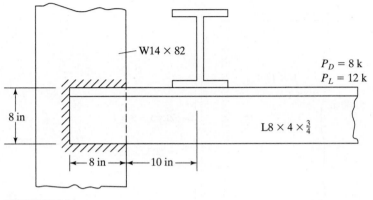

FIGURE P14-31

14-32. Determine the value of the loads ϕP_n and P_n/Ω that can be applied to the connection shown in the accompanying illustration if 3/8-in fillet welds are used. E70. SAW. Use the elastic method.

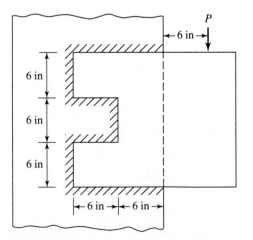

FIGURE P14-32

14-33. Using LRFD and ASD, determine the length of 1/4-in SMAW E70 fillet welds 12 in on center required to connect the cover plates for the section shown in the accompanying illustration at a point where the external shear V_u is 80 k and V_a = 55 k. E70. (*Ans.* 2.5 in both LRFD & ASD)

PL$\frac{1}{2}$ × 12

W21 × 62

PL$\frac{1}{2}$ × 12

FIGURE P14-33

14-34. The welded girder shown in the accompanying illustration has an external shear V_D = 300 k and V_L = 350 k at a particular section. Determine the fillet weld size required to fasten the plates to the web if the SMAW process is used. E70. Use LRFD and ASD.

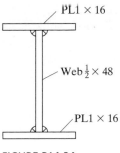

PL1 × 16

Web $\frac{1}{2}$ × 48

PL1 × 16

FIGURE P14-34

14-35. (a) Using both LRFD and ASD procedures and assuming the A36 plates in Fig. 14.29 are 12 in wide and 1/2 in thick, determine their design tensile strength and their ASD allowable strength if a full-penetration groove weld is used. Use E70 electodes.

(b) Repeat if a 5/16-in partial-penetration groove weld is used on one side.
(*Ans.* (a) 194.4 k (LRFD); 129.3 k (ASD), (b) 75.6 k (LRFD); 50.3 k (ASD))

14-36. (a) If full-penetration groove welds formed with E70 electrodes are used to splice together the two halves of a W24 × 117, determine the shear strength capacity of the splice using both the LRFD and ASD procedures. Use SMAW weld process, F_y = 50 ksi, F_u = 65 ksi.

(b) Repeat part (a) if two vertical partial-penetration groove welds (45° Bevel) with 1/4 in throat thickness are used.

CHAPTER 15

Building Connections

15.1 SELECTION OF TYPE OF FASTENER

This chapter is concerned with the actual beam-to-beam and beam-to-column connections commonly used in steel buildings. Under present-day steel specifications, three types of fasteners are permitted for these connections: *welds, unfinished bolts, and high-strength bolts*.

Selection of the type of fastener or fasteners to be used for a particular structure usually involves many factors, including requirements of local building codes, relative economy, preference of designer, availability of good welders, loading conditions (as static or fatigue loadings), preference of fabricator, and equipment available. It is impossible to list a definite set of rules from which the best type of fastener can be selected for any given structure. We can give only a few general statements that may be helpful in making a decision:

1. Unfinished bolts are often economical for light structures subject to small static loads and for secondary members (such as purlins, girts, bracing, etc.) in larger structures.
2. Field bolting is very rapid and involves less skilled labor than welding. The purchase price of high-strength bolts, however, is rather high.
3. If a structure is later to be disassembled, welding probably is ruled out, leaving the job open to selection of bolts.
4. For fatigue loadings, slip-critical high-strength bolts and welds are very good.
5. Notice that special care has to be taken to properly install high-strength, slip-critical bolts.
6. Welding requires the smallest amounts of steel, probably provides the most attractive-looking joints, and also has the widest range of application to different types of connections.
7. When continuous and rigid, fully moment-resisting joints are desired, welding probably will be selected.

8. Welding is almost universally accepted as being satisfactory for shopwork. For fieldwork, it is very popular in most areas of the United States, while in a few others it is stymied by the idea that field inspection is rather questionable.

9. To use welds for very thick members requires a great deal of extra care, and bolted connections may very well be used instead. Furthermore, such bolted connections are far less susceptible to brittle fractures.

15.2   TYPES OF BEAM CONNECTIONS

All connections have some restraint—that is, some resistance to changes of the original angles between intersecting members when loads are applied. Depending on the amount of restraint, the AISC Specification (B3.6) classifies connections as being fully restrained (Type FR) and partially restrained (Type PR). These two types of connections are described in more detail as follows:

1. Type FR connections are commonly referred to as rigid or continuous frame connections. They are assumed to be sufficiently rigid or restrained to keep the original angles between members virtually unchanged under load.

2. Type PR connections are those that have insufficient rigidity to keep the original angles virtually unchanged under load. Included in this classification are simple and semirigid connections, as described in detail in this section.

A *simple connection* is a Type PR connection for which restraint is ignored. It is assumed to be completely flexible and free to rotate, thus having no moment resistance. A *semirigid connection*, or *flexible moment connection*, is also a Type PR connection whose resistance to angle change falls somewhere between the simple and rigid types.

As there are no perfectly rigid connections nor completely flexible ones, all connections really are partly restrained, or PR, to one degree or another. The usual practice in the past was to classify connections based on a ratio of the moment developed for a particular connection to the moment that would theoretically be developed by a completely rigid connection. A rough rule was that simple connections had 0–20 percent rigidity, semirigid connections had 20–90 percent rigidity, and rigid connections had 90–100 percent rigidity. Figure 15.1 shows a set of typical moment-rotation curves for these connections. Notice that the lines are curved because as the moments become larger, the rotations increase at a faster rate.

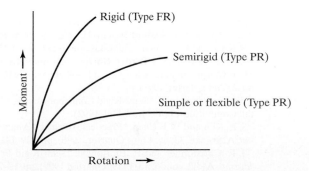

FIGURE 15.1

Typical moment-rotation curves for connections.

During the past few years, quite a few investigators around the world have been trying to develop empirical formulas for describing the rotational characteristics of connections.[1-6] Though they have made some progress, the only accurate method for developing such information today involves the actual fabrication of connections, followed by load testing. It is very difficult to include in a formula the effects of such things as poor fit, improper tightening of bolts, and so on.

Each of these three general types of connections is briefly discussed in this section, with little mention of the specific types of connectors used. The remainder of the chapter is concerned with detailed designs of these connections, using specific types of fasteners. *In this discussion, the author probably overemphasizes the semirigid and rigid type connections, because a very large percentage of the building designs with which the average designer works will be assumed to have simple connections.* A few descriptive comments are given in the paragraphs that follow concerning each of these three types of connections.

Simple connections (Type PR) are quite flexible and are assumed to allow the beam ends to be substantially free to rotate downward under load, as true simple beams should. Although simple connections do have some moment resistance (or resistance to end rotation), it is taken to be negligible, and they are assumed to be able to resist shear only. Several types of simple connections are shown in Fig. 15.2. More detailed descriptions of each of these connections and their assumed behavior under load are given in later sections of this chapter. In this figure, most of the connections are shown as being made entirely with the same type of fastener—that is, all bolted or all welded—while in actual practice two types of fasteners are often used for the same connection. For example, a very common practice is to shop-weld the web angles to the beam web and field-bolt them to the column or girder.

Semirigid connections or flexible moment connections (Type PR) are those that have appreciable resistance to end rotation, thus developing appreciable end moments. In design practice, it is quite common for the designer to assume that all connections are either simple or rigid, with no consideration given to those situations in between, thereby simplifying the analysis. Should he or she make such an assumption for a true semirigid connection, he or she may miss an opportunity for appreciable moment reductions. To understand this possibility, the reader is referred to the moment diagrams shown in Fig. 15.3 for a group of uniformly loaded beams supported with connections having different percentages of rigidity. This figure shows that the maximum moments in a beam vary greatly with different types of end connections. For example, the maximum moment in the semirigid connection of part (d) of the figure is only 50 percent of the

[1]R. M. Richard, "A Study of Systems Having Conservative Non-Linearity," Ph.D. Thesis, Purdue University, 1961.

[2]N. Kishi and W. F. Chen, "Data Base of Steel Beam-to-Column Connections," no. CE-STR-86-26, Vols. I and II (West Lafayette, IN: Purdue University, School of Engineering, July 1986).

[3]L. F. Geschwindner, "A Simplified Look at Partially Restrained Beams," *Engineering Journal*, AISC, vol. 28, no. 2 (2nd Quarter, 1991), pp. 73–78.

[4]Wai-Fah Chen et al., "Semi-Rigid Connections in Steel Frames," Council on Tall Buildings and Urban Habitat, Committee 43 (McGraw Hill, 1992).

[5]S. E. Kim and W. F. Chen, "Practical Advanced Analysis for Semi-Rigid Frame Design," *Engineering Journal*, AISC, vol. 33, no. 4 (4th Quarter, 1996), pp. 129–141.

[6]J. E. Christopher and R. Bjorhovde, "Semi-Rigid Frame Design for Practicing Engineers," *Engineering Journal*, AISC, vol. 36, no. 1 (1st Quarter, 1999), pp. 12–28.

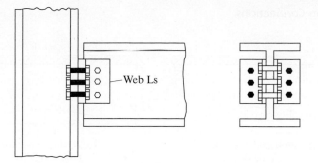

(a) Framed simple connection

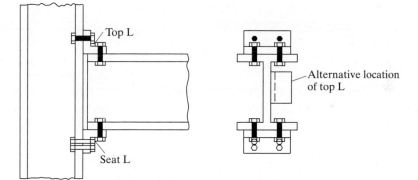

(b) Seated simple connection

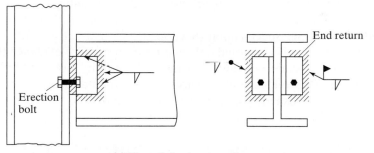

(c) Framed simple connection

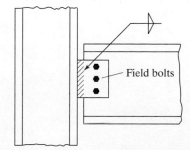

(d) Single-plate or shear tab simple connection

FIGURE 15.2

Some simple connections. Notice how these connections are placed up towards the top flanges so that they provide lateral stability at the compression flanges at the beam supports. (a) Framed simple connection. (b) Seated simple connection. (c) Framed simple connection. (d) Single-plate or shear tab simple connection.

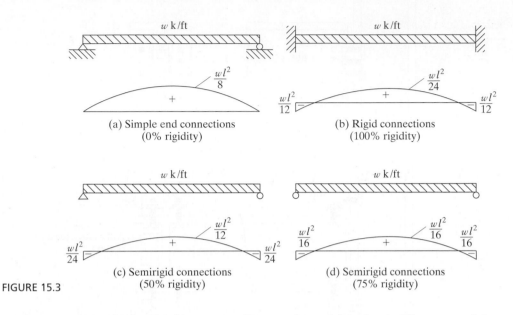

w k/ft

$\dfrac{wl^2}{8}$

$+$

(a) Simple end connections
(0% rigidity)

w k/ft

$\dfrac{wl^2}{24}$

$+$

$\dfrac{wl^2}{12}$ $\dfrac{wl^2}{12}$

(b) Rigid connections
(100% rigidity)

w k/ft

$\dfrac{wl^2}{12}$

$+$

$\dfrac{wl^2}{24}$ $\dfrac{wl^2}{24}$

(c) Semirigid connections
(50% rigidity)

w k/ft

$\dfrac{wl^2}{16}$

$+$

$\dfrac{wl^2}{16}$ $\dfrac{wl^2}{16}$

(d) Semirigid connections
(75% rigidity)

FIGURE 15.3

maximum moment in the simply supported beam of part (a) and only 75 percent of the maximum moment in the rigidly supported beam of part (b).

Actual semirigid connections are used fairly often, but usually no advantage is taken of their moment-reducing possibilities in the calculations. Perhaps one factor that keeps the design professional from taking advantage of them more often is the statement of the AISC Specification (Section B3.6b) that consideration of a connection as being semirigid is permitted only upon presentation of evidence that it is capable of providing a certain percentage of the end restraint furnished by a completely rigid connection. This evidence must consist of documentation in the technical literature, or must be established by analytical or empirical means.

A typical framing angle shop-welded to a beam. It will be field-bolted to another member. (Courtesy of CMC South Carolina Steel.)

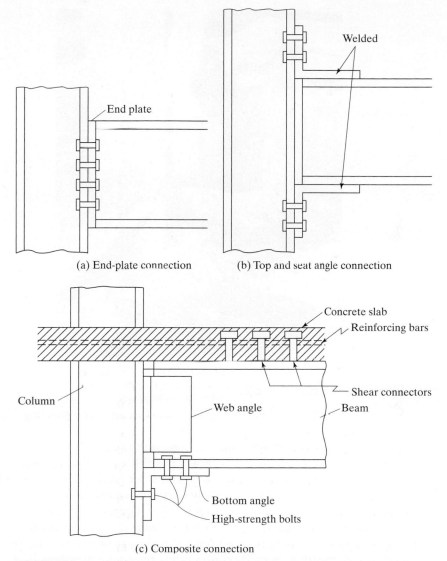

(a) End-plate connection   (b) Top and seat angle connection

(c) Composite connection

FIGURE 15.4

Some semirigid or flexible moment connections. (Additional types are shown in Part 11 of the AISC Manual.)

Three practical semirigid, or PR, connections capable of providing considerable moment resistance are shown in Fig. 15.4. If the end-plate connection shown in part (a) of the figure is extended above the beam and more bolts are installed, the moment resistance of the connection can be appreciably increased. Part (c) of the figure shows a semirigid connection that is proving to be quite satisfactory for steel-concrete composite floors. Moment resistance in this connection is provided by reinforcing bars placed in the concrete slab above the beam and by the horizontal leg of the seat angle.[7]

---

[7]D. J. Ammerman and R.T. Leon, "Unbraced Frames with Semirigid Composite Connections," *Engineering Journal*, AISC, vol. 27, no. 1 (1st Quarter 1990), pp. 12–21.

A semirigid beam-to-column connection, Ainsley Building, Miami, FL. (Courtesy of the Lincoln Electric Company.)

Another type of semirigid connection is illustrated in an accompanying photograph from the Lincoln Electric Company.

The use of partially restrained connections with roughly 60 to 75 percent rigidities is gradually increasing. When it becomes possible to accurately predict the percentages of rigidity for various connections, and when better design procedures are available, this type of design will probably become even more common.

Rigid connections (Type FR) are those which theoretically allow no rotation at the beam ends and thus transfer close to 100 percent of the moment of a fixed end. Connections of this type may be used for tall buildings in which wind resistance is developed. The connections provide continuity between the members of the building frame. Several Type FR connections that provide almost 100 percent restraint are shown in Fig. 15.5. It will be noticed in the figure that column web stiffeners may be required for some of these connections to provide sufficient resistance to rotation. The design of these stiffeners is discussed in Section 15.12.[8]

The moment connection shown in part (d) of Fig. 15.5 is rather popular with steel fabricators, and the end-plate connection of part (e) also has been frequently used in recent years.[9]

---

[8]J. D. Griffiths "End-Plate Moment Connections—Their Use and Misuse," *Engineering Journal*, AISC, vol. 21, no. 1 (1st Quarter, 1984), pp. 32–34.
[9]AISC, "Seismic Provisions for Structural Steel Buildings," ANSI/AISC 341–05, American Institute of Steel Construction, Inc. (Chicago, IL).

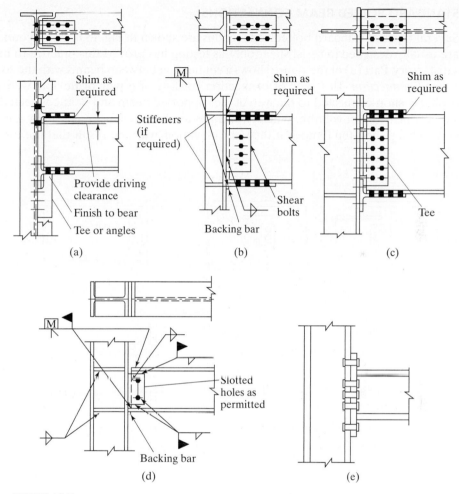

FIGURE 15.5

Moment-resisting connections.

You will note the use of *shims* in parts (a) to (c). Shims are thin strips of steel that are used to adjust the fit at connections. They can be one of two types: conventional shims or finger shims. *Conventional shims* are installed with the bolts passing through them, while *finger shims* can be installed after the bolts are in place. You should realize that there is some variation in the depths of beams as they come from the steel mills. (See Table 1-22 in Part 1 of the AISC Manual for permissible tolerances.) To provide for such variations, it is common to make the distance between flange plates or angles larger than the nominal beam depths given in the Manual.[10]

---

[10]W. T. Segui, *Fundamentals of Structural Steel Design*, 4th ed. (Boston: PWS-Kent, 2007), p. 487.

15.3   STANDARD BOLTED BEAM CONNECTIONS

Several types of standard bolted connections are shown in Fig. 15.6. These connections are usually designed to resist shear only, as testing has proved this practice to be quite satisfactory. Part (a) of the figure shows a connection between beams with the so-called *framed connection*. This type of connection consists of a pair of flexible web angles, probably shop-connected to the web of the supported beam and field-connected to the supporting beam or column. When two beams are being connected, usually it is necessary to keep their top flanges at the same elevation, with the result that the top flange

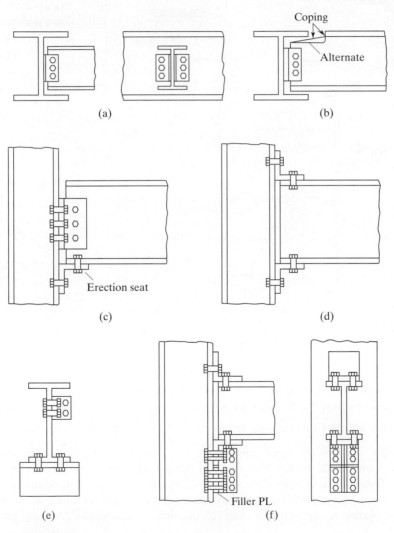

FIGURE 15.6

(a) Framed connection. (b) Framed connection. (c) Framed connection. (d) Seated connection. (e) Seated connection. (f) Seated connection with stiffener angles.

of one will have to be cut back (called *coping*), as shown in part (b) of the figure. For such connections, we must check block shear, as discussed in Section 3.7 of this text. Coping is an expensive process and should be avoided where possible.

Simple connections of beams to columns can be either framed or seated, as shown in Fig. 15.6. In part (c) of the figure, a framed connection is shown in which two web angles are connected to the beam web in the shop, after which bolts are placed through the angles and column in the field. It is often convenient to use an angle called an *erection seat* to support the beam during erection. Such an angle is shown in the figure.

The seated connection has an angle under the beam similar to the erection seat just mentioned, which is shop-connected to the column. In addition, there is another angle — probably on top of the beam — that is field-connected to the beam and column. A seated connection of this type is shown in part (d) of the figure. Should space prove to be limited above the beam, the top angle may be placed in the optional location shown in part (e) of the figure. The top angle at either of the locations mentioned is very helpful in keeping the top flange of the beam from being accidentally twisted out of place during construction.

The amount of load that can be supported by the types of connections shown in parts (c), (d), and (e) of Fig. 15.6 is severely limited by the flexibility or bending strength of the horizontal legs of the seat angles. For heavier loads, it is necessary to use stiffened seats, such as the one shown in part (f) of the figure.

The designer selects most of these connections by referring to standard tables. The AISC Manual has excellent tables for selecting bolted or welded beam connections of the types shown in Fig. 15.6. After a rolled-beam section has been selected, it is quite convenient for the designer to refer to these tables and select one of the standard connections, which will be suitable for the vast majority of cases.

In order to make these standard connections have as little moment resistance as possible, the angles used in making up the connections are usually light and flexible. To qualify as simple end supports, the ends of the beams should be as free as possible to rotate downward. Figure 15.7 shows the manner in which framed and seated end connections will theoretically deform as the ends of the beams rotate downward. The designer does not want to do anything that will hamper these deformations if he or she is striving for simple supports.

For the rotations shown in Fig. 15.7 to occur, there must be some deformation of the angles. As a matter of fact, if end slopes of the magnitudes that are computed for simple ends are to occur, the angles will actually bend enough to be stressed beyond their yield points. If this situation occurs, they will be permanently bent and the connections will quite closely approach true simple ends. The student should now see why it is desirable to use rather thin angles and large gages for the bolt spacing if flexible simple end connections are the goal of the designer.

These connections do have some resistance to moment. When the ends of the beam begin to rotate downward, the rotation is certainly resisted to some extent by the tension in the top bolts, even if the angles are quite thin and flexible. Neglecting the moment resistance of these connections will cause conservative beam sizes. If moments of any significance are to be resisted, more rigid-type joints need to be provided than are available with the framed and seated connections.

The beam on the right has a coped top flange, while the beam on the left has both flanges coped because its depth is very close to that of the supporting girder. (Courtesy of CMC South Carolina Steel.)

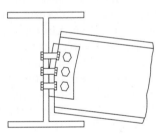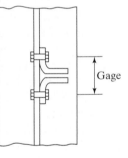

(a)

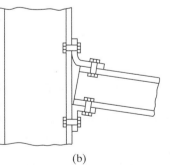

(b)

FIGURE 15.7

(a) Bending of framed-beam connection.
(b) Bending of seated-beam connection.

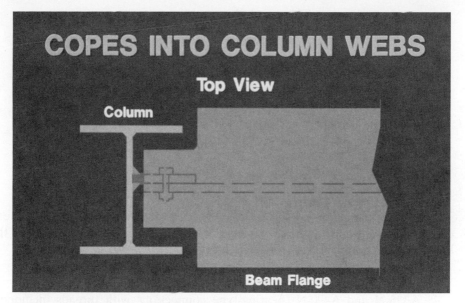

Top view showing a beam framing into a column web where the beam flange is wider than the opening between the column flanges. Thus, both top and bottom flanges are coped by flame-cutting.

15.4   AISC MANUAL STANDARD CONNECTION TABLES

In Part 10 of the AISC Manual, a series of tables is presented that the designer may use to select several different types of standard connections. There are tables for bolted or welded two-angle framed connections, seated beam connections, stiffened seated beam connections, eccentrically loaded connections, single-angle framed connections, and others.

In the next few sections of this chapter (15.5 through 15.8), a few standard connections are selected from the tables in the Manual. The author hopes that these examples will be sufficient to introduce the reader to the AISC tables and to enable him or her to make designs by using the other tables with little difficulty.

Sections 15.9 to 15.11 present design information for some other types of connections.

15.5   DESIGNS OF STANDARD BOLTED FRAMED CONNECTIONS

For small and low-rise buildings (and that means most buildings), simple framed connections of the types previously shown in parts (a) and (b) of Fig. 15.6 are usually used to connect beams to girders or to columns. The angles used are rather thin (1/2 in is the arbitrary maximum thickness used in the AISC Manual), so they will have the necessary flexibility shown in Fig. 15.7. The angles will develop some small moments (supposedly, not more than 20 percent of full fixed-end conditions), but they are neglected in design.

The framing angles extend out from the beam web by 1/2 in, as shown in Fig. 15.8. This protrusion, which is often referred to as the *setback*, is quite useful in fitting members together during steel erection.

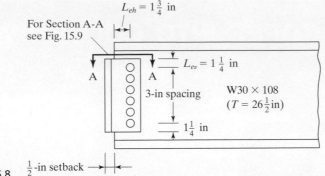

For Section A-A
see Fig. 15.9

$L_{eh} = 1\frac{3}{4}$ in

$L_{ev} = 1\frac{1}{4}$ in

A

A

3-in spacing

W30 × 108

$(T = 26\frac{1}{2}$ in$)$

$1\frac{1}{4}$ in

FIGURE 15.8 $\frac{1}{2}$-in setback

In this section, several standard bolted framed connections for simple beams are designed, with use of the tables provided in Part 10 of the AISC Manual. In these tables, the following abbreviations for different bolt conditions are used:

1. A325-SC and A490-SC (slip-critical connections)
2. A325-N and A490-N (bearing-type connections with threads included in the shear planes)
3. A325-X and A490-X (bearing-type connections with threads excluded from shear planes)

It is thought that the minimum depth of framing angles should be at least equal to one-half of the distance between the web toes of the beam fillets (called the T distances and given in the properties tables of Part 1 of the Manual). This minimum depth is used so as to provide sufficient stability during steel erection.

Example 15-1 presents the design of standard framing angles for a simply supported beam using bearing-type bolts in standard-size holes. In this example, the design strengths of the bolts and the angles are taken from the appropriate tables.

Example 15-1

Select an all-bolted double-angle framed simple end connection for the uncoped W30 × 108 ($t_w = 0.545$ in) shown in Fig. 15.8 if $R_D = 50$ k and $R_L = 70$ k and if the connection frames into the flange of a W14 × 61 column ($t_f = 0.645$ in). Assume that $F_y = 36$ ksi and $F_u = 58$ ksi for the framing angles and 50 ksi and 65 ksi, respectively, for the beam and column. Use 3/4-in A325-N bolts (bearing-type threads included in shear plane) in standard-size holes.

Solution

LRFD	ASD
$R_u = (1.2)(50) + (1.6)(70) = 172$ k	$R_a = 50 + 70 = 120$ k

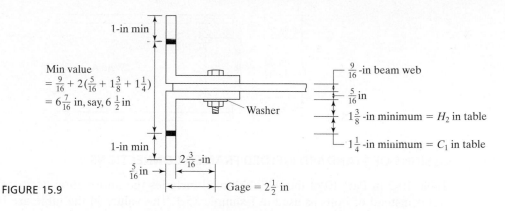

FIGURE 15.9

Using a double-angle connection with $F_y = 36$ ksi and $F_u = 58$ ksi, we look in the tables for the least number of rows of bolts that can be used with a W30 section. It's 5 rows (page 10-20, Part 10 of the AISC Manual), but the connection for 3/4-in A325-N bolts will not support a reaction that large. Thus, we move to a 6-row connection (on page 10-19 in the Manual) and try a connection with an angle thickness of 5/16 in and a length $L = 15 + (2)\left(1\frac{1}{4}\right) = 17.5$ in. (See Fig. 15.8.) This length seems satisfactory compared with the T of 26 1/2 in for this shape that is given in the tables of Part 1 of the Manual.

LRFD	ASD
$\phi R_n = 187$ k > 172 k **OK**	$\dfrac{R_n}{\Omega} = 124$ k > 120 k **OK**

To select the lengths of the angle legs, it is necessary to study the dimensions given in Fig. 15.9, which is a view along Section A-A in Fig. 15.8. The lower right part of the figure shows the minimum clearances needed for insertion and tightening of the bolts. These are the H_2 and C_1 distances and are obtained for 3/4-in bolts from Table 7.16 in Part 7 of the AISC Manual.

For the legs bolted to the beam web, a $2\frac{1}{2}$-in gage is used. Using a minimum edge distance of 1 in here, we will make these angle legs $3\frac{1}{2}$ in. For the outstanding legs, the minimum gage is $\frac{5}{16} + 1\frac{3}{8} + 1\frac{1}{4} = 2\frac{15}{16}$ in—say, 3 in. We will make this a 4-in angle leg.

Use 2Ls $4 \times 3\frac{1}{2} \times \frac{5}{16} \times 1$ **ft** $- 5\frac{1}{2}$**-in A36.**

From the tables of Part 10 of the AISC Manual, framed bolted connections can be easily selected where one or both flanges of a beam are coped. Coping may substantially reduce the design strength of beams and require the use of larger members or the addition of web reinforcing. Elastic section moduli values for coped sections are provided in Table 9.2 in the Manual.

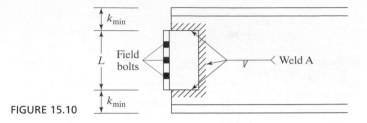

FIGURE 15.10

15.6   DESIGNS OF STANDARD WELDED FRAMED CONNECTIONS

Table 10-2 in Part 10 of the AISC Manual includes the information necessary to use welds instead of bolts as used in Example 15-1. The values in the table are based on E70 electrodes. The table is normally used where the angles are welded to the beams in the shop and then field-bolted to the other member. Should the framing angles be welded to both members, the weld values provided in Table 10-3 in Part 10 of the AISC Manual would be applied.

In the tables, the weld used to connect the angles to the beam web is called Weld A, as shown in Fig. 15.10. If a weld is used to connect the beam to another member, that weld is called Weld B.

For the usual situations, $4 \times 3\frac{1}{2}$-in angles are used with the $3\frac{1}{2}$-in legs connected to the beam webs. The 4-in outstanding legs will usually accommodate the standard gages for the bolts going into the other members. The angle thickness selected equals the weld size plus 1/16 in or the minimum value given in Table 10-1 for the bolts. The angle lengths are the same as those used for the nonstaggered bolt cases (that is, $5\frac{1}{2}$ through $35\frac{1}{2}$ in).

The design strengths of the welds to the beam webs (Weld A) given in Table 10-2 in Part 10 of the AISC Manual were computed by the instantaneous center of rotation method, which we briefly described in Chapter 14. To select a connection of this type, the designer picks a weld size from Table 10-2 and then goes to Table 10-1 to determine the number of bolts required for connection to the other member. This procedure is illustrated in Example 15-2.

Example 15-2

Design a framed beam connection to be welded (SMAW) to a W30 $\times$ 90 beam ($t_w = 0.470$ in and $T = 26\frac{1}{2}$ in) and then bolted to another member. The value of the reaction R_D is 75 k, while R_L is 70 k. The steel for the angles is A36, the weld is E70, and the bolts are $\frac{3}{4}$-in A325-N.

Solution. Table 10-2, Weld A Design Strength, AISC Manual.

LRFD	ASD
$R_u = (1.2)(75) + (1.6)(70) = 202$ k	$R_a = 75 + 70 = 145$ k

From this table (10-2), one of several possibilities is a 3/16-in weld $20\frac{1}{2}$ in long with an LRFD design strength of $228\text{ k} > 202\text{ k}$ and an ASD allowable strength of $152\text{ k} > 145\text{ k}$. The depth or length is compatible with the T value of $26\frac{1}{2}$ in. For a 3/16-in weld, the minimum angle thickness is 1/4 in, as per AISC Specification J2.2b. The minimum web thickness for shear given in Table 10-2 is 0.286 in for $F_y = 36$ ksi, and this is less than the 0.470 in furnished.

Table 10-1, "Selection of Bolts," AISC Manual.

The angle length selected is $20\frac{1}{2}$ in, and this corresponds to a 7-row bolted connection in Table 10-1. From this table, we find that 7 rows of 3/4-in A325-N bolts have an LRFD design strength of 174 k if the angle thickness is 1/4-in but R_u is 208 k. If we use a 5/16-in thickness, ϕR_n is 217 k; and R_n/Ω is 145 k which is equal to $R_a = 145$ k

Use 2Ls $4 \times 3\frac{1}{2} \times \frac{5}{16} \times 1$ ft $8\frac{1}{2}$ in **A36 steel for LRFD and ASD.**

Table 10-3 in Part 10 of the AISC Manual provides the information necessary for designing all-welded standard framed connections—that is, with Welds A and B as shown in Fig. 15.11. The design strengths for Weld A were determined by the ultimate strength, instantaneous center, or rotation method, while the values for Weld B were determined by the elastic method. A sample design is presented in Example 15-3.

Example 15-3

Select an all-welded double-angle framed simple end connection for fastening a W30 × 108 beam ($t_w = 0.545$ in) to a W14 × 61 column ($t_f = 0.645$ in). Assume that $F_y = 50$ ksi and $F_u = 65$ ksi for the members, with $R_D = 60$ k and $R_L = 80$ k.

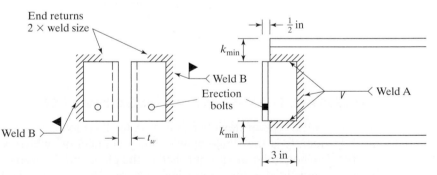

FIGURE 15.11

Solution

LRFD	ASD
$R_u = (1.2)(60) + (1.6)(80) = 200$ k	$R_a = 60 + 80 = 140$ k

Select Weld A for beam

From Table 10-3 in the AISC Manual, one possibility is a 3/16-in weld 20 in long. For a beam with $F_y = 50$ ksi, the minimum web thickness is 0.286 in, which is $< t_w$ of 0.545 in of the beam.

LRFD	ASD
$\phi R_n = 223$ k > 200 k OK	$\dfrac{R_n}{\Omega} = 149$ k > 140 k OK

Select Weld B for column

The 1/4-in weld will not provide sufficient capacity. $\therefore$ Use 5/16-in weld. For a column with $F_y = 50$ ksi, the minimum flange t is 0.238 in $< t_f$ of 0.545 in of the column.

LRFD	ASD
$\phi R_n = 226$ k > 200 k **OK**	$\dfrac{R_n}{\Omega} = 151$ k > 140 k **OK**

$$\text{Minimum angle thickness} = \frac{5}{16} + \frac{1}{16} = \frac{3}{8} \text{ in.}$$

On page 10-12 in the Manual, the AISC says to use 4×3 angles when angle length ≥ 18 in. Use two angles 3×3 otherwise.

Use 2 Ls $4 \times 3 \times \dfrac{3}{8} \times 1$ ft 8 in **A36**.

15.7 SINGLE-PLATE, OR SHEAR TAB, FRAMING CONNECTIONS

An economical type of flexible connection for light loads that is being used more and more frequently is the single-plate framing connection, which was illustrated in Fig. 15.2(d). The bolt holes are prepunched in the plate and the web of the beam. The plate is then shop-welded to the supporting beam or column; lastly, the beam is bolted to the plate in the field. Steel erectors like this kind of connection because of its simplicity. They are particularly pleased with it when there is a beam connected to each side of a girder, as shown in Fig. 15.12(a). All they have to do is bolt the beam webs to the single plate on each side of the girder. Should web clip angles be used for such a connection, the bolts will have to pass through the angles on each side of the girder, as well as the girder web, as shown in part (b) of the figure. This is a slightly more difficult field

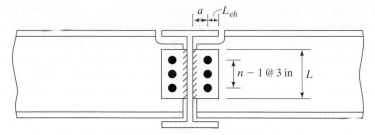

(a) Single-plate framing connection

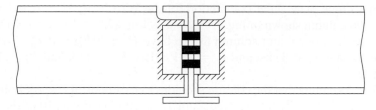

(b) Simple connection with web angles

FIGURE 15.12

(a) Single-plate framing connection. (b) Simple connection with web angles.

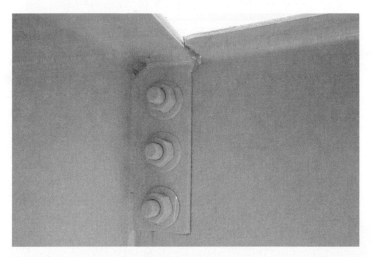

A shear tab, or single-plate, framing connection. (Picture taken by the author.)

operation. Frequently, one side will have one extra bolt row so that each beam can be erected separately.

With the single-plate connection, the reaction or shear load is assumed to be distributed equally among the bolts passing through the beam web. It is also assumed that relatively free rotation occurs between the end of the member and the supporting beam or column. Because of these assumptions, this type of connection often is referred to as a "shear tab" connection. Various studies and tests have shown that these connections can develop some end moments, depending on the number and size of the bolts and their arrangement, the thicknesses of the plate and beam web, the span-to-depth ratio of the beam, the type of loading, and the flexibility of the supporting element.

Table 10-9 for shear tab connections is presented in the AISC Manual. An empirical design procedure that uses service loads is presented by R. M. Richard et al.[11]

Example 15-4

Design a single-plate shear connection for the W16 × 50 (t_w = 0.380 in) beam to the W14 × 90 column shown in Fig. 15.13. Use 3/4-in A325-N high-strength bolts and E70 electrodes. The beam and column are to have F_y = 50 ksi and F_u = 65 ksi, while the plate is to have F_y = 36 ksi and F_u = 58 ksi. Use R_D = 15 k and R_L = 20 k.

Solution. Assuming that the column provides rigid support, 4 rows of bolts, a 1/4-in plate, and 3/16-in fillet welds are selected from Table 10-9(a) in Part 10 of the AISC Manual.

LRFD	ASD
$R_U = (1.2)(15) + (1.6)(20) = 50$ k	$R_a = 15 + 20 = 35$ k

LRFD	ASD
$\phi R_n = 52.2$ k > 50 k **OK**	$\dfrac{R_n}{\Omega} = 34.8$ k ≈ 35 k **OK**

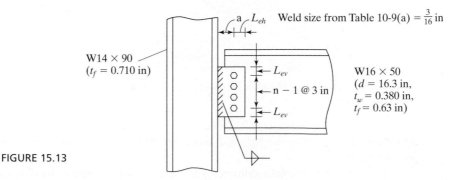

FIGURE 15.13

---

[11]R. M. Richard et al., "The Analysis and Design of Single-Plate Framing Connections," *Engineering Journal*, AISC, vol. 17, no. 2 (2nd Quarter, 1980), pp. 38–52.

Checking supported beam web

From AISC Table 10-9(a) for 4 rows of bolts with $L_{ev} = 1\frac{1}{4}$ in, and $L_{eh} = 1\frac{3}{4}$ in, and an uncoped beam. For beam web design strength, reference is made to AISC Table 10-1 for a 4-row connection and an uncoped beam.

LRFD	ASD
$\phi R_n = (0.380)(351) = 133.4 \text{ k} > 50 \text{ k}$ **OK**	$\dfrac{R_n}{\Omega} = (0.380)(234) = 88.9 \text{ k} > 35 \text{ k}$ **OK**

Use PL $\frac{1}{4} \times 5 \times$ 0 ft $11\frac{1}{2}$ in A36 steel with 4 rows of 3/4-in A325 N bolts and 3/16-in E70 fillet welds. The 5 in dimension was determined from page 10-102 in the Manual.

"a", see Fig. 15.13, must be less than or equal to $3\frac{1}{2}$ in. If L_{eh} is $1\frac{3}{4}$ in, then maximum plate width is $5\frac{1}{4}$ in.

Should the beam be coped, it will be necessary to also check flexural yielding and local web buckling at the cope.

15.8   END-PLATE SHEAR CONNECTIONS

Another type of connection is the *end-plate* connection. It consists of a plate shop-welded flush against the end of a beam and field-bolted to a column or another beam. To use this type of connection, it is necessary to carefully control the length of the beam and the squaring of its ends, so that the end plates are vertical. Camber must also be considered as to its effect on the position of the end plate. After a little practice in erecting members with end-plate connections, fabricators seem to like to use them. Nonetheless, there still is trouble in getting the dimensions just right, and end-plate connectors are not as commonly used as the single-plate connectors.

Part (a) of Fig. 15.14 shows an end-plate connection that is satisfactory for PR situations. End-plate connections are illustrated in Fig. 12-6 of the AISC Manual. Should the end plate be extended above and below the beam, as shown in part (b) of Fig. 15.14, appreciable moment resistance will be achieved.

Table 10-4 of the AISC Manual provides tables and a procedure for designing extended end-plate connections. These connections may be designed as FR, for statically loaded structures, for buildings in areas of low seismicity. Their design is described in Part 12 of the Manual.

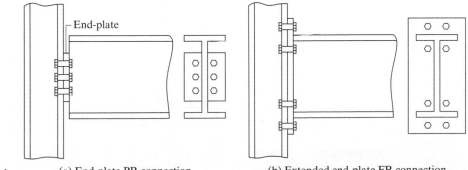

FIGURE 15.14   (a) End-plate PR connection   (b) Extended end-plate FR connection

15.9 DESIGNS OF WELDED SEATED BEAM CONNECTIONS

Another type of fairly flexible beam connection is achieved by the use of a beam seat, such as the one shown in Fig. 15.15. Beam seats obviously offer an advantage to the worker performing the steel erection. The connections for these angles may be bolts or welds, but only the welded type is considered here. For such a situation, the seat angle would usually be shop-welded to the column and field-welded to the beam. When welds are used, the seat angles, also called *shelf angles*, may be punched for erection bolts, as shown in the figure. These holes can be slotted, if desired, to permit easy alignment of the members.

A seated connection may be used only when a top angle is used, as shown in Fig. 15.15. This angle provides lateral support for the beam and may be placed on top of the beam, or at the optional location shown on the side of the beam in part (a) of the figure. As the top angle is not usually assumed to resist any of the load, its size probably is selected by judgment. Fairly flexible angles are used that will bend away from the column or girder to which they are connected when the beam tends to rotate downward

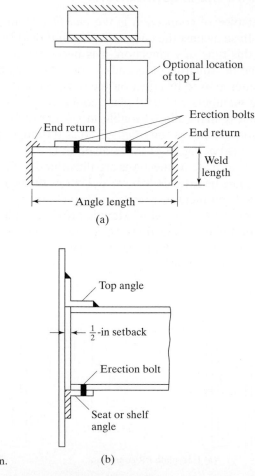

(a)

(b)

FIGURE 15.15

Seated beam connection.

under load. This desired situation is illustrated in Fig. 15.7(b). A common top angle size selected is 4 × 4 × 1/4.

As will be seen in AISC Tables 10-5 through 10-8, unstiffened seated beam connections of practical sizes can support only fairly light factored loads. For such light loads, two vertical end welds on the seat are sufficient. The top angle is welded only on its toes, so when the beam tends to rotate, this thin flexible angle will be free to pull away from the column.

The seat design strengths given in the AISC tables were developed for seat angles with either 3 1/2- or 4-in outstanding legs. The steel used for the angles is A36 with F_y = 36 ksi and F_u = 58 ksi.

The design strengths in the tables were obtained with consideration given to both shear and flexural yielding of the outstanding leg of the seat angle, and crippling of the beam web as well. The values were calculated on the basis of a 3/4-in setback rather than the nominal 1/2 in used for web framing angles. This larger value was used to provide for possible mill underrun in the beam lengths. Example 15-5 illustrates the use of the Manual tables for designing a welded unstiffened seated beam connection. Other tables are included in the Manual for bolted seated connections, as well as bolted or welded stiffened seated connections.

Example 15-5

Design an all-welded unstiffened seated connection with E70 electrodes to support the reactions $R_D = 20$ k and $R_L = 30$ k from a W24 × 55 beam (d = 23.6 in, t_w = 0.395 in, t_f = 0.505 in, and k = 1.01 in). The connection is to be made to the flange of a W14 × 68 column (t_f = 0.720 in). The angles are A36, while the beam and column have an F_y = 50 ksi and an F_u = 65 ksi.

Solution. Design seat angle and welds.

LRFD	ASD
$R_u = (1.2)(20) + (1.6)(30) = 72$ k	$R_a = 20 + 30 = 50$ k

Checking local web yielding, assuming that N $= 3\frac{1}{2}$ in for outstanding leg

Using AISC Equation J10-2 and AISC Table 9-4

LRFD	ASD
$\phi R_1 = 49.6$ and $\phi R_2 = 19.8$	$\dfrac{R_1}{\Omega} = 33.1$ and $\dfrac{R_2}{\Omega} = 13.2$
$\phi R_n = \phi R_1 + N\,(\phi R_2)$	$\dfrac{R_n}{\Omega} = \dfrac{R_1}{\Omega} + N\!\left(\dfrac{R_2}{\Omega}\right)$
$72 = 49.6 + (N)(19.8)$	$50 = 33.1 + N(13.2)$
$N_{req'd} = 1.13$ in	$N_{req'd} = 1.28$ in

Checking web crippling

LRFD	ASD
$\phi R_3 = 63.7$ and $\phi R_4 = 5.61$	$\dfrac{R_3}{\Omega} = 42.5$ and $\dfrac{R_4}{\Omega} = 3.74$
$\dfrac{N}{d} = \dfrac{3.5}{23.6} = 0.148 < 0.2$	$\dfrac{N}{d} = \dfrac{3.5}{23.6} = 0.148 < 0.2$
$\therefore$ Use AISC Equation J10-5a	$\therefore$ Use AISC Equation J10-5a
$\phi R_n = \phi R_3 + N(\phi R_4)$	$\dfrac{R_n}{\Omega} = \dfrac{R_3}{\Omega} + N\left(\dfrac{R_4}{\Omega}\right)$
$72 = 63.7 + N(5.61)$	$50 = 42.5 + (N)(3.74)$
$N_{req'd} = 1.48$ in	$N_{req'd} = 2.00$ in

Use 4-in angle leg with 1/2 in nominal setback. $\therefore$ Bearing length, $N = 3.5$ in. Using AISC Table 10-6 for an 8×4 angle, determine t

LRFD	ASD
With $N_{req} = 1.48$ in, say $1\frac{1}{2}$ in	With $N_{req} = 2.00$ in
From Table 10.6 upper portion, an 8 in angle length with 3/4-in thickness will provide:	From Table 10.6 upper portion, an 8 in angle length with 3/4-in thickness will provide:
$\phi R_n = 97.2$ k > 72 k **OK**	$\phi R_n = 38.8$ k > 50 k **N.G.** Increase angle thickness to $\frac{7}{8}$-in.
From Table 10.6 lower portion, an 8×4 angle with a 3/8-in weld will provide:	From Table 10.6 lower portion, an 8×4 angle with a 3/8-in weld will provide:
$\phi R_n = 80.1$ k > 72 k **OK**	$\phi R_n = 53.4$ k > 50 k **OK**
Use L8 $\times$ 4 $\times$ 3/4 $\times$ 0 ft 8 in with 3/8-in weld.	**Use L8 $\times$ 4 $\times$ 7/8 $\times$ 0 ft 8 in with 3/8-in weld.**

15.10   DESIGNS OF STIFFENED SEATED BEAM CONNECTIONS

When beams are supported by seated connections and when the factored reactions become fairly large, it is necessary to stiffen the seats. These larger reactions cause moments in the outstanding or horizontal legs of the seat angles that cannot be supported by standard thickness angles unless they are stiffened in some manner. Typical stiffened seated connections are shown in Fig. 15.6(f) and in Fig. 15.16.

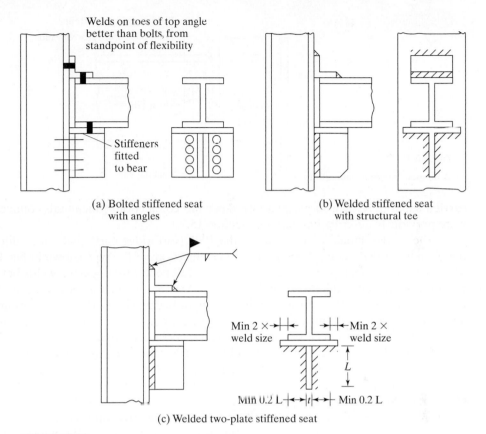

Welds on toes of top angle
better than bolts, from
standpoint of flexibility

Stiffeners
fitted
to bear

(a) Bolted stiffened seat
with angles

(b) Welded stiffened seat
with structural tee

Min 2 × weld size

Min 2 × weld size

L

Min 0.2 L   t   Min 0.2 L

(c) Welded two-plate stiffened seat

FIGURE 15.16

(a) Bolted stiffened seat with angles. (b) Welded stiffened seat with structural tee. (c) Welded two-plate stiffened seat.

Stiffened seats may be bolted or welded. Bolted seats may be stiffened with a pair of angles, as shown in part (a) of Fig. 15.16. Structural tee stiffeners either bolted or welded may be used. A welded one is shown in part (b) of the same figure. Welded two-plate stiffeners, such as the one shown in part (c) of the figure, also are commonly used. AISC Table 10-8 provides information for designing stiffened seated connections.

15.11   DESIGNS OF MOMENT-RESISTING FR MOMENT CONNECTIONS

In this section, a brief introduction to moment-resisting connections is presented. It is not the author's intention to describe in detail all of the possible arrangements of bolted and welded moment-resisting connections at this time, nor to provide a complete design. Instead, he attempts to provide the basic theory of transferring shear and moment from a beam to another member. One numerical example is included. This theory is very easy to understand and should enable the reader to design other moment-resisting connections—regardless of their configurations.

One moment-resisting connection that is popular with many fabricators is shown in Fig. 15.17. There, the flanges are groove-welded to the column, while the shear is

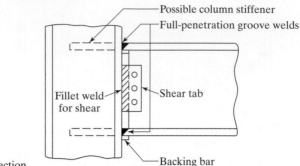

FIGURE 15.17

A moment-resisting connection.

carried separately by a single plate or shear tab connection. (Shear tab connections were previously described in detail in Section 15.7.)

The reader should realize that, at the 1994 Northridge earthquake in California, quite a few brittle fractures were initiated in connections of the type shown in Fig. 15.17. These fractures apparently began at or near the full penetration groove welds between the bottom flanges and the column flanges. Among the factors involved in these failures were notch effects caused by the backup or backing bars, which were commonly left in place. Other factors were welds with porosity and slag inclusions, inconsistent strength and deformation capacities of the steel sections, and so on.

Detailed information on these problems at Northridge was provided in FEMA 267 Report Number SAC-95-02, entitled "Interim Guidelines, Evaluation, Repair, Modification and Design of Steel Moment Frames," dated August 1995. In "Interim Guidelines Advisory No. 1 Supplement to FEMA 267," published in March 1997, several recommendations were presented for correcting the problems. These included the removal of backup bars and weld tabs, the incorporation of full-scale inelastic testing of joints of the types used, and several others.

To design a moment-resisting connection, the first step is to compute the magnitude of the internal compression and tension forces, C and T. These forces are assumed to be concentrated at the centers of the flanges, as shown in Fig. 15.18.

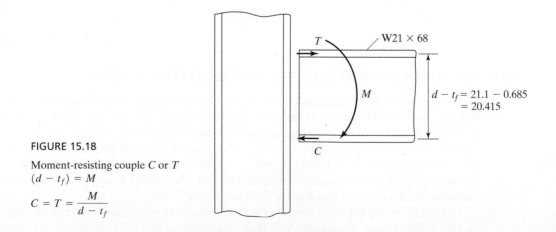

FIGURE 15.18

Moment-resisting couple C or T

$(d - t_f) = M$

$$C = T = \frac{M}{d - t_f}$$

Next, the areas of the full-penetration welds up against the column are determined. They equal the magnitude of C or T divided by the design stress of a full penetration groove weld, as provided in Table 14.1 (AISC Table J2.5), with $\phi = 0.9$.

$$\text{Area reqd.} = \frac{C_u \text{ or } T_u}{\phi F_y} \text{ or } \frac{C_a \text{ or } T_a}{F_y / \Omega}$$

With this procedure, it is theoretically possible to have a weld area larger than the cross-sectional area of the flange. It would then be theoretically necessary to use an auxiliary plate on the flange to resist the extra force. (We may just transfer all of the forces with plates on the flanges. Sometimes, the beam flanges are groove-welded flush with the column on one end and connected to the beam on the other end with the auxiliary plates just described. This could help us solve our problems of fit. The forces are transferred from the beam to the plate with fillet welds and from the plate to the column by groove welds.)

Recent research at the University of California and Lehigh University has shown that the full plastic moment capacity of a beam can be developed with full-penetration welds made only to the flanges.

Example 15-6 illustrates the design of a moment-resisting connection with flange full-penetration groove welds. The reader should understand that this example is not quite complete. It is necessary also to design the shear plate, seat angle, or whatever is used to transfer the shear, and to check the column for the concentrated T or C force.

Example 15-6

Design a moment-resisting connection for the W21 × 68 beam shown in Fig. 15.18, with the flanges groove-welded to a column. The beam, which consists of 50 ksi steel, has end reactions $R_D = 20$ k and $R_L = 20$ k, along with moments $M_D = 60$ ft-k and $M_L = 90$ ft-k. Use E70 electrodes.

Solution

Using a W21 × 68 ($d = 21.1$ in, $b_f = 8.27$ in, $t_f = 0.685$ in, $T = 18\frac{3}{8}$ in)

Design of moment welds

LRFD $\phi = 0.90$	ASD $\Omega = 1.67$
$M_u = (1.2)(60) + (1.6)(90) = 216$ ft-k	$M_a = 60 + 90 = 150$ ft-k
$C_u = T_u = \dfrac{(12)(216)}{21.1 - 0.685} = 127$ k	$C_a = T_a = \dfrac{(12)(150)}{21.1 - 0.685} = 88.17$ k
A of groove weld $= \dfrac{127}{(0.9)(50)} = 2.82$ in^2	A of groove weld $= \dfrac{88.17}{50/1.67} = 2.94$ in^2
width reqd $= \dfrac{2.82}{t_f} = \dfrac{2.82}{0.685} = 4.12$ in $< b_f$	width reqd $= \dfrac{2.94}{t_f} = \dfrac{2.94}{0.685} = 4.29$ in $< b_f$
Use 5-in-wide E70 full-penetration groove welds.	**Use 5-in-wide E70 full-penetration groove welds.**

Design of shear welds

Try 1/4-in fillet welds on shear tabs (or on seat angle or on beam web)

$$R_n \text{ of weld per in} = F_{nw}A_{we} = (0.60 \times 70)\left(\frac{1}{4} \times 0.707\right) = 7.42 \text{ k/in}$$

LRFD $\phi = 0.75$	ASD $\Omega = 2.00$
$R_u = (1.2)(20) + (1.6)(20) = 56 \text{ k}$	$R_a = 20 + 20 = 40 \text{ k}$
$\phi R_n = (0.75)(7.42) = 5.56 \text{ k/in}$	$\dfrac{R_n}{\Omega} = \dfrac{7.42}{2.00} = 3.71 \text{ k/in}$
weld length reqd $= \dfrac{56}{5.56} = 10.07 \text{ in}$	weld length $= \dfrac{40}{3.71} = 10.78 \text{ in}$
Use $\frac{1}{4}$-in fillet welds $5\frac{1}{2}$ in long each side.	Use $\frac{1}{4}$-in fillet welds $5\frac{1}{2}$ in long each side.

Note: The design is incomplete, as it is necessary to design the shear tab, seat L, or whatever is used, and to check the column for the calculated shear force (C or T) from the beam.

Figure 15.19 shows a moment-resisting connection where the C and T forces are carried by cover plates on the top and bottom of a W section. The moment to be resisted is divided by the distance between the centers of gravity of the top and bottom parts of the couple (C and T), and then welds or bolts are selected that will provide the necessary design strengths so determined. Next, a shear tab, a pair of framing angles, or a beam seat is selected to resist the shear force. Finally, as described in the next section, it may be necessary to provide stiffeners for the column web, or to select a larger column section.

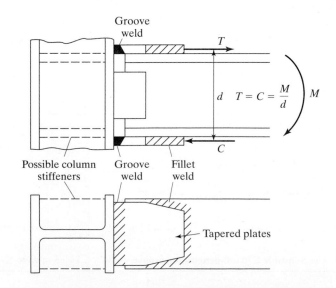

FIGURE 15.19

In this particular connection, the T and C values are transferred by fillet welds into the plates and by groove welds from the plates to the columns. For easier welding, these plates may be tapered as shown in the bottom of the figure. You may have noticed such tapered plates used for facilitating welding in other situations.

For a rigid or continuous connection of the type shown in Fig. 15.19, we must be careful to check the strength of the top and bottom plates. Should the plates be bolted, this check involves the tensile strength of the top plate, including the effect of the bolt holes and block shear. The design compressive strength of the other plate must also be checked.

15.12 COLUMN WEB STIFFENERS

If a column to which a beam is being connected bends appreciably at the connection, the moment resistance of the connection will be reduced, regardless of how good the connection may be. Furthermore, if the top connection plate, in pulling away from the column, tends to bend the column flange as shown in part (a) of Fig. 15.20, the middle part of the weld may be greatly overstressed (like the prying action for bolts discussed in Chapter 13).

When there is a danger of the column flange bending, as described here, we must make sure that the desired moment resistance of the connection is provided. We may do this either by using a heavier column with stiffer flanges or by introducing column web stiffener plates, as shown in part (b) of Fig. 15.20. *It is almost always desirable to use a heavier column, because column web stiffener plates are quite expensive and a nuisance to use.*

Column web stiffener plates are somewhat objectionable to architects, who find it convenient to run pipes and conduits inside their columns; this objection can

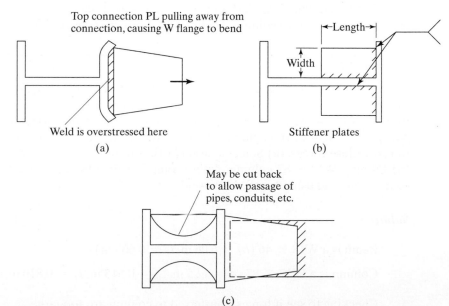

FIGURE 15.20

easily be overcome, however. First, if the connection is to only one column flange, the stiffener does not have to run for more than half the column depth, as shown in part (b) of Fig. 15.20. If connections are made to both column flanges, the column stiffener plates may be cut back to allow the passage of pipes, conduits, etc., as shown in part (c) of the figure.

Should the LRFD or ASD forces applied from the beam flange to the column be greater than any of the values given by the AISC equations for flange local bending, web local yielding, web crippling, and web compression buckling, it will be necessary to use column stiffeners or doubler plates for the column web or to select a column with a thicker flange. The equations for these items were previously presented in Chapter 10 of this text. Their application is again illustrated in the solution of Example 15-7.

The AISC Manual presents a set of suggested rules for the design of column web stiffeners. These are as given in AISC specificaion J10.

1. The width of the stiffener plus one-half of the column web thickness should not be less than one-third the width of the beam flange or of the moment connection plate that applies the concentrated force.
2. The stiffener thickness should not be less than $t_f/2$ or half the thickness of the moment connection plate delivering the concentrated load and not less than the width divided by 16.
3. If there is a moment connection applied to only one flange of the column, the length of the stiffener plate does not have to exceed one-half the column depth.
4. The stiffener plate should be welded to the column web with a sufficient strength to carry the force caused by the unbalanced moment on the opposite sides of the column.

For the column given in Example 15-7, it is necessary to use column web stiffeners or to select a larger column. These alternatives are considered in the solution.

Example 15-7

It is assumed that a particular column is a W12 × 87 consisting of 50 ksi steel and subjected to $C_D = T_D = 60$ k and $C_L = T_L = 90$ k transferred by an FR-type connection from a W18 × 46 beam on one side of the column. The connection is located a distance $>d$ from the end of the column. It will be found that this column is not satisfactory to resist these forces. (a) Select a larger W12 column section that will be satisfactory. (b) Using a W12 × 87 column, design column web stiffeners, including the stiffener connections and using E70 SMAW welds.

Solution

Beam is a W18 × 46 ($b_f = 6.06$ in, $t_f = 0.605$ in)

Column is a W12 × 87 ($d = 12.5$ in, $t_w = 0.515$ in, $t_f = 0.810$ in, $k = 1.41$ in)

Checking to see if forces transferred to column are too large.

LRFD	ASD
$C_u = (1.2)(60) + (1.6)(90) = 216$ k	$C_a = 60 + 90 = 150$ k

Flange local bending

$$R_n = (6.25)(0.810 \text{ in})^2(50 \text{ ksi}) = 205 \text{ k} \qquad \text{(AISC Equation J10-1)}$$

LRFD $\phi = 0.90$	ASD $\Omega = 1.67$
$\phi R_n = (0.90)(205) = 184.5$ k < 216 k **N.G.**	$\dfrac{R_n}{\Omega} = \dfrac{205}{1.67} = 122.8$ k < 150 k **N.G.**

$\therefore$ Must use a larger column or a pair of transverse stiffeners.

Web local yielding

$$R_n = (5 \times 1.41 \text{ in} + 6.06 \text{ in})(50 \text{ ksi})(0.515 \text{ in}) = 337.6 \text{ k} \quad \text{(AISC Equation J10-2)}$$

LRFD $\phi = 1.00$	ASD $\Omega = 1.5$
$\phi R_n = (1.00)(337.6) = 337.6 > 216$ k **OK**	$\dfrac{R_n}{\Omega} = \dfrac{337.6}{1.5} = 225.1$ k > 140 k **OK**

Web crippling

$$R_n = (0.80)(0.515 \text{ in})^2 \left[1 + 3\left(\frac{6.06 \text{ in}}{12.5 \text{ in}}\right)\left(\frac{0.515 \text{ in}}{0.810 \text{ in}}\right)^{1.5}\right]\sqrt{\frac{(29 \times 10^3 \text{ ksi})(50 \text{ ksi})(0.810 \text{ in})}{0.515 \text{ in}}}$$

$$= 556.7 \text{ k} \qquad \text{(AISC Equation J10-4)}$$

LRFD $\phi = 0.75$	ASD $\Omega = 2.00$
$\phi R_n = (0.75)(556.7) = 417.5$ k > 216 k **OK**	$\dfrac{R_n}{\Omega} = \dfrac{556.7}{2.00} = 278.3$ k > 150 k **OK**

(a) selecting a larger column

Try W12 $\times$ 96 ($t_f = 0.900$)

Flange local buckling

$$R_n = (6.25)(0.900 \text{ in})^2(50 \text{ ksi}) = 253.1 \text{ k} \qquad \text{(AISC Equation J10-1)}$$

LRFD $\phi = 0.90$	ASD $\Omega = 1.67$
$\phi R_n = (0.90)(253.1) = 227.8 \text{ k} > 216 \text{ k } \textbf{OK}$	$\dfrac{R_n}{\Omega} = \dfrac{253.1}{1.67} = 151.6 \text{ k} > 150 \text{ k } \textbf{OK}$

Use W12 × 96 column.

(b) Design of web stiffeners using a W12 × 87 column and the suggested rules presented before this example. The author shows only the LRFD solution for this part of problem.

$$\text{Reqd stiffener area} = \frac{216 \text{ k} - 184.5 \text{ k}}{50 \text{ ksi}} = 0.63 \text{ in}^2$$

$$\text{Min width} = \frac{1}{3}b_f - \frac{t_w}{2} = \frac{6.06}{3} - \frac{0.515}{2} = 1.76 \text{ in}$$

$$\text{Min } t \text{ of stiffeners} = \frac{0.63 \text{ in}^2}{1.76 \text{ in}} = 0.358 \text{ in } \textbf{say, 3/8 in}$$

$$\text{Reqd width} = \frac{0.63 \text{ in}^2}{0.375 \text{ in}} = 1.68 \text{ in } \textbf{say, 4 in for practical purposes}$$

$$\text{Minimum length} = \frac{d}{2} - t_f = \frac{12.5}{2} - 0.810 = 5.45 \text{ in } \textbf{say, 6 in}$$

Design of welds for stiffener plates

Minimum weld size as reqd by AISC Table J-2.4

$$= \frac{3}{16} \text{ in based on the column web } t_w = 0.515 \text{ in}$$

$$\text{Reqd length of weld} = \frac{216 \text{ k} - 184.5 \text{ k}}{(0.75)(0.60 \times 70 \text{ ksi})(0.707)\left(\dfrac{3}{16} \text{ in}\right)} = 7.54 \text{ in } \textbf{say, 8 in}$$

15.13 PROBLEMS FOR SOLUTION

For Problems 15-1 through 15-15 use the tables of Part 10 of the AISC Manual.

15-1.   Determine the maximum end reaction that can be transferred through the A36 web angle connection shown in the accompanying illustration. Solve by LRFD and ASD. The beam steel is 50 ksi, and the bolts are 3/4-in A325-N and are used in standard-size holes. The beam is connected to the web of a W30 × 90 girder with A36 angles. (*Ans.* 126 k, 83.9 k)

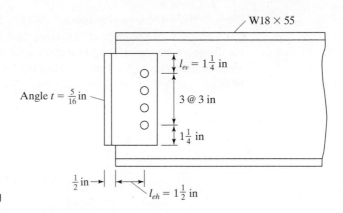

FIGURE P15-1

15-2.   Repeat Prob. 15-1 if the bolts are 1-in A325-X.

15-3.   Repeat Prob. 15-1 if the bolts are 7/8-in A325-N. (*Ans.* 122 k, 81.6 k).

15-4.   Using the AISC Manual, select a pair of bolted standard web angles (LRFD and ASD) for a W33 × 141 connected to the flange of a W14 × 120 column with a dead load service reaction of 75 k and a live load service reaction of 50 k. The bolts are to be 7/8-in A325-N in standard-size holes, and the steel is A36 for the angles and A992 for the W shapes.

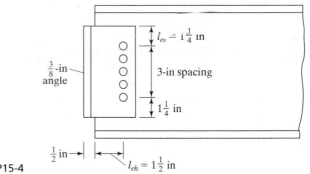

FIGURE P15-4

15-5.   Repeat Prob. 15-4 if 1-in A325-N bolts are to be used. (*Ans.* 6 row conn. with 2Ls 4 × $3\frac{1}{2}$ × $\frac{3}{8}$ × 1 ft − $5\frac{1}{2}$ in)

15-6.   Repeat Prob. 15-1 if 3/4-in A325 SC Class A bolts are to be used.

15-7.   Design framed beam connections for a W27 × 84 connected to a W30 × 116 girder web to support a dead load reaction of 40 k and a live load reaction of 50 k, using the LRFD and ASD methods. The bolts are to be 7/8-in A325 SC Class A in standard-size holes. Angles are A36 steel, while beams are A992. The edge distances and bolt spacings are the same as those shown in the sketch for Prob. 15-4. (*Ans.* 5 row connection with 2Ls 5 × $3\frac{1}{2}$ × $\frac{3}{8}$ × 1 ft − $2\frac{1}{2}$ in)

15-8.   Repeat Prob. 15-1 if the bolts are 3/4-in A490-SC Class A and are used in $1\frac{1}{16}$ × $1\frac{5}{16}$ in short slots with long axes perpendicular to the transmitted force. Angle *t* is 1/2 in.

15-9.   Design a framed beam connection for a W18 × 50 to support a dead load reaction of 30 k and a live load reaction of 20 k, using the LRFD and ASD methods. The beam's top flange is to be coped for a 2-in depth, and 7/8-in A325-X bolts in standard-size holes are to be used. The beam is connected to a W27 × 146 girder. Connection is A36, while W shapes are A992. (*Ans.* 4 row connection 2Ls 5 × $3\frac{1}{2}$ × $\frac{1}{4}$ × 0 ft − $11\frac{1}{2}$ in)

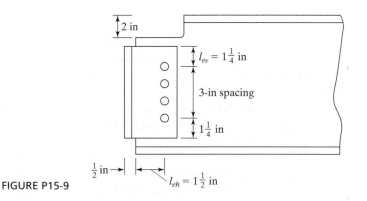

2 in

$l_{ev} = 1\frac{1}{4}$ in

3-in spacing

$1\frac{1}{4}$ in

$\frac{1}{2}$ in

$l_{eh} = 1\frac{1}{2}$ in

FIGURE P15-9

15-10. Repeat Prob. 15-7 if the dead load reaction is 80 k and the live load reaction is 110 k, and if 1-in A325-N bolts and A572 steel (F_y = 50 ksi and F_u = 65 ksi) are to be used.

15-11. Select a framed beam connection, using LRFD and ASD, for a W33 × 130 beam (A992 steel) with a dead load reaction = 45 k and a live load reaction = 65 k. It is to be connected to the flange of a W36 × 150 column. The A36 web angles are to be welded with E70 electrodes (weld A in the Manual) and are to be field-connected to the girder with 3/4-in A325-N bolts. (*One ans.* 2Ls 4 × $3\frac{1}{2}$ × $\frac{5}{16}$ × 1 ft $5\frac{1}{2}$ in, $\frac{3}{16}$-in weld A and 6-row bolt connection to girder.)

15-12. Repeat Prob. 15-11 using SMAW shop and field welds (welds A and B in the Manual).

15-13. Select an A36 framed beam connection from the AISC Manual (LRFD and ASD) for a W30 × 124 consisting of 50 ksi steel, using SMAW E70 shop and field welds. The dead load reaction is 60 k, while the live load one is 80 k. The beam is to be connected to the flange of a 50 ksi W14 × 145 column. (*Ans.* 2Ls 4 × 3 × $\frac{3}{8}$ × 1 ft − 8 in weld A = $\frac{3}{16}$ in, weld B = $\frac{5}{16}$ in)

15-14. Repeat Prob. 15-13 if R_D = 90 k and R_L = 100 k.

15-15. Select unstiffened A36 seated beam connections (LRFD and ASD) bolted with 7/8-in A325-N bolts in standard-size holes for the following data: Beam is W16 × 67, column is W14 × 82, both consisting of 50 ksi steel, R_D = 25 k, R_L = 30 k, and the column gage is $5\frac{1}{2}$ in. (*Ans.* 1L 6 × 4 × $\frac{3}{4}$ × 0 ft − 8 in)

15-16. Design SMAW welded moment-resisting connections LRFD and ASD for the ends of a W24 × 76 to resist R_D = 30 k, R_L = 60 k, M_D = 60 ft-k, and M_L = 80 ft-k. Use A36 steel and E70 electrodes. Assume that the column flange is 14 in wide. The moment is to be resisted by full-penetration groove welds in the flanges, and the shear is to be resisted by welded clip angles along the web. Assume that the beam was selected for bending with $0.9F_y$.

15-17.   The beam shown in the accompanying illustration is assumed to be attached at its ends with moment-resisting connections. Select the beam, assuming full lateral support, and E70 SMAW electrodes. Use a connection of the type used in Prob. 15-16. $F_y = 50$ ksi. Use ASD and LRFD. (*Ans.* W21 × 55, shear welds $10\frac{1}{2}$ in each side LRFD, 11 in ASD.)

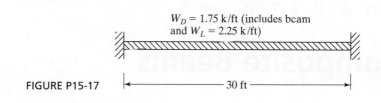

$W_D = 1.75$ k/ft (includes beam and $W_L = 2.25$ k/ft)

FIGURE P15-17 |← ———————— 30 ft ———————— →|

C H A P T E R   1 6

Composite Beams

16.1 COMPOSITE CONSTRUCTION

When a concrete slab is supported by steel beams, and there is no provision for shear transfer between the two, the result is a noncomposite section. Loads applied to noncomposite sections obviously cause the slabs to deflect along with the beams, resulting in some of the load being carried by the slabs. Unless a great deal of bond exists between the two (as would be the case if the steel beam were completely encased in concrete, or where a system of mechanical steel anchors is provided), the load carried by the slab is small and may be neglected.

For many years, steel beams and reinforced-concrete slabs were used together, with no consideration made for any composite effect. In recent decades, however, it has been shown that a great strengthening effect can be obtained by tying the two together to act as a unit in resisting loads. Steel beams and concrete slabs joined together compositely can often support 33 to 50 percent or more load than could the steel beams alone in noncomposite action.

Composite construction for highway bridges was given the green light by the adoption of the 1944 AASHTO Specifications, which approved the method. Since about 1950, the use of composite bridge floors has rapidly increased, until today they are commonplace all over the United States. In these bridges, the longitudinal shears are transferred from the stringers to the reinforced-concrete slab or deck with steel anchors (described in Section 16.5), causing the slab or deck to assist in carrying the bending moments. This type of section is shown in part (a) of Fig. 16.1.

The first approval for composite building floors was given by the 1952 AISC Specification; today, they are very common. These floors may either be encased in concrete (very rare due to expense) as shown in part (b) of Fig. 16.1, or be nonencased with shear connectors as shown in part (c) of the figure. Almost all composite building floors being built today are of the nonencased type. If the steel sections are encased in concrete, the shear transfer is made by bond and friction between the beam and the

562

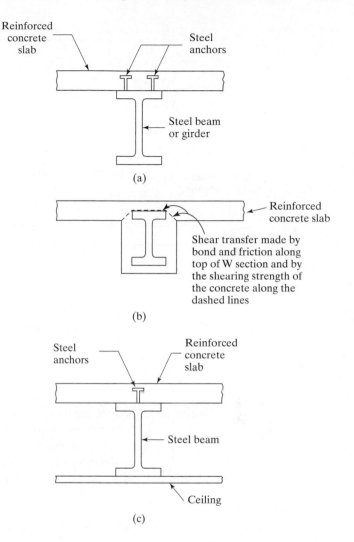

FIGURE 16.1

(a) Composite bridge floor with steel anchors. (b) Encased section for building floors. (c) Building floors with steel anchors.

concrete and by the shearing strength of the concrete along the dashed lines shown in part (b) of Fig. 16.1.

Today, formed steel deck (illustrated in Fig. 16.2) is used for almost all composite building floors. The initial examples in this chapter, however, pertain to the calculations for composite sections where formed steel deck is not used. Sections that make use of formed steel deck are described later in the chapter.

16.2 ADVANTAGES OF COMPOSITE CONSTRUCTION

The floor slab in composite construction acts not only as a slab for resisting the live loads, but also as an integral part of the beam. It actually serves as a large cover plate for the upper flange of the steel beam, appreciably increasing the beam's strength.

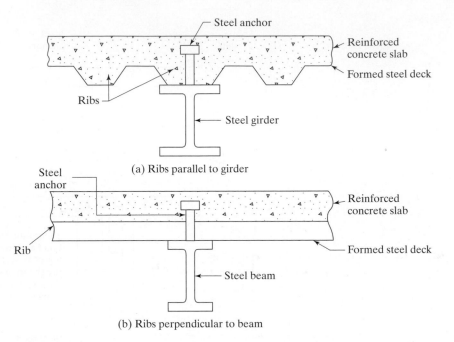

FIGURE 16.2

Composite sections using formed steel deck.

A particular advantage of composite floors is that they make use of concrete's high compressive strength by putting a large part of the slab in compression. At the same time, a larger percentage of the steel is kept in tension (also advantageous) than is normally the case in steel-frame structures. The result is less steel tonnage required for the same loads and spans (or longer spans for the same sections). Composite sections have greater stiffness than noncomposite sections, and they have smaller deflections—perhaps only 20 to 30 percent as large. Furthermore, tests have shown that the ability of a composite structure to take overload is decidedly greater than for a noncomposite structure.

An additional advantage of composite construction is the possibility of having smaller overall floor depths—a fact of particular importance for tall buildings. Smaller floor depths permit reduced building heights, with the consequent advantages of smaller costs for walls, plumbing, wiring, ducts, elevators, and foundations. Another important advantage available with reduced beam depths is a saving in fireproofing costs, because a coat of fireproofing material is provided on smaller and shallower steel shapes.

It occasionally is necessary to increase the load-carrying capacity of an existing floor system. Often, this can be handled quite easily for composite floors by welding cover plates onto the bottom flanges of the beams.

A disadvantage for composite construction is the cost of furnishing and installing the steel anchors. This extra cost usually will exceed the cost reductions mentioned when spans are short and lightly loaded.

16.3   DISCUSSION OF SHORING

After the steel beams are erected, the concrete slab is placed on them. The formwork, wet concrete, and other construction loads must therefore be supported by the beams or by temporary shoring. Should no shoring be used, the steel beams must support all of these loads as well as their own weights. Most specifications say that after the concrete has gained 75 percent of its 28-day strength, the section has become composite and all loads applied thereafter may be considered to be supported by the composite section. When shoring is used, it supports the wet concrete and the other construction loads. It does not really support the weight of the steel beams unless they are given an initial upward deflection (which probably is impractical). When the shoring is removed (after the concrete gains at least 75 percent of its 28-day strength), the weight of the slab is transferred to the composite section, not just to the steel beams. The student can see that if shoring is used, it will be possible to use lighter, and thus cheaper, steel beams. The question then arises, "Will the savings in steel cost be greater than the extra cost of shoring?" The answer probably is no. The usual decision is to use heavier steel beams and to do without shoring for several reasons, including the following:

1. Apart from reasons of economy, the use of shoring is a tricky operation, particularly where settlement of the shoring is possible, as is often the case in bridge construction.
2. Both theory and load tests indicate that the ultimate strengths of composite sections of the same sizes are the same, whether shoring is used or not. If lighter steel beams are selected for a particular span because shoring is used, the result is a smaller ultimate strength.
3. Another disadvantage of shoring is that after the concrete hardens and the shoring is removed, the slab will participate in composite action in supporting the dead loads. The slab will be placed in compression by these long-term loads and will have substantial creep and shrinkage parallel to the beams. The result will be a great decrease in the stress in the slab, with a corresponding increase in the steel stresses.

Floor framing for Glen Oaks School, Bellerose, NY. (Courtesy of CMC South Carolina Steel.)

The probable consequence is that most of the dead load will be supported by the steel beams anyway, and composite action will really apply only to the live loads, as though shoring had not been used.

4. Also, in shored construction cracks occur over the steel girders, necessitating the use of reinforcing bars. In fact, we should use reinforcing over the girders in unshored construction, too. Although cracks will be smaller there, they are going to be present nonetheless, and we need to keep them as small as possible.

Nevertheless, shored construction does present some advantages compared with unshored construction. First, deflections are smaller because they are all based on the properties of the composite section. (In other words, the initial wet concrete loads are not applied to the steel beams alone, but rather to the whole composite section.) Second, it is not necessary to make a strength check for the steel beams for this wet load condition. This is sometimes quite important for situations in which we have low ratios of live to dead loads.

The deflections of unshored floors due to the wet concrete sometimes can be quite large. If the beams are not cambered, additional concrete (perhaps as much as 10 percent or more) will be used to even up the floors. If, on the other hand, too much camber is specified, we may end up with slabs that are too thin in those areas where wet concrete deflections aren't as large as the camber.

16.4   EFFECTIVE FLANGE WIDTHS

There is a problem involved in estimating how much of the slab acts as part of the beam. Should the beams be rather closely spaced, the bending stresses in the slab will be fairly uniformly distributed across the compression zone. If, however, the distances between beams are large, bending stresses will vary quite a bit nonlinearly across the flange. The further a particular part of the slab or flange is away from the steel beam, the smaller will be its bending stress. Specifications attempt to handle this problem by replacing the actual slab with a narrower or effective slab that has a constant stress. This equivalent slab is deemed to support the same total compression as is supported by the actual slab. The effective width of the slab b_e is shown in Fig. 16.3.

The portion of the slab or flange that can be considered to participate in the composite beam action is controlled by the specifications. AISC Specification I3.1a states that the effective width of the concrete slab on each side of the beam center line shall

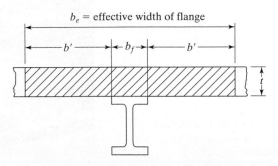

FIGURE 16.3

not exceed the least of the values to follow. The following set of rules applies, whether the slab exists on one or both sides of the beam:

1. One-eighth of the span of the beam measured center-to-center of supports for both simple and continuous spans.
2. One-half of the distance from the beam center line to the center line of the adjacent beam.
3. The distance from the beam center line to the edge of the slab.

The AASHTO requirements for determining effective flange widths are somewhat different. The maximum total flange width may not exceed one-fourth of the beam span, twelve times the least thickness of the slab, or the distance center-to-center of the beams. Should the slab exist on only one side of the beam, its effective width may not exceed one-twelfth of the beam span, six times the slab thickness, or one-half of the distance from the center line of the beam to the center line of the adjacent beam.

16.5 SHEAR TRANSFER

The concrete slabs may rest directly on top of the steel beams, or the beams may be completely encased in concrete for fireproofing purposes. This latter case, however, is very expensive and thus is rarely used. The longitudinal shear can be transferred between the two by bond and shear (and possibly some type of shear reinforcing), if needed, when the beams are encased. When not encased, mechanical connectors must transfer the load. Fireproofing is not necessary for bridges, and the slab is placed on top of the steel beams. Bridges are subject to heavy impactive loads, and the bond between the beams and the deck, which is easily broken, is considered negligible. For this reason, steel anchors are designed to resist all of the shear between bridge slabs and beams.

Various types of steel anchors have been tried, including spiral bars, channels, zees, angles, and studs. Several of these types of connectors are shown in Fig. 16.4. Economic considerations have usually led to the use of round studs welded to the top flanges of the beams. These studs are available in diameters from 1/2 to 1 in and in lengths from 2 to 8 in, but the AISC Specification (I8.2) states that their length may not be less than 4 stud diameters. This specification also permits the use of hot-rolled steel channels, but not spiral connectors.

The studs actually consist of rounded steel bars welded on one end to the steel beams. The other end is upset or headed to prevent vertical separation of the slab from the beam. These studs can be quickly attached to the steel beams through the steel decks with stud-welding guns by semiskilled workers. The AISC Commentary (I3.2d) describes special procedures needed for 16-gage and thicker decks and for decks with heavy galvanized coatings (>1.25 ounces per sq ft).

A rather interesting practical method is used by many engineers in the field to check the adequacy of the welds used to connect the studs to the steel beams. They take a 5 or 6 lb hammer and hit occasional studs a sufficient number of times to cause them to bend over roughly 25 or 30 degrees. If the studs don't break loose during this hammering, the welds are considered to be satisfactory and the studs are left in their bent positions, which is OK because they will later be encased in the concrete. Should the welds be poor, as where they were made during wet conditions, they may break loose and have to be replaced.

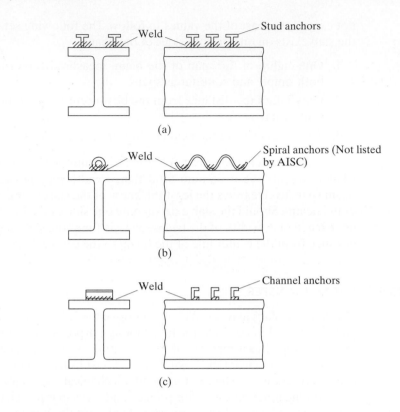

FIGURE 16.4

Steel anchors.

Shop installation of steel anchors initially is more economical, but there is a growing tendency to use field installation. There are two major reasons for this trend: The anchors may easily be damaged during transportation and setting of the beams, and they serve as a hindrance to the workers walking along the top flanges during the early phases of construction.

When a composite beam is being tested, failure will probably occur with a crushing of the concrete. It seems reasonable to assume that at that time the concrete and steel will both have reached a plastic condition.

For the discussion to follow, reference is made to Fig. 16.5. Should the plastic neutral axis (PNA) fall in the slab, the maximum horizontal shear (or horizontal force on the plane between the concrete and the steel) is said to be $A_s F_y$; and if the plastic neutral axis is in the steel section, the maximum horizontal shear is considered to equal $0.85 f'_c A_c$, where A_c is the effective area of the concrete slab. (For the student unfamiliar with the strength design theory for reinforced concrete, the average stress at failure on the compression side of a reinforced concrete beam is usually assumed to be $0.85 f'_c$.)

From this information, expressions for the shear to be taken by the anchors can be determined: The AISC (I3.2d) says that, for composite action, the total horizontal shear between the points of maximum positive moment and zero moment is to be taken as the least of the following, where ΣQ_n is the total nominal strength of the steel anchors provided,

a. For concrete crushing

$$V' = 0.85 f'_c A_c \qquad \text{(AISC Equation I3-1a)}$$

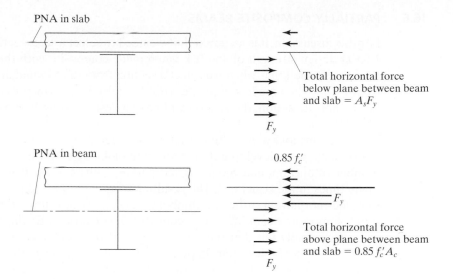

PNA in slab

Total horizontal force
below plane between beam
and slab = $A_s F_y$

F_y

PNA in beam

$0.85 f_c'$

F_y

Total horizontal force
above plane between beam
and slab = $0.85 f_c' A_c$

F_y

FIGURE 16.5

b. For tensile yielding of the steel section (for hybrid beams, this yield force must be calculated separately for each of the components of the cross section)

$$V' = F_y A_s \qquad \text{(AISC Equation I3-1b)}$$

c. For strength of steel anchors

$$V' = \Sigma Q_n \qquad \text{(AISC Equation I3-1c)}$$

Composite floor framing, North Charleston, SC. (Courtesy of CMC South Carolina Steel.)

16.6 PARTIALLY COMPOSITE BEAMS

For this discussion, it is assumed that we need to select a steel section that will have an LRFD design strength of 450 ft-k when made composite with the concrete slab. It is further assumed that when we select a section from the Manual, it has an $\phi_b M_n$ equal to 510 ft-k (when made composite with the slab). If we now provide steel anchors for full composite action, the section will have a design strength of 510 ft-k. But we need only 450 ft-k.

It seems logical to assume that we may decide to provide only a sufficient number of anchors to develop a design strength of 450 ft-k. In this way we can reduce the number of anchors and reduce costs (perhaps substantially if we repeat this section many times in the structure). The resulting section is a *partially composite section*, one that does not have a sufficient number of anchors to develop the full flexural strength of the composite beam. We will encounter this situation in Examples 16-3 and 16-4.

It is usually felt that the total strength of the steel anchors used in a particular beam should not be less than 25 percent of the shearing strength required for full composite action ($A_s F_y$). Otherwise, our calculations may not accurately depict the stiffness and strength of a composite section.

16.7 STRENGTH OF STEEL ANCHORS

For composite sections, it is permissible to use normal weight stone concrete (made with aggregates conforming to ASTM C33) or lightweight concrete weighing not less than 90 lb/ft³ (made with rotary kiln-produced aggregates conforming to ASTM C330).

The AISC Specification provides strength values for headed steel studs not less than 4 diameters in length after installation and for hot-rolled steel channels. *They do not, however, give resistance factors for the strength of steel anchors.* This is because they feel that the factor used for determining the flexural strength of the concrete is sufficient to account for variations in concrete strength, including those variations that are associated with steel anchors.

16.7.1 Steel Headed Stud Anchors

The nominal shear strength in kips of one stud steel anchors embedded in a solid concrete slab is to be determined with an expression from AISC Specification I8.2a. In this expression, A_{sa} is the cross-sectional area of the shank of the anchors in square inches and f'_c is the specified compressive stress of the concrete in ksi. E_c is the modulus of elasticity of the concrete in ksi (MPa) and equals $w^{1.5}\sqrt{f'_c}$ in which w is the unit weight of the concrete in lb/ft³, while F_u is the specified minimum tensile strength of the steel stud in ksi (MPa). R_g is a coefficient used to account for the group effect of the anchors, while R_p is the position effect of the anchors. Values of these latter two factors are given in AISC Specification I8.2a. Here is the expression for the normal shear strength:

$$Q_n = 0.5 A_{sa}\sqrt{f'_c E_c} \leq R_g R_p A_{sa} F_u \qquad \text{(AISC Equation I8-1)}$$

Values of Q_n are listed in Table 3-21 in the AISC Manual. These values are given for different stud diameters, for 3 and 4 ksi normal and lightweight concrete weighing 110 lbs/ft^3, and for composite sections with or without steel decking.

16.7.2 Steel Channel Anchors

The nominal shear strength in kips for one channel steel anchors is to be determined from the following expression from AISC Specification I8.2b, in which t_f and t_w are, respectively, the flange and web thicknesses of the channel and l_a is its length in in (mm):

$$Q_n = 0.3(t_f + 0.5t_w)l_a\sqrt{f'_c E_c} \qquad \text{(AISC Equation I8-2)}$$

16.8 NUMBER, SPACING, AND COVER REQUIREMENTS FOR SHEAR CONNECTORS

The number of steel anchors to be used between the point of maximum moment and each adjacent point of zero moment equals the horizontal force to be resisted divided by the nominal strength of one anchor Q_n.

16.8.1 Spacing of Anchors

Tests of composite beams with steel anchors spaced uniformly, and of composite beams with the same number of anchors spaced in variation with the statical shear, show little difference as to ultimate strengths and deflections at working loads. This situation prevails as long as the total number of anchors is sufficient to develop the shear on both sides of the point of maximum moment. As a result, the AISC Specification (I8.2c and I8.2d) permits uniform spacings of anchors on each side of the maximum moment points. However, the number of anchors placed between a concentrated load and the nearest point of zero moment must be sufficient to develop the maximum moment at the concentrated load.

16.8.2 Maximum and Minimum Spacings

Except for formed steel decks, the minimum center-to-center spacing of steel anchors along the longitudinal axes of composite beams permitted by AISC Specification (I8.2d) is 6 diameters, while the minimum value transverse to the longitudinal axis is 4 diameters. Within the ribs of formed steel decks, the minimum permissible spacing is 4 diameters in any direction. The maximum spacing may not exceed 8 times the total slab thickness, or 36 in.

When the flanges of steel beams are rather narrow, it may be difficult to achieve the minimum transverse spacings described here. For such situations, the studs may be staggered. Figure 16.6 shows possible arrangements.

If the deck ribs are parallel to the axis of the steel beam and more anchors are required than can be placed within the rib, the AISC Commentary (I8.2d) permits the splitting of the deck so that adequate room is made available.

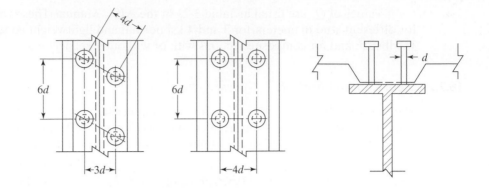

FIGURE 16.6

Anchor arrangements.

Steel anchors must be capable of resisting both horizontal and vertical movement, because there is a tendency for the slab and beam to separate vertically, as well as to slip horizontally. The upset heads of stud steel anchors help to prevent vertical separation.

16.8.3 Cover Requirements

The AISC Specification (I8.2d) requires that there be at least 1 in of lateral concrete cover provided for steel anchors. This rule does not apply to anchors used within the ribs of formed steel decks, because tests have shown that strengths are not reduced, even when studs are placed as close as possible to the ribs.

When studs are not placed directly over beam webs, they have a tendency to tear out of the beam flanges before their full shear capacity is reached. To keep this situation from occurring, the AISC Specification (I8.1) requires that the diameter of the studs may not be greater than 2.5 times the flange thickness of the beam to which they are welded, unless they are located over the web.

When formed steel deck is used, the steel beam must be connected to the concrete slab with steel headed stud anchors with diameters not larger than 3/4 in. These may be welded through the deck or directly to the steel beam. After their installation, they must extend for at least 1 1/2 in above the top of the steel deck, and the concrete slab thickness above the steel deck may not be less than 2 in (AISC Specification I3.2c(1)).

16.8.4 Strong and Weak Positions for Steel Headed Stud Anchors

If we examine Table 3-21 in the AISC Manual, which provides the calculated strengths of steel headed studs for different situations, we will see that if formed steel decks are used with their ribs placed perpendicular to the longitudinal direction of the steel beam, the studs are referred to as being either *strong* or *weak*. Most composite steel floor decks today have a stiffening rib in the middle of the deck flutes. This means that the shear studs will have to be placed on one side or the other of the ribs. Figure C-I8.1 of the AISC Specification Commentary shows strong and weak positions for placing the studs. The strong position is on the side away from the direction from which the shear is applied. Making sure that the studs are placed in the strong positions in the field is not an easy task, and some designers assume conservatively that the studs will always be placed in the weak positions, with their smaller shear strengths, as shown in AISC Table 3-21.

16.9   MOMENT CAPACITY OF COMPOSITE SECTIONS

The nominal flexural strength of a composite beam in the positive moment region may be controlled by the plastic strength of the section, by the strength of the concrete slab, or by the strength of the steel anchors. Furthermore, if the web is very slender and if a large portion of the web is in compression, web buckling may limit the nominal strength of the member.

Little research has been done on the subject of web buckling for composite sections, and for this reason the AISC Specification (I3.2) has conservatively applied the same rules to composite section webs as to plain steel webs. The positive nominal flexural strength, M_n, of a composite section is to be determined, assuming a plastic stress distribution if $h/t_w \leq 3.76\sqrt{E/F_{yf}}$. In this expression, h is the distance between the web toes of the fillet (that is, $d - 2k$), t_w is the web thickness, and F_{yf} is the yield stress of the beam flange. All of the rolled W, S, M, HP, and C shapes in the Manual meet this requirement for F_y values up to 65 ksi. (For built-up sections, h is the distance between adjacent lines of fasteners or the clear distance between flanges when welds are used.)

If h/t_w is greater than $3.76\sqrt{E/F_{yf}}$, the value of M_n with $\phi_b = 0.90$ and $\Omega = 1.67$ is to be determined by superimposing the elastic stresses. The effects of shoring must be considered for these calculations.

The nominal moment capacity of composite sections as determined by load tests can be estimated very accurately with the plastic theory. With this theory, the steel section at failure is assumed to be fully yielded, and the part of the concrete slab on the compression side of the neutral axis is assumed to be stressed to $0.85\, f'_c$. If any part of the slab is on the tensile side of the neutral axis, it is assumed to be cracked and incapable of carrying stress.

The plastic neutral axis (PNA) may fall in the slab or in the flange of the steel section or in its web. Each of these cases is discussed in this section.

16.9.1   Neutral Axis in Concrete Slab

The concrete slab compression stresses vary somewhat from the PNA out to the top of the slab. For convenience in calculations, however, they are assumed to be uniform, with a value of $0.85\, f'_c$ over an area of depth a and width b_e, determined as described in Section 16.4. (This distribution is selected to provide a stress block having the same total compression C and the same center of gravity for the total force as we have in the actual slab.)

The value of a can be determined from the following expression, where the total tension in the steel section is set equal to the total compression in the slab:

$$A_s F_y = 0.85 f'_c a b_e$$

$$a = \frac{A_s F_y}{0.85 f'_c b_e}$$

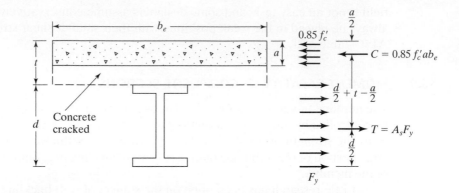

FIGURE 16.7

Plastic neutral axis
(PNA) in the slab.

If a is equal to or less than the slab thickness, the PNA will fall in the slab and the nominal or plastic moment capacity of the composite section may be written as the total tension T or the total compression C times the distance between their centers of gravity. Reference is made here to Fig. 16.7.

Example 16-1 illustrates the calculation of $\phi_b M_p = \phi_b M_n$ and $\dfrac{M_n}{\Omega}$ for a composite section where the PNA falls within the slab.

Example 16-1

Compute $\phi_b M_n$ and $\dfrac{M_n}{\Omega_b}$ for the composite section shown in Fig. 16.8 if $f'_c = 4$ ksi and $F_y = 50$ ksi.

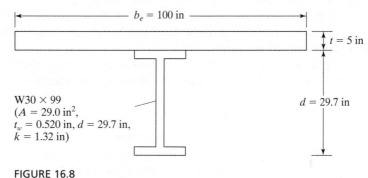

W30 × 99
($A = 29.0$ in^2,
$t_w = 0.520$ in, $d = 29.7$ in,
$k = 1.32$ in)

$b_e = 100$ in
$t = 5$ in
$d = 29.7$ in

FIGURE 16.8

Solution. Determining M_n

$$h = d - 2k = 29.7 \text{ in} - (2)(1.32 \text{ in}) = 27.06 \text{ in}$$

$$\frac{h}{t_w} = \frac{27.06 \text{ in}}{0.520 \text{ in}} = 52.04 < 3.76\sqrt{\frac{E}{F_{yf}}} = 3.76\sqrt{\frac{29 \times 10^3}{50}} = 90.55$$

∴ OK to determine M_n from the plastic stress distribution on the composite section for the limit state of yielding (plastic moment).

Locate PNA

$$a = \frac{A_s F_y}{0.85 f'_c b_e} = \frac{(29.0 \text{ in}^2)(50 \text{ ksi})}{(0.85)(4 \text{ ksi})(100 \text{ in})} = 4.26 \text{ in} < 5 \text{ in} \quad \therefore \text{ PNA is in slab}$$

$$\therefore M_n = M_p = A_s F_y \left(\frac{d}{2} + t - \frac{a}{2} \right)$$

$$= (29.0 \text{ in}^2)(50 \text{ ksi}) \left(\frac{29.7 \text{ in}}{2} + 5 \text{ in} - \frac{4.26 \text{ in}}{2} \right)$$

$$= 25{,}694 \text{ in-k} = 2141.2 \text{ ft-k}$$

LRFD $\phi_b = 0.90$	ASD $\Omega_b = 1.67$
$\phi_b M_n = (0.90)(2141.2) = 1927.1$ ft-k	$\dfrac{M_n}{\Omega_b} = \dfrac{2141.2}{1.67} = 1282.2$ ft-k

Note: If the reader refers to Part 3 of the AISC Manual, he or she can determine the ϕM_n and $\dfrac{M_n}{\Omega}$ values for this composite beam; see Fig. 16.9. To use the Manual composite tables, we assume that the PNA is located at the top of the steel flange (TFL) or down in the steel shape. In the AISC tables, Y1 represents the distance from the PNA to the top of the beam flange while Y2 represents the distance from the centroid of the effective concrete flange force to the top flange of the beam $(Y_{con} - a/2)$.

The reader should realize that the PNA locations will be different for LRFD and ASD if the live load over the dead load is not equal to 3. The AISC has selected the PNA locations that require the most shear transfer. The result is a slight variation in the calculated moment values and Table 3-19 values.

With the PNA for the preceding example being located at the top of the beam flange, from page 3-170 of the Manual, with Y2 = 5 − 4.28/2 = 2.86 in and Y1 = 0, and for a W30 × 99, the value of $\phi M_n = \phi_b M_p$ by interpolation is 1926 ft-k.

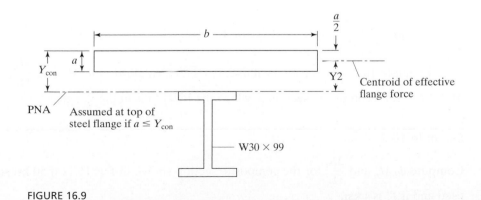

FIGURE 16.9

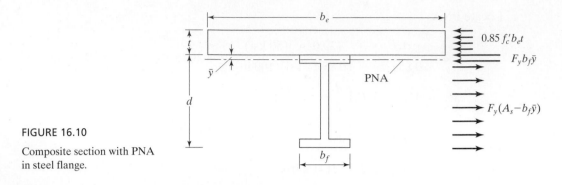

FIGURE 16.10

Composite section with PNA
in steel flange.

16.9.2 Neutral Axis in Top Flange of Steel Beam

If a is calculated as previously described and is greater than the slab thickness t, the PNA will fall down in the steel section. If this happens, it will be necessary to find out whether the PNA is in the flange or below the flange. Suppose we assume that it's at the base of the flange. We can calculate the total compressive force C above the PNA $= 0.85\, f'_c b_e t + A_f F_y$, where A_f is the area of the flange, and the total tensile force below $T = F_y(A_s - A_f)$. If $C > T$, the PNA will be in the flange. If $C < T$, the PNA is below the flange.

Assuming that we find that the PNA is in the flange, we can determine its location, letting $\bar{y}$ be the distance to the PNA measured from the top of the top flange, by equating C and T as follows:

$$0.85\, f'_c b_e t + F_y b_f \bar{y} = F_y A_s - F_y b_f \bar{y}$$

From this, $\bar{y}$ is

$$\bar{y} = \frac{F_y A_s - 0.85\, f'_c b_e t}{2 F_y b_f}$$

Then the nominal or plastic moment capacity of the section can be determined from the expression to follow, with reference being made to Fig. 16.10. Taking moments about the PNA, we get

$$M_p = M_n = 0.85\, f'_c b_e t \left(\frac{t}{2} + \bar{y} \right) + 2 F_y b_f \bar{y} \left(\frac{\bar{y}}{2} \right) + F_y A_s \left(\frac{d}{2} - \bar{y} \right)$$

Example 16-2, which follows, illustrates the calculation of $\phi_b M_n$ and $\dfrac{M_n}{\Omega_b}$, where $M_n = M_p$, for a composite section in which the PNA falls in the flange.

Example 16-2

Compute $\phi_b M_n$ and $\dfrac{M_n}{\Omega_b}$ for the composite section shown in Fig. 16.11 if 50 ksi steel is used and if f'_c is 4 ksi.

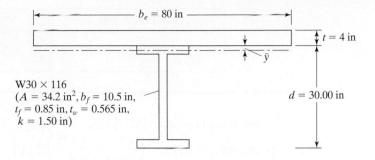

W30 × 116
($A = 34.2$ in², $b_f = 10.5$ in,
$t_f = 0.85$ in, $t_w = 0.565$ in,
$k = 1.50$ in)

FIGURE 16.11

Solution. Determining M_n

$$h = d - 2k = 30.00 - (2)(1.50) = 27.00 \text{ in}$$

$$\frac{h}{t_w} = \frac{27.00 \text{ in}}{0.565 \text{ in}} = 47.78 < 3.76\sqrt{\frac{29 \times 10^3}{50}} = 90.55$$

$$\therefore M_n = M_p$$

Is PNA located at top of steel flange?

$$a = \frac{A_s f_y}{0.85 f'_c b_e} = \frac{(34.2 \text{ in}^2)(50 \text{ ksi})}{(0.85)(4 \text{ ksi})(80 \text{ in})} = 6.29 \text{ in} > 4.00 \text{ in}$$

$\therefore$ PNA is located down in steel section.

Is PNA in flange or in web? Here, we assume it is at base of steel flange.

$$C = 0.85 f'_c b_e t + F_y b_f t_f = (0.85)(4 \text{ ksi})(80 \text{ in})(4 \text{ in}) + $$
$$(50 \text{ ksi})(10.5 \text{ in})(0.850 \text{ in}) = 1534 \text{ k}$$

$$T = F_y(A_s - b_f t_f) = (50 \text{ ksi})(34.2 \text{ in}^2 - 10.5 \text{ in} \times 0.850 \text{ in}) = 1264 \text{ k}$$

Since $C > T$, the PNA falls in the steel flange and can be located as follows:

$$\bar{y} = \frac{F_y A_s - 0.85 f'_c b_e t}{2 F_y b_f} = \frac{(50 \text{ ksi})(34.2 \text{ in}^2) - (0.85)(4 \text{ ksi})(80 \text{ in})(4 \text{ in})}{(2)(50 \text{ ksi})(10.5 \text{ in})} = 0.592 \text{ in}$$

Then, using the moment expression given just before this example, we have

$$M_n = M_p = (0.85)(4 \text{ ksi})(80 \text{ in})(4 \text{ in})\left(\frac{4}{2} \text{ in} + 0.592 \text{ in}\right)$$

$$+ (2)(50 \text{ ksi})(10.5 \text{ in})(0.592 \text{ in})\left(\frac{0.592}{2} \text{ in}\right)$$

$$+ (50 \text{ ksi})(34.2 \text{ in}^2)\left(\frac{30.00}{2} \text{ in} - 0.592 \text{ in}\right)$$

$$= 27{,}650 \text{ in-k} = 2304 \text{ ft-k}$$

LRFD $\phi_b = 0.90$	ASD $\Omega_b = 1.67$
$\phi_b M_n = (0.90)(2304) = 2074$ ft-k	$\dfrac{M_n}{\Omega_b} = \dfrac{2304}{1.67} = 1380$ ft-k

By interpolation in AISC Table 3-19, page 3-170, with Y1 = 0.592 in and Y2 = 2.0 in, we get $2110 - \left(\dfrac{0.592 - 0.425}{0.638 - 0.425} \right)(2110 - 2060) = 2070$ ft-k for LRFD, and in a similar manner for ASD, we get 1376.5 ft-k.

If we have a partially composite section with ΣQ_n less than $A_s F_y$, the PNA will be down in the shape; if in the flange, the value of $\phi_b M_n$ can be determined with the equation used in Example 16-2. In the composite design tables presented in the Manual, values of ΣQ_n and $\phi_b M_n$ are shown for seven different PNA locations—top of the flange, quarter points in the flange, bottom of the flange, and two points down in the web. Straight-line interpolation may be used for numbers in between the tabulated values.

16.9.3   Neutral Axis in Web of Steel Section

If for a particular composite section we find that a is larger than the slab thickness, and if we then assume that the PNA is located at the bottom of the steel flange and we calculate C and T and find T is larger than C, the PNA will fall in the web. We can go through calculations similar to the ones we used for the case in which the PNA was located in the flange. Space is not taken to show such calculations, because the Composite Design tables in Part 3 of the Manual cover most common cases.

16.10   DEFLECTIONS

Deflections for composite beams may be calculated by the same methods used for other types of beams. The student must be careful to compute deflections for the various types of loads separately. For example, there are dead loads applied to the steel section alone (if no shoring is used), dead loads applied to the composite section, and live loads applied to the composite section.

The long-term creep effect in the concrete in compression causes deflections to increase with time. These increases, however, are usually not considered significant for the average composite beam. This is usually true, unless long spans and large permanent live loads are involved (AISC Commentary I3.2.4).

Should lightweight concrete be used, the actual modulus of elasticity of that concrete E_c (which may be rather small) should be used in calculating the transformed section moment of inertia I_{tr} for deflection computations. For stress calculations, we use E_c for normal-weight concrete.

Generally speaking, shear deflections are neglected, although on occasion they can be quite large.[1] The steel beams can be cambered for all or some portion of deflec-

---

[1] L. S. Beedle et al., *Structural Steel Design* (New York: Ronald Press, 1964), p. 452.

tions. It may be feasible in some situations to make a floor slab a little thicker in the middle than on the edges to compensate for deflections.

The designer may want to control vibrations in composite floors subject to pedestrian traffic or other moving loads. This may be the case where we have large open floor areas with no damping furnished by partitions, as in shopping malls. For such cases, dynamic analyses should be made.[2]

When the AISC Specification is used to select steel beams for composite sections, the results often will be some rather small steel beams and thus some quite shallow floors. Such floors, when unshored, frequently will have large deflections when the concrete is placed. For this reason, designers will often require cambering of the beams. Other alternatives include the selection of larger beams or the use of shoring.[3,4]

The beams selected must, of course, have sufficient $\phi_b M_n$ or $\dfrac{M_n}{\Omega}$ values to support themselves and the wet concrete. Nevertheless, their sizes are often dictated more by wet concrete deflections than by moment considerations. It is considered to be good practice to limit these deflections to maximum values, of about 2 1/2 in. Larger deflections than this value tend to cause problems with the proper placement of the concrete.

An alternative solution for these problems involves the use of partly restrained or semirigid PR connections (discussed in Chapter 15). When these connections are used, the midspan deflections and moments are appreciably reduced, enabling us to use smaller girders. Furthermore, there are reductions in the annoying vibrations that are a problem in shallow composite floors.

When semirigid PR connections are used, negative moments will develop at the supports. In Section 16.12 of this text, it is shown that the AISC Specification permits the use of negative design moment strength for composite floors, provided that certain requirements are met as to shear connectors and development of slab reinforcing in the negative moment region.

For unshored composite construction, the final deflections will equal the initial deflections caused by the wet concrete calculated with the moments of inertia of the steel beams, plus the deflections due to the loads applied after the concrete hardens, calculated with the moments of inertia of the composite sections. Should shored construction be used, all deflections will be calculated with moments of inertia of the composite sections. These latter moments of inertia, which are referred to as lower bound moments of inertia, are discussed later in the next section of this chapter.

16.11   DESIGN OF COMPOSITE SECTIONS

Composite construction is of particular advantage economically when loads are heavy, spans are long, and beams are spaced at fairly large intervals. For steel building frames, composite construction is economical for spans varying roughly from 25 to 50 ft, with particular advantage in the longer spans. For bridges, simple spans have been economically

[2]Thomas M. Murray, "Design to Prevent Floor Vibrations," *Engineering Journal*, AISC, vol. 12, no. 3 (3rd Quarter, 1975), pp. 82–87.

[3]R. Leon, "Composite Semi-Rigid Connections," *Modern Steel Construction*, vol. 32, no. 9 (AISC, Chicago: October 1992), pp. 18–23.

[4]"Innovative Design Cuts Costs," *Modern Steel Construction*, vol. 33, no. 4 (AISC, Chicago: April 1993), pp. 18–21.

constructed up to approximately 120 ft—and continuous spans 50 or 60 ft longer. Composite bridges are generally economical for simple spans greater than about 40 ft and for continuous spans greater than about 60 ft.

Occasionally, cover plates are welded to the bottom flanges of steel beams, with improved economy. One can see that with the slab acting as part of the beam, there is a very large compressive area available and that by adding cover plates to the tensile flange, a slightly better balance is obtained.

In tall buildings where headroom is a problem, it is desirable to use the minimum possible overall floor thicknesses. For buildings, minimum depth-span ratios of approximately 1/24 are recommended if the loads are fairly static and 1/20 if the loads are of such a nature as to cause appreciable vibration. The thicknesses of the floor slabs are known (from the concrete design), and the depths of the steel beams can be fairly well estimated from these ratios.

Before we attempt some composite designs, several additional points relating to lateral bracing, shoring, estimated steel beam weights, and lower bound moments of inertia are discussed in the paragraphs to follow.

16.11.1 Lateral Bracing

After the concrete slab hardens, it will provide sufficient lateral bracing for the compression flange of the steel beam. However, during the construction phase before the concrete hardens, lateral bracing may be insufficient and its design strength may have to be reduced, depending on the estimated unbraced length. When steel-formed decking or concrete forms are attached to the beam's compression flange, they usually will provide sufficient lateral bracing. The designer must very carefully consider lateral bracing for fully encased beams.

16.11.2 Beams with Shoring

If beams are shored during construction, we will assume that all loads are resisted by the composite section after the shoring is removed.

16.11.3 Beams without Shoring

If temporary shoring is not used during construction, the steel beam alone must be able to support all the loads before the concrete is sufficiently hardened to provide composite action.

Without shoring, the wet concrete loads tend to cause large beam deflections, which may lead us to build thicker slabs where the beam deflections are larger. This situation can be counteracted by cambering the beams.

The AISC Specification does not provide any extra margin against yield stresses occurring in beams during construction of unshored composite floors. Assuming that satisfactory lateral bracing is provided, the Specification (F2) states that the maximum factored moment may not exceed $0.90 F_y Z$. The 0.90 in effect limits the maximum factored moment to a value about equal to the yield moment $F_y S$.

To calculate the moment to be resisted during construction, it makes sense to count the wet concrete as a live load and also to include some extra live load (perhaps 20 psf) to account for construction activities.

16.11.4 Estimated Steel Beam Weight

As illustrated in Example 16-3, it sometimes may be useful to make an estimate of the weight of the steel beam. The third edition of the LRFD Manual provided the following empirical formula for this purpose in its Part 5 (page 5-26):

$$\text{Estimated beam weight} = \left[\frac{12M_u}{(d/2 + Y_{con} - a/2)\phi F_y} \right] 3.4$$

Here,

M_u = required flexural strength of composite section, ft-k

d = nominal steel beam depth, in

Y_{con} = distance from top of steel beam to top of concrete slab, in

a = effective concrete slab thickness, in (which can be conservatively estimated as somewhere in the range of about 2 in)

ϕ = 0.85

16.11.5 Lower Bound Moment of Inertia

To calculate the service load deflections for composite sections, a table of lower bound moment of inertia values is presented in Part 3 of the Manual (Table 3-20). These values are computed from the area of the steel beam and an equivalent concrete area of $\Sigma Q_n/F_y$. The remainder of the concrete flange is not used in these calculations. This means that if we have partially composite sections, the value of the lower bound moment of inertia will reflect this situation because ΣQ_n will be smaller. The lower bound moment of inertia is computed with the expression that follows. See Fig. 16.12 which is Figure 3-5 from Part 3 of the AISC Manual. Terms are defined in AISC commentary I3.

$$I_{LB} = I_s + A_s(Y_{ENA} - d_3)^2 + \left(\frac{\Sigma Q_n}{F_y} \right)(2d_3 + d_1 - Y_{ENA})^2$$

(AISC Commentary Equation C-I3-1)

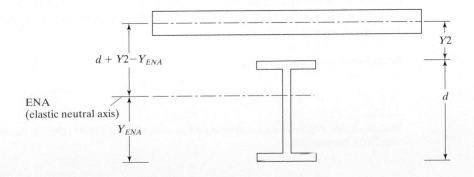

FIGURE 16.12

Here,

I_{LB} = lower bound moment of inertia, in^4

I_s = moment of inertia of steel section, in^4

d_1 = distance from the compression force in the concrete to the top of the steel section, in

d_3 = distance from the resultant steel tension force for full section tension yield to the top of the steel, in

Y_{ENA} = the distance from the bottom of the beam to the elastic neutral axis (ENA), in

$$= \left[\left(A_s d_3 + \left(\frac{\Sigma Q_n}{F_y} \right)(2d_3 + d_1) \right) \Big/ \left(A_s + \frac{\Sigma Q_n}{F_y} \right) \right] \qquad \text{(C-I3-2)}$$

16.11.6 Extra Reinforcing

For building design calculations, the spans often are considered to be simply support-ed, but the steel beams generally do not have perfectly simple ends. The result is that some negative moment may occur at the beam ends, with possible cracking of the slab above. To prevent or minimize cracking, some extra steel can be placed in the top of the slab, extending 2 or 3 ft out into the slab. The amount of steel added is in addition to that needed to meet the temperature and shrinkage requirements specified by the American Concrete Institute.[5]

16.11.7 Example Problems

Examples 16-3 and 16-4 illustrate the designs of two unshored composite sections.

Example 16-3

Beams 10 ft on center with 36-ft simple spans are to be selected to support a 4-in-deep lightweight concrete slab on a 3-in-deep formed steel deck with no shoring. The ribs for the steel deck, which are perpendicular to the beam center lines, have average widths of 6 in. If the service dead load (including the beam weight) is to be 0.78 k/ft of length of the beams and the service live load is 2 k/ft, (a) select the beams, (b) determine the number of 3/4-in-diameter headed studs required, (c) compute the service live load deflection, and (d) check the beam shear. Other data are as follows: 50 ksi steel, f_c = 4 ksi, and concrete weight 110 lb/ft^3.

Solution. Loads and moments

---

[5]*Building Code Requirements for Reinforced Concrete*, ACI std. 318-05 (Detroit: American Concrete Insti-tute, 2005), Section 7.12.

LRFD	ASD
$w_u = (1.2)(0.78) + (1.6)(2.0) = 4.14$ k/ft	$w_a = 0.78 + 2.0 = 2.78$ k/ft
$M_u = \dfrac{(4.14)(36)^2}{8} = 670.7$ ft-k	$M_a = \dfrac{(2.78)(36)^2}{8} = 450.4$ ft-k

Effective flange width b_e

$$b_e = (2)\left(\tfrac{1}{8} \times 36 \times 12\right) = 108 \text{ in} \ \leftarrow$$
$$b_e = (2)(5 \times 12) = 120 \text{ in}$$

a.   Select W section

Y_{con} = distance from top of slab to top of steel flange = $4 + 3 = 7$ in

Assume $a = 2$ in < 4 in slab thickness (It's usually quite small, particularly for relatively light sections.)

Y1 is distance from PNA to top flange = 0 in

Y2 is the distance from the center of gravity of the concrete flange force to the top flange of the beam = $7 - a/2 = 7 - 2/2 = 6$ in

Looking through the composite tables of the Manual, with $M_u = 670.7$ ft-k, Y1 = 0, and Y2 = 6 in, we can see that several W 18s (46 lb, 50 lb, and 55 lb) seem reasonable.

Try W18 × 46 ($A = 13.5$ in^2, $I_x = 712$ in^4);

Check deflection due to wet concrete plus beam wt

$$w = \left(\frac{4}{12}\right)(110)(10) + 46 = 413 \text{ lbs/ft}$$

$$M = \frac{(0.413)(36)^2}{8} = 66.9 \text{ ft-k}$$

$$C_1 = 161 \text{ from Fig. 3-2 in AISC Manual}$$

$$\Delta = \frac{ML^2}{C_1 I_x} = \frac{(66.9)(36)^2}{(161)(712)} = 0.76 \text{ in} < 2.5 \text{ in} \textbf{ OK}$$

Assume $\Sigma Q_n = A_s F_y = (13.5)(50) = 675$ k ($\Sigma Q_n = 677$ k from AISC Table 3-19)

Thus, a required $= \dfrac{\Sigma Q_n}{0.85 f'_c b_e} = \dfrac{675}{(0.85)(4)(108)} = 1.84$ in < 4 in

$$Y1 = 0$$
$$Y2 = 7.00 - \frac{1.84}{2} = 6.08 \text{ in}$$

$\phi_b M_n$ from Manual, by interpolation,

$$= 763 + \left(\frac{0.08}{0.50}\right)(788 - 763) = 767 \text{ ft-k} > 670.7 \text{ ft-k} \textbf{ OK}$$

This moment for which ΣQ_n is 677 k is somewhat on the high side. With this same W18 × 46, we can go to the case in AISC Table 3-19 where Y1 is the largest possible, to provide a $\phi_b M_n$ of about 671 ft-k with Y2 = approximately 6 in. The result will be that ΣQ_n will be smaller, and fewer shear connectors will be necessary. This will occur when Y1 = 0.303 in and ΣQ_n = 494 ft-k (see AISC Table 3-19) with Y2 = 6 in.

$$a_{reqd} = \frac{\Sigma Q_n}{0.85 f'_c b_e} = \frac{494}{(0.85)(4)(108)} = 1.35 \text{ in}$$

$$Y2 = 7.00 - \frac{1.35}{2} = 6.33 \text{ in}$$

LRFD	ASD
$\phi_b M_n = 678 + \left(\dfrac{0.33}{0.50}\right)(697 - 678) = 690.5$ ft-k	$\dfrac{M_n}{\Omega} = 451 + \left(\dfrac{0.33}{0.50}\right)(464 - 451)$
> 670.7 ft-k **OK**	$= 459.6$ ft-k > 450.4 ft-k

Use W18 × 46, F_y = 50 ksi.

b. Design of steel headed stud anchors

The strengths (or Q_n values) of individual studs are provided in AISC Table 3-21. The author selected from this table a value of 21.2 k for 3/4-in headed studs enclosed in 4 ksi lightweight concrete weighing 110 lbs/ft³. To obtain this value, he also made the assumptions that only one stud would be placed in each rib and that the studs would be in the strong position described in the AISC Commentary Section I8.2a.

Number of anchors required $= \dfrac{\Sigma Q_n}{Q_n} = \dfrac{494}{21.2} = 23.3$

Use 24 3/4-in studs on each side of point of maximum moment (which is L_c here).

c. Compute LL deflection

Assume maximum permissible LL deflection

$$\frac{1}{360} \text{ span} = \left(\frac{1}{360}\right)(12 \times 36) = 1.2 \text{ in}$$

$$C_1 = 161 \text{ from Fig. 3-2 in AISC Manual}$$

$$M_L = \frac{(2.0 \text{ k/ft})(36 \text{ ft})^2}{8} = 324 \text{ ft-k}$$

I_{LB} = lower bound moment of inertia from AISC Table 3-20
using straight-line interpolation

$$= 2000 + \left(\frac{0.33}{0.50}\right)(2090 - 2000) = 2059 \text{ in}^4$$

$$\Delta_L = \frac{ML^2}{C_1 I_{LB}} = \frac{(324)(36)^2}{(161)(2059)} = 1.27 \text{ in} > 1.2 \text{ in} \text{ **A little high**}$$

d.   Check beam shear for steel section

LRFD	ASD
$V_u = \dfrac{4.14(36)}{2} = 74.5 \text{ k}$	$V_a = \dfrac{2.78(36)}{2} = 50.0 \text{ k}$
$\phi V_n = 195$ k from Table 3-2 in Manual **OK**	$\dfrac{V_n}{\Omega} = 130$ k from Table 3-2 in Manual **OK**

Use W18 × 46 with forty-eight 3/4-in headed studs.

Example 16-4

Using the same data as for Example 16-3, except that $w_L = 1.2$ k/ft and $f'_c = 3$ ksi, perform the following tasks:

a.   Select steel beam for composite action.
b.   If the studs are placed in the weak position with no more than one stud per rib, determine the number of 3/4-in headed studs required.
c.   Check the beam strength before the concrete hardens.
d.   Compute service load deflection before concrete hardens. Assume a construction live load of 20 psf.
e.   Determine the service live load deflection after composite action is available.
f.   Check shear.
g.   Select a steel section to carry all the loads if no shear connectors are used, and compute its service live load deflection.

Solution

LRFD	ASD
$w_u = (1.2)(0.78) + (1.6)(1.2) = 2.86$ k/ft	$w_a = 0.78 + 1.2 = 1.98$ k/ft
$M_u = \dfrac{(2.86)(36)^2}{8} = 463.3$ ft-k	$M_a = \dfrac{(1.98)(36)^2}{8} = 320.8$ ft-k

a.   Select W section

$$Y_{con} = 4 + 3 = 7 \text{ in}$$

Assume $a = 2$ in

$$Y1 = 0$$

$$Y2 = 7 - \frac{2}{2} = 6 \text{ in}$$

Try W16 × 31 ($A = 9.13$ in^2, $d = 15.9$ in, $t_w = 0.275$ in)

Assume $\Sigma Q_n = (9.13)(50) = 456.5$ k or 456 k from AISC Manual Table 3-19

$$a = \frac{456.5}{(0.85)(3)(108)} = 1.66 \text{ in} < 4 \text{ in}$$

$$Y2 = 7 - \frac{1.66}{2} = 6.17 \text{ in}$$

$\phi_b M_n$ from AISC Table 3-19 by interpolation

$$= 477 + \left(\frac{0.17}{0.50}\right)(494 - 477) = 482.8 \text{ ft-k} > 463.3 \text{ ft-k} \quad \textbf{OK}$$

b. Design of studs

Q_n from AISC Table 3-21 $= 17.2$ k

We cannot go down in the W16 × 31 values to reduce, because $\phi_b M_n$ values are insufficient

$$\Sigma Q_n = 456.5 \text{ kips}$$

Number of stud anchors required $= \dfrac{456.5}{17.2} = 26.5$

Use twenty-seven 3/4-in stud anchors each side of ℄.

c. Check strength of W section before concrete hardens.

Assume that the wet concrete is a LL during construction, and also add a 20 psf construction live load.

Concrete wt = wt of slab + wt of ribs

$$= \left(\frac{4}{12}\right)(10)(110) + [(3)(6)/144](110)(10) = 504 \text{ lbs/ft}$$

Other dead loads = Deck wt (assume = 2 psf) + Beam wt

$$= 2(10) + 31 = 51 \text{ lbs/ft}$$

LRFD	ASD
$w_u = (1.2)(0.051) + (1.6)(0.020 \times 10 + 0.504) = 1.19$ k/ft	$w_a = 0.051 + 0.704 = 0.76$ k/ft
$M_u = \dfrac{(1.19)(36)^2}{8} = 192.8$ ft-k	$M_a = \dfrac{(0.76)(36)^2}{8} = 123.1$ ft-k
Assume metal deck provides lateral bracing	$\dfrac{M_n}{\Omega} = 135$ ft-k from AISC Table 3-2
$\phi M_n = 203$ ft-k from AISC Table 3-2 > 192.8 ft-k **OK**	> 123.1 ft-k **OK**

(d) Service load deflection before concrete hardens. I_x for W16 × 31 $= 375$ in^4 (not lower bound I)

Use $w_D = 0.76$ k/ft

$$M_D = \frac{(0.76)(36)^2}{8} = 123.1 \text{ ft-k}$$

$$\Delta_{DL} = \frac{(123.1)(36)^2}{(161)(375)} = 2.64 \text{ in} > 2.50 \text{ in}$$

(We might camber beam for this deflection and/or use PR connections.)

e.   Service LL deflection after composite action is available

$$M_L = \frac{(1.2)(36)^2}{8} = 194.4 \text{ ft-k}$$

Lower bound I from AISC Table 3-20 with Y1 = 0 and Y2 = 6.17 in

$$I = 1260 + \left(\frac{0.17}{0.50}\right)(1320 - 1260) = 1284 \text{ in}^4$$

$$\Delta_L = \frac{(194.4)(36)^2}{(161)(1284)} = 1.22 \text{ in} > \frac{L}{360} = 1.2 \text{ in (Probably OK)}$$

f.   Check shear

LRFD	ASD
$w_u = (1.2)(0.78) + (1.6)(1.2) = 2.86$ k/ft	$w_a = 0.78 + 1.2 = 1.98$ k/ft
$V_u = \dfrac{(2.86)(36)}{2} = 51.5$ k	$V_a = \dfrac{(1.98)(36)}{2} = 35.64$ k
$\phi V_n = 131$ k from AISC Table 3-2 > 51.4 k **OK**	$\dfrac{V_n}{\Omega} = 87.3$ k from AISC Table 3-2 > 35.64 k **OK**

g.   Selecting steel section, no composite action

LRFD	ASD
$M_u = 463.3$ ft-k	$M_a = 320.8$ ft-k
Select W21 × 55 from AISC Table 3-2	Select W24 × 55 from AISC Table 3-2
$\phi M_n = 473$ ft-k > 463.3 ft-k	$\dfrac{M_n}{\Omega} = 334$ ft-k > 320.8 ft-k

I_x for W21 × 55 = 1140 in^4

$M_L = 194.4$ ft-k from part e of this problem

$$\text{Service live load deflection} = \frac{(194.4)(36)^2}{(161)(1140)} = 1.37 \text{ in} > \frac{L}{360} = 1.2 \text{ in}$$

$$\text{Min } I_x \text{ to limit deflection to 1.2 in} = \left(\frac{1.37}{1.2}\right)(1140) = 1302 \text{ in}^4$$

(A W24 × 55 would be required by Table 3-3 in the Manual to provide such an I_x, or a W21 × 62 to keep depth approximately the same.)

Continuous-welded plate girders
in Henry Jefferson County, IA.
(Courtesy of the Lincoln Electric
Company.)

16.12 CONTINUOUS COMPOSITE SECTIONS

The AISC Specification (I3.2b) permits the use of continuous composite sections. The flexural strength of a composite section in a negative moment region may be considered to equal $\phi_b M_n$ for the steel section alone, or it may be based upon the plastic strength of a composite section considered to be made up of the steel beam and the longitudinal reinforcement in the slab. For this latter method to be used, the following conditions must be met:

1. The steel section must be compact and adequately braced.
2. The slab must be connected to the steel beams in the negative moment region with shear connectors.
3. The longitudinal reinforcing in the slab parallel to the steel beam and within the effective width of the slab must have adequate development lengths. (*Development length* is a term used in reinforced-concrete design and refers to the length that reinforcing bars have to be extended or embedded in the concrete in order to properly anchor them or develop their stresses by means of bond between the bars and the concrete.)

For a particular beam, the total horizontal shear force between the point of zero moment and the point of maximum negative moments is to be taken as the smaller of $A_{sr}F_{ysr}$ and ΣQ_n, where A_{sr} is the cross-sectional area of the properly developed reinforcing and F_{ysr} is the yield stress of the bars. The plastic stress distribution for negative moment in a composite section is illustrated in Fig. 16.13.

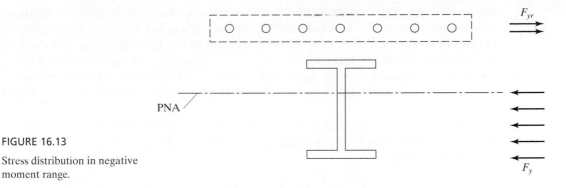

FIGURE 16.13

Stress distribution in negative moment range.

16.13   DESIGN OF CONCRETE-ENCASED SECTIONS

For fireproofing purposes, it is possible to completely encase in concrete the steel beams used for building floors. This practice definitely is not economical, because light-weight spray-on fire protection is so much cheaper. Furthermore, encased beams may increase the floor system dead load by as much as 15 percent.

For the rare situation in which encased beams are used, steel anchors shall be provided.

The nominal flexual strength, M_n, shall be determined using one of several methods, see AISC Specification I3.3.

1   With one method, the design strength of the encased section may be based on the plastic moment capacity $\phi_b M_p$ or $\dfrac{M_p}{\Omega_b}$ of the steel section alone, with $\phi_b = 0.90$ and $\Omega_b = 1.67$.

2.   By another method, the design strength is based on the first yield of the tension flange, assuming composite action between the concrete, which is in compression, and the steel section. Again, $\phi_b = 0.90$ and $\Omega_b = 1.67$.

If the second method is used and we have unshored construction, the stresses in the steel section caused by the wet concrete and other construction loads are calculated. Then the stresses in the composite section caused by loads applied after the concrete hardens are computed. These stresses are superimposed on the first set of stresses. If we have shored construction, all of the loads may be assumed to be supported by the composite section and the stresses computed accordingly. For stress calculations, the properties of a composite section are computed by the transformed area method. In this method, the cross-sectional area of one of the two materials is replaced or transformed into an equivalent area of the other. For composite design, it is customary to replace the concrete with an equivalent area of steel, whereas the reverse procedure is used in the working stress design method for reinforced-concrete design.

In the transformed area procedure, the concrete and steel are assumed to be bonded tightly together so that their strains will be the same at equal distances from the neutral axis. The unit stress in either material can then be said to equal its strain times its modulus of elasticity (ϵE_c for the concrete or ϵE_s for the steel). The unit stress in the steel is then $\epsilon E_s/\epsilon E_c = E_s/E_c$ times as great as the corresponding unit stress in the concrete. The E_s/E_c ratio is referred to as the modular ratio n; therefore, n in^2 of concrete are required to resist the same total stress as 1 in^2 of steel; and the cross-sectional area of the slab (A_c) is replaced with a transformed or equivalent area of steel equal to A_c/n.

The American Concrete Institute (ACI) Building Code states that the following expression may be used for calculating the modulus of elasticity of concrete weighing from 90 to 155 lb/ft^3:

$$E_c = w_c^{1.5} 33 \sqrt{f_c'}$$

In this expression, w_c is the weight of the concrete in pounds per cubic foot, and f_c' is the 28-day compressive strength in pounds per square inch.

> In SI units with w_c varying from 1500 to 2500 kg/m^3 and with
> f_c' in N/mm^2 or MP_a $E_c = w_c^{1.5}(0.043)\sqrt{f_c'}$

There are no slenderness limitations required by the AISC Specification for either of the two methods, because the encasement is effective in preventing both local and lateral buckling.

In Example 16-5, which follows, stresses are computed by the elastic theory, assuming composite action as described for the second method. Notice that the author has divided the effective width of the slab by n to transform the concrete slab into an equivalent area of steel.

Example 16-5

Review the encased beam section shown in Fig. 16.14 if no shoring is used and the following data are assumed:

$$\text{Simple span} = 36 \text{ ft}$$

$$\text{Service dead load} = 0.50 \text{ k/ft before concrete hardens plus an}$$

$$\text{additional } 0.25 \text{ k/ft after concrete hardens}$$

$$\text{Construction live loads} = 0.2 \text{ k/ft}$$

$$\text{Service live load} = 1.0 \text{ k/ft after concrete hardens}$$

$$\text{Effective flange width } b_e = 60 \text{ in and } n = 9$$

$$F_y = 50 \text{ ksi}$$

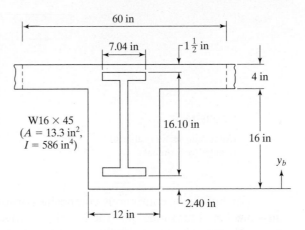

FIGURE 16.14

Solution. Calculated Properties of Composite Section: Neglecting concrete area below the slab.

$$A = 13.3 \text{ in}^2 + \frac{(4 \text{ in})(60 \text{ in})}{9} = 39.96 \text{ in}^2$$

$$y_b = \frac{(13.3 \text{ in}^2)(10.45 \text{ in}) + (26.66 \text{ in}^2)(18 \text{ in})}{39.96} = 15.50 \text{ in}$$

$$I = 586 \text{ in}^4 + (13.3 \text{ in}^2)(5.05 \text{ in})^2 + \left(\tfrac{1}{12}\right)\left(\tfrac{60}{9} \text{ in}\right)(4 \text{ in})^3 + (26.66 \text{ in}^2)(2.5 \text{ in})^2$$
$$= 1127 \text{ in}^4$$

Stresses before concrete hardens
   Assume that wet concrete is a live load

$$w_u = (1.6)(0.5 \text{ k/ft} + 0.2 \text{ k/ft}) = 1.12 \text{ k/ft}$$
$$M_u = \frac{(1.12 \text{ k/ft})(36 \text{ ft})^2}{8} = 181.4 \text{ ft-k}$$

   Assume beam only properties

$$f_t = \frac{(12 \text{ in/ft})(181.4 \text{ ft-k})(8.05 \text{ in})}{586 \text{ in}^4} = 29.90 \text{ ksi}$$
$$< \phi_b F_y = (0.9)(50) = 45 \text{ ksi} \qquad\qquad \text{(OK)}$$

Stresses after concrete hardens

$$w_u = (1.2)(0.25 \text{ k/ft}) + (1.6)(1.0 \text{ k/ft}) = 1.9 \text{ k/ft}$$
$$M_u = \frac{(1.9 \text{ k/ft})(36 \text{ ft})^2}{8} = 307.8 \text{ ft-k}$$

$$f_t = \frac{(12 \text{ in/ft})(307.8 \text{ ft-k})(15.50 \text{ in} - 2.40 \text{ in})}{1127 \text{ in}^4} = 42.93 \text{ ksi}$$

Total $f_t = 29.90 + 42.93 = 72.83 \text{ ksi} > 0.9F_y = 45 \text{ ksi}$ \qquad (NG)

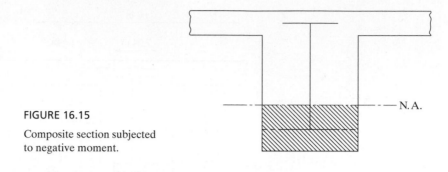

FIGURE 16.15

Composite section subjected
to negative moment.

For buildings, continuous composite construction with encased sections is permissible. For continuous construction, the positive moments are handled exactly as has been illustrated by the preceding example. For negative moments, however, the transformed section is taken as shown in Fig. 16.15. The crosshatched area represents the concrete in compression, and all concrete on the tensile side of the neutral axis (that is, above the neutral axis) is neglected.

16.14   PROBLEMS FOR SOLUTION

Use LRFD and ASD methods for Problems 16-1 through 16-19.

16-1.   Determine $\phi_b M_n$ and $\dfrac{M_n}{\Omega_b}$ for the section shown, assuming that sufficient steel anchors are provided to ensure full composite section. Solve by using the procedure presented in Section 16.9 and check answers with tables in the Manual. $F_y = 50$ ksi, $f'_c = 3$ ksi. (*Ans.* 366.3 ft-k, LRFD; 243.7 ft-k, ASD)

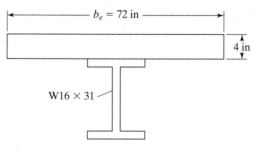

FIGURE P16-1

16-2.   Repeat Prob. 16-1 if a W18 × 55 is used.

16-3.   Repeat Prob. 16-2, using tables in Manual if it is considered to be partially composite and if ΣQ_n is 454 k. (*Ans.* 637.8 ft-k, 424.2 ft-k)

16-4.   Determine $\phi_b M_n$ and $\dfrac{M_n}{\Omega_b}$ for the section shown, if 50 ksi steel and sufficient steel anchors are used to guarantee full composite action. Use formulas and check with Manual. $f'_c = 4$ ksi.

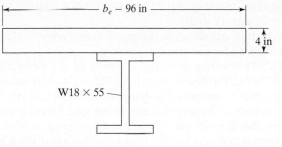

FIGURE P16-4

16-5.   Repeat Prob. 16-4 if a W16 × 36 is used. (*Ans.* 442.7 ft-k, 294.5 ft-k)

16-6.   Compute $\phi_b M_n$ and $\dfrac{M_n}{\Omega_b}$ for the composite section shown, if 50 ksi steel is used and sufficient steel anchors are used to provide full composite action. A 3-in concrete slab is supported by 2-in-deep composite metal deck ribs perpendicular to beam; $f'_c = 4$ ksi. Check answers with Manual.

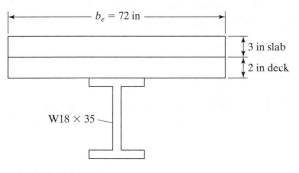

FIGURE P16-6

16-7.   Repeat Prob. 16-6 using Manual tables if ΣQ_n of the anchors is 387 k. (*Ans.* 462.9 ft-k, 308.2 ft-k)

16-8.   Using the Composite Design Tables of the AISC Manual, 50 ksi steel, a 145 lb/ft^3 concrete slab with $f'_c = 4$ ksi and shored construction, select the steel section, design 3/4-in headed studs, calculate live load service deflection, and check the shear if the service live load is 100 psf. Refer to the accompanying figure.

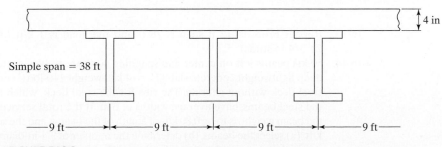

FIGURE P16-8

16-9.   Repeat Prob. 16-8 if span is 32 ft and live load is 80 psf. (*Ans.* For LRFD and ASD W14 × 22 with 19 studs)

16-10.   For Prob. 16-9, calculate the deflection during construction for wet concrete, plus 20 psf live load for construction activities.

16-11.   Select a 50 ksi section to support a service dead load of 200 psf and a service live load of 100 psf. The beams are to have 37.5-ft simple spans and are to be spaced 8 ft 6 in on center. Construction is shored, concrete weighs 110 lb/ft^3, f'_c is 3.5 ksi, and a metal deck with ribs perpendicular to the steel beams is used together with a 4-in concrete slab. The ribs are 3-in deep and have average widths of 6 in. Design 3/4-in headed studs and calculate live load deflection. (*Ans.* For LRFD W18 × 46 with 61 studs)

16-12.   Using the AISC Manual and 50 ksi steel, design a nonencased unshored composite section for the simple span beams shown in the accompanying figure if a 4-in concrete slab (145 lb/ft^3) with $f'_c = 4$ ksi is used. The total service dead load, including the steel beam, is 0.6 k/ft of length of the beam, and the service live load is 1.25 k/ft. Assume construction $LL = 20$ psf and $L_b = 0$.

    a.   Select the beams.

    b.   Determine the number of 3/4-in-diameter headed studs required.

    c.   Compute the service live load deflection.

    d.   Check the beam shear.

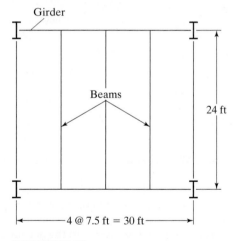

FIGURE P16-12

16-13.   Repeat Prob. 16-12 if span is 28 ft and live load is 1 k/ft (*Ans.* W14 × 22 with 22 3/4-in studs)

16-14.   50 ksi beams 9 ft on center and spanning 40 ft are to be selected to support a 4-in-deep lightweight concrete slab ($f'_c = 4$ ksi, weight 110 lb/ft^3) on a 3-in-deep formed steel deck without shoring. The ribs for the steel deck, which are perpendicular to the steel beams, have average widths of 6 in. If the total service dead load, including the beam weight, is to be 0.80 k/ft of length of the beams, and the service live load is 1.25 k/ft, (a) select the beams, (b) determine the number of 3/4-in-diameter headed studs required, (c) compute the service live load deflection, and (d) check the beam shear.

16-15. Repeat Prob. 16-14 if spans are 45 ft. (*Ans.* W21 × 44 with 68 studs LRFD)

16-16. Repeat Prob. 16-14 if spans are 32 ft.

16-17. Repeat Prob. 16-16 if spans are 34 ft and the live load is 2 k/ft. (*Ans.* For LRFD W18 × 40 with 60 3/4-in studs)

16-18. Using the same data as for Prob. 16-14, except that unshored construction is to be used for a 45 ft span, perform the following tasks:

    a. Select the steel beam.

    b. If the stud reduction factor for metal decks is 1.0, determine the number of 3/4-in headed studs required assuming deck ribs are perpendicular to the beams.

    c. Check the beam strength before the concrete hardens.

    d. Compute service load deflection before the concrete hardens, assuming a construction live load of 25 psf.

    e. Determine the service load deflection after composite action is available.

    f. Check shear.

16-19. Repeat Prob. 16-18 if span is 35 ft and live load is 1.60 k/ft. (*Ans.* W18 × 35 with 54 studs LRFD)

16-20. Using the transformed area method, compute the stresses in the encased section shown in the accompanying illustration if no shoring is used. The section is assumed to be used for a simple span of 30 ft and to have a service dead uniform load of 30 psf applied after composite action develops and a service live uniform load of 120 psf. Assume $n = 9$, $F_y = 50$ ksi, $f'_c = 4$ ksi, and concrete weighing 150 lb/ft^3.

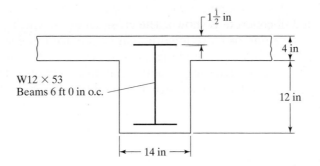

FIGURE P16-20

C H A P T E R   1 7

Composite Columns

17.1   INTRODUCTION

Composite columns are constructed with rolled or built-up steel shapes encased in concrete, or with concrete placed inside HSS or pipe sections. The resulting members are able to support significantly higher loads than reinforced-concrete columns of the same sizes.

Several composite columns are shown in Fig. 17.1. In part (a) of the figure, a *W* shape embedded in concrete is shown. The cross sections, which usually are square or rectangular, have one or more longitudinal bars placed in each corner. In addition, lateral ties are wrapped around the longitudinal bars at frequent vertical intervals. Ties are effective in increasing column strengths. They prevent the longitudinal bars from being displaced during construction, and they resist the tendency of these

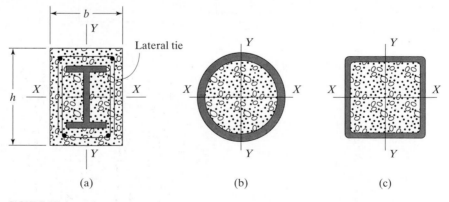

FIGURE 17.1

Composite columns.

same bars to buckle outward under load, which would cause breaking or spalling off of the outer concrete cover. Notice that these ties are all open and U-shaped. Otherwise they could not be installed, because the steel column shapes always will have been erected at an earlier time. In parts (b) and (c) of the figure, steel pipe and HSS sections filled with concrete are shown.

17.2   ADVANTAGES OF COMPOSITE COLUMNS

For a number of decades, structural steel shapes have been used in combination with plain or reinforced concrete. Originally, the encasing concrete was used to provide only fire and corrosion protection for the steel, with no consideration given to its strengthening effects. More recently, however, the development and increasing popularity of composite frame construction has encouraged designers to include the strength of the concrete in their calculations.[1,2]

Composite columns may be practically used for low-rise and high-rise buildings. For the low-rise warehouses, parking garages, and so on, the steel columns are often encased in concrete for the sake of appearance or for protection from fire, corrosion, and (in garages) vehicles. If we are going to encase the steel in concrete anyway, we may as well take advantage of the concrete and use smaller steel shapes.

For high-rise buildings, the sizes of composite columns often are considerably smaller than is required for reinforced-concrete columns to support the same loads. The results with composite designs are appreciable savings of valuable floor space. Closely spaced composite steel-concrete columns connected with spandrel beams may be used around the outsides of high-rise buildings to resist lateral loads by the tubular concept (described in Chapter 19). Very large composite columns are sometimes placed on the corners of high-rise buildings to increase lateral resisting moments. Also, steel sections embedded within reinforced-concrete shear walls may be used in the central core of high-rise buildings. This also ensures a greater degree of precision in the construction of the core.

With composite construction, the bare steel sections support the initial loads, including the weights of the structure, the gravity, lateral loads occurring during construction, and the concrete later cast around the W shapes or inside the tube shapes. The concrete and steel are combined in such a way that the advantages of both materials are used in the composite sections. For instance, the reinforced concrete enables the building frame to more easily limit swaying or lateral deflections. At the same time, the light weight and strength of the steel shapes permit the use of smaller and lighter foundations.[3]

---

[1] D. Belford, "Composite Steel Concrete Building Frame," *Civil Engineering* (New York: ASCE, July 1972), pp. 61–65.
[2] Fazlur R. Kahn, "Recent Structural Systems in Steel for High Rise Buildings," BCSA Conference on Steel in Architecture, (London, November 24–26, 1969).
[3] L. G. Griffis, "Design of Encased W-shape Composite Columns," Proceedings 1988 National Steel Construction Conference (AISC, Chicago, June 8–11, 1988), pp. 20-1–20-28.

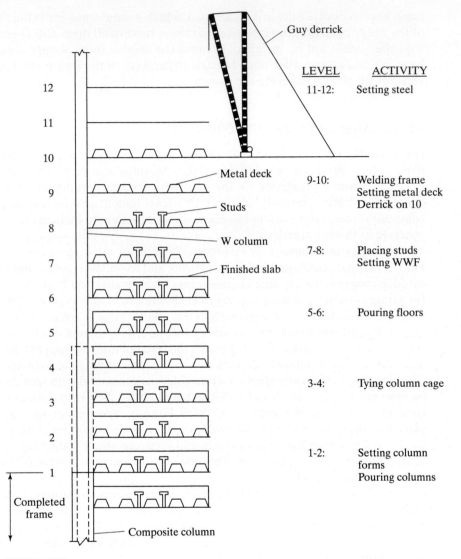

Guy derrick

LEVEL	ACTIVITY
11-12:	Setting steel
9-10:	Welding frame Setting metal deck Derrick on 10
7-8:	Placing studs Setting WWF
5-6:	Pouring floors
3-4:	Tying column cage
1-2:	Setting column forms Pouring columns

Metal deck

Studs

W column

Finished slab

Composite column

Completed frame

FIGURE 17.2

Sequence of construction operations for a composite building frame. (Courtesy of AISC.)

   Composite high-rise structures are erected in a rather efficient manner. There is quite a vertical spread of construction going on at any one time, with numerous trades working simultaneously. This situation, which is pictured in Fig. 17.2, is briefly described here.[4]

---

[4]Griffis, op.cit.

1. One group of workers may be erecting the steel beams and columns for one or two stories on top of the frame.

2. Two or three stories below, another group will be setting the metal decking for the floors.

3. A few stories below that, another group may be placing the concrete for the floor slabs.

4. This continues as we go down the building, with one group tying the column reinforcing bars in cages, while below them others are placing the column forms, placing the column concrete, and so on.

17.3 DISADVANTAGES OF COMPOSITE COLUMNS

As described in the preceding section, composite columns have several important advantages. They also have a few disadvantages. One particular problem with their use in high-rise buildings is the difficulty of controlling their rates and amounts of shortening in relation to shear walls and, perhaps, adjacent plain steel columns. The accurate estimation of these items is made quite difficult by the different types and stages of construction activities going on simultaneously over a large number of building stories.

If composite columns are used around the outside of a high-rise building, and plain steel sections are used in the building core (or if we have shear walls), the creep in the composite sections can be a problem. The results may be concrete floors that are not very level. Some erectors make very careful elevation measurements at column splices and then try to make appropriate adjustments with steel shims to try to even out the differences between measured elevations and computed elevations.

Another problem with composite columns is the lack of knowledge available concerning the mechanical bond between the concrete and the steel shapes. This is particularly important for the transfer of moments through beam–column joints. It is feared that if large cyclical strain reversals were to occur at such a joint (as in a seismic area), there could be a severe breakdown of the joint.[5]

17.4 LATERAL BRACING

Resistance to lateral loads for the usual structural steel or reinforced-concrete high-rise building is provided as the floors are being constructed. For instance, diagonal bracing or moment-resisting joints may be provided for each floor as a structural steel building frame is being constructed. In a similar manner, the needed lateral strength of a reinforced-concrete frame may be provided by the moment resistance obtained with monolithic construction of its members and/or by shear walls.

For composite construction, the desired lateral strength of a building is not obtained until the concrete has been placed around or inside the erected steel members and has sufficiently hardened. This situation is probably being achieved 10 to 18 stories behind the steel erection (see Fig. 17.2).

---

[5]Griffis, op.cit.

As we have mentioned, the steel fabricator is used to erecting a steel frame and providing the necessary wind bracing as the floors are erected. The steel frames used for high-rise composite buildings, however, do not usually have such bracing, and the frames will not have the desired lateral strength. This strength will be achieved only after the concrete is placed and cured for many building stories. Thus, the engineer of record for a composite high-rise building must clearly state the lateral force conditions and what is to be done about them during erection.[6]

17.5   SPECIFICATIONS FOR COMPOSITE COLUMNS

Composite columns theoretically can be constructed with cross sections that are square, rectangular, round, triangular, or any other shape. Practically, however, they usually are square or rectangular, with one reinforcing bar in each column corner. This arrangement enables us to use reasonably simple connections from the exterior spandrel beams and floor beams to the steel shapes in the columns, without unduly interfering with the vertical reinforcing.

The AISC Specification does not provide detailed requirements for reinforcing bar spacings, splices, and so on. Therefore, it seems logical that the requirements in this regard of the ACI 318 Code[7] should be followed for situations not clearly covered by the AISC Specification.

Sections I1 and I2 of the AISC Specification provide detailed requirements pertaining to cross-sectional areas of steel shapes, concrete strengths, tie areas, and spacings for the vertical reinforcing bars, and so on. This information is listed and briefly discussed in the paragraphs to follow.

For encased composite columns

1. *The total cross-sectional area of the steel section or sections may not be less than 1 percent of the gross column area.* If the steel percentage is less than 1 percent, the member is classified as a reinforced-concrete column, and its design must adhere to the *Building Code Requirements for Reinforced Concrete* of the American Concrete Institute.

2. *Concrete encasement of the steel core shall be reinforced with continuous longitudinal bars and lateral ties or spirals.*

    Where lateral ties are used, a minimum of either a No. 3 bar spaced at a maximum of 12 in on center, or a No. 4 bar or larger spaced at a maximum of 16 in on center shall be used. Deformed wire or welded wire reinforcement of equivalent area are permitted.

    Maximum spacing of lateral ties shall not exceed 0.5 times the least column dimension.

3. The minimum reinforcing ratio for this steel is $\rho_{sr} = A_{sr}/A_g = 0.004$

    where A_{sr} = area of continuous reinforcing bars, in^2
    A_g = gross area of composite member, in^2

---

[6]Griffis, op. cit.
[7]American Concrete Institute, *Building Code Requirements for Reinforced Concrete*, ACI 318-08 (Detroit: 2008).

4. *It is necessary to use steel anchors to resist the shear force in Section 14.1 of the AISC Specification. Steel anchors utilized to transfer longitudinal shear shall be distributed within the* load introduction length, *which shall not exceed a distance of two times the minimum transverse dimension of the encased composite member above and below the load transfer region. Anchors utilized to transfer longitudinal shear shall be placed on at least two faces of the steel shape in a generally symmetric configuration about the steel shape axes.*

    Steel anchor spacing, both within and outside of the load introduction length, shall conform to Section I8.3e.

5. *Should two or more steel shapes be used in the composite section, they must be connected together with lacing, tie plates, batten plates, or similar components.* Their purpose is to prevent the buckling of individual shapes before the concrete sets.

6. *There must be at least 1.5 in clear cover of concrete outside of any steel (ties or longitudinal bars).* The cover is needed for the protection of the steel from fire or corrosion. The amount of longitudinal and transverse reinforcing required in the encasement is thought to be sufficient to prevent severe spalling of the concrete surface from occurring during a fire.

7. *The specified compression strength of the concrete f'_c must be at least 3 ksi (21 MPa), but not more than 10 ksi if normal-weight concrete is used. For lightweight concrete, it may not be less than 3 ksi or more than 6 ksi.* The upper limits are provided, because sufficient test data are not available for composite columns with higher-strength concretes at this time. The lower limits of 3 ksi were specified for the purpose of ensuring the use of good-quality, but readily available, concrete and for the purpose of making sure that adequate quality control is used. This might not be the case if a lower grade of concrete were specified. The upper limit of 10 ksi for normal-weight concrete was specified because of the lack of data available for higher-strength concretes and because of the changes in behavior that have been observed in such concretes. The upper limit of 6 ksi for lightweight concrete is to ensure the use of readily available material. Higher-strength concretes may be used for calculating the modulus of elasticity for stiffness calculations, but may not be used for strength calculations, unless such use is justified by testing and analysis.

8. *The yield stresses of the steel sections and reinforcing bars used may not be greater than 75 ksi (525 MPa),* unless higher strengths are justified by testing and analysis.

The original reason for limiting the value of F_y is given here. One major objective in composite design is the prevention of local buckling of the longitudinal reinforcing bars and the contained steel section. To achieve this objective, the covering concrete must not be allowed to break or spall. It was assumed by the writers of previous LRFD Specifications that such concrete was in danger of breaking or spalling if its strain reached 0.0018. If we take this strain and multiply it by F_s, we get $(0.0018)(29,000) \approx 55$ ksi. Hence, that value was specified as the maximum useable yield stress.

Recent research has shown that, due to concrete confinement effects, the 55 ksi value is conservative, and it has been raised to 75 ksi in the specification.

For filled composite columns

1. The cross-sectional area of the HSS section must make up no less than 1 percent of the total composite member cross section.

2. Filled composite columns are classified as compact, non-compact, or slender (AISC I1.4). They are compact if the width-to-thickness ratio does not exceed λp. If the ratio exceeds λp but does not exceed λ_r the section is non-compact. If the ratio exceeds λ_r the section is slender. The maximum permitted width-to-thickness ratios for both rectangular HSS (b/t) and round filled sections (D/t) are specified in Table I1.1A in the AISC Specification.

17.6   AXIAL DESIGN STRENGTHS OF COMPOSITE COLUMNS

If a composite column were perfectly axially loaded and fully braced laterally, its nominal strength would equal the sum of the axial strengths of the steel shape, the concrete, and the reinforcing bars, as given by

$$P_{no} = A_s F_y + A_{sr} F_{ysr} + 0.85 f'_c A_c \quad \text{(AISC Equation I2-4)}$$

in which

A_s = area of steel section, in^2

A_{sr} = area of continuous reinforcing bars, in^2

F_{ysr} = specified minimum yield strength of reinforcing bars, ksi

A_c = area of concrete, in^2

Unfortunately, these ideal conditions are not present in practical composite columns. The contribution of each component of a composite column to its overall strength is difficult, if not impossible, to determine. The amount of flexural concrete cracking varies throughout the height of the column. The concrete is not nearly as homogeneous as the steel, and furthermore, the modulus of elasticity of the concrete varies with time and under the action of long-term or sustained loads. The effective lengths of composite columns in the rigid monolithic structures in which they are frequently used cannot be determined very well. The contribution of the concrete to the total stiffness of a composite column varies, depending on whether it is placed inside a tube or it is on the outside of a W section where its stiffness contribution is less.

The preceding paragraph presented some of the reasons it is difficult to develop a useful theoretical formula for the design of composite columns. As a result, one set of empirical equations is presented by the AISC Specification for concrete-encased sections (AISC I2.1), and another set is presented for concrete-filled sections (AISC I2.2).

Concrete-encased sections

In the expressions to follow, the following terms are used:

P_{no} = nominal compressive strength of the column
without consideration of its length

$$= A_s F_y + A_{sr} F_{ysr} + 0.85 A_c f'_c \quad \text{(AISC Equation I2-4)}$$

$$C_1 = 0.1 + 2\left(\frac{A_s}{A_c + A_s}\right) \le 0.3 \qquad \text{(AISC Equation I2-7)}$$

I_s = moment of inertia of steel shape, in^4

I_{sr} = moment of inertia of reinforcing bars, in^4

EI_{eff} = effective stiffness of composite column, kip-in^2

$$= E_s I_s + 0.5 E_s I_{sr} + C_1 E_c I_c \qquad \text{(AISC Equation I2-6)}$$

P_e = elastic buckling load, kips

$$= \frac{\pi^2 (EI_{eff})}{(KL)^2} \qquad \text{(AISC Equation I2-5)}$$

The available compressive strength $\phi_c P_n$, with $\phi_c = 0.75$, and the allowable compressive strength P_n/Ω_c, with $\Omega_c = 2.00$, of doubly symmetric axially loaded encased composite columns are to be determined with the following expressions:

When $\dfrac{P_{no}}{P_e} \le 2.25$

$$P_n = P_{no}\left[0.658^{\left(\frac{P_{no}}{P_e}\right)}\right] \qquad \text{(AISC Equation I2-2)}$$

When $\dfrac{P_{no}}{P_e} > 2.25$

$$P_n = 0.877 P_e \qquad \text{(AISC Equation I2-3)}$$

Concrete-filled composite columns

(a) For compact sections:

$$P_{no} = P_p \qquad \text{(AISC Equation I2-9a)}$$

$$P_p = A_s F_y + C_2 f'_c\left[A_c + A_{sr}\left(\frac{E_s}{E_c}\right)\right] \qquad \text{(AISC Equation I2-9b)}$$

$C_2 = 0.85$ for rectangular sections and 0.95 for circular ones

For all sections:

$$EI_{eff} = E_s I_s + E_s I_{sr} + C_3 E_c I_c \qquad \text{(AISC Equation I2-12)}$$

$$C_3 = 0.6 + 2\left(\frac{A_s}{A_c + A_s}\right) \le 0.9 \qquad \text{(AISC Equation I2-13)}$$

P_e and P_n are determined with AISC Equations I2-2, I2-3 and I2-5, as in concrete-encased sections.

(b) For non-compact sections:

$$P_{no} = P_p - \frac{P_p - P_y}{(\lambda_r - \lambda_p)^2}(\lambda - \lambda_p)^2 \qquad \text{(AISC Equation I2-9c)}$$

$\lambda, \lambda_p, \lambda_r$ are slenderless ratio from Table I1.1a

P_p is from Equation I2-9b

$$P_y = A_s F_y + 0.7 f_c' \left(A_c + A_{sr} \left(\frac{E_s}{E_c} \right) \right) \qquad \text{(AISC Equation I2-9d)}$$

(c) For slender sections:

$$P_{no} = A_s F_{cr} + 0.7 f_c' \left[A_c + A_{sr} \left(\frac{E_s}{E_c} \right) \right] \qquad \text{(AISC Equation I2-9e)}$$

For rectangular-filled sections: $F_{cr} = \dfrac{9 E_s}{(b/t)^2}$ \qquad (AISC Equation I2-10)

or

For round-filled sections: $F_{cr} = \dfrac{0.72 F_y}{\left[\left(\dfrac{D}{t} \right) \left(\dfrac{F_y}{E_s} \right) \right]^{0.2}}$ \qquad (AISC Equation I2-11)

Example 17-1

Compute the values of $\phi_c P_n$ and P_n / Ω_c for the axially loaded encased composite column shown in Fig. 17.3 if KL = 12.0 ft, F_y = 50 ksi, and f_c' = 3.5 ksi. The concrete weighs 145 lb/ft^3.

Solution

Using a W12 × 72(A_s = 21.1 in^2, I_{sx} = 597 in^4, I_{sy} = 195 in^4)

A_c = (20 in)(20 in) − 21.1 in^2 − (4)(1.0 in^2) = 374.9 in^2

P_{no} = $A_s F_y + A_{sr} F_{ysr} + 0.85 A_c f_c'$ \qquad (AISC Equation I2-4)

= (21.1 in^2)(50 ksi) + (4.0 in^2)(60 ksi) + (0.85)(374.9 in)(3.5 ksi) = 2410 k

C_1 = $0.1 + 2 \left(\dfrac{A_s}{A_c + A_s} \right) \leq 0.3$ \qquad (AISC Equation I2-7)

= $0.1 + 2 \left(\dfrac{21.1}{374.9 + 21.1} \right)$ = 0.2066 < 0.3 \qquad **OK**

E_c = $w_c^{1.5} \sqrt{f_c'}$ = $145^{1.5} \sqrt{3.5}$ = 3.267 × 10^3 ksi

I_c = $\left(\dfrac{1}{12} \right) (20)(20)^3 - 195$ = 13,138 in^4

EI_{eff} = $E_s I_s + 0.5 E_s I_{sr} + C_1 E_c I_c$ \qquad (AISC Equation I2-6)

= (29 × 10^3)(195) + (0.5)(29 × 10^3)(4 × 1.0 × 7.5^2)

\quad + (0.2066)(3.267 × 10^3)(13,138) = 17.785 × 10^6 k-in^2

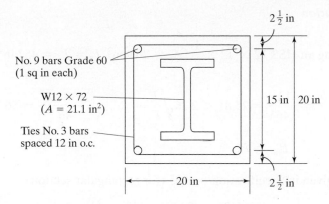

No. 9 bars Grade 60
(1 sq in each)

W12 × 72
($A = 21.1$ in^2)

Ties No. 3 bars
spaced 12 in o.c.

$2\frac{1}{2}$ in

15 in | 20 in

20 in

$2\frac{1}{2}$ in

FIGURE 17.3

$$P_e = \frac{(\pi^2)(EI_{eff})}{(KL)^2}$$ (AISC Equation I2-5)

$$= \frac{(\pi)^2(17.785 \times 10^6)}{(12 \times 12)^2} = 8465 \text{ k}$$

$$\frac{P_{no}}{P_c} = \frac{2410}{8465} = 0.28 \leq 2.25$$

∴ Must use AISC Equation I2-2 for P_n.

$$P_n = P_{no}\left[0.658^{\frac{P_{no}}{P_e}}\right] = 2410\left[0.658^{\frac{2410}{8465}}\right] = 2139 \text{ k}$$

LRFD $\phi_c = 0.75$	ASD $\Omega_c = 2.00$
$\phi_c P_n = (0.75)(2139) = 1604$ k	$\dfrac{P_n}{\Omega_c} = \dfrac{2139}{2.00} = 1070$ k

Note: The W12 × 72 alone has $\phi_c P_n = 807$ k and $P_n/\Omega_c = 537$ k.

Example 17-2

Determine the LRFD design strength $\phi_c P_n$ and the ASD allowable strength P_n/Ω_c of a concrete-filled 46 ksi HSS 12 × 12 × 1/2 section filled with 4 ksi concrete weighing 145 lb/ft^3. $(KL)_x = (KL)_y = 16$ ft.

Solution

Using an HSS $12 \times 12 \times \dfrac{1}{2}$ ($A_s = 20.9$ in^2, $I_x = I_y = 457$ in^4, $t = 0.465$ in)

$$\frac{b}{t} = \frac{12}{0.465} = 25.81 < 2.26\sqrt{\frac{E}{F_y}} = 2.26\sqrt{\frac{29 \times 10^3}{46}} = 56.7 \therefore \text{Compact section}$$

$$P_{no} = P_p \qquad\qquad \text{(AISC Equation I2-9a)}$$

C_2 given in specification = 0.85 (for rectangular section)

$$A_c = (12)(12) - (20.9) = 123.1 \text{ in}^2$$

$$P_p = A_s F_y + C_2 f_c' \left[A_c + A_{sr}\left(\frac{E_s}{E_c}\right) \right] \qquad \text{(AISC Equation I2-9b)}$$

$$= (20.9)(46) + 0.85(4)[123.1 + 0] = 1380 \text{ k} = P_p = P_{no}$$

$$C_3 = 0.6 + 2\left(\frac{A_s}{A_c + A_s}\right) \le 0.9 \qquad \text{(AISC Equation I2-13)}$$

$$= 0.6 + 2\left(\frac{20.9}{(12 \times 12) + 20.9}\right) = 0.85 < 0.9 \qquad \textbf{OK}$$

$$E_c = w_c^{1.5}\sqrt{f_c'} = (145)^{1.5}\sqrt{4} = 3.492 \times 10^3 \text{ ksi}$$

$$I_c = \left(\frac{1}{12}\right)(12)(12)^3 - 457 = 1271 \text{ in}^4$$

$$EI_{eff} = E_s I_s + E_s I_{sr} + C_3 E_c I_c \qquad \text{(AISC Equation I2-12)}$$

$$= (29 \times 10^3)(457) + 0 + (0.85)(3.492 \times 10^3)(1271)$$

$$= 17.026 \times 10^6 \text{ k-in}^2$$

$$P_e = \frac{\pi^2 EI_{eff}}{(KL)^2} \qquad\qquad \text{(AISC Equation I2-5)}$$

$$= \frac{(\pi^2)(17.026 \times 10^6)}{(12 \times 16)^2} = 4558 \text{ k}$$

$$\frac{P_{no}}{P_e} = \frac{1380}{4558} = 0.30 \le 2.25$$

$\therefore$ Use AISC Equation I2-2.

$$P_n = P_{no}\left[0.658^{\frac{P_{no}}{P_e}} \right] = 1380\left[0.658^{\frac{1380}{4558}} \right] = 1216 \text{ k}$$

LRFD $\phi_c = 0.75$	ASD $\Omega_c = 2.00$
$\phi_c P_n = (0.75)(1216) = 912 \text{ k}$	$\dfrac{P_n}{\Omega_c} = \dfrac{1216}{2.00} = 608 \text{ k}$

From AISC Table 4-15, the values are $\phi_c P_n = 911$ k and $P_n/\Omega_c = 607$ k. In computing the properties of the column, the author did not account for the fillets on the inside corners of the HSS. For this reason, his results vary a little from those given in the AISC tables.

17.7   SHEAR STRENGTH OF COMPOSITE COLUMNS

Section I4 of the AISC Specification states that the shear strength of composite columns may be calculated based on one of the following:

1. available shear strength of the steel section alone per AISC Spec. Chapter G.
2. available shear strength of the reinforced concrete portion (concrete plus steel reinforcement) alone per ACI 318 with $\phi_v = 0.75$ (LRFD) or $\Omega_v = 2.00$ (ASD).
3. nominal shear strength of the steel section per AISC Spec. Chapter G plus the nominal strength of the reinforcing steel per ACI 318 with $\phi_v = 0.75$ (LRFD) or $\Omega_v = 2.00$ (ASD).

The shear strength for method 2 is determined using ACI 318 Chapter 11 equation:

$$V_n = V_c + V_s$$

where $V_c = 2\sqrt{f'_c}\, bd$

$V_s = A_{st}\, F_{yt}\, \dfrac{d}{s}$

The shear strength for method 3 is determined using the following expression:

$$V_n = 0.6F_y A_w + A_{st}\, F_{yt}\, \dfrac{d}{s}$$

where A_w represents the steel section area. For square and rectangular HSS sections and for box sections, it equals $2ht$ (AISC Specification G5), where h is the clear distance between the member flanges less the inside corner radius on each side. Should this radius not be available, the designer may assume that h equals the outside dimension minus three times the flange thickness t.

Example 17-3 presents the calculation of the shear strength of an HSS section filled with concrete.

Example 17-3

It is assumed that the HSS 12 × 12 × 1/2 column of Example 17-2 is filled with 4 ksi concrete and is subjected to the end shear forces $V_D = 50$ k and $V_L = 100$ k. Does the member possess sufficient strength to resist these forces if $F_y = 46$ ksi?

Solution

Calculating required shear strength using method 1.

LRFD	ASD
$V_u = (1.2)(50) + (1.6)(100) = 220$ k	$V_a = 50 + 100 = 150$ k

Using an HSS 12 × 12 × $\dfrac{1}{2}$ (d = 12.00 in, t_w = 0.465 in)

$$h = d - 3t = 12.00 - (3)(0.465) = 10.605 \text{ in}$$

$$A_w = 2ht = (2)(10.605)(0.465) = 9.86 \text{ in}^2$$

$$V_n = 0.6F_y A_w = (0.6)(46)(9.86) = 272 \text{ k}$$

LRFD $\phi_v = 0.90$	ASD $\Omega_v = 1.67$
$\phi_v V_n = (0.90)(272) = 244.8$ k > 220 k **OK**	$\dfrac{V_n}{\Omega_v} = \dfrac{272}{1.67} = 162.9$ k > 150 k **OK**

17.8 LRFD AND ASD TABLES

In Part 4 of the Manual, a series of tables is presented for HSS sections and steel pipe sections filled with concrete. These tables, numbered 4-13 to 4-20, are set up in exactly the same fashion as the tables for axially loaded plain steel columns, which are also presented in Section 4 of the Manual. The axial strengths are given with respect to the minor axis for a range of $(KL)_y$ values.

Included are values for composite HSS square and rectangular sections ($F_y = 46$ ksi), for round HSS sections ($F_y = 42$ ksi), and for steel pipe sections ($F_y = 35$ ksi). The tables cover steel sections filled with 4 and 5 ksi concretes. For other grades of concrete, and for steel shapes made composite with encasing concrete, the formulas presented earlier in this chapter may be used to determine $\phi_c P_n$ values.

Examples 17-4 and 17-5 show how the tables can be used to directly determine design strengths for square and rectangular composite HSS sections.

Example 17-4

Determine the LRFD design strength and the ASD allowable strength of a 46 ksi HSS $10 \times 10 \times 3/8$ filled with 4 ksi concrete if $(KL)_x = (KL)_y = 15$ ft.

Solution. From Table 4-15 in the Manual, for $(KL)_y = 15$ ft.

LRFD	ASD
$\phi_c P_n = 575$ k	$\dfrac{P_n}{\Omega_c} = 383$ k

Example 17-5

Determine $\phi_c P_n$ and $\dfrac{P_n}{\Omega_c}$ for a concrete-filled HSS $20 \times 12 \times 5/8$ $(F_y = 46$ ksi$)$ if $f'_c = 5$ ksi, $(KL)_x = 24$ ft, and $(KL)_y = 12$ ft.

Solution. From the AISC Manual, Table 4-14, we obtain $\dfrac{r_{mx}}{r_{my}} = 1.54$. Then, determining the controlling unbraced length yields

$$(KL)_y = 12 \text{ ft}$$

$$(KL)_{y\,\text{EQUIV}} = \frac{(KL)_x}{r_{mx}/r_{my}} = \frac{24}{1.54} = 15.58 \text{ ft} > 12 \text{ ft}$$

$$\therefore (KL)_y = 15.58 \text{ ft controls}$$

By interpolation, values to follow are found.

LRFD	ASD
$\phi_c P_n = 1652.6$ k	$\dfrac{P_n}{\Omega_c} = 1098.4$ k

17.9   LOAD TRANSFER AT FOOTINGS AND OTHER CONNECTIONS

A small steel base plate usually is provided at the base of a composite column. Its purpose is to accommodate the anchor bolts needed to anchor the embedded steel shape to the footing for the loads occurring during the erection of the structure before the encasing concrete hardens and composite action is developed. This plate should be sufficiently small to be out of the way of the dowels needed for the reinforced-concrete part of the column.[8]

The AISC Specification does not provide details for the design of these dowels, but a procedure similar to the one provided by Section 10.14 of the ACI 318 Code seems to be a good one. If the column P_u exceeds $\phi_c(0.85 f_c' A_1)(\sqrt{A_2/A_1})$ the excess load should be resisted by dowels. If P_u does not exceed the equation above, it would appear that no dowels are needed. For such a situation, the ACI Code (Section 15.8.2.1) states that a minimum area of dowels equal to 0.005 times the cross-sectional area of the column must be used, and those dowels may not be larger than No. 11 bars. This diameter requirement ensures sufficient tying together of the column and footing over the whole contact area. The use of a very few large dowels spaced far apart might not do this very well.

17.10   TENSILE STRENGTH OF COMPOSITE COLUMNS

The tensile design strength, or the allowable design, strength of composite sections may be needed when uplift forces are present and perhaps for some beam column interaction situations. The AISC Specification (I2.1c and I2.2c) provides the nominal tensile strength for such situations by the following expression, for which $\phi_t = 0.90$ and $\Omega_t = 1.67$:

$$P_n = A_s F_y + A_{sr} F_{ysr} \qquad \text{(AISC Equations I2-8 and I2-14)}$$

17.11   AXIAL LOAD AND BENDING

To determine the required strengths for composite columns subject to both axial load and bending, it is necessary (as for steel beam columns) to include second-order effects in the analysis. The AISC Specification does not provide specific equations to assess the available strength of such members. Section I5 of the Specification does provide information with which interaction curves for the forces may be constructed, much as they are done in reinforced concrete design. Furthermore, a suggested procedure for doing this is presented in Section I5 of the AISC Commentary, and a numerical example is provided on the CD enclosed with the Manual.

17.12   PROBLEMS FOR SOLUTION

For all problems, use 145 lb/ft^3 concrete and 50 ksi steel shapes.

17-1 to 17-3. *Using the AISC equations, compute $\phi_c P_n$ and P_n/Ω_c for each encased concrete (5 ksi) section shown. F_y = 50 ksi, F_{yr} = 60 ksi, w_c = 145 lbs/ft^3.*

---

[8]Griffis, op. cit.

17-1.

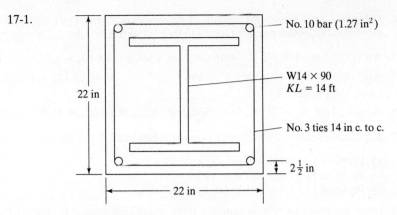

No. 10 bar (1.27 in²)

W14 × 90
KL = 14 ft

No. 3 ties 14 in c. to c.

22 in

22 in

$2\frac{1}{2}$ in

FIGURE P17-1   (*Ans.* 2343 k, 1562 k)

17-2.

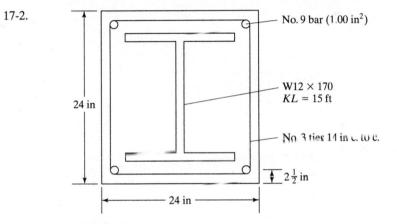

No. 9 bar (1.00 in²)

W12 × 170
KL = 15 ft

No. 3 ties 14 in c. to c.

24 in

24 in

$2\frac{1}{2}$ in

FIGURE P17-2

17-3.

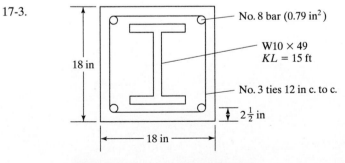

No. 8 bar (0.79 in²)

W10 × 49
KL = 15 ft

No. 3 ties 12 in c. to c.

18 in

18 in

$2\frac{1}{2}$ in

FIGURE P17-3   (*Ans.* 1262.6 k, 841.7 k)

17-4 to 17-6. *Using the appropriate AISC equations, determine $\phi_c P_n$ and P_n/Ω_c for each of the given sections, which are filled with* 145 lb/ft^3 *concrete.*

17-4. An HSS $14 \times 14 \times 1/2$, $F_y = 46$ ksi, $f_c' = 4$ ksi, and $(KL)_x = (KL)_y = 14$ ft.

17-5. An HSS $14 \times 10 \times 1/2$, $F_y = 46$ ksi, $f_c' = 4$ ksi, and $(KL)_x = (KL)_y = 12$ ft. (*Ans.* 930.4 k, 620.2 k)

17-6. A pipe 12 std, $F_y = 35$ ksi, $f_c' = 5$ ksi, and $(KL)_x = (KL)_y = 15$ ft.

17-7. Repeat the following problems, using the tables in Part 4 of the AISC Manual:

(a) Problem 17-4 (*Ans.* 1200 k, 797 k)
(b) Problem 17-5 (*Ans.* 928 k, 619 k)
(c) Problem 17-6 (*Ans.* 671 k, 448 k)

17-8. Select the lightest available concrete-filled round HSS column to support $P_D = 80$ k and $P_L = 120$ k. $F_y = 42$ ksi. $f_c' = 4$ ksi. $KL = 16$ ft.

C H A P T E R   1 8

Cover-Plated Beams and Built-up Girders

18.1   COVER-PLATED BEAMS

Should the largest available W section be insufficient to support the loads anticipated for a certain span, several possible alternatives may be taken. Perhaps the most economical solution involves the use of a higher-strength steel W section. If this is not feasible, we may make use of one of the following: (1) two or more regular W sections side-by-side (an expensive solution), (2) a cover-plated beam, (3) a built-up girder, or (4) a steel truss. This section discusses the cover-plated beams, while the remainder of the chapter is concerned with built-up girders.

In addition to being practical for cases in which the moments to be resisted are slightly in excess of those that can be supported by the deepest W sections, there are other useful applications for cover-plated beams. On some occasions, the total depth may be so limited that the resisting moments of W sections of the specified depth are too small. For instance, the architect may show a certain maximum depth for beams in his or her drawings for a building. In a bridge, beam depths may be limited by clearance requirements. Cover-plated beams frequently will be a very satisfactory solution for situations like these. Furthermore, there may be economical uses for cover-plated beams where the depth is not limited and where there are standard W sections available to support the loads. A smaller W section than required by the maximum moment can be selected and have cover plates attached to its flanges. These plates can be cut off where the moments are smaller, with resulting saving of steel. Applications of this type are quite common for continuous beams.

Should the depth be fixed and a cover-plated beam seem to be a feasible solution, the usual procedure will be to select a standard section with a depth that leaves room for top and bottom cover plates. Then the cover plate sizes can be selected.

For this discussion, reference is made to Fig. 18.1. For the derivation to follow, Z is the plastic modulus for the entire built-up section, Z_W is the plastic modulus for the

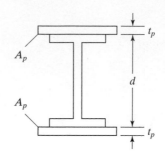

FIGURE 18.1

W section, d is the depth of the W section, t_p is the thickness of one cover plate, and A_p is the area of one cover plate.

The expressions to follow are written for LRFD design. Similar expressions could be developed for ASD. In Example 18-1, the author selects a cover-plated beam and then computes its LRFD design strength and its ASD allowable strength.

An expression for the required area of one flange cover plate can be developed as follows:

$$Z_{\text{reqd}} = \frac{M_u}{\phi_b F_y}$$

The total Z of the built-up section must at least equal the Z required. It will be furnished by the W shape and the cover plates as follows:

$$Z_{\text{reqd}} = Z_W + Z_{\text{plates}}$$
$$= Z_W + 2A_p \left(\frac{d}{2} + \frac{t_p}{2} \right)$$
$$A_p = \frac{Z_{\text{reqd}} - Z_W}{d + t_p}$$

Example 18-1 illustrates the design of a cover-plated beam. Quite a few satisfactory solutions involving different W sections and varying-size cover plates are available, other than the one made in this problem.

Example 18-1

Select a beam limited to a maximum depth of 29.50 in for the loads and span of Fig. 18.2. A 50 ksi steel is used, and the beam is assumed to have full lateral bracing for its compression flange.

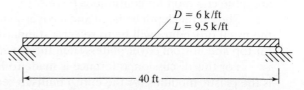

FIGURE 18.2

Solution

Assume beam wt = 350 lb/ft

LRFD	ASD
$w_u = (1.2)(6.35) + (1.6)(9.5) = 22.82$ k/ft	$w_a = 6.35 + 9.5 = 15.85$ k/ft
$M_u = \dfrac{(22.82)(40)^2}{8} = 4564$ ft-k	$M_a = \dfrac{(15.85)(40)^2}{8} = 3170$ ft-k

$$Z_{reqd} = \frac{(12)(4564)}{(0.9)(50)} = 1217 \text{ in}^3$$

The only W sections listed in the Manual with depths $<$ 29.50 in and Z values $>$ 1217 in^3 are the impractical, extremely heavy and expensive, W14 $\times$ 605, W14 $\times$ 665, and W14 $\times$ 730. As a result, the author decided to use a lighter W section with cover plates. He assumes the plates are each 1 in thick.

Try W27 $\times$ 146 ($d = 27.4$ in, $Z_x = 464$ in^3, $b_f = 14.0$ in)

Total depth = 27.4 + (2)(1.00) = 29.4 in < 29.5 in   **OK**

Area of 1 cover plate for each flange

$$A_p = \frac{Z_{reqd} - Z_w}{d + t_p} = \frac{1217 - 464}{27.4 + 1.00} = 26.51 \text{ in}^2$$

Try a 1 $\times$ 28 in **cover plate each flange**

$$Z_{furnished} = 464 + (1)(28)(2)\left(\frac{27.4}{2} + 0.5\right)$$

$$= 1259.2 \text{ in}^3 > 1217 \text{ in}^3 \quad \textbf{OK}$$

$$M_n = \frac{F_y Z}{12} = \frac{(50)(1259.2)}{12} = 5246.7 \text{ ft-k}$$

LRFD $\phi_b = 0.9$	ASD $\Omega_b = 1.67$
$\phi_b M_n = (0.9)(5246.7) = 4722$ ft-k > 4564 ft-k	$\dfrac{M_n}{\Omega_b} = \dfrac{5246.7}{1.67} = 3142$ ft-k < 3170 ft-k
OK	**Not quite**

A check of the b/t ratios for the plates, web, and flanges show them to be satisfactory.

Use W27 × 146 with one 1 × 28 in each flange for LRFD (slightly larger plate needed for ASD).

Wt of steel for LRFD design $= 146 + \left(\dfrac{2 \times 1 \times 28}{144}\right)(490) = 337$ lb/ft $<$ estimated

350 lb/ft   **OK**

18.2 BUILT-UP GIRDERS

Built-up I-shaped girders, frequently called plate girders, are made up with plates and perhaps with rolled sections. They usually have design moment strengths somewhere between those of rolled beams and steel trusses. Several possible arrangements are shown in Fig. 18.3. Rather obsolete bolted girders are shown in parts (a) and (b) of the figure, while several welded types are shown in parts (c) through (f). *Since nearly all built-up girders constructed today are welded (although they may make use of bolted field splices), this chapter is devoted almost exclusively to welded girders.*

The welded girder of part (d) of Fig. 18.3 is arranged to reduce overhead welding, compared with the girder of part (c), but in so doing may be creating a slightly worse corrosion situation if the girder is exposed to the weather. A box girder, illustrated in part (g), occasionally is used where moments are large and depths are quite limited. Box girders also have great resistance to torsion and lateral buckling. In addition, they make very efficient curved members because of their high torsional strengths.

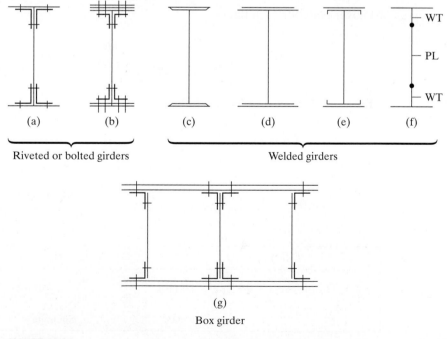

FIGURE 18.3

Built-up girders.

Plates and shapes can be arranged to form built-up girders of almost any reasonable proportions. This fact may seem to give them a great advantage for all situations, but for the smaller sizes the advantage usually is canceled out by the higher fabrication costs. For example, it is possible to replace a W36 with a built-up girder roughly twice as deep that will require considerably less steel and will have much smaller deflections; however, the higher fabrication costs will almost always rule out such a possibility.

Most steel highway bridges built today for spans of less than about 80 ft are steel-beam bridges. For longer spans, the built-up girder begins to compete very well economically. Where loads are extremely large, such as for railroad bridges, built-up girders are competitive for spans as small as 45 or 50 ft.

The upper economical limits of built-up girder spans depend on several factors, including whether the bridge is simple or continuous, whether a highway or railroad bridge is involved, and what is the largest section that can be shipped in one piece.

Generally speaking, built-up girders are very economical for railroad bridges in spans of 50 to 130 ft (15 to 40 m), and for highway bridges in spans of 80 to 150 ft (24 to 46 m). However, they are often very competitive for much longer spans, particularly when continuous. In fact, they are actually common for 200-ft (61-m) spans and have been used for many spans in excess of 400 ft (122 m). The main span of the continuous Bonn-Beuel built-up girder bridge over the Rhine River in Germany is 643 ft.

Built-up girders are not only used for bridges. They also are fairly common in various types of buildings to support heavy concentrated loads. Frequently, a large ballroom or dining room with no interfering columns is desired on a lower floor of a multistory building. Such a situation is shown in Fig. 18.4. The girder shown must support some tremendous column loads for many stories above. The usual building girder

Buffalo Bayou Bridge, Houston, TX—a 270-ft span. (Courtesy of the Lincoln Electric Company.)

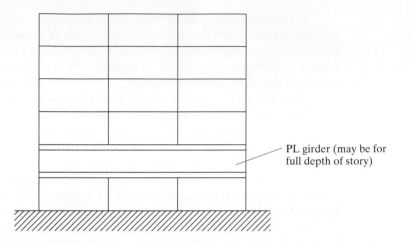

FIGURE 18.4

PL girder (may be for full depth of story)

of this type is simple to analyze because it probably does not have moving loads, although some building girders may be called upon to support traveling cranes.

The usual practical alternative to built-up girders in the spans for which they are economical is the truss. In general, plate girders have the following advantages, particularly compared with trusses:

1. The pound price for fabrication is lower than for trusses, but it is higher than for rolled beam sections.
2. Erection is cheaper and faster than for trusses.
3. Due to the compactness of built-up girders, vibration and impact are not serious problems.
4. Built-up girders require smaller vertical clearances than trusses.
5. The built-up girder has fewer critical points for stresses than do trusses.
6. A bad connection here or there is not as serious as in a truss, where such a situation could spell disaster.
7. There is less danger of injury to built-up girders in an accident, compared with trusses. Should a truck run into a bridge plate girder, it would probably just bend it a little, but a similar accident with a bridge truss member could cause a broken member and, perhaps, failure.
8. A built-up girder is more easily painted than a truss.

On the other hand, built-up girders usually are heavier than trusses for the same spans and loads, and they have a further disadvantage in the large number of connections required between webs and flanges.

18.3 BUILT-UP GIRDER PROPORTIONS

18.3.1 Depth

The depths of built-up girders vary from about 1/6 to 1/15 of their spans, with average values of 1/10 to 1/12, depending on the particular conditions of the job. One condition that may limit the proportions of the girder is the largest size that can be fabricated in the shop and shipped to the job. There may be a transportation problem such as clearance requirements that limit maximum depths to 10 or 12 ft along the shipping route.

Shallow girders probably will be used when loads are light, and the deeper ones when very large concentrated loads need to be supported, as from the columns in a tall building. If there are no depth restrictions for a particular girder, it will probably pay for the designer to make rough designs and corresponding cost estimates to arrive at a depth decision. (Computer solutions will be very helpful in preparing these alternative designs.)

18.3.2   Web Size

After the total girder depth is estimated, the general proportions of the girder can be established from the maximum shear and the maximum moment. As previously described for I-shaped sections in Section 10.2, the web of a beam carries nearly all of the shearing stress; this shearing stress is assumed by the AISC Specification to be uniformly distributed throughout the web. The web depth can be closely estimated by taking the total girder depth and subtracting a reasonable value for the depths of the flanges (roughly 1 to 2 in each). The web depths usually are selected to the nearest even inch, because these plates are not stocked in fractional dimensions.

As a plate girder bends, its curvature creates vertical compression in the web, as illustrated in Fig. 18.5. This is due to the downward vertical component of the compression flange bending stress and the upward vertical component of the tension flange bending stress.

The web must have sufficient vertical buckling strength to withstand the squeezing effect shown in Fig. 18.5. This problem is handled in Section G of the AISC Specification. There, the nominal shearing strength of the webs of stiffened or unstiffened built-up I-shaped girders is presented. In the following equation, A_w is the depth of the girder web times its thickness $= dt_w$, while C_v is a web coefficient, values of which are given after the equation:

$$V_n = 0.6F_y A_w C_v \qquad \text{(AISC Equation G2-1)}$$

$$\phi_v = 0.90 \quad \Omega_v = 1.67$$

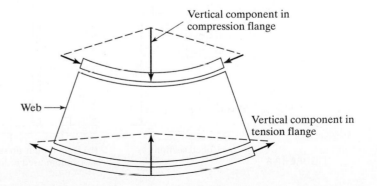

Vertical component in
compression flange

Web

Vertical component in
tension flange

FIGURE 18.5

Squeezing of plate girder web.

Values of C_v are as follows:

1. For $h/t_w \leq 1.10\sqrt{k_v \, E/F_y}$

$$C_v = 1.0 \qquad \text{(AISC Equation G2-3)}$$

2. For $1.10\sqrt{k_v \, E/F_y} < \dfrac{h}{t_w} \leq 1.37\sqrt{k_v \, E/F_y}$

$$C_v = \frac{1.10\sqrt{k_v \, E/F_y}}{h/t_w} \qquad \text{(AISC Equation G2-4)}$$

3. For $h/t_w > 1.37\sqrt{k_v \, E/F_y}$

$$C_v = \frac{1.51 E k_v}{(h/t_w)^2 F_y} \qquad \text{(AISC Equation G2-5)}$$

In the preceding C_v expressions, h is equal to (a) the clear distance between flanges less two times the fillet or corner radius for rolled shapes, (b) the distance between adjacent lines of fasteners for built-up sections, or (c) the clear distance between flanges for built-up sections when welds are used. These values are illustrated in Fig. 18.6.

The term k_v is a web plate-buckling coefficient $= 5 + \dfrac{5}{(a/h)^2}$ except that it is to be 5.0 when $a/h > 3.0$ or $> \left[\dfrac{260}{(h/t_w)}\right]^2$.

From a corrosion standpoint, the usual practice is to use some absolute minimum web thickness. For bridge girders, 3/8 in is a common minimum, while 1/4 or 5/16 in are the minimum values used for the more sheltered building girders.

18.3.3 Flange Size

After the web dimensions are selected, the next step is to select an area of the flange so that it will not be overloaded in bending. The total bending strength of a plate girder

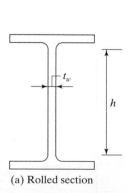

(a) Rolled section

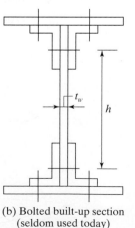

(b) Bolted built-up section
(seldom used today)

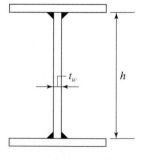

(c) Welded built-up section

FIGURE 18.6

equals the bending strength of the flange plus the bending strength of the web. As almost all of the bending strength is provided by the flange, an approximate expression can be developed to estimate the flange area as follows:

$$Z_{reqd} = \frac{M_u}{\phi_b F_y}$$

$$Z_{furnished} = 2A_f\left(\frac{h + t_f}{2}\right) + (2)\left(\frac{h}{2}\right)(t_w)\left(\frac{h}{4}\right)$$

Equating $Z_{reqd} = Z_{furnished}$ and solving for A_f

$$\frac{M_u}{\phi_b F_y} = A_f(h + t_f) + (2)\left(\frac{h}{2}\right)(t_w)\left(\frac{h}{4}\right)$$

$$A_f = \frac{M_u}{\phi_b F_y(h + t_f)} - \frac{t_w h^2}{4(h + t_f)}$$

In Example 18-2, the web and flanges are proportioned for a built-up I-shaped girder such that transverse stiffeners are not required for the web.

Example 18-2

Select trial proportions for a 60-in deep welded built-up I-shaped section with a 70 ft simple span to support a service dead load (not including the beam weight) of 1.1 k/ft and a service live load of 3 k/ft. The A36 section will be assumed to have full lateral bracing for its compression flange, and an unstiffened web is to be used.

Solution

Trial proportions

Try 60 in depth $\approx l/14$

$$\text{Estimated beam weight: 2 Flanges} = 2(1.0)(15) = 30 \text{ in}^2$$

$$\text{Web} = (0.75)(58) = 43.5 \text{ in}^2$$

$$\overline{A \text{ total} = 73.5 \text{ in}^2}$$

$$\text{wt (plf)} = \frac{73.5 \text{ in}^2}{144 \text{ in}^2/\text{ft}^2}\left(490 \frac{\text{lbs}}{\text{ft}^3}\right) = 250.1 \text{ plf}$$

Assume beam wt = 250 lbs/ft

Maximum moment and shear

LRFD	ASD
$w_u = (1.2)(1.1 + 0.250) + (1.6)(3) = 6.42$ k/ft	$w_a = 1.1 + 0.250 + 3 = 4.35$ k/ft
$R_u = \left(\dfrac{70}{2}\right)(6.42) = 224.7$ k	$R_a = \left(\dfrac{70}{2}\right)(4.35) = 152.2$ k
$M_u = \dfrac{(6.42)(70)^2}{8} = 3932$ ft-k	$M_a = \dfrac{(4.35)(70)^2}{8} = 2664$ ft-k

Design for a *compact web* and *flange*

$$Z_{\text{reqd}} = \frac{M_u}{\phi F_y} = \frac{(12)(3932)}{(0.9)(36)} = 1456 \text{ in}^3$$

Trial web size

For web to be compact by AISC Table B4.1 $\dfrac{h}{t_w}$ must be $\leq 3.76\sqrt{\dfrac{E}{F_y}}$

$$= 3.76\sqrt{\frac{29 \times 10^3}{36}} = 106.7 \qquad\qquad \text{(Case 15 AISC Table B4.1b)}$$

Assuming h to be 60 in $- \ 2(1.0 \text{ in}) = 58$ in

$$Min \ t_w = \frac{58}{106.7} = 0.544 \text{ in, Say, } \frac{9}{16} \text{ in } (0.563 \text{ in})$$

Try $\frac{9}{16} \times 58$ web

$$\frac{h}{t_w} = \frac{58}{\frac{9}{16}} = 103.1$$

Since 103.1 is $> 2.46\sqrt{\dfrac{E}{F_y}} = 2.46\sqrt{\dfrac{29 \times 10^3}{36}} = 69.82$

transverse stiffeners may be required, states AISC Specification G2.2.

But the same specification states that stiffeners are not required if the necessary shear strength for the web is less than or equal to its available shear strength, as stipulated by AISC Specification G2.1, using $k_v = 5.0$.

$$\frac{h}{t_w} = 103.1 > 1.37\sqrt{\frac{k_v E}{F_y}} = 1.37\sqrt{\frac{(5)(29 \times 10^3)}{36}}$$

$$= 86.94$$

$$\therefore C_v = \frac{1.51 E k_v}{\left(\dfrac{h}{t_w}\right)^2 F_y} = \frac{(1.51)(29 \times 10^3)(5.0)}{(103.1)^2(36)} = 0.572$$

$$V_n = 0.6 F_y A_w C_v = (0.6)(36)\left(58 \times \frac{9}{16}\right)(0.572)$$

$$= 403.1 \text{ k}$$

Available shear strength without stiffeners

LRFD $\phi_v = 0.90$	ASD $\Omega_v = 1.67$
$\phi_v V_n = (0.90)(403.1) = 362.8 \text{ k}$	$\dfrac{V_n}{\Omega_v} = \dfrac{403.1}{1.67} = 241.4 \text{ k}$
$> 224.7 \text{ k}$	$> 152.2 \text{ k}$
$\therefore$ **Stiffeners are not required.**	$\therefore$ **Stiffeners are not required.**

Trial flange size

Assume 1-in plates ($t_f = 1.0$ in)

$$A_f = \frac{M_u}{\phi_b F_y(h + t_f)} - \frac{t_w h^2}{4(h + t_f)}$$

$$= \frac{(12)(3932)}{(0.9)(36)(58 + 1)} - \frac{\left(\dfrac{9}{16}\right)(58)^2}{4(58 + 1)} = 16.66 \text{ in}^2$$

Try 1×18 plate each flange. Are they compact by AISC Table B4.1b (Case 11)?

$$\frac{b_f}{2t_f} = \frac{18.00}{(2)(1.00)} = 9.00 < 0.38\sqrt{\frac{E}{F_y}} = 0.38\sqrt{\frac{29,000}{36}} = 10.79 \text{ (Yes, is compact)}$$

Check Z of section

$$Z = (2)\left(\frac{58}{2}\right)\left(\frac{9}{16}\right)\left(\frac{58}{4}\right) + (2)(1 \times 18)\left(\frac{58}{2} + \frac{1}{2}\right)$$
$$= 1535 \text{ in}^3 > 1456 \text{ in}^3 \quad \textbf{OK}$$

Check girder wt

$$\text{wt} = \frac{\left(\dfrac{9}{16}\right)(58) + (2)(1 \times 18)}{144}(490) = 233.5 \text{ lb} < 250 \text{ lb estimated } \textbf{OK}$$

Trial Section $\frac{9}{16} \times 58$ web with 1×18 PL each flange. (See Fig. 18.7.)

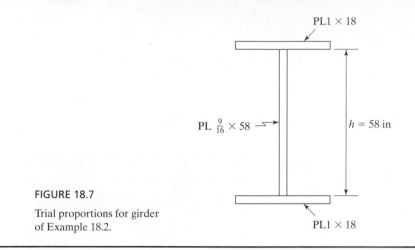

FIGURE 18.7

Trial proportions for girder
of Example 18.2.

18.4 FLEXURAL STRENGTH

The *nominal* flexural strength, M_n, of a plate girder bent about its major axis is based on one of the limit states as defined in Chapter F of the AISC Specification, Section F2 to F5. These limit states include yielding (Y), lateral-torsional buckling (LTB), compression flange local buckling (FLB), compression flange yielding (CFY), and tension flange yielding (TFY). This strength, M_n, is the lowest value obtained according to these limit states. The application of the limit states defined in F2 to F5 are based on whether plate girder has compact, non-compact, or slender flanges and webs as defined in Specification Section B4.1 for flexure and the unbraced length of the compression flange, L_b. Table F1.1 in the specification summarizes the application of the Chapter F sections.

In Example 18-2, the doubly symmetrical I-shaped plate girder was proportioned so that both the flanges and web were compact. With this condition, Section F2 was applicable and the limit states of yielding (Y) and lateral-torsional buckling (LTB) would need to be checked to determine M_n. In the example, the member was assumed to have full lateral bracing for its compression flange. Therefore, $L_b = 0$ and the limit state of lateral-torsional buckling does not apply. The yielding limit state was used to determine the *nominal* flexural strength.

Section F3 applies to doubly symmetrical I-shaped members having compact webs and non-compact or slender flanges. The *nominal* flexural strength, M_n, shall be the lower value obtained from the limit states of LTB and FLB. Section F4 applies to doubly symmetrical I-shaped members with non-compact webs and singly symmetrical I-shaped members with compact or non-compact webs. The *nominal* flexural strength, M_n, shall be the lowest value obtained from the limit states of CFY, LTB, FLB and TFY. Section F5 applies to doubly symmetric and singly symmetric I-shaped members with slender webs. The *nominal* flexural strength, M_n, shall be the lowest value obtained from the limit states of CFY, LTB, FLB and TFY.

The *design* flexural strength, $\Phi_b M_n$, and the *allowable* flexural strength, M_n/Ω_b, shall be determined using $\Phi_b = 0.90$ (LRFD) and $\Omega_b = 1.67$ (ASD).

Example 18-3

Determine the *design* flexural strength, $\Phi_b M_n$, and the *allowable* flexural strength, M_n/Ω_b, of the following welded I-shaped plate girder. The flanges are 1 1/4 in $\times$ 15 in, the web is 1/4 in $\times$ 50 in, and the member is uniformly loaded and simply-supported. Use A36 steel and assume the girder has continuous bracing for its compression flange.

Solution

Determine if flange is compact, non-compact, or slender? *Case 11*. Table B4.1b

$$\frac{b}{t_f} = \frac{b_f/2}{t_f} = \frac{15/2}{1.25} = 6.0 < 0.38\sqrt{\frac{E}{F_y}} = 0.38\sqrt{\frac{29,000}{36}} = 10.79$$

Compact — Flange

Determine if web is compact, non-compact, or slender? *Case 15*. Table B4.1b

$$\frac{h}{t_w} = \frac{50}{0.25} = 200 > 5.70\sqrt{\frac{E}{F_y}} = 5.70\sqrt{\frac{29,000}{36}} = 161.78$$

Slender — Web

$\therefore$ (F5) Doubly symmetric section with *slender web* and *compact flange* bent about their major axis.

$\phi_b M_n$ is lowest value of Y, LTB, FLB, TFY

$\boxed{\text{LTB}}$ — Since $L_b = 0$, limit state of LTB does not apply.

$\boxed{\text{FLB}}$ — Since flange is compact, limit state of FLB does not apply

$\boxed{\text{TFY}}$ — Since member is symmetric about x–x axis $S_{xt} = S_{xc}$, limit state of TFY does not apply.

$\boxed{\text{Y}}$ — Compression flange yielding.

$$M_n = R_{pg} F_y S_{xc} \qquad \text{(AISC Equation F5-1)}$$

$$a_w = \frac{h_c t_w}{b_{fc} t_{fc}} = \frac{50(1/4)}{15(1.25)} \qquad \text{(AISC Equation F4-12)}$$

$$a_w = 0.667 < 10 \text{ (upper limit)}$$

$$R_{pg} = 1 - \frac{a_w}{1200 + 300 a_w}\left[\frac{h_c}{t_w} - 5.7\sqrt{\frac{E}{F_y}}\right] \le 1.0 \qquad \text{(AISC Equation F5-6)}$$

$$R_{pg} = 1 - \frac{0.667}{1200 + 300(0.667)}\left[200 - 5.7\sqrt{\frac{29,000}{36}}\right] \le 1.0$$

$$R_{pg} = 0.982$$

$$F_y = 36 \text{ ksi}$$

$$I_{xc} = \frac{1}{12}\left(\frac{1}{4}\right)(50)^3 + 2\left(\frac{1}{12}\right)(15)(1.25)^3 + 2(1.25)(15)(25.625)^2$$

$$I_{xc} = 27{,}233 \text{ in}^4$$

$$S_{xc} = \frac{I}{c} = \frac{27{,}233}{26.25} = 1037.4 \text{ in}^3$$

$$M_n = R_{pg}F_yS_{xc} = \frac{(0.982)(36 \text{ ksi})(1037.4 \text{ in}^3)}{12 \text{ in /ft}}$$

$$M_n = 3056 \text{ ft-k}$$

LRFD ϕ = 0.90	ASD Ω = 1.67
$\phi M_n = 0.9(3056)$ ft-k	$M_n/\Omega = 3056$ ft-k/1.67
$\phi M_n = 2750$ ft-k	$M_n/\Omega = 1830$ ft-k

Example 18-4

Determine the *design* flexural strength, $\Phi_b M_n$, and the *allowable* flexural strength, M_n/Ω_b, of the following welded I-shaped plate girder. The flanges are 1 in × 24 in, the web is 5/16 in × 45 in, and the member is uniformly loaded and has a simply-supported 100 ft. span. Use A36 steel and the unbraced length of the compression flange is 20 ft.

Solution

Determine if flange is compact, non-compact, or slender? *Case 11*. Table B4.1b

$$\frac{b}{t_f} = \frac{b_f/2}{t_f} = \frac{24/2}{1} = 12.00 > 0.38\sqrt{\frac{E}{F_y}} = 0.38\sqrt{\frac{29,000}{36}} = 10.79$$

$$k_c = \frac{4}{\sqrt{h/t_w}} = \frac{4}{\sqrt{45/0.3125}} = 0.333$$

$$\frac{b}{t_f} = 12.00 < 0.95\sqrt{\frac{k_c E}{F_y}} = 0.95\sqrt{\frac{0.333(29,000)}{36}} = 15.56$$

Non-Compact—Flange

Determine if web is compact, non-compact, or slender? *Case 15.* Table B4.1b

$$\frac{h}{t_w} = \frac{45}{0.3125} = 144 > 3.76\sqrt{\frac{E}{F_y}} = 3.76\sqrt{\frac{29,000}{36}} = 106.72$$

$$< 5.70\sqrt{\frac{E}{F_y}} = 5.70\sqrt{\frac{29,000}{36}} = 161.78$$

Non-Compact—Web

∴ (F4) Doubly symmetric I-shaped members with *non-compact webs* bent about their major axis.

$\phi_b M_n$ is lowest value of Y, LTB, FLB, TFY

$\boxed{\text{TFY}}$ — Since member is symmetric about *x-x* axis, $S_{xt} = S_{xc}$, limit state of TFY does not apply.

$\boxed{\text{Y}}$ — Compression flange yielding

$$M_n = R_{pc} F_y S_{xc} \qquad\qquad \text{(AISC Equation F4-1)}$$

$$\text{Since } \frac{h}{t_w} = 144 \geq \lambda_{pw} = 106.72 - 3.76\sqrt{\frac{E}{F_y}}$$

$$R_{pc} = \left[\frac{M_p}{M_{yc}} - \left(\frac{M_p}{M_{yc}} - 1\right)\left(\frac{\lambda - \lambda_{pw}}{\lambda_{rw} - \lambda_{pw}}\right)\right] \leq \frac{M_p}{M_{yc}} \qquad \text{(AISC Equation F4-9b)}$$

$$\frac{M_p}{M_{yc}} = \frac{Z}{S}$$

where: $\quad Z = 2(1)(24)(22.5 + 0.5) + 0.3125(2)(22.5)(11.25)$

$\qquad\quad Z = 1262 \text{ in}^3$

$$S = \frac{2\left(\frac{1}{12}\right)(24)(1)^3 + \left(\frac{1}{12}\right)(0.3125)(45)^3 + 2(24)(1)(22.5 + 0.5)^2}{23.5}$$

$\qquad\quad S = 1182 \text{ in}^3$

$$\frac{M_p}{M_{yc}} = \frac{Z}{S} = \frac{1262}{1182} = 1.068$$

$$R_{pc} = \left[1.068 - (1.068-1)\left(\frac{144 - 106.72}{161.78 - 106.72} \right) \right] \le 1.068$$

$$R_{pc} = 1.022$$

$$M_n = \frac{(1.022)(36 \text{ ksi})(1182 \text{ in}^3)}{12 \text{ in /ft}} = 3624 \text{ ft-k} \quad \boxed{Y}$$

LTB — Lateral Torsional Buckling check unbraced length of 20 ft,

$$L_b = 20 \text{ ft or } 240 \text{ in}$$

$$L_p = 1.1\, r_t \sqrt{\frac{E}{F_y}} \qquad \text{(AISC Equation F4-7)}$$

$$r_t = \frac{b_{fc}}{\sqrt{12\left(\dfrac{h_o}{d} + \dfrac{1}{6} a_w \dfrac{h^2}{h_o d} \right)}} \qquad \text{(AISC Equation F4-11)}$$

$$a_w = \frac{h_c t_w}{b_{fc} t_{fc}} = \frac{45(0.3125)}{24(1.0)} = 0.586 \qquad \text{(AISC Equation F4-12)}$$

$$b_{fc} = 24 \text{ in}, \quad h_o = 46 \text{ in}, \quad d = 47 \text{ in}, \quad h = 45 \text{ in}$$

$$r_t = \frac{24}{\sqrt{12\left(\dfrac{46}{47} + \dfrac{1}{6}(0.586)\left(\dfrac{45^2}{46(47)} \right) \right)}} = 6.70 \text{ in}$$

$$L_p = 1.1(6.70 \text{ in })\sqrt{\frac{29{,}000}{36}} = 209.1 \text{ in} = 17.42 \text{ ft}$$

$$L_r = 1.95\, r_t \frac{E}{F_L} \sqrt{\frac{J}{S_{xc} h_o} + \sqrt{\left(\frac{J}{S_{xc} h_o} \right)^2 + 6.76\left(\frac{F_L}{E} \right)^2}} \qquad \text{(AISC Equation F4-8)}$$

Since $\dfrac{S_{xt}}{S_{xc}} = 1.0 \ge 0.7$

$$\therefore F_L = 0.7\, F_y = 0.7\,(36) = 25.2 \text{ ksi} \qquad \text{(AISC Equation F4-6a)}$$

$$J = \sum \frac{1}{3} b t^3 = 2\left(\frac{1}{3} \right)(24)(1)^3 + \left(\frac{1}{3} \right)(45)(0.3125)^3 = 16.46 \text{ in}^3$$

$$L_r = 1.95(6.70)\frac{29{,}000}{25.2} \sqrt{\frac{16.46}{1182(46)} + \sqrt{\left(\frac{16.46}{1182(46)} \right)^2 + 6.76\left(\frac{25.2}{29{,}000} \right)^2}}$$

$$L_r = 764.0 \text{ in} = 63.67 \text{ ft}$$

$$M_n = C_b \left[R_{pc}M_{yc} - (R_{pc}M_{yc} - F_L S_{xc}) \left[\frac{L_b - L_p}{L_r - L_p} \right] \right] \leq R_{pc}M_{yc} \text{ (AISC Equation F4-2)}$$

$$R_{pc}M_{yc} = \frac{1.022(36)(1182)}{12} = 3624 \text{ ft-k}$$

$$M_n = 1.0 \left[3624 - \left(3624 - \frac{25.2(1182)}{12} \right) \left[\frac{20 - 17.42}{63.67 - 17.42} \right] \right] \leq 3624$$

$$M_n = 3560 \text{ ft-k} \quad \boxed{\text{LTB}}$$

$\boxed{\text{FLB}}$ – Flange Local Buckling

$$M_n = \left[R_{pc}M_{yc} - (R_{pc}M_{yc} - F_L S_{xc}) \left[\frac{\lambda - \lambda_{pf}}{\lambda_{rf} - \lambda_{pf}} \right] \right] \qquad \text{(AISC Equation F4-12)}$$

For sections with non-compact flanges

$$\lambda = \frac{b_f/2}{t_f} = 12.00 \quad \lambda_{pf} = 10.79 = 0.38\sqrt{\frac{E}{F_y}}$$

$$\lambda_{rf} = 15.56 = 0.95\sqrt{\frac{kE}{F_y}}$$

$$M_n = \left[3624 - \left(3624 - \frac{25.2(1182)}{12} \right) \left(\frac{12.00 - 10.79}{15.56 - 10.79} \right) \right]$$

$$M_n = 3334 \text{ ft-k} \quad \boxed{\text{FLB}}$$

M_n is controlled by least of Y, LTB, FLB

$$\therefore M_n = 3334 \text{ ft-k} \quad \boxed{\text{FLB}}$$

LRFD $\phi = 0.90$	ASD $\Omega = 1.67$
$\phi M_n = 0.9 \ (3334 \text{ ft-k})$	$M_n/\Omega = 3334 \text{ ft-k}/1.67$
$\phi M_n = 3001 \text{ ft-k}$	$M_n/\Omega = 1996 \text{ ft-k}$

18.5   TENSION FIELD ACTION

The AISC Specification for built-up I-shaped girders permits their design on the basis of postbuckling strength. Designs on this basis provide a more realistic idea of the actual strength of a girder. (Such designs, however, do not necessarily result in better economy, because stiffeners are required.) Should a girder be loaded until initial buckling occurs, it will not then collapse, because of a phenomenon known as *tension field action*.

The panels of a built-up I-shaped girder located between suitably designed vertical stiffeners will resist much larger shear forces than the theoretical buckling strength of the girder's web. Once the shearing forces reach the theoretical buckling strength of the web, the girder will be displaced by a small and insignificant amount.

If transverse stiffeners have been properly designed, membrane forces or diagonal tension fields will develop in the web between the stiffeners, as illustrated in Fig. 18.8. These diagonal tension stresses are caused by those shearing forces which are larger than the shearing forces theoretically required to buckle the web. For these excess forces, the girder will behave much like a Pratt truss, with parts of the web acting as tension diagonals and with the stiffeners acting as compression verticals, as shown in Fig. 18.8.

Students who have studied reinforced-concrete design will notice that tension field action is somewhat like the behavior of reinforced-concrete beams with web reinforcing (according to the Ritter-Morsch theory), as the beams resist shearing forces. Actually, there the beam behavior, according to this theory, is rather like a Warren truss with the concrete "diagonals" being in compression and the web reinforcing serving as tension verticals.

The stiffeners of the built-up I-girders keep the flanges from coming together, and the flanges keep the stiffeners from coming together. The intermediate stiffeners, which before initial buckling were assumed to resist no load, will after buckling resist compressive loads (or will serve as the compression verticals of a truss) due to diagonal tension. The result is that a plate-girder web probably can resist loads equal to two or three times those present at initial buckling before complete collapse will occur.

Until the web buckles initially, deflections are relatively small. However, after initial buckling, the girder's stiffness decreases considerably, and deflections may increase to several times the values estimated by the usual deflection theory.

The estimated total or ultimate shear that a panel (a part of the girder between a pair of stiffeners) can withstand equals the shear initially causing web buckling plus the shear that can be resisted by tension field action. The amount of tension field action is dependent on the proportions of the panels.

The capability necessary for the girders to develop tension field action is based on the ability of the stiffeners to resist compression from both sides of a panel. You can see that there is a panel on only one side of an end panel, so tension field action is not to be considered in such panels. Also, it's not allowed when the panels have very large aspect ratios. The aspect ratio, a, is the ratio of the clear distance between stiffeners in a panel to the height of the panel. According to AISC Specification G3.1,

FIGURE 18.8

Tension field action in plate-girder web. (Note that end panels cannot develop tension field action.)

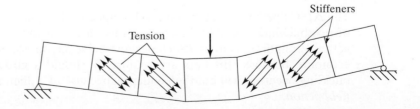

the *a/h* ratios may not be larger than 3.0 or $\left[\dfrac{260}{h/t_w}\right]^2$. Tension field action also may not

be considered if $\left(\dfrac{2A_w}{A_{fc} + A_{ft}}\right) > 2.5$ or if h/b_{fc} or $h/b_{ft} > 6.0$. (Here, A_{fc} and A_{ft} are

the areas of the compression and tension flanges, respectively, while b_{fc} and b_{ft} are the widths of those same flanges.

Example 18-5

The built-up A36 I-shaped girder shown in Fig. 18.9 has been selected for a 65-ft simple span to support the loads $w_D = 1.1$ k/ft (not including the beam weight) and $w_L = 2$ k/ft. Select transverse stiffeners as needed.

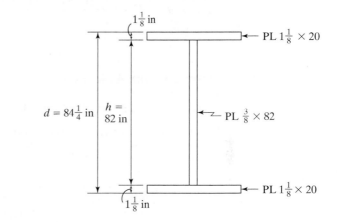

FIGURE 18.9

Computing girder weight

$$A = (2)\left(1\frac{1}{8}\text{ in}\right)(20\text{ in}) + \left(\frac{3}{8}\text{ in}\right)(82\text{ in}) = 75.75\text{ in}^2$$

$$\text{wt per ft} = \left(\frac{75.75\text{ in}^2}{144\text{ in}^2/\text{ft}^2}\right)\left(490\text{ lb/ft}^3\right) = 258\text{ lb/ft}$$

Computing the required shear strength at the support

LRFD	ASD
$w_u = (1.2)(1.1 + 0.258) + (1.6)(2) = 4.83$ k/ft	$w_a = 1.1 + 0.258 + 2 = 3.358$ k/ft
$R_u = \left(\dfrac{65}{2}\right)(4.83) = 156.98$ k	$R_a = \left(\dfrac{65}{2}\right)(3.358) = 109.14$ k

Are stiffeners needed?

$$A_w = dt_w = (84.25 \text{ in})\left(\frac{3}{8} \text{ in}\right) = 31.59 \text{ in}^2$$

$$\frac{h}{t_w} = \frac{82}{0.375} = 219 < 260 \therefore k_v = 5.0, \text{ says AISC Section G2.1b.}$$

C_v from same AISC section

$$219 > 1.37\sqrt{k_v E/F_y} = 1.37\sqrt{\frac{(5)(29 \times 10^3)}{36}} = 86.95$$

$\therefore$ Must use AISC Equation G2-5.

$$C_v = \frac{1.51Ek_v}{\left(\dfrac{h}{t_w}\right)^2 F_y} = \frac{(1.51)(29 \times 10^3)(5)}{(219)^2(36)} = 0.1268$$

Computing V_n with AISC Equation G2-1

$$V_n = 0.6F_y A_w C_v = (0.6)(36 \text{ k/in}^2)(31.59 \text{ in}^2)(0.1268) = 86.52 \text{ k}$$

Calculating shear strengths without stiffeners

LRFD $\phi_v = 0.90$	ASD $\Omega_v = 1.67$
$\phi_v V_n = (0.90)(86.52) = 77.87 \text{ k}$	$\dfrac{V_n}{\Omega_v} = \dfrac{86.52}{1.67} = 51.81 \text{ k}$
$< 156.98 \text{ k} \therefore$ **Stiffeners are required.**	$< 109.14 \text{ k} \therefore$ **Stiffeners are required.**

Can we use tensile field action? (AISC Specification G3)

a.   Not in end panels with transverse stiffeners.

b.   Not in members where $\dfrac{a}{h} > 3.0$ or $\left[260\left(\dfrac{h}{t_w}\right)\right]^2$.

c.   Not if $\left(\dfrac{2A_w}{A_{fc} + A_{ft}}\right) > 2.5$. Here, A_{fc} = the area of compression flange and

A_{ft} = the area of the tension flange.

d.   Not if $\dfrac{h}{b_{fc}}$ or $\dfrac{h}{b_{ft}} > 6.0$.

Select stiffener spacing for end panel.

By (a), tension field action may not be used

LRFD	ASD
$\dfrac{\phi V_n}{A_w} = \dfrac{V_u}{A_w} = \dfrac{156.98}{31.59} = 4.97$ ksi	$\dfrac{V_n}{\Omega_c A_w} = \dfrac{V_a}{A_w} = \dfrac{109.14}{31.59} = 3.45$ ksi

Referring to AISC Table 3-16a, which provides the available shear stress (tension field action not included). Entering left margin with $h/t_w = 219$ and moving horizontally from that value to the $\phi V_n / A_w$ curve $= 4.97$ ksi (actually interpolating between the curves in table). There, move down vertically to base and read 1.00. This is the a/h value that can be used.

$$\therefore a = (1.00)(82) = 82 \text{ in.}$$

Using the ASD values $h/t_w = 219$ and $V_n/\Omega_v A_w = 3.45$ ksi and entering AISC Table 3-16a, we read at the base the value 0.98. Thus, $a = (0.98)(82) = 80$ in.

Select stiffeners for second panel, noting that tension field action is permitted, since it's not an end panel.

Required shear strength needed for 2nd panel

LRFD (82 in out in span)	ASD (80 in out in span)
$V_u = 156.98 - \left(\dfrac{82}{12}\right)(4.83)$	$V_a = 109.14 - \left(\dfrac{80}{12}\right)(3.358)$
$= 123.97$ k	$= 86.75$ k

Computing the available shear strength without stiffeners

LRFD $\phi_v = 0.90$	ASD $\Omega_v = 1.67$
$\phi_v V_n = (0.90)(86.52) = 77.87$ k	$\dfrac{V_n}{\Omega_v} = \dfrac{86.52}{1.67} = 51.81$ k
< 123.97 k	< 86.75 k
∴ More stiffeners reqd.	**∴ More stiffeners reqd.**
$\dfrac{\phi V_n}{A_w} = \dfrac{123.97}{31.59} = 3.92$ ksi	$\dfrac{V_n}{\Omega_v A_w} = \dfrac{V_a}{A_w} = \dfrac{86.75}{31.59} = 2.75$ ksi

For LRFD with $\phi_v V_n / A_w = 3.92$ ksi and $h/t_w = 219$, we use Table 3-16b, tension field action is included. The stress does not intersect the h/t_w value, so we read the maximum value $a/h = 1.4$. This is obtained by moving horizontally from $h/t_w = 219$ then pivoting on the bold line and moving down vertically to the base and read 1.40.

$$a = (1.4)(82) = 114.8 \text{ in}$$

ASD results are the same with $a = 114.8$ in.

18.6 DESIGN OF STIFFENERS

As previously indicated, it usually is necessary to stiffen the high thin webs of built-up girders to keep them from buckling. If the girders are bolted, the stiffeners probably will consist of a pair of angles bolted to the girder webs. If the girders are welded (the normal situation), the stiffeners probably will consist of a pair of plates welded to the girder webs. Figure 18.10 illustrates these types of stiffeners.

Stiffeners are divided into two groups: *bearing stiffeners*, which transfer heavy reactions or concentrated loads to the full depth of the web, and *intermediate* or *nonbearing stiffeners*, which are placed at various intervals along the web to prevent buckling due to diagonal compression. An additional purpose of bearing stiffeners is the transfer of heavy loads to plate girder webs without putting all the loads on the flange connections.

As described in Sections G2.2 and J10.8 and the corresponding commentaries for those sections in the AISC Specification, transverse stiffeners may be either single or double. They do not have to be connected to the flanges, except in the following situations:

1. Where bearing strength is needed to transmit concentrated loads or reactions.
2. Where single stiffeners are used and the flange of the girder consists of a rectangular plate. For such a situation, the stiffener must be attached to the flange to resist any possible uplift tendency that may be caused by torsion in the flange.
3. Where lateral bracing is attached to a stiffener or stiffeners. Such a stiffener must be connected to the compression flange with a strength sufficient to transmit no less than 1 percent of the total flange stress, unless the flange is composed only of angles.

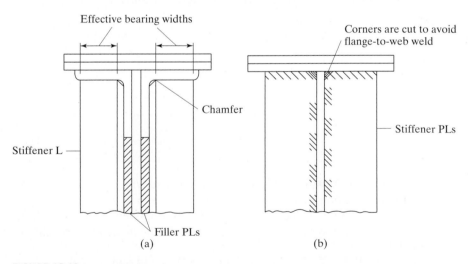

FIGURE 18.10

(a) Angle-bearing stiffeners. (b) Plate-bearing stiffeners.

Welds used to attach stiffeners to girder webs must be terminated at distances not less than 4 times the web thickness nor more than 6 times the web thickness from the near toe of the flange to the web weld.

If bolts are used to connect stiffeners to the web, they may not be spaced farther apart than 12 inches. If intermittent fillet welds are used, the clear distance between them may not exceed the lesser of 16 times the web thickness or 10 inches.

Stiffeners that are located at concentrated loads or reactions have some special requirements because of the possibility of web crippling or compression buckling of the web. In these situations, the stiffeners need to be designed as columns. Should the load or reaction be tensile, it will be necessary to weld the stiffeners to the loaded flange. If the force is compressive, the stiffener may either bear against the loaded flange or be welded to it.

18.6.1 Bearing Stiffeners

Bearing stiffeners are placed in pairs on the webs of built-up girders at unframed girder ends and where required for concentrated loads. They should fit tightly against the flanges being loaded and should extend out toward the edges of the flange plates or angles as far as possible. If the load normal to the girder flange is tensile, the stiffeners must be welded to the loaded flange. If the load is compressive, it is necessary to obtain a snug fit—that is, a good bearing between the flange and the stiffeners. To accomplish this goal, the stiffeners may be welded to the flange or the outstanding legs of the stiffeners may be milled.

A bearing stiffener is a special type of column that is difficult to accurately analyze, because it must support the load in conjunction with the web. The amount of support provided by these two elements is difficult to estimate. The AISC Specification (J10.8) states that the factored load, or reaction, may not exceed the design strength of a column consisting of the stiffener effective area plus a portion of the web equal to $12t_w$ at girder ends and $25t_w$ at interior concentrated loads. Only the part of the stiffeners outside of the fillets of the flange angles or the parts outside of the flange to web welds (see Fig. 18.10) are to be considered effective to support the bearing loads. The effective length of these *bearing stiffener columns* is assumed by the AISC Specification (J10.8) to be equal to $0.75\,h$. The welds made to the girder flange must be designed for the difference between the required strength and the applicable limit state strength. Several recommentations for the details of bearing stiffeners are given in AISC Commentary J10.8.

At an unframed girder end, an end-bearing stiffener is required if the factored reaction R_u is larger than ϕR_n or if $R_a > R_n/\Omega$ If an interior load or reaction is larger than the same values, an interior bearing stiffener is needed.

If the concentrated force is applied at a distance from the member end greater than the member depth d,

$$R_n = (5k + l_b)F_{yw}t_w \qquad \text{(AISC Equation J10-2)}$$

If the concentrated force is applied at a distance from the member end less than or equal to the member depth d,

$$R_n = (2.5k + l_b)F_{yw}t_w \qquad \text{(AISC Equation J10-3)}$$

In these expressions,

$$F_{yw} = \text{specified minimum yield stress of web, ksi}$$
$$k = \text{distance from outer face of flange to web toe of fillet, in}$$
$$l_b = \text{length of bearing (not less than } k \text{ for end beam reactions), in}$$
$$t_w = \text{web thickness, in}$$
$$\phi = 1.0 \text{ and } \Omega = 1.50$$

18.6.2 Intermediate Stiffeners

Intermediate or non-bearing stiffeners are also called stability or transverse intermediate stiffeners. These stiffeners are not required by the AISC Specification Section G2.2 where $h/t_w \leq 2.46\sqrt{\frac{E}{F_{yw}}}$ or where the available strength provided in accordance with Section G2.1 for $k_v = 5$ is greater than the required shear strength.

If $1.10\sqrt{\dfrac{k_v E}{F_y}} < \dfrac{h}{t_w} \leq 1.37\sqrt{\dfrac{k_v E}{F_y}}$

$$C_v = \frac{1.10\sqrt{k_v E/F_y}}{h/t_w} \qquad \text{(AISC Equation G2-4)}$$

If $\dfrac{h}{t_w} > 1.37\sqrt{\dfrac{k_v E}{F_y}}$

$$C_v = \frac{1.51 E k_v}{(h/t_w)^2 F_y} \qquad \text{(AISC Equation G2-5)}$$

The AISC Specification imposes some arbitrary limits on panel aspect ratios (a/h) for plate girders, even where shear stresses are small. The purpose of these limitations is to facilitate the handling of girders during fabrication and erection.

$$V_n = 0.6 F_y A_w C_v \qquad \text{(AISC Equation G2-1)}$$

You should note here that, as the intermediate stiffener spacing is decreased, C_v will become larger, as will the shear capacity of the girder.

The AISC Specification (G2.2) states that the moment of inertia of a transverse intermediate stiffener about an axis at the center of the girder web if a pair of stiffeners is used, or about the face in contact with the web when single stiffeners are used, may not be less than

$$I_{st} \text{ minimum} \geq b t_w^3 j \quad \text{where } b \text{ is the smaller of } a \text{ and } h$$

AISC Specification G3 states that transverse stiffeners subject to tension field action must meet the following limitations in which $(b/t)_{st}$ is the width thickness ratio of the stiffener:

$$\left(\frac{b}{t}\right)_{st} \leq 0.56\sqrt{\frac{E}{F_{yst}}} \qquad \text{(AISC Equation G3-3)}$$

$$I_{st} \geq I_{st1} + (I_{st2} - I_{st1})\left[\frac{V_r - V_{c1}}{V_{c2} - V_{c1}}\right] \qquad \text{(AISC Equation G3-4)}$$

I_{st} is the moment of inertia of the transverse stiffeners about an axis in the web center for stiffener pairs, or about the face in contact with the web plate for single stiffeners. I_{st1} is the minimum moment of inertia of transverse stiffeners required for development of the web shear buckling resistance in Section G2.2. I_{st2} is the *minimum* moment of inertia of the transverse stiffeners required for development of the full web shear buckling plus the web tension field resistance, $V_r = V_{c2}$.

$$I_{st2} = \frac{h^4 \rho_{st}^{1.3}}{40}\left[\frac{F_{yw}}{E}\right]^{1.5} \qquad \text{(AISC Equation G3-5)}$$

V_r is the larger of the required shear strengths in the adjacent web panels. V_{c1} is the smaller of the available shear strength in the adjacent web panels with V_n as defined in Section G2.1 V_{c2} is the smaller of the available shear strength in the adjacent web panels with V_n as defined in Section G 3.2. ρ_{st} is the larger of F_{yw}/F_{yst} and 1.0.

Equation G3-4 is the same requirement as specified in AASHTO (2007).[1]

18.6.3   Longitudinal Stiffeners

Longitudinal stiffeners, though not as effective as transverse ones, frequently are used for bridge plate girders, because many designers feel they are more attractive. As such stiffeners are seldom used for plate girders in buildings, they are not covered in this textbook.

Highway bypass bridge, Stroudsburg, PA. (Courtesy of Bethlehem Steel Corporation.)

---

[1]AASHTO LRFD Bridge Design Specifications, 2007 (Washington, DC: American Association of State Highway and Transportation Officials).

Example 18-6

Design bearing and intermediate stiffeners for the plate girder of Example 18-5, which is not framed at its ends.

Solution

a.   Are end bearing stiffeners required?

Assume point bearing (conservative) of the end reaction (that is, $l_b = 0$) and assume a 5/16-in web-to-flange fillet weld.

Check local web yielding

$$k = \text{distance from outer edge of flange to web end of fillet weld}$$

$$= 1\frac{1}{8} + \frac{5}{16} = 1.44 \text{ in}$$

$$R_n = (2.5\,k + l_b)F_y t_w \qquad \text{(AISC Equation J10-3)}$$

$$= (2.5 \times 1.44 + 0)(36)\left(\frac{3}{8}\right) = 48.6 \text{ k}$$

LRFD $\phi = 1.00$	ASD $\Omega = 1.50$
$\phi R_n = (1.00)(48.6) = 48.6 \text{ k} < 156.98 \text{ k}$	$\dfrac{R_n}{\Omega} = \dfrac{48.6}{1.50} = 32.4 \text{ k} < 109.14 \text{ k}$
$\therefore$ End bearing stiffener is reqd.	$\therefore$ End bearing stiffener is reqd.

If local web yielding check had been satisfactory, it would have been necessary to also check the web crippling criteria set forth in AISC Section J10.3 before we could definitely say that end bearing stiffeners were not required.

b.   Design of end-bearing stiffeners

Try two plate stiffeners $\frac{5}{8} \times 9$, as shown in Fig. 18.11.

Check width–thickness ratio (AISC Table B4.1a).

$$\frac{9.00}{0.875} = 10.29 < 0.56\sqrt{\frac{29{,}000}{36}} = 15.89 \qquad \textbf{OK}$$

Check column strength of stiffener, which is shown crosshatched in Fig. 18.11.

$$I \approx \left(\frac{1}{12}\right)\left(\frac{5}{8}\right)(18.375)^3 = 323 \text{ in}^4 \text{ (neglecting the almost infinitesimal contribu-}$$
tion of the web)

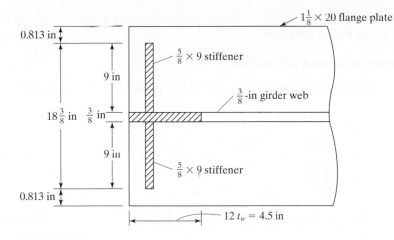

FIGURE 18.11

End bearing stiffeners.

$$A \text{ of column} = (2)\left(\frac{5}{8}\right)(9) + (4.5)\left(\frac{3}{8}\right) = 12.94 \text{ in}^2$$

$$r = \sqrt{\frac{323}{12.94}} = 5.00 \text{ in}$$

$$kL = (0.75)(82) = 61.5 \text{ in}$$

$$\frac{kL}{r} = \frac{61.5}{5.00} = 12.30$$

$$\phi_c F_{cr} = 32.17 \text{ ksi and } \frac{F_{cr}}{\Omega} = 21.39 \text{ ksi from AISC Table 4-22}$$

LRFD	ASD
$\phi_c P_n = (32.17)(12.94) = 416.3 \text{ k} > 156.98 \text{ k }$ **OK**	$\dfrac{P_n}{\Omega_c} = (21.39)(12.94) = 276.8 \text{ k} > 109.14 \text{ k }$ **OK**

Check bearing criterion

$$P_n = 1.8 F_y A_{pd} = (1.8)(36)(2)(9 - 0.5)\left(\frac{5}{8}\right) = 688.5 \text{ k}$$

LRFD $\phi = 0.75$	ASD $\Omega = 2.00$
$\phi P_n = (0.75)(688.5) = 516.4 \text{ k} > 156.98 \text{ k }$ **OK**	$\dfrac{P_n}{\Omega} = \dfrac{688.5}{2.00} = 344.2 \text{ k} > 109.14 \text{ k }$ **OK**

Use 2 PLS $\frac{5}{8} \times 9 \times 6$ ft $9\frac{3}{4}$ in for bearing stiffeners (A depth of $81\frac{3}{4}$ in is used instead of 82 in, for fitting purposes.)

Design of intermediate stiffeners

The first intermediate stiffener (LRFD) is placed 82 in from the end.

$$\frac{a}{h} = \frac{82}{82} = 1.00$$

$$j = \frac{2.5}{\left(\dfrac{a}{h}\right)^2} - 2 = \frac{2.5}{(1.0)^2} - 2 = 0.5 \qquad \text{(AISC Equation G2-8)}$$

$$\text{Min } I_{st} = bt_w^3 j = (82)\left(\frac{3}{8}\right)^3 (0.5) = 2.16 \text{ in}^4 \qquad \text{(AISC Equation G2-7)}$$

There are many possible satisfactory stiffener sizes, but only two sizes are tried here.

Try $\dfrac{1}{4} \times 6$ single plate stiffener

$$I_{st} = \left(\frac{1}{12}\right)\left(\frac{1}{4}\right)(6)^3 = 4.5 \text{ in}^4 > 2.16 \text{ in}^4$$

Try a pair of $\dfrac{1}{4} \times 4$ plate stiffeners

$$I_{st} = \left(\frac{1}{12}\right)\left(\frac{1}{4}\right)\left(2 \times 4 + \frac{3}{8}\right)^3 = 12.24 \text{ in}^4 > 2.16 \text{ in}^4$$

use $\dfrac{1}{4} \times 6$ single plate stiffener

18.7 PROBLEMS FOR SOLUTION

All problems are to be solved by both the LRFD and ASD procedures, UNO.

18-1. Select a cover-plated W section limited to a maximum depth of 21.00 in to support the service loads shown in the accompanying figure. Use A36 steel and assume the beam has full lateral bracing for its compression flange. (*Ans.* one solution W18 × 86 with PL7/8 × 14 each flange LRFD, W18 × 86 with PL7/8 × 16, ASD)

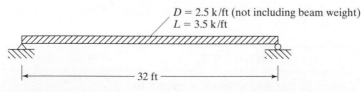

$D = 2.5$ k/ft (not including beam weight)
$L = 3.5$ k/ft

32 ft

FIGURE P18-1

18-2. The architects specify that a cover-plated beam no greater than 16.00 in in depth be designed to support a dead service uniform load of 4 klf, not including the beam weight, and a live service uniform load of 6 klf for a 24-ft simple span. A572 steel with F_y = 50 ksi is to be used, and full lateral bracing for the compression flange is to be provided. Use a W14 × 53 plus cover plates.

18-3. Use LRFD method for the following welded shape: The flanges are 7/8 in × 12 in, the web is 3/8 in × 46 in, and the member is simply supported, uniformly loaded, and has continuous lateral support of the compression flange.
   (a)   Determine the design flexural strength if A572 – Grade 50 steel is used. (*Ans.* ΦM_n = 2383.7 ft-k, LRFD)
   (b)   If the span is 40 ft and supports the loads w_D = 4.6 k/ft (not including the beam weight) and w_L = 3.7 k/ft, verify the moment capacity and select transverse stiffeners as needed. (*Ans.* M_u = 2320 ft-k < ΦM_n = 2383.7 ft-k; 1st stiffener panel, a = 57 in)

18-4. Use LRFD method for the following welded shape: The flanges are 1 in × 10 in, the web is 5/16 in × 50 in, and the member is simply supported, uniformly loaded. Lateral support of the compression flange is provided at the ends and at the midspan.
   (a)   Determine the design flexural strength if A572 – Grade 50 steel is used.
   (b)   If the span is 34 ft and supports the loads w_D = 4.5 k/ft (not including the beam weight) and w_L = 4.5 k/ft, verify the moment capacity and select transverse stiffeners as needed.

18-5. Design a 48-in-deep welded built-up wide-flange section with no intermediate stiffeners for a 50-ft simple span to support a service dead load of 1 klf (not including the beam weight) and a service live load of 1.8 klf. The section is to be framed between columns and is to have full lateral bracing for its compression flange. The design is to be made with A36 steel. (*Ans.* One solution $\frac{7}{16}$ × 46 -in web, $\frac{5}{8}$ × 10 in flange plates)

18-6. Repeat Prob. 18-5 if the span is to be 60 ft and F_y = 50 ksi.

18-7. Design, using the LRFD method, a simply supported plate girder to span 50 ft and support the service loads shown in the figure below. The maximum permissible depth of the girder is 58 in. Use A36 steel and E70XX electrodes and assume that the girder has continuous lateral support of the compression flange. The ends have a bearing-type connection (with the bearing length, l_b = 6 in). Use 16 in wide flanges and a 5/16 in thick web. Also, select transverse stiffeners as needed. Design end bearing stiffener, first intermediate stiffener, the welded connection of the web to the flange, and if a bearing stiffener is required at the concentrated loads. (*Ans.* 1 1/2 × 16 flanges, 5/16 × 55 web; M_u = 3414 ft-k < ΦM_n = 3943 ft-k; 1st stiffener panel, a = 35 in)

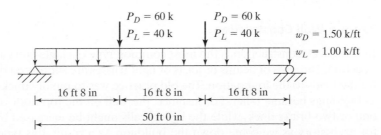

P_D = 60 k
P_L = 40 k
P_D = 60 k
P_L = 40 k
w_D = 1.50 k/ft
w_L = 1.00 k/ft

16 ft 8 in   16 ft 8 in   16 ft 8 in

50 ft 0 in

FIGURE P18-7

C H A P T E R 1 9

Design of Steel Buildings

19.1 INTRODUCTION TO LOW-RISE BUILDINGS

The material in this section and the next pertains to the design of low-rise steel buildings up to several stories in height, while Sections 19.3 to 19.10 provide information concerning common types of building floors, and Section 19.11 presents common types of roof construction. Sections 19.12 and 19.13 are concerned with walls, partitions, and fireproofing. The sections thereafter present general information relating to high-rise or multistory buildings.

The low-rise buildings considered include apartment houses, office buildings, warehouses, schools, and institutional buildings that are not very tall with respect to their least lateral dimensions.

19.2 TYPES OF STEEL FRAMES USED FOR BUILDINGS

Steel buildings usually are classified as being in one of four groups according to their type of construction: *bearing-wall* construction, *skeleton* construction, *long-span* construction, and *combination steel and concrete framing*. More than one of these construction types can be used in the same building. We discuss each of these types briefly in the paragraphs that follow.

19.2.1 Bearing-Wall Construction

Bearing-wall construction is the most common type of single-story light commercial construction. The ends of beams or joists or light trusses are supported by the walls that transfer the loads to the foundation. The old practice was to rapidly thicken load-bearing walls as buildings became taller. For instance, the wall on the top floor of a building might be one or two bricks thick, while the lower walls might be increased by one brick thickness for each story as we come down the building. As a result, this type of construction was usually thought to have an upper economical limit of about two or three stories, although

Coliseum in Spokane, WA. (Courtesy of Bethlehem Steel Corporation.)

some load-bearing buildings were much higher. The tallest load-bearing building built in the United States in the nineteenth century was the 17-story Monadnock building in Chicago. This building, completed in 1891, had 72-in-thick walls on its first floor. A great deal of research has been conducted for load-bearing construction in recent decades, and it has been discovered that thin load-bearing walls may be quite economical for many buildings up to 10, 20, or even more stories.

The average engineer is not very well versed in the subject of bearing-wall construction, with the result that he or she may often specify complete steel or reinforced-concrete frames where bearing-wall construction might have been just as satisfactory and more economical. Bearing-wall construction is not very resistant to seismic loadings, and has an erection disadvantage for buildings of more than one story. For such cases, it is necessary to place the steel floor beams and trusses floor by floor as the masons complete their work below, thus requiring alternation of the masons and ironworkers.

Bearing plates usually are necessary under the ends of the beams, or light trusses, that are supported by the masonry walls, because of the relatively low bearing strength of the masonry. Although, theoretically, the beam flanges may on many occasions provide sufficient bearing without bearing plates, plates are almost always used—particularly where the members are so large and heavy that they must be set by a steel erector. The plates usually are shipped loose and set in the walls by the masons. Setting them in their correct positions and at the correct elevation is a critical part of the construction. Should they not be properly set, there will be some delay in correcting their positions. If a steel erector is used, he or she probably will have to make an extra trip to the job.

When the ends of a beam are enclosed in a masonry wall, some type of wall anchor is desirable to prevent the beam from moving longitudinally with respect to the

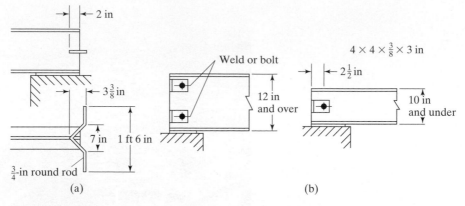

FIGURE 19.1

(a) Government anchor. (b) Angle wall anchors. From American Institute Steel Construction, *Manual of Steel Construction Load & Resistance Factor Design*, 2d ed. (Chicago: AISC, 1994), p. 12–24. "Copyright © American Institute of Steel Construction. Reprinted with permission. All rights reserved."

wall. The usual anchors consist of bent steel bars passing through beam webs. These so-called *government anchors* are shown in Fig. 19.1(a). Occasionally, clip angles attached to the web are used instead of government anchors. These are shown in Fig. 19.1(b). Should longitudinal loads of considerable magnitude be anticipated, regular vertical anchor bolts may be used at the beam ends.

For small commercial and industrial buildings, bearing-wall construction is quite economical when the clear spans are not greater than roughly 35 or 40 ft. If the clear spans are much greater, it becomes necessary to thicken the wall and use pilasters to ensure stability. For these cases, it may often be more economical to use intermediate columns if permissible.

19.2.2 Skeleton Construction

In skeleton construction, the loads are transmitted to the foundations by a framework of steel beams and columns. The floor slabs, partitions, exterior walls, and so on, all are supported by the frame. This type of framing, which can be erected to tremendous heights, often is referred to as beam-and-column construction.

In beam-and-column construction, the frame usually consists of columns spaced 20, 25, or 30 ft apart, with beams and girders framed into them from both directions at each floor level. One very common method of arranging the members is shown in Fig. 19.2. The more heavily loaded girders are placed in the short direction between the columns, while the comparatively lightly loaded beams are framed between the girders in the long direction. It is common practice to orient columns in a way that will minimize eccentric loading. Many engineers will orient columns so that the girders will frame into the web and beams will frame into the flange as shown for the interior bay in Fig. 19.2. With various types of floor construction, other arrangements of beams and girders may be used.

For skeleton framing, the walls can be supported by the steel frame and generally are referred to as *nonbearing* or *curtain walls*. The beams supporting the exterior walls

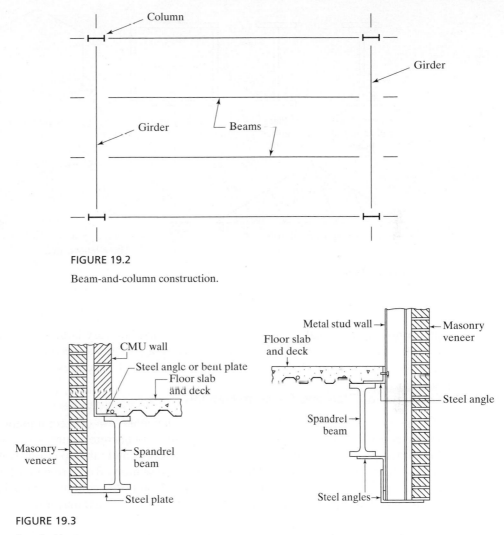

FIGURE 19.2

Beam-and-column construction.

FIGURE 19.3

Spandrel beams.

are called *spandrel beams*. These beams, illustrated in Fig. 19.3, can be placed so that they will serve as the lintels for the windows or other openings.

19.2.3 Long-Span Steel Structures

When it becomes necessary to use very large spans between columns—such as for gymnasiums, auditoriums, theaters, hangars, or hotel ballrooms—the usual skeleton construction may not be sufficient. Should the ordinary rolled *W* sections be insufficient, it may be necessary to use cover-plated beams, built-up I-shaped girders, box girders, large trusses, arches, rigid frames, and the like. When depth is limited, cover-plated beams, plate girders, or box girders may be called upon to do the job. Should

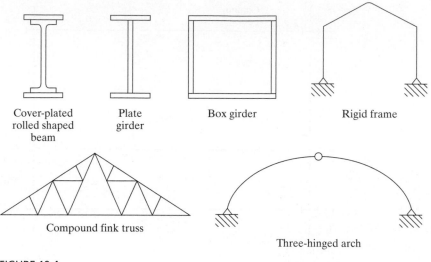

Cover-plated
rolled shaped
beam

Plate
girder

Box girder

Rigid frame

Compound fink truss

Three-hinged arch

FIGURE 19.4

Long-span structures.

depth not be so critical, trusses may be satisfactory. For very large spans, arches and rigid frames often are used. These various types of structures are referred to as *long-span structures*. Figure 19.4 shows a few of these types of structures.

19.2.4 Combination Steel and Concrete Framing

A large percentage of the buildings erected today make use of a combination of reinforced concrete and structural steel. When reinforced-concrete columns are used in very tall buildings, they are rather large on the lower floors and take up considerable space. Steel column shapes surrounded by and bonded to reinforced concrete may be used and are referred to as *encased* or *composite* columns. Composite columns consisting of HSS members filled with concrete (called filled composite columns) may also be used.

19.3 COMMON TYPES OF FLOOR CONSTRUCTION

Concrete floor slabs of one type or another are used almost universally for steel-frame buildings. They are strong and have excellent fire ratings and good acoustic ratings. On the other hand, appreciable time and expense are required to provide the formwork necessary for most slabs. Concrete floors are heavy, they must include some type of reinforcing bars or mesh, and there may be a problem involved in making them watertight. The following are some of the types of concrete floors used today for steel-frame buildings:

1. Concrete slabs supported with open-web steel joists (Section 19.4)
2. One-way and two-way reinforced-concrete slabs supported on steel beams (Section 19.5)

3. Concrete slab and steel beam composite floors (Section 19.6)
4. Concrete-pan floors (Section 19.7)
5. Steel-decking floors (Section 19.8)
6. Flat slab floors (Section 19.9)
7. Precast concrete slab floors (Section 19.10)

Among the several factors to be considered in selecting the type of floor system to be used for a particular building are loads to be supported; fire rating desired; sound and heat transmission; dead weight of floor; ceiling situation below (to be flat or to have beams exposed); facility of floor for locating conduits, pipes, wiring, and so on; appearance; maintenance required; time required to construct; and depth available for floor.

You can get a lot of information about these and other construction practices by referring to various engineering magazines and catalogs, particularly *Sweet's Catalog File*, published by McGraw-Hill Information Systems Company. The author cannot make too strong a recommendation for these books to help the student see the tremendous amount of data available. The sections to follow present brief descriptions of the floor systems mentioned in this section, along with some discussions of their advantages and chief uses.

19.4   CONCRETE SLABS ON OPEN-WEB STEEL JOISTS

Perhaps the most common type of floor slab in use for small steel-frame buildings is the slab supported by open-web steel joists. The joists are small parallel chord trusses whose members usually are made from bars (hence, the common name *bar joist*) or small angles or other rolled shapes. Steel forms or decks are usually attached to the joists by welding or self-drilling or self-tapping screws; then concrete slabs are poured on top. This is one of the lightest types of concrete floors and also one of the most economical. A sketch of an open-web joist floor is shown in Fig. 19.5.

Open-web joists are particularly well suited to building floors with relatively light loads and for structures where there is not too much vibration. They have been used a great deal for fairly tall buildings, but generally speaking, they are better suited for shorter buildings. Open-web joists are very satisfactory for supporting floor and roof slabs for schools, apartment houses, hotels, office buildings, restaurant buildings, and similar low-level structures. These joists generally are not suitable for supporting concentrated loads, however, unless they are especially detailed to carry such loads.

Open-web joists must be braced laterally to prevent them from twisting or buckling and to keep the floors from becoming too springy. Lateral support is provided by *bridging*, which consists of continuous horizontal rods fastened to the top and bottom chords of the joists, or of diagonal cross bracing. Bridging is desirably used at spaces not exceeding 7 ft on centers.

Open-web joists are easy to handle, and they are quickly erected. If desired, a ceiling can be attached to the bottom or suspended from the joists. The open spaces in the webs are well suited for placing conduits, ducts, wiring, piping, and so on. The joists should be either welded to supporting steel beams, or well-anchored in the masonry

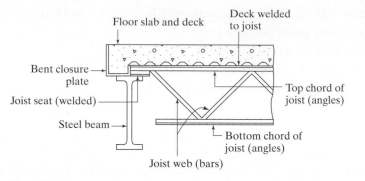

FIGURE 19.5

Open-web joists.

walls. When concrete slabs are placed on top of the joists, they are usually from 2 to 2 1/2 in thick. Nearly all of the many concrete slabs cast in place or precast on the market today can be used successfully on top of open-web joists.

Detailed information concerning open-web joists is not included in the AISC Manual. Such data can be obtained from a publication by the Steel Joist Institute, entitled *Standard Specifications, Load Tables, and Weight Tables for Steel Joists and Joist Girders.*[1] The load tables and specifications are based on an allowable stress method (ASD) and load and resistance factor design (LRFD) method.

Three categories of joists are presented in *Standard Specifications: Open-Web Joists* (called the K-series), the *Longspan Steel Joists* (LH-series), and the *Deep Longspan Steel Joists* (DLH-series). These three types of joists are designed as simply supported trusses which themselves support uniform loads on their top chords. Should concentrated loads need to be supported, a special analysis should be requested from the manufacturer.

One page of the Steel Joist Institute's Standard ASD Load Table for the K-series joists is presented as Table 19.1 of this chapter. As an illustration of the use of this table, look across the top headings of the columns until you reach the 12K3 joists. This member has a nominal depth of 12 in and an approximate weight of 5.7 lb/ft. If we now move vertically down the column under the 12K3 heading until we reach a 20-ft span, as shown on the left side of the table, we can see that this joist can support a total service load equal to 302 lb/ft. The number given below the 302 is 177 lb/ft. It represents the live uniform load that will cause this joist to have an approximate deflection equal to 1/360th of the span. The joists given in the Institute tables run all the way from the 8K1 to the 72DLH19. This latter joist will support a total load of 497 lb/ft for a 144 ft span.

The Steel Joist Institute actually presents a fourth category of joists called *joist girders.* These are quite large joists that may be used to support open-web joists.

---

[1]Myrtle Beach, SC: Steel Joist Institute, 2005.

TABLE 19.1   Standard Load Table Open Web Steel Joists, K-Series Based on a Maximum Allowable Tensile Stress of 30,000 psi

Joist Designation	8K1	10K1	12K1	12K3	12K5	14K1	14K3	14K4	14K6	16K2	16K3	16K4	16K5	16K6	16K7	16K9
Depth (in)	8	10	12	12	12	14	14	14	14	16	16	16	16	16	16	16
Approx. Wt. (lb/ft)	5.1	5.0	5.0	5.7	7.1	5.2	6.0	6.7	7.7	5.5	6.3	7.0	7.5	8.1	8.6	10.0
Span (ft) ↓																
8	550 / 550															
9	550 / 550															
10	550 / 480	550 / 550														
11	532 / 377	550 / 542														
12	444 / 288	550 / 455	550 / 550	550 / 550	550 / 550											
13	377 / 225	479 / 363	550 / 510	550 / 510	550 / 510											
14	324 / 179	412 / 289	500 / 425	550 / 463	550 / 463	550 / 550	550 / 550	550 / 550	550 / 550							
15	281 / 145	358 / 234	434 / 344	543 / 428	550 / 434	511 / 475	550 / 507	550 / 507	550 / 507							
16	246 / 119	313 / 192	380 / 282	476 / 351	550 / 396	448 / 390	550 / 467	550 / 467	550 / 467	550 / 550	550 / 550	550 / 550	550 / 550	550 / 550	550 / 550	550 / 550
17		277 / 159	336 / 234	420 / 291	550 / 366	395 / 324	495 / 404	550 / 443	550 / 443	512 / 488	550 / 526	550 / 526	550 / 526	550 / 526	550 / 526	550 / 526
18		246 / 134	299 / 197	374 / 245	507 / 317	352 / 272	441 / 339	530 / 397	550 / 408	456 / 409	508 / 456	550 / 490	550 / 490	550 / 490	550 / 490	550 / 490
19		221 / 113	268 / 167	335 / 207	454 / 269	315 / 230	395 / 287	475 / 336	550 / 383	408 / 347	455 / 386	547 / 452	550 / 455	550 / 455	550 / 455	550 / 455
20		199 / 97	241 / 142	302 / 177	409 / 230	284 / 197	356 / 246	428 / 287	525 / 347	368 / 297	410 / 330	493 / 386	550 / 426	550 / 426	550 / 426	550 / 426
21			218 / 123	273 / 153	370 / 198	257 / 170	322 / 212	388 / 248	475 / 299	333 / 255	371 / 285	447 / 333	503 / 373	548 / 405	550 / 406	550 / 406
22			199 / 106	249 / 132	337 / 172	234 / 147	293 / 184	353 / 215	432 / 259	303 / 222	337 / 247	406 / 289	458 / 323	498 / 351	550 / 385	550 / 385
23			181 / 93	227 / 116	308 / 150	214 / 128	268 / 160	322 / 188	395 / 226	277 / 194	308 / 216	371 / 252	418 / 282	455 / 307	507 / 339	550 / 363
24			166 / 81	208 / 101	282 / 132	196 / 113	245 / 141	295 / 165	362 / 199	254 / 170	283 / 189	340 / 221	384 / 248	418 / 269	465 / 298	550 / 346
25						180 / 100	226 / 124	272 / 145	334 / 175	234 / 150	260 / 167	313 / 195	353 / 219	384 / 238	428 / 263	514 / 311
26						166 / 88	209 / 110	251 / 129	308 / 156	216 / 133	240 / 148	289 / 173	326 / 194	355 / 211	395 / 233	474 / 276
27						154 / 79	193 / 98	233 / 115	285 / 139	200 / 119	223 / 132	268 / 155	302 / 173	329 / 188	366 / 208	439 / 246
28						143 / 70	180 / 88	216 / 103	265 / 124	186 / 106	207 / 118	249 / 138	281 / 155	306 / 168	340 / 186	408 / 220
29										173 / 95	193 / 106	232 / 124	261 / 139	285 / 151	317 / 167	380 / 198
30										161 / 86	180 / 96	216 / 112	244 / 126	266 / 137	296 / 151	355 / 178
31										151 / 78	168 / 87	203 / 101	228 / 114	249 / 124	277 / 137	332 / 161
32										142 / 71	158 / 79	190 / 92	214 / 103	233 / 112	259 / 124	311 / 147

Short-span joists in the Univac Center
of Sperry-Rand Corporation, Blue Bell,
PA. (Courtesy of Bethlehem Steel
Corporation.)

19.5   ONE-WAY AND TWO-WAY REINFORCED-CONCRETE SLABS

19.5.1   One-Way Slabs

A very large number of concrete floor slabs in old office and industrial buildings consisted of one-way slabs about 4 in thick, supported by steel beams 6 to 8 ft on centers.
These floors were often referred to as *concrete arch* floors, because, at one time, brick
or tile floors were constructed in approximately the same shape—that is, in the shape
of arches with flat tops.

A one-way slab is shown in Fig. 19.6. The slab spans in the short direction shown by
the arrows in the figure. One-way slabs usually are used when the long direction is two or
more times the short direction. In such cases, the short span is so much stiffer than the long
span that almost all of the load is carried by the short span. The short direction is the main
direction of bending and will be the direction of the main reinforcing bars in the concrete,
but temperature-and-shrinkage steel is needed in the other direction.

A typical cross section of a one-way slab floor with supporting steel beams is
shown in Fig. 19.7. When steel beams or joists are used to support reinforced-concrete
floors, it may be necessary to encase them in concrete or other materials to provide the
required fire rating. Such a situation is very expensive.

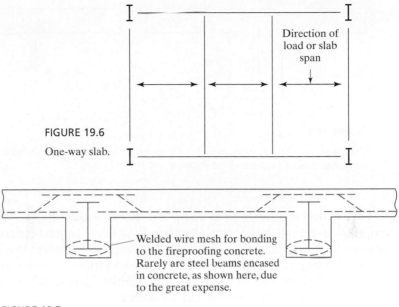

FIGURE 19.6

One-way slab.

FIGURE 19.7

Cross section of one-way slab floor.

It may be necessary to leave steel lath protruding from the bottom flanges or soffits of the beam for the purpose of attaching plastered ceilings. Should such ceilings be required to cover the beam stems, this floor system will lose a great deal of its economy.

One-way slabs have an advantage when it comes to formwork, because the forms can be supported entirely by the steel beams, with no vertical shoring needed. They have a disadvantage in that they are much heavier than most of the newer lightweight floor systems. The result is that they are not used as often as formerly for lightly loaded floors. Should, however, a rigid floor, a floor to support heavy loads, or a durable floor be needed, the one-way slab may be the appropriate selection.

19.5.2 Two-Way Concrete Slabs

The two-way concrete slab is used when the slabs are square or nearly so, with supporting beams under all four edges. The main reinforcing runs in both directions. Other characteristics are similar to those of the one-way slab.

19.6   COMPOSITE FLOORS

Composite floors, previously discussed in Chapter 16, have steel beams (rolled sections, cover-plated beams, or built-up members) bonded together with concrete slabs in such a manner that the two act as a unit in resisting the total loads that the beam sections would otherwise have to resist alone. The steel beams can be smaller when composite floors are used, because the slabs act as part of the beams.

FIGURE 19.8

Composite floors. (a) Steel beam encased in concrete (very expensive). (b) Steel beam bonded to concrete slab with steel anchors.

A particular advantage of composite floors is that they utilize concrete's high compressive strength by keeping all or nearly all of the concrete in compression, and at the same time stress a larger percentage of the steel in tension than is normally the case with steel-frame structures. The result is less steel tonnage in the structure. A further advantage of composite floors is an appreciable reduction in total floor thickness, which is particularly important in taller buildings.

Two types of composite floor systems are shown in Fig. 19.8. The steel beam can be completely encased in the concrete and the horizontal shear transferred by friction and bond (plus some shear reinforcement if necessary). *This type of composite floor is usually quite uneconomical.* The usual type of composite floor is shown in part (b), where the steel beam is bonded to the concrete slab with some type of steel anchors. Various types of steel anchors have been used during the past few decades, including spiral bars, channels, angles, studs, and so on, but economic considerations usually lead to the use of round studs welded to the top flanges of the beams in place of the other types mentioned. Typical studs are 1/2 to 3/4 in in diameter and 2 to 4 in in length.

Cover plates may occasionally be welded to the bottom flanges of the rolled steel sections used for composite floors. The student can see that, with the slab acting as a part of the beam, there is quite a large area available on the compressive side of the beam. By adding plates to the tensile flange, a little better balance is obtained.

19.7   CONCRETE-PAN FLOORS

There are several types of pan floors that are constructed by placing concrete in removable pan molds. (Some special light corrugated pans also are available that can be left in place.) Rows of the pans are arranged on wooden floor forms, and the concrete is placed over the top of them, producing a floor cross section such as the one shown in Fig. 19.9. Joists are formed between the pans, giving a tee-beam-type floor.

These floors, which are suitable for fairly heavy loads, are appreciably lighter than the one-way and two-way concrete slab floors. They require a good deal of formwork, including appreciable shoring underneath the stems. Labor is thus higher than for many floors, but savings due to weight reduction and the reuse of standard-size pans may make pan floor design economically competitive. If suspended ceilings are required, pan floors will have a decided economic disadvantage.

The 104-ft joists for the Bethlehem Catholic High School, Bethlehem, PA. (Courtesy of Bethlehem Steel Corporation.)

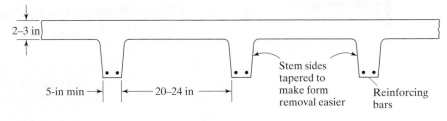

FIGURE 19.9

Concrete-pan floor.

Two-way construction is available—that is, with ribs or stems running in both directions. Pans with closed ends are used, and the result is a waffle-type floor. This type of floor usually is used when the floor panels are square or nearly so. Two-way construction can be obtained for reasonably economical prices and yields a very attractive ceiling below with fairly good acoustical properties.

19.8   STEEL FLOOR DECK

Typical cross sections of steel-decking floors are shown in Fig. 19.10; several other variations are available. *Today, formed steel decking with a concrete topping is by far the most common type of floor system used for office and apartment buildings.* It also is popular for hotels and other buildings where the loads are not very large.

One-piece metal dome pans. (Courtesy of Gateway Erectors, Inc.)

Temple Plaza parking facility, Salt Lake City, UT. (Courtesy of Ceco Steel Products Corporation.)

A particular advantage of steel-decking floors is that the decking immediately forms a working platform. The corrugated steel floor decks are quite strong. Due to the strength of the decking, the concrete does not have to be particularly strong, which permits the use of lightweight concrete, often as thin as 2 or 2 1/2 in, depending on the space of the supporting beams or joists.

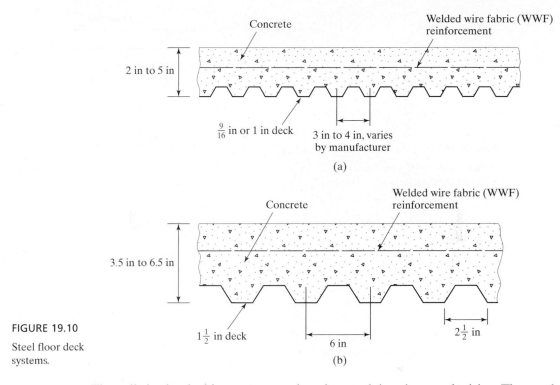

FIGURE 19.10

Steel floor deck systems.

The cells in the decking are convenient for conduits, pipes, and wiring. The steel usually is galvanized, and if exposed underneath, it can be left as it comes from the manufacturer or painted as desired. Should fire resistance be necessary, a suspended ceiling with metal lath and plaster may be used.

19.9   FLAT SLAB FLOORS

Formerly, flat slab floors were limited to reinforced-concrete buildings, but today it is possible to use them in steel-frame buildings. A flat slab is reinforced in two or more directions and transfers its loads to the supporting columns without the use of beams and girders protruding below. The supporting concrete beams and girders are made so wide that they are the same depth as the slab.

Flat slabs are of great value when the panels are approximately square, when more headroom is desired than is provided with the normal beam and girder floors, when heavy loads are anticipated, and when we want to place the windows as near to the tops of the walls as possible. Another advantage is the flat ceiling produced for the floor below. Although the large amounts of reinforcing steel required increased costs, the simple formwork cuts expenses decidedly. The significance of simple formwork will be understood when it is realized that over one-half of the cost of the average poured reinforced-concrete floor slab is in the formwork.

For some reinforced-concrete frame buildings with flat slab floors, it is necessary to flare out the tops of the columns, forming column capitals, and perhaps thicken the slab around the column with the so-called *drop panels*. These items, which are shown in

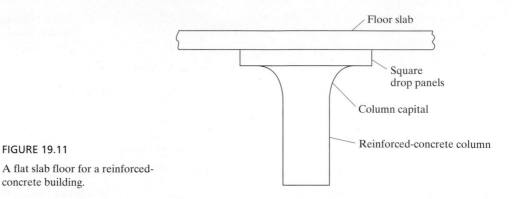

FIGURE 19.11

A flat slab floor for a reinforced-concrete building.

Fig. 19.11, may be necessary to prevent shear or punching failures in the slab around the column.

It is possible in steel-frame buildings to use short steel cantilever beams connected to the steel columns and embedded in the slabs. These beams serve the purposes of the flared columns and drop panels in ordinary flat slab construction. This arrangement often is called a *steel grillage* or *column head*.

The flat slab is not a very satisfactory type of floor system for the usual tall building where lateral forces (wind or earthquake) are appreciable, because protruding beams and girders are desirable to serve as part of the lateral bracing system.

19.10 PRECAST CONCRETE FLOORS

Precast concrete sections are more commonly associated with roofs than they are with floor slabs, but their use for floors is increasing. They are quickly erected and reduce the need for formwork. Lightweight aggregates are often used in the concrete, making the sections light and easy to handle. Some of the aggregates used make the slabs nailable and easily cut and fitted on the job. For floor slabs with their fairly heavy loads, the aggregates should be of a quality that will not greatly reduce the strength of the resulting concrete.

The reader again is referred to *Sweet's Catalog File*. In these catalogs, a great amount of information is available on the various types of precast floor slabs on the market today. Some of the common types available are listed here, and a cross section of each type mentioned is presented in Fig. 19.12. Due to slight variations in the upper

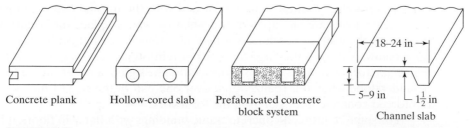

Concrete plank   Hollow-cored slab   Prefabricated concrete block system   Channel slab

FIGURE 19.12

Precast concrete roof and floor slabs.

101 Hudson St., Jersey City, NJ. (Courtesy of Owen Steel Company, Inc.)

Interns' living quarters, Good
Samaritan Hospital, Dayton, OH.
(Courtesy of Flexicore Company, Inc.)

surfaces of precast sections, it usually is necessary to use a mortar topping of 1 to 2 in before asphalt tile or other floor coverings can be installed.

1. *Precast concrete planks* are roughly 2 to 3 in thick and 12 to 24 in wide and are placed on joists up to 5 or 6 ft on center. They usually have tongue-and-groove edges.
2. *Hollow-cored slabs* are sections roughly 6 to 8 in deep and 12 to 18 in wide and have hollow, perhaps circular, cores formed in a longitudinal direction, thus reducing their weights by approximately 50 percent (as compared with solid slabs of the same dimensions). These sections, which can be used for spans of roughly 10 to 25 ft, may be prestressed.
3. *Prefabricated concrete block systems* are slabs made by tying together precast concrete blocks with steel rods (may also be prestressed). These seldom-used floors vary from 4 to 8 in deep and can be used for spans of roughly 8 to 32 ft.
4. *Channel slabs* are used for spans of approximately 8 to 24 ft. Rough dimensions of these slabs are given in Fig. 19.12.

19.11 TYPES OF ROOF CONSTRUCTION

The types of roof construction commonly used for steel-frame buildings include concrete slabs on open-web joists, steel roof deck, and various types of precast concrete slabs. For industrial buildings, the roof system in predominant use today involves the use of a cold-rolled steel deck. In addition, the other types mentioned in the preceding sections on floor types are occasionally used, but they usually are unable to compete economically. Among the factors to be considered in selecting the specific types of roof construction are strength, weight, span, insulation, acoustics, appearance below, and type of roof covering to be used.

The major differences between floor-slab selection and roof-slab selection probably occur in the considerations of strength and insulation. The loads applied to roofs generally are much smaller than those applied to floor slabs, thus permitting the use of many types of lightweight aggregate concretes that may be appreciably weaker than floor material. Roof slabs should have good insulation properties, or they will have to have insulation materials placed on them and covered by the roofing. Among the many types of lightweight aggregates used are wood fibers, zonolite, foams, sawdust, gypsum, and expanded shale. Although some of these materials decidedly reduce concrete strengths, they provide very light roof decks with excellent insulating properties.

Precast slabs made with these aggregates are light, are quickly erected, have good insulating properties, and usually can be sawed and nailed. For poured concrete slabs on open-web joists, several lightweight aggregates work very well (zonolite, foams, gypsum, etc.), and the resulting concrete easily can be pumped up to the roofs, thereby facilitating construction. By replacing the aggregates with certain foams, concrete can be made so light it will float in water (for a while). Needless to say, the strength of the resulting concrete is quite low.

Steel decking with similar thin slabs of lightweight and insulating concrete placed on top make very good economical roof systems. A competitive variation consists of steel decking with rigid insulation board placed on top, followed by the regular roofing material. The other concrete-slab types are hard pressed to compete economically with these types, for lightly loaded roofs. Other types of poured concrete decks require much more labor.

19.12 EXTERIOR WALLS AND INTERIOR PARTITIONS

19.12.1 Exterior Walls

The purposes of exterior walls are to provide resistance to atmospheric conditions, including insulation against heat and cold; satisfactory sound-absorption and lightrefraction characteristics; sufficient strength; and acceptable fire ratings. They also should look good and yet be reasonably economical.

For many years, exterior walls were constructed of some type of masonry, glass, or corrugated sheeting. Recently, however, the number of satisfactory materials available for exterior walls has increased tremendously. Precast concrete panels, insulated metal sheeting, and many other prefabricated units are commonly used today. Of increasing popularity are the light prefabricated sandwich panels consisting of three layers. The exterior surface is made from aluminum, stainless steel, ceramics, plastics, and other materials. The center of the panel consists of some type of insulating material, such as fiberglass or fiberboard, while the interior surface is made from metal, plaster, masonry, or some other attractive material.

19.12.2 Interior Partitions

The main purpose of interior partitions is to divide the inside space of a building into rooms. Partitions are selected for appearance, fire rating, weight, and acoustical properties. Partitions may be loadbearing or nonloadbearing.

Bearing partitions support gravity loads in addition to their own weights and are thus permanently fixed in position. They can be constructed from wood or steel studs or from masonry units and faced with plywood, plaster, wallboard, or other material.

Nonbearing partitions do not support any loads in addition to their own weights, and thus may be fixed or movable. Selection of the type of material to be used for a partition is based on the answers to the following questions: Is the partition to be fixed or movable? Is it to be transparent or opaque? Is it to extend all the way to the ceiling? Is it to be used to conceal piping and electrical conduits? Are there fire-rating and acoustical requirements? The more common types are made of metal, masonry, or concrete. For design of the floor slabs in a building that has movable partitions, some allowance should be given to the fact that the partitions may be moved from time to time. Probably the most common practice is to increase the floor-design live load by 15 or 20 psf over the entire floor area.

19.13 FIREPROOFING OF STRUCTURAL STEEL

Although structural steel members are incombustible, their strength is tremendously reduced at temperatures normally reached in fires when the other materials of a building burn. Many disastrous fires have occurred in empty buildings where the only fuel was the buildings themselves. Steel is an excellent heat conductor, and nonfireproofed steel members may transmit enough heat from one burning compartment of a building to ignite materials with which they are in contact in adjoining sections of the building.

The fire resistance of structural steel members can be greatly increased by coating them with fire-protective covers such as concrete, gypsum, mineral fiber sprays, special paints, and other materials. The thickness and kind of fireproofing used depends on the type of structure, the degree of fire hazard, and economics.

In the past, concrete was commonly used for fire protection. Although concrete is not a particularly good insulation material, it is very satisfactory when applied in thicknesses of 1 1/2 to 2 in or more, due to its mass. Furthermore, the water in concrete (16 to 20 percent when fully hydrated) improves its fireproofing qualities appreciably. The boiling off of the water from the concrete requires a great deal of heat. As the water or steam escapes, it greatly reduces the concrete temperature. This is identical to the behavior of steel steam boilers, where the steel temperature is held to maximum values of only a few hundred degrees Fahrenheit due to the escape of the steam. It is true, however, that in very intense fires the boiling off of the water may cause severe cracking and spalling of the concrete.

Although concrete is an everyday construction material, and in mass it is a quite satisfactory fireproofing material, its installation cost is extremely high and its weight is large. As a result, for most steel construction, spray-on fireproofing materials have almost completely replaced concrete.

The spray-on materials usually consist of either mineral fibers or cementious fireproofing materials. The mineral fibers formerly used were made of asbestos. Due to the health hazards associated with this material, its use has been discontinued, and other fibers now are used by manufacturers. The cementious fireproofing materials are composed of gypsum, perlite, vermiculite, and others. Sometimes, when plastered ceilings are required in a building, it is possible to hang the ceilings and light-gage furring channels by wires from the floor systems above and to use the plaster as the fireproofing.

The cost of fireproofing structural steel buildings is high and hurts steel in its economic competition with other materials. As a result, a great deal of research is being conducted by the steel industry on imaginative new fireproofing methods. Among these ideas is the coating of steel members with expansive and insulative paints. When heated to certain temperatures, these paints will char, foam, and expand, forming an insulative shield around the members. These swelling, or extumescent, paints are very expensive.

Other techniques involve the isolation of some steel members outside of the building, where they will not be subject to damaging fire exposure, and the circulation of liquid coolants inside box- or tube-shaped members in the building. It is probable that major advances will be made with these and other fireproofing methods in the near future.[2]

19.14   INTRODUCTION TO HIGH-RISE BUILDINGS

In this introductory text, we do not discuss the design of multistory buildings in detail, but the material of the next few sections gives the student a general idea of the problems involved in the design of such buildings, without presenting elaborate design examples.

---

[2]W. A. Rains, "A New Era in Fire Protective Coatings for Steel," *Civil Engineering* (New York: ASCE, September 1976), pp. 80–83.

Office buildings, hotels, apartment houses, and other buildings of many stories are quite common in the United States, and the trend is toward an even larger number of tall buildings in the future. Available land for building in our heavily populated cities is becoming scarcer and scarcer, and costs are becoming higher and higher. Tall buildings require a smaller amount of this expensive land to provide required floor space. Other factors contributing to the increased number of multistory buildings are new and better materials and construction techniques.

On the other hand, several factors that may limit the heights to which buildings will be erected in the future include the following:

1. Certain city building codes prescribe maximum heights for buildings.
2. Foundation conditions may not be satisfactory for supporting buildings of many stories.
3. Floor space may not be rentable above a certain height. Someone will always be available to rent the top floor or two of a 250-story building, but floors numbered 100 through 248 may not be so easy to rent.
4. There are several cost factors that tend to increase with taller buildings. Among these factors are elevators, plumbing, heating and air-conditioning, glazing, exterior walls, and wiring.

Whether a multistory building is used for an office building, a hospital, a school, an apartment house, or something else, the problems of design generally are the same. The construction usually is of the skeleton type in which the loads are transmitted to the foundation by a framework of steel beams and columns. The floor slabs, partitions, and exterior walls all are supported by the frame. This type of framing, which can be erected to tremendous heights, may also be referred to as *beam-and-column* construction. Bearing-wall construction is not often used for buildings of more than a few stories, although it has on occasion been used for buildings up to 20 or 25 stories. The columns of a skeleton frame usually are spaced 20, 25, or 30 ft on centers, with beams and girders framing into them from both directions. (See Fig. 19.2.) On some of the floors, however, it may be necessary to have much larger open areas between columns for dining rooms, ballrooms, and so on. For such cases, very large beams (perhaps built-up I girders) are needed to support column loads for many floors above.

It usually is necessary in multistory buildings to fireproof the members of the frame with concrete, gypsum, or some other material. The exterior walls often are constructed with concrete or masonry units, although an increasing number of modern buildings are erected with large areas of glass in the exterior walls.

For these tall and heavy buildings, the usual spread footings may not be sufficient to support the loads. If the bearing strength of the soil is high, steel grillage footings may be sufficient; for poor soil conditions, pile or pier foundations may be necessary.

For multistory buildings, the beam-to-column systems are superimposed on top of each other, story by story or tier by tier. The column sections can be fabricated for one, two, or more floors, with the two-story lengths probably being the most common. Theoretically, column sizes can be changed at each floor level, but the costs of the splices involved usually would more than cancel any savings in column weights. Columns of three or more stories in height are difficult to erect. The two-story heights work out very well most of the time.

19.15   DISCUSSION OF LATERAL FORCES

For tall buildings, lateral forces as well as gravity forces must be considered. High wind pressures on the sides of tall buildings produce overturning moments. These moments probably are resisted without difficulty by the axial strengths of the columns, but the horizontal shears produced on each level may be of such magnitude as to require the use of special bracing or moment-resisting connections.

Unless they are fractured, the floors and walls provide a large part of the lateral stiffness of tall buildings. Although the amount of such resistance may be several times that provided by the lateral bracing, it is difficult to estimate and may not be reliable. Today, so many modern buildings have light, movable interior partitions, glass exterior walls, and lightweight floors, that only the steel frame should be assumed to provide the required lateral stiffness.

Not only must a building be sufficiently braced laterally to prevent architectual failure, but it must also be prevented from deflecting so much as to damage its various architectural parts. Another item of importance is the provision of sufficient bracing to give the occupants a feeling of safety. This is commonly referred to as the serviceability limit state. They might not have this feeling of safety in tall buildings that have a great deal of lateral movement in times of high winds. There have actually been tales of occupants of the upper floors of tall buildings complaining of seasickness on very windy days.

The horizontal deflection of a multistory building due to wind or seismic loading is called *drift*. It is represented by Δ in Fig. 19.13. Drift is measured by the *drift index*, Δ/h, where h is the height or distance to the ground.

The usual practice in the design of multistory steel buildings is to provide a structure with sufficient lateral stiffness to keep the drift index between about 0.0015 and 0.0030 radians for the worst storms that occur in a period of approximately 10 years.

Broadview Apartments, Baltimore, MD. (Courtesy of Lincoln Electric Company.)

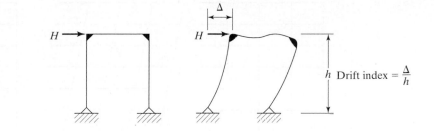

FIGURE 19.13

Lateral drift.

These so-called 10-year storms might have wind in the range of about 90 mph, depending on location and weather records. It is felt that when the index does not exceed these values, the users of the building will be reasonably comfortable.

In addition, multistory buildings should be designed to withstand 50-year storms safely. For such cases, the drift index will be larger than the range mentioned, with the result that the occupants will suffer some discomfort. The 1450-ft twin towers of the World Trade Center in New York City, which were destroyed in 2001, theoretically would deflect or sway about 3 ft in 10-year storms (drift index = 0.0021), while in hurricane winds they theoretically would sway about 7 ft (drift index = 0.0048).

Most buildings can be designed, with little extra expense, to withstand the forces caused during an earthquake of fairly severe intensity. On the other hand, earthquakes during recent years have clearly shown that the average building not designed for earthquake forces can be destroyed by earthquakes which are not particularly severe. The usual practice is to design buildings for additional lateral loads (representing the estimate of the earthquake forces) that are equal to some percentage of the weight of the building and its contents.

The lateral forces require the use of more steel, even though load factors are reduced and safety factors increased for wind and earthquake forces. Additional steel will be used for bracing or moment-resisting connections, as described in the next section.

19.16   TYPES OF LATERAL BRACING

A steel building frame with no lateral bracing is represented in Fig. 19.14(a). Should the beams and columns shown be connected with the standard ("simple beam") connections, the frame would have little resistance to the lateral forces shown. Assuming that the joints act as frictionless pins, the frame would be laterally deflected, as shown in Fig. 19.14(b), eventually collapsing because the structure is unstable.

To resist these lateral deflections, the simplest method, from a theoretical standpoint, is the insertion of full diagonal bracing, as shown in Fig. 19.14(c). From a practical standpoint, however, the student can easily see that in the average building, full diagonal bracing will often be in the way of doors, windows, and other wall openings. Also, many buildings have movable interior partitions, and the presence of interior cross bracing would greatly reduce this flexibility. Usually, diagonal bracing is convenient in solid walls in and around elevator shafts, stairwells, and other walls in which few or no openings are planned.

Another method of providing resistance to lateral forces involves the use of moment-resisting connections, as illustrated in Fig. 19.15. This *moment frame* may be

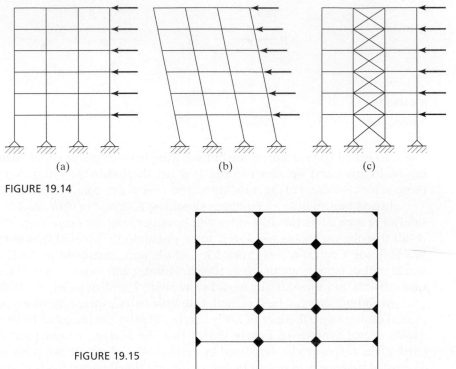

FIGURE 19.14

FIGURE 19.15

Moment frame

used economically to provide lateral bracing for lower-rise buildings. As buildings become taller, however, moment frame is not very economical, nor is it very satisfactory in limiting lateral deflections.

Some ways in which we can transmit lateral forces to the ground for buildings in the 20- to 60-story range are shown in Fig. 19.16. The X bracing system of part (a) works very well, except that it might get in the way. Furthermore, as compared with the systems shown in Fig. 19.16(b) and (c), the floor beams have longer spans and will have to be larger.

The K bracing system (b) provides more freedom for the placing of openings than does the full X bracing. K braces are connected at midspan, with the result that the floor beams will have smaller moments. The K bracing system (like the knee bracing system) uses less material than does X bracing. If we need more room than is available with the K bracing system, we can go to the full-story knee bracing system shown in Fig. 19.16(c).[3]

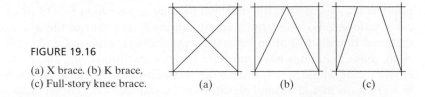

FIGURE 19.16

(a) X brace. (b) K brace.
(c) Full-story knee brace.

(a)   (b)   (c)

[3]E. H. Gaylord, Jr., and C. N. Gaylord, *Structural Engineering Handbook*, 2d ed. (New York: McGraw-Hill, 1979), pp. 19–77 to 19–111.

Construction of the twin Petronas Towers in Kuala Lumpur, Malaysia. (Courtesy of
Corbis/Sergio Dolzntes.)

In the usual building, the floor system (beams and slabs) is assumed to be rigid in the horizontal plane, and the lateral loads are assumed to be concentrated at the floor levels. Floor slabs and girders acting together provide considerable resistance to lateral forces. Investigation of steel buildings that have withstood high wind forces has shown that the floor slabs distribute the lateral forces so that all of the columns on a particular floor have essentially equal deflections, as long as twisting of the structure does not occur. When rigid floors are present, they spread the lateral shears to the columns or walls in the building. When lateral forces are particularly large, as in very tall buildings or where seismic forces are being considered, certain specially designed walls may be used to resist large parts of the lateral forces. These walls are called *shear walls*.

It is not necessary to brace every panel in a building. Usually, the bracing can be placed in the outside walls with less interference than in the inside walls, where movable partitions may be desired. Probably, the bracing of the outside panels alone is not enough, and some interior panels may need to be braced. It is assumed that the floors and beams are sufficiently rigid to transfer the lateral forces to the braced panels. Three possible arrangements of braced panels are shown in Fig. 19.17 for lateral forces in one direction. A symmetrical arrangement probably is desirable to prevent uneven lateral deflection, and thus torsion, in the building.

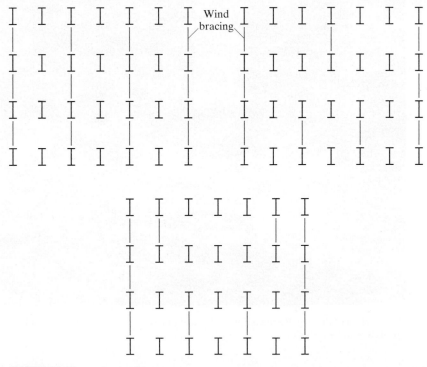

FIGURE 19.17

Possible locations for lateral bracing.

Bracing around elevator shafts and stairwells usually is permissible, while for other locations it often will interfere with windows, doors, movable partitions, glass exterior walls, open spaces, and so on. Should bracing be used around an elevator shaft, as shown in Fig. 19.18(a), and should calculations show that the drift index is too large, it may be possible to use a *hat truss* on the top floor, as shown in part (b). Such a truss will substantially reduce drift. If a hat truss is not feasible because of interference with other items, one or more *belt trusses* may be possible, as shown in Fig. 19.18(c). A belt truss will appreciably reduce drift, although not as much as a hat truss.

The bracing systems described so far are not efficient for buildings with more than approximately 60 stories. These taller buildings have very large lateral wind and, perhaps, seismic forces applied hundreds of feet above the ground. The designer needs to develop a system that will resist these loads without failure and in such a manner that lateral deflections are not so great as to frighten the occupants. The bracing methods used for such buildings usually are based on a tubular frame concept.

With the tubular system, a vertical tubular cantilever frame is created, much like the tube of Fig. 19.19. The tube consists of the building columns and girders in both the longitudinal and transverse directions of the building. The idea is to create a tube that will act like a continuous chimney or stack.

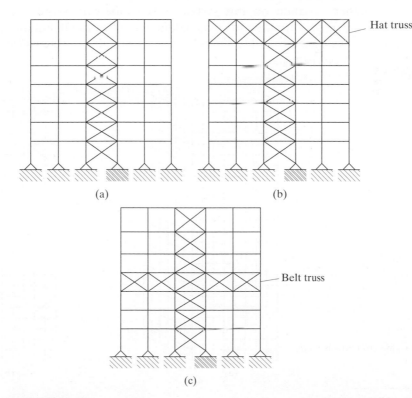

FIGURE 19.18
Bracing systems.

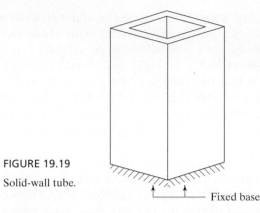

FIGURE 19.19

Solid-wall tube.

Fixed base

    To build the tube, the exterior columns are spaced close together—from 3 or 4 ft up to 10 or 12 ft on center. They are connected with spandrel beams at the floor levels, as shown in Fig. 19.20(a).

    Further improvements of the tubular system can be made by cross bracing the frame with X bracing over many stories, as illustrated in Fig. 19.20(b). This latter system is very stiff and efficient, and is very helpful in distributing gravity loads rather uniformly over the exterior columns.

    Another variation on this system is the tube-within-a-tube system. Interior columns and girders are used to create additional tubes. In tall buildings, it is common to group elevator shafts, utility shafts, and stairs, and these systems can include shear walls and braced bents.[4]

FIGURE 19.20

(a) Tubular frame, (b) Cross-braced tubular frame.

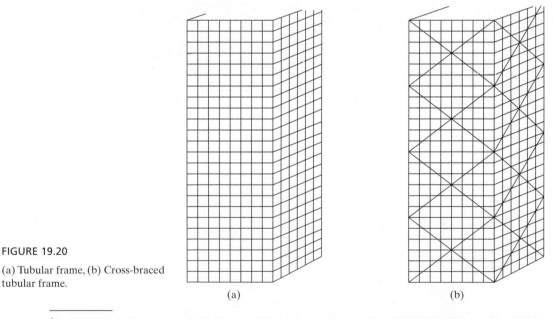

(a)

(b)

[4]R. N. White, P. Gergely, and R. G. Sexsmith, *Structural Engineering*, vol. 3 (New York: Wiley, 1974), pp. 537–546.

19.17   ANALYSIS OF BUILDINGS WITH DIAGONAL WIND BRACING FOR LATERAL FORCES

Full diagonal cross bracing has been described as being an economical type of wind bracing for tall buildings. Where this type of bracing cannot practically be used due to interference with windows, doors, or movable partitions, moment-resisting brackets often are used. However, should very tall narrow buildings (with height–least-width ratios of 5.0 or greater) be constructed, lateral deflections may become a problem with moment-resisting joints. The joints can satisfactorily resist the moments, but the deflections may be excessive.

Should maximum wind deflections be kept under 0.002 times the building height, there is little chance of injury to the building, according to ASCE Subcommittee 31.[5] The deflection is to be computed neglecting any resistance supplied by floors and walls. To keep lateral deflections within this range, it is necessary to use deep knee bracing, K bracing, or full diagonal cross bracing when the height–least-width ratio is about 5.0; and for greater values of the ratio, full diagonal cross bracing or some other method such as the tubular frame is needed.[6]

The student often may see bracing used in buildings for which he or she would think wind stresses were negligible. Such bracing stiffens up a building appreciably and serves the useful purpose of plumbing the steel frame during erection. Before bracing is installed, the members of a steel building frame may be twisted in many directions. Connecting the diagonal bracing should pull members into their proper positions.

When diagonal cross bracing is used, it is desirable to introduce initial tension into the diagonals. This prestressing will make the building frame tight and reduce its lateral deflection. Furthermore, these light diagonal members can support compressive stress due to their pretensioning. Since the members can resist compression, the horizontal shear to be resisted will be assumed to split equally between the two diagonals. For buildings with several bay widths, equal shear distribution usually is assumed for each bay.

Should the diagonals not be initially tightened, rather stiff sections should be used so that they will be able to resist appreciable compressive forces. The direct axial forces in the girders and columns can be found for each joint from the shear forces assumed in the diagonals. Usually, the girder axial forces so computed are too small to consider, but the values for columns can be quite important. The taller the building becomes, the more critical are the column axial forces caused by lateral forces.

For types of bracing other than cross bracing, similar assumptions can be made for analysis. The student is referred to pages 333–339 of *Theory of Modern Steel Structures*, by L. E. Grinter (New York: Macmillan, 1962), for a discussion of this subject.

---

[5]"Wind Bracing in Steel Buildings," *Transactions ASCE* 105 (1940), pp. 1713–1739.
[6]L. E. Grinter, *Theory of Modern Steel Structures* (New York: Macmillan, 1962), p. 326.

Transfer truss, 150 Federal Street, Boston, MA. (Courtesy of Owen Steel Company, Inc.)

19.18   MOMENT-RESISTING JOINTS

For a large percentage of buildings under eight to ten stories, the beams and girders are connected to each other and to the columns with simple end-framed connections of the types described in Chapter 15. As buildings become taller, it is absolutely necessary to use a definite wind-bracing system or moment-resisting joints. Moment-resisting joints also may be used in lower buildings where it is desired to take advantage of continuity and the consequent smaller beam sizes and depths and shallower floor construction. Moment-resisting brackets also may be necessary in some locations for loads that are applied eccentrically to columns.

Two types of moment-resisting connections that may be used as wind bracing are shown in Fig. 19.21. The design of connections of these types was also presented in Chapter 14. The average design company through the years probably will develop a file of moment-resisting connections from their previous designs. When they have a wind moment of a certain value, they will refer to their file and select one of their former designs that would provide the required moment resistance.

These are the two most commonly used moment-resisting connections today. Most fabricators select the type shown in 19.21(a) as being the most economical. Single-plate (or shear tab) connectors are shown, but framing angles may be used instead. Column stiffeners (shown as dashed lines in the figures) sometimes are required by AISC Specification J10—or local flange bending, local web yielding, sidesway buckling of the web, and so on. Stiffeners are a nuisance and cost an appreciable amount of money. As a result, we will, when the specification shows they are needed, try to avoid them by increasing column sizes. (The reader is again reminded of the troubles experienced with this type of connection during the Northridge earthquake in California as described in Section 15.11 of this text.)

A very satisfactory variation of these last two connections involves an end plate, as shown in Fig. 12-6 in Part 12 of the AISC Manual. This type of connection may, however, be used only for static load situations.

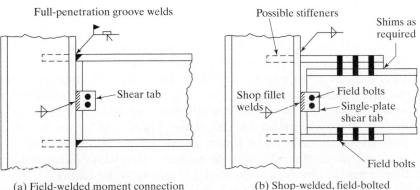

(a) Field-welded moment connection   (b) Shop-welded, field-bolted
                                                moment connection

FIGURE 19.21

Popular moment-resisting connections.

Chase Manhattan Bank Building, New
York City. (Courtesy of Bethlehem Steel
Corporation.)

19.19 DESIGN OF BUILDINGS FOR GRAVITY LOADS

19.19.1 Simple Framing

If *simple framing* is used, the design of the girders is less complex, because the shears
and moments in each girder can be determined by statics. The gravity loads applied to
the columns are relatively easy to estimate, but the column moments may be a little
more difficult. If the girder reactions on each side of the interior column of Fig. 19.22
are equal, then, theoretically, no moment will be produced in the column at that level.
This situation probably is not realistic, however, because it is highly possible for the
live load to be applied on one side of the column and not on the other (or at least be
unequal in magnitude). The results will be column moments. If the reactions are unequal, the moment produced in the column will equal the difference between the reactions times the distances to the center of gravity of the column.

Exterior columns often may have moments due to spandrel beams opposing the
moments caused by the floor loads on the inside of the column. Nevertheless, gravity
loads generally will cause the exterior columns to have larger moments than the interi-
or columns.

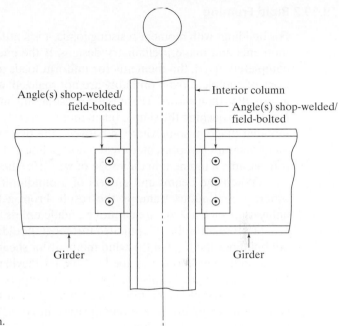

Angle(s) shop-welded/
field-bolted

Interior column

Angle(s) shop-welded/
field-bolted

Girder

Girder

FIGURE 19.22

Simple framing of interior column.

To estimate the moment applied to a column above a certain floor, it is reasonable to assume that the unbalanced moment at that level splits evenly between the columns above and below. In fact, such an assumption is often on the conservative side, as the column below may be larger than the one above.

"Topping out" of Blue Cross–Blue Shield Building in Jacksonville, FL. (Courtesy of Owen Steel Company, Inc.) The Christmas tree is an old north European custom used to ward off evil spirits. It also is used today to show that the steel frame was erected with no lost-time accidents to personnel.

19.19.2 Rigid Framing

For buildings with moment-resisting joints, it is a little more difficult to estimate the girder moments and make preliminary designs. If the ends of each girder are assumed to be completely fixed, the moments for uniform loads are as shown in Fig. 19.23(a). Conditions of complete fixity probably are not realized, with the result that the end moments are smaller than shown. There is a corresponding increase in the positive centerline moments approaching the simple beam moment $(wL^2/8)$ shown in Fig. 19.23(b). Probably, a moment diagram somewhere in between the two extremes is more realistic. Such a moment diagram is represented by the dotted line of Fig. 19.23(a). A reasonable procedure is to assume a moment in the range of $wL^2/10$, where L is the clear span.

When the beams and girders of a building frame are rigidly connected to each other, a continuous frame is the result. From a theoretical standpoint, an accurate analysis of such a structure cannot be made unless the entire frame is handled as a unit.

Before this subject is pursued further, it should be realized that in the upper floors of tall buildings the total of the wind moments or shear above will be small, and the framed and seated connections described in Chapter 15 will provide sufficient moment resistance. For this reason, the beams and girders of the upper floors may very well be designed on the basis of simple beam moments, while those of the lower floors may be designed as continuous members with moment-resistant connections due to the larger total of the wind moments or shear above.

From a strictly theoretical viewpoint, there are several live-load conditions that need to be considered to obtain maximum shears and moments at various points in a continuous structure. For the building frame shown in Fig. 19.24, it is desired to place live loads to cause maximum positive moment in span AB. A qualitative influence line for positive moment at the centerline of this span is shown in part (a). This influence line shows that, to obtain maximum positive moment at the centerline of span AB, the live loads should be placed as shown in Fig. 19.24(b).

To obtain maximum negative moment at point B, or maximum positive moment in span BC, other loading situations need to be considered. With the availability of computers, more detailed analyses are being done every day. The student can see, however, that unless the designer uses a computer, he or she probably will not have sufficient time to go through all of these theoretical situations. Furthermore, it is doubtful if the accuracy of our analysis methods would justify all of the work anyway. Nonetheless, it often is feasible to take out two stories of the building at a time as a free body and analyze that part by one of the "exact" methods such as the successive correction method of moment distribution.

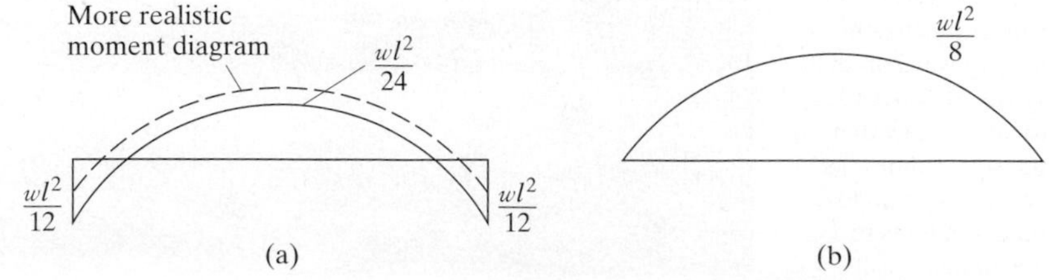

FIGURE 19.23

(a) Fixed-end beam.
(b) Simple beam.

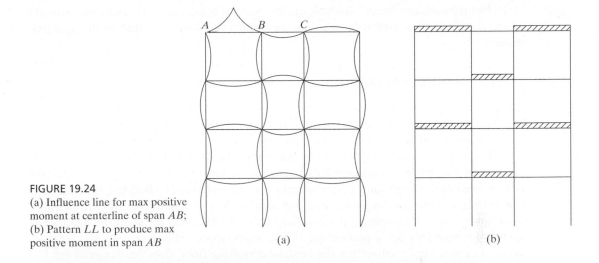

FIGURE 19.24
(a) Influence line for max positive moment at centerline of span AB; (b) Pattern LL to produce max positive moment in span AB

(a)

(b)

Construction of bridge spanning the Piscataqua River linking Kittery, Maine, with Portsmouth New Hampshire. (Courtesy of Getty Images/Hilton Archive.)

Before such an "exact" analysis can be made, it is necessary to make an estimate of the member sizes, probably based on the results of analysis by one of the approximate methods.

19.20   SELECTION OF MEMBERS

The girders can be designed for the load combinations mentioned in Section 19.19. For the top several floors the second condition will not control design, but farther down in the building it will control the sizes for the reasons described in the next paragraph.

Each of the members of a building frame must be designed for all of the dead loads that it supports, but it may be possible to design some members for lesser loads than their full theoretical live-load values. For example, it seems unlikely in a building frame of several stories that the absolutely maximum-design live load will occur on every floor at the same time. The lower columns in a building frame are designed for all of the dead loads above, but probably for a percentage of live loads appreciably less than 100 percent. Some specifications require that the beams supporting floor slabs be designed for full dead and live loads, but permit the main girders to be designed under certain conditions for reduced live loads. (This reduction is based on the improbability that a very large area of a floor would be loaded to its full live-load value at any one time.) The ASCE Standard 7 provides some very commonly used reduction expressions.[7]

Should simple framing be used, the girders will be proportioned for simple beam moments plus the moments caused by the lateral loads. For continuous framing, the girders will be proportioned for $wL^2/10$ (for uniform loads) plus the moments caused by the lateral loads. An interesting comparison of designs by the two methods is shown in pages 717–719 of Beedle et al., *Structural Steel Design* (New York: Ronald, 1964). It may be necessary to draw the moment diagram for the two cases (gravity loads and lateral loads) and add them together to obtain the maximum positive moment out in the span. The value so computed may very well control the girder size in some of the members.

A major part of the design of a multistory building is involved in setting up a column schedule that shows the loads to be supported by the various columns story by story. The gravity forces can be estimated very well, while the shears, axial forces, and moments caused by lateral forces can be roughly approximated for the preliminary design from some approximate method (portal, cantilever, factor, or other).

If both axes of the columns are free to sway, the column effective lengths theoretically must be calculated for both axes, as described in Chapter 5. If diagonal bracing is used in one direction, thus preventing sidesway, the K value will be less than 1.0 in that direction. If these frames are braced also in the narrow direction, it is reasonable to use $K = 1.0$ for both axes.

After the sizes of the girders and columns are tentatively selected for a two-story height, an "exact" analysis can be performed and the members redesigned. The two stories taken out for analysis often are referred to as a *tier*. This process can be continued tier by tier down through the building. Many tall buildings have been designed and are performing satisfactorily in which this last step (the two-story "exact" analysis) was omitted.

---

[7]*American Society of Civil Engineers Minimum Design Loads for Buildings and Other Structures.* ASCE 7–10 (New York: ASCE), Section 4.8.

APPENDIX A

Derivation of the Euler Formula

The Euler formula is derived in this section for a straight, concentrically loaded, homogeneous, long, slender, elastic, and weightless column with rounded ends. It is assumed that this perfect column has been laterally deflected by some means, as shown in Fig. A.1 and that, if the concentric load P were removed, the column would straighten out completely.

The x and y axes are located as shown in the figure. As the bending moment at any point in the column is $-Py$, the equation of the elastic curve can be written as

$$EI\frac{d^2y}{dx^2} = -Py$$

For convenience in integration, both sides of the equation are multiplied by $2dy$:

$$EI2\frac{dy}{dx}d\frac{dy}{dx} = -2Pydy$$

$$EI\left(\frac{dy}{dx}\right)^2 = -Py^2 + C_1$$

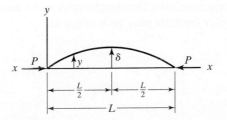

FIGURE A.1

When $y = \delta$, $dy/dx = 0$, and the value of C_1 will equal $P\delta^2$ and

$$EI\left(\frac{dy}{dx}\right)^2 = -Py^2 + P\delta^2$$

The preceding expression is arranged more conveniently as follows:

$$\left(\frac{dy}{dx}\right)^2 = \frac{P}{EI}(\delta^2 - y^2)$$

$$\frac{dy}{dx} = \sqrt{\frac{P}{EI}}\sqrt{\delta^2 - y^2}$$

$$\frac{dy}{\sqrt{\delta^2 - y^2}} = \sqrt{\frac{P}{EI}}dx$$

Integrating this expression, the result is

$$\text{arc sin}\frac{y}{\delta} = \sqrt{\frac{P}{EI}}x + C_2$$

When $x = 0$ and $y = 0$, $C_2 = 0$. The column is bent into the shape of a sine curve expressed by the equation

$$\text{arc sin}\frac{y}{\delta} = \sqrt{\frac{P}{EI}}x$$

When $x = L/2$, $y = \delta$, resulting in

$$\frac{\pi}{2} = \frac{L}{2}\sqrt{\frac{P}{EI}}$$

In this expression, P is the *critical buckling load*, or the maximum load that the column can support before it becomes unstable. Solving for P, we have

$$P = \frac{\pi^2 EI}{L^2}$$

This expression is the Euler formula, but usually it is written in a little different form involving the slenderness ratio. Since $r = \sqrt{I/A}$ and $r^2 = I/A$ and $I = r^2 A$, the Euler formula may be written as

$$\frac{P}{A} = \frac{\pi^2 E}{(L/r)^2} = F_e$$

APPENDIX B

Slender Compression Elements

Section B4 of the AISC Specification is concerned with the local buckling of compression elements. In that section, compression elements are classified as non-slender element or slender element sections. As non-slender elements have been discussed previously, this appendix is concerned only with slender elements. A brief summary of the AISC method for determining design stresses for such members is presented in the next few paragraphs.

When b/t ratios exceed the values given in AISC Table B4.1a, those elements will be classified as being slender, and their critical, or F_{cr}, stresses will have to be reduced. The design strength of an axially loaded compression member with slender elements will be reduced by multiplying it by a reduction factor Q.

The value of Q is equal to the product of two reduction factors Q_s and Q_a. Their values are dependent on whether the member consists of stiffened and/or unstiffened elements. Q_a is a reduction factor for slender stiffened compression elements, while Q_s is a reduction factor for slender unstiffened compression elements. Two cases are considered in AISC Specification E7:

1. For members consisting of unstiffened elements only, Q_s is to be determined with the appropriate formulas presented in AISC Section E7.1, and $Q_a = 1.0$.
2. For members consisting of stiffened elements only, $Q_s = 1.0$, and Q_a is to be determined with the formulas presented in AISC Section E7.2.

Equations for computing Q are given in AISC Section E7 for the following types of members: (a) single angles; (b) flanges, angles, and plates projecting from rolled columns or other compression members; (c) flanges, angles, and plates projecting from built-up I-shaped columns or other compression members; and (d) stems of tees. In these equations, the following terms are used:

$$b = \text{width of unstiffened compression element,}$$
$$\text{as defined in section B4.1, in}$$

$$t = \text{thickness of unstiffened compression element, in}$$
$$F_y = \text{specified minimum yield stress, ksi}$$
$$d = \text{full nominal depth of tee, in}$$

Only the Q_s expressions for case (d), stems of tees, are presented here,

When　　$\dfrac{d}{t} \leq 0.75\sqrt{\dfrac{E}{F_y}}$

$$Q_s = 1.0 \qquad\qquad \text{(AISC Equation E7-13)}$$

When　　$0.75\sqrt{\dfrac{E}{F_y}} < \dfrac{d}{t} \leq 1.03\sqrt{\dfrac{E}{F_y}}$

$$Q_s = 1.908 - 1.22\left(\dfrac{d}{t}\right)\sqrt{\dfrac{E}{F_y}} \qquad\qquad \text{(AISC Equation E7-14)}$$

When　　$\dfrac{d}{t} > 1.03\sqrt{\dfrac{E}{F_y}}$

$$Q_s = \dfrac{0.69E}{F_y\left(\dfrac{d}{t}\right)^2} \qquad\qquad \text{(AISC Equation E7-15)}$$

In Example B-1, the value of Q_s is computed for a WT section that is used as a compression member. As the member consists only of unstiffened elements, $Q_a = 1.0$. The computed value for Q_s can be checked in the WT shape tables of Part 1 (Table 1-8) of the Manual. In Appendix C, this same WT shape member is further considered as to lateral-torsional buckling, and the Q_s determined here is used there.

Example B-1

A WT10.5 × 31, shown in Fig. B.1, is used as a compression member. Compute Q_s for this member, which is assumed to have an $F_y = 50$ ksi.

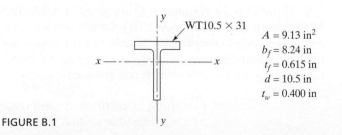

$A = 9.13$ in^2
$b_f = 8.24$ in
$t_f = 0.615$ in
$d = 10.5$ in
$t_w = 0.400$ in

FIGURE B.1

Solution

For tee flange, use AISC Section E7.1a, with

$$\frac{b}{t} = \frac{b_f/2}{t_f} = \frac{8.24/2}{0.615} = 6.70 \le 0.56\sqrt{\frac{E}{F_y}} = 0.56\sqrt{\frac{29,000}{50}} = 13.49$$

$$\therefore Q_s = 1.0 \qquad \text{(AISC Equation E7-4)}$$

For tee stem, use AISC Section E7.1 (d), with

$$\frac{d}{t} = \frac{10.5}{0.400} = 26.25 > 1.03\sqrt{\frac{E}{F_y}} = 1.03\sqrt{\frac{29,000}{50}} = 24.81$$

$$\therefore Q_s = \frac{0.69E}{F_y\left(\dfrac{d}{t}\right)^2} = \frac{0.69(29,000)}{50(26.25)^2} \qquad \text{(AISC Equation E7-15)}$$

$$Q_s = 0.581 \leftarrow$$

Checks with value in WT shapes tables of AISC (1-8) where $Q_s = 0.581$, for $F_y = 50$ ksi.

If we have slender elements in a compression member, its design compression strength is to be computed as follows:

$$\text{For} \qquad \frac{KL}{r} \le 4.71\sqrt{\frac{E}{QF_y}} \qquad \left(\text{or } \frac{QF_y}{F_e} \le 2.25\right)$$

$$F_{cr} = Q\left[0.658^{\frac{QF_y}{F_e}}\right]F_y \qquad \text{(AISC Equation E7-2)}$$

$$\text{For} \qquad \frac{KL}{r} > 4.71\sqrt{\frac{E}{QF_y}} \qquad \left(\text{or } \frac{QF_y}{F_e} > 2.25\right)$$

$$F_{cr} = 0.877F_e \qquad \text{(AISC Equation E7-3)}$$

Where F_e is the elastic buckling stress, calculated using Equations E3-4 and E4-5 for singly symmetric members and $Q = Q_sQ_a$.

APPENDIX C

Flexural–Torsional Buckling of Compression Members

Usually, symmetrical members such as W sections are used as columns. Torsion will not occur in such sections if the lines of action of the lateral loads pass through their shear centers. The *shear center* is that point in the cross section of a member through which the resultant of the transverse loads must pass so that no torsion will occur. The calculations necessary to locate shear centers were presented in Chapter 10. The shear centers of the commonly used doubly symmetrical sections occur at their centroids. This is not necessarily the case for other sections, such as channels and angles. Shear center locations for several types of sections are shown in Fig. C.1. Also shown in the figure are the coordinates x_0 and y_0 for the shear center of each section with respect

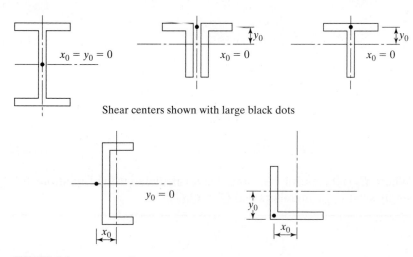

Shear centers shown with large black dots

FIGURE C.1

Shear center locations for some common column sections.

to its centroid. These values are needed to solve the flexural-torsional formulas, presented later in this section.

Even though loads pass through shear centers, torsional buckling still may occur. If we load any section through its shear center, no torsion will occur, but we still compute torsional buckling strength for these members—that is, buckling load does not depend on the nature of the axial or transverse loading; rather, it depends on the cross-section properties, column length, and support conditions.

The average designer does not consider the torsional buckling of symmetrical shapes or the flexural-torsional buckling of unsymmetrical shapes. He or she usually thinks that these conditions don't control the critical column loads, or at least that they don't affect them much. If, however, we have unsymmetrical columns or even symmetrical columns made up of thin plates, we will find that torsional buckling or flexural-torsional buckling may significantly reduce column capacities.

In Section E of the AISC Specification, a long list of formulas is presented for computing the flexural-torsional strength of column sections. The values given for column design strengths ($\phi_c P_n$ and P_n/Ω_c values) for double angles, single angles, and tees in Part 4 of the AISC Manual make use of these formulas.

For flexural-torsion, $P_u \leq \phi_c P_n = \phi_c A_g F_{cr}$, with $\phi_c = 0.90$ and F_{cr} to be determined from the formulas to follow from the specification. A list of definitions that are needed for using these formulas also is provided. When the column is defined as a slender-element compression member, the critical stress, F_{cr}, is found from equations in Section E7.

When $\dfrac{KL}{r} < 4.71 \sqrt{\dfrac{E}{QF_y}}$ $\left(\text{or } \dfrac{QF_y}{F_e} \leq 2.25 \right),$

$$F_{cr} = Q\left[0.658^{\frac{QF_y}{F_e}} \right] F_y \qquad \text{(AISC Equation E7-2)}$$

and when $\dfrac{KL}{r} > 4.71 \sqrt{\dfrac{E}{QF_y}}$ $\left(\text{or } \dfrac{QF_y}{F_e} > 2.25 \right),$

$$F_{cr} = 0.877 F_e, \qquad \text{(AISC Equation E7-3)}$$

When the member meets the width-thickness ratio, λ_r, of AISC Section B4.1, it is classified as a non-slender section. In this case, $Q = 1.0$ and F_{cr} is determined from Equations E3-2 and E3-3 in Section E3.

For either case, F_e, the critical flexural-torsional elastic bucking stress is calculated based on the formulas in Section E4.

For doubly symmetric shapes,

$$F_e = \left[\frac{\pi^2 E C_w}{(K_z L)^2} + GJ \right] \frac{1}{I_x + I_y}. \qquad \text{(AISC Equation E4-4)}$$

For singly symmetric shapes where y is the axis of symmetry,

$$F_e = \frac{F_{ey} + F_{ez}}{2H} \left[1 - \sqrt{1 - \frac{4F_{ey}F_{ez}H}{(F_{ey} + F_{ez})^2}} \right].$$ (AISC Equation E4-5)

For unsymmetrical sections, F_e is the lowest root of the following cubic equation:

$$(F_e - F_{ex})(F_e - F_{ey})(F_e - F_{ez}) - F_e^2(F_e - F_{ey})\left(\frac{x_0}{\overline{r}_0}\right)^2$$

$$- F_e^2(F_e - F_{ex})\left(\frac{y_0}{\overline{r}_0}\right)^2 = 0$$ (AISC Equation E4-6)

The following is a more convenient form of AISC Equation E4-6:

$$HF_e^3 + \left[\frac{1}{\overline{r}_0^2}(y_0 F_{ex} + x_0^2 F_{ey}) - (F_{ex} + F_{ey} + F_{ez}) \right] F_e^2$$

$$+ (F_{ex}F_{ey} + F_{ex}F_{ez} + F_{ey}F_{ez})F_e - F_{ex}F_{ey}F_{ez} = 0$$

Here,

r_o = polar radius of gyration about the shear center (in)

K_z = effective length factor for torsional buckling

G = shear modulus of elasticity of steel = 11,200 ksi

C_w = warping constant (in^6)

J = torsional constant (in^4)

$$\overline{r}_0^2 = x_0^2 + y_0^2 + \frac{I_x + I_y}{A_g}$$ (AISC Equation E4-11)

$$H = 1 - \left(\frac{x_0^2 + y_0^2}{\overline{r}_0^2}\right)$$ (AISC Equation E4-10)

$$F_{ex} = \frac{\pi^2 E}{(KL/r)_x^2}$$ (AISC Equation E4-7)

$$F_{ey} = \frac{\pi^2 E}{(KL/r)_y^2}$$ (AISC Equation E4-8)

$$F_{ez} = \left[\frac{\pi^2 E C_w}{(K_z L)^2} + GJ \right] \frac{1}{A_g \overline{r}_0^2}$$ (AISC Equation E4-9)

The values of C_w, J, $\overline{r}_0$, and H are provided for many sections in the "Flexural-Torsional Properties" tables of Part 1 of the Manual.

In Example C-1, the authors have gone through all of these formulas for a WT shape column. The resulting values for $\phi_c P_n$ is shown to agree with the values given in the Manual.

Example C-1

Determine (a) the flexural buckling strength and (b) the flexural-torsional buckling strength of an 18-ft pinned-end column consisting of A992 Grade 50 steel. The cross-section and other properties of the member are shown in Fig. C.2; $G = 11,500$ ksi, and $K = 1.0$.

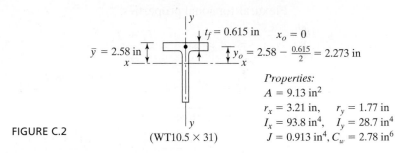

FIGURE C.2

(WT10.5 × 31)

$t_f = 0.615$ in $\quad x_o = 0$

$\bar{y} = 2.58$ in

$\bar{y}_o = 2.58 - \frac{0.615}{2} = 2.273$ in

Properties:
$A = 9.13$ in^2
$r_x = 3.21$ in, $\quad r_y = 1.77$ in
$I_x = 93.8$ in^4, $\quad I_y = 28.7$ in^4
$J = 0.913$ in^4, $C_w = 2.78$ in^6

Solution. Using WT1.50 × 31

(a) Flexural buckling about the x-axis (perpendicular to the axis of symmetry)

$$\left(\frac{KL}{r}\right)_x = \frac{(1.0)(12 \times 18)}{3.21} = 67.29$$

From Example B-1, $Q_s = 0.581$

$$Q_a = 1.0$$

$$\therefore Q = Q_s Q_a = 0.581 \,(1.0) = 0.581$$

when $\dfrac{KL}{r} = 67.29 \le 4.71\sqrt{\dfrac{E}{QF_y}} = 4.71\sqrt{\dfrac{29,000}{0.581(50)}} = 148.81$

Then $$F_{cr} = Q\left[0.658^{\frac{QF_y}{F_e}}\right]F_y \qquad \text{(AISC Equation E7-2)}$$

where $F_e = \dfrac{\pi^2 E}{\left(\dfrac{KL}{r}\right)^2} = \dfrac{\pi^2(29,000)}{(67.29)^2} = 63.21$ ksi

Therefore, $F_{cr} = 0.581 \left[0.658^{\frac{0.581(50)}{63.21}} \right] 50 = 23.97 \text{ ksi}$

$$\phi_c P_n = \phi_c F_{cr} A_g = 0.9 \,(23.97)\,(9.13) = 197 \text{ kips}$$

Checks with value in Table 4-7 (Available strength in Axial Compression – WT shapes) where for x-x axis, $\phi_c P_n = 197$ kips.

(b) Flexural-torsional buckling with respect to the y-axis passing through the shear center of the section.

Flexural-torsional properties:

$$\bar{r}_0^2 = x_0^2 + y_0^2 + \frac{I_x + I_y}{A_g} \qquad \text{(AISC Equation E4-11)}$$

$$\bar{r}_0^2 = (0)^2 + (2.273)^2 + \frac{93.8 + 28.7}{9.13} = 18.584 \text{ in}^2$$

$$\bar{r}_0 = \sqrt{18.58} = 4.31 \text{ in}$$

$$H = 1 - \frac{x_0^2 + y_0^2}{\bar{r}_0^2} = 1 - \frac{(0)^2 + (2.273)^2}{18.58} = 0.722 \text{ in}$$

For tee shaped compression member the critical stress, F_{cr}, is determined from AISC Equation E4-2.

$$F_{cr} = \left[\frac{F_{cry} + F_{crz}}{2H} \right] \left[1 - \sqrt{1 - \frac{4 F_{cry} F_{crz} H}{(F_{cry} + F_{crz})^2}} \right]$$

where F_{cry} is taken as F_{cr} from Equation E3-2 or E3-3 and $\dfrac{KL}{r} = \left(\dfrac{KL}{r}\right)_y$

Since $\left(\dfrac{KL}{r}\right)_y = \dfrac{1.0(12)(18)}{1.77} = 122.03 < 4.71\sqrt{\dfrac{E}{QF_y}} = 4.71\sqrt{\dfrac{29,000}{0.581(50)}} = 148.81$

where $Q = Q_s Q_a = 0.581\,(1.0) = 0.581$

$$F_e = \frac{\pi^2 E}{\left(\dfrac{KL}{r}\right)_y^2} = \frac{\pi^2 (29,000)}{(122.03)^2} = 19.22 \text{ ksi} \qquad \text{(AISC Equation E3-4)}$$

Therefore, $F_{cry} = Q\left[0.658^{\frac{QF_y}{F_e}}\right]F_y$ (AISC Equation E3.2)

$$F_{cry} = 0.581\left[0.658^{\frac{0.581(50)}{19.22}}\right]50 = 15.43 \text{ ksi}$$

and where $F_{crz} = \dfrac{GJ}{A_g \bar{r}_0^2}$ (AISC Equation E4-3)

$$F_{crz} = \frac{11,200(0.913)}{9.13(18.584)} = 60.27 \text{ ksi}$$

Therefore,

$$F_{cr} = \left[\frac{15.43 + 60.27}{2(0.722)}\right]\left[1 - \sqrt{1 - \frac{4(15.43)(60.27)(0.722)}{(15.43 + 60.27)^2}}\right]$$

$F_{cr} = 14.21 \text{ ksi.}$

$\phi_c P_n = \phi_c F_{cr} A_g = 0.9(14.21)(9.13)$

$\phi P_n = 117 \text{ kips} \leftarrow$

Checks with value in Table 4-7 (Available strength in Axial Compression – WT shapes) where for y-y axis, $\phi_c P_n = 118$ kips.

APPENDIX D

Moment-Resisting Column Base Plates

Column bases frequently are designed to resist bending moments as well as axial loads. An axial load causes compression between a base plate and the supporting footing, while a moment increases the compression on one side and decreases it on the other side. For small moments, the forces may be transferred to the footing through flexure of the base plate. When they are very large, stiffened or booted connections may be used. For a small moment, the entire contact area between the plate and the supporting footing will remain in compression. This will be the case if the resultant load falls within the middle third of the plate length in the direction of bending.

Figures D.1(a) and (b) show base plates suitable for resisting relatively small moments. For these cases, the moments are sufficiently small to permit their transfer to the footings by bending of the base plates. The anchor bolts may or may not have calculable stresses, but they are nevertheless considered necessary for good construction practice. They definitely are needed to hold the columns firmly in place and upright during the initial steel erection process. Temporary guy cables are also necessary during erection. The anchor bolts should be substantial and capable of resisting unforeseen erection forces. Sometimes these small plates are attached to the columns in the shop, and other times they are shipped loose to the job and carefully set to the correct elevations in the field.

Should the eccentricity ($e = M/P$) be sufficiently large that the resultant falls outside the middle third of the plate, there will be an uplift on the other side of the column, putting the anchor bolts on that side in tension.

The moment will be transferred from the column into the footing by means of the anchor bolts, embedded a sufficient distance into the footing to develop the anchor bolt forces. The embedment should be calculated as required by reinforced-concrete design methods.[1] The booted connection shown in Fig. D.1(c) is assumed to be welded

---

[1]*Building Code Requirements for Reinforced Concrete (ACI 318-05) and Commentary (ACI 318R-05)* (Detroit: American Concrete Institute, 2005), pp. 196–200.

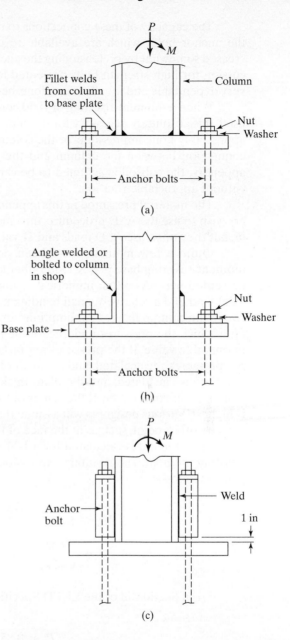

Fillet welds
from column
to base plate

Column

Nut
Washer

Anchor bolts

(a)

Angle welded or
bolted to column
in shop

Nut
Washer

Base plate →

Anchor bolts

(b)

Anchor
bolt

Weld

1 in

FIGURE D.1

Moment-resisting column base plates.

(c)

to the column. The boots generally are made of angles or channels and are not, as a rule, connected directly to the base plate. Rather, the tensile force component induced by the moment is transmitted from the column to the foundation by means of the anchor bolts. When booted connections are used, the base plates normally are shipped loose to the job and carefully set to the correct elevations in the field.

The capacity of these connections to resist rotation is dependent on the lengths of the anchor bolts, which are available to deform elastically. This capacity can be increased somewhat by pretensioning the anchor bolts. (This is similar to the prestress discussion for high-strength bolts presented in Section 13.4.) Actually, prestressing is not very dependable and usually is not done, because of the long-term creep in the concrete.

When a moment-resisting or rigid connection between a column and its footing is used, it is absolutely necessary for the supporting soil or rock beneath the footing to be appreciably noncompressible, or the column base will rotate. If this happens, the rigid connection between the column and the footing is useless. For the purpose of this appendix, the subsoil is assumed to be capable of resisting the moment applied to it without appreciable rotation.

The material presented in this appendix applies to LRFD designs. Should designers wish to use the ASD procedure, they may follow exactly the same steps given herein, but they must use ASD loads and Ω values.

Quite a few methods have been developed through the years for designing moment-resisting base plates. One rather simple procedure used by many designers is presented here. As a first numerical example, a column base plate is designed for an axial load and a relatively small bending moment such that the resultant load falls between the column flanges. Assumptions are made for the width and length of the plate, after which the pressures underneath the plate are calculated and compared with the permissible value. If the pressures are unsatisfactory, the dimensions are changed and the pressures recalculated, and so on, until the values are satisfactory. The moment in the plate is calculated, and the plate thickness is determined. The critical section for bending is assumed to be at the center of the flange on the side where the compression is highest. Various designers will assume that the point of maximum moment is located at some other point, such as at the face of the flange or the center of the anchor bolt.

The moment is calculated for a 1-in-wide strip of the plate and is equated to its resisting moment. The resulting expression is solved for the required thickness of the plate as follows:

$$M_u \leq \phi_b M_n = \frac{\phi_b F_y I}{c} = \frac{\phi_b F_y \left(\frac{1}{12}\right)(1)(t)^3}{t/2}$$

$$t \geq \sqrt{\frac{6M_u}{\phi_b F_y}} \quad \text{with} \quad \phi_b = 0.9$$

From Section J8 of the LRFD Specification,

$$P_p = 0.85 f'_c A_1 \sqrt{\frac{A_2}{A_1}} \qquad \text{(AISC Equation J8-2)}$$

If we assume $\sqrt{\dfrac{A_2}{A_1}} \geq 2$, then

$$P_p = 1.7 f'_c A_1$$

$$\phi_c P_p = \phi_c 1.7 f'_c A_1 \quad \text{with} \quad \phi_c = 0.65 \quad (\Omega_c = 2.31)$$

Example D-1

Design a moment-resisting base plate to support a W14 × 120 column with an axial load P_u of 620 k and a bending moment M_u of 225 ft-k. Use A36 steel with $F_y = 36$ ksi and a concrete footing with $f'_c = 3.0$ ksi. $\phi_c F_p = (0.65)(1.7)(3.0) = 3.32$ ksi.

Solution. Using a W14 × 120($d = 14.5$ in, $t_w = 0.590$ in, $b_f = 14.70$ in, $t_f = 0.940$ in),

$$e = \frac{(12)(225)}{620} = 4.35 \text{ in}$$

∴ The resultant falls between the column flanges and within the middle third of the plate.

Try a 20 × 28 in plate (after a few trials)

$$f = -\frac{P_u}{A} \pm \frac{P_u ec}{I} = -\frac{620}{(20)(28)} \pm \frac{(620)(4.35)(14)}{\left(\frac{1}{12}\right)(20)(28)^3}$$

$$= -1.107 \pm 1.032 \begin{cases} -2.139 < \phi_c P_n = 3.32 \text{ ksi} \\ -0.075 \text{ ksi (still compression)} \end{cases} \qquad \text{(OK)}$$

Taking moments to right at center of right flange (see Fig. D.2):

$$M_u = (1.606)(7.22)\left(\frac{7.22}{2}\right) + (2.139 - 1.606)(7.22)\left(\frac{2}{3} \times 7.22\right) = 51.12 \text{ in-k}$$

$$t \geq \sqrt{\frac{6M_u}{\phi_b F_y}} = \sqrt{\frac{(6)(51.12)}{(0.9)(36)}} = 3.08 \text{ in}$$

Checking bending in transverse direction

$$n = \frac{B - 0.8b_f}{2} = \frac{20 - (0.80)(14.7)}{2} = 4.12 \text{ in}$$

$$\text{Average } f_p = \frac{0.075 + 2.139}{2} = 1.107 \text{ ksi}$$

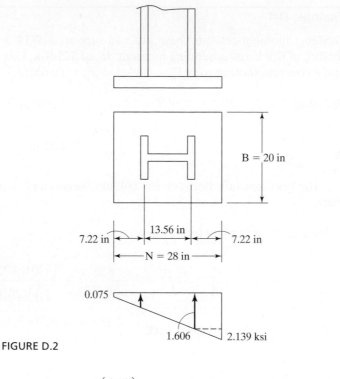

FIGURE D.2

$$M_u = (1.107)(4.12)\left(\frac{4.12}{2}\right) = 9.40 \text{ in-k} < 51.19 \text{ in-k} \qquad \text{(OK)}$$

Use PL$3\frac{1}{4}$ × 20 × 2 ft 4 in A36

The moment considered in Example D-2 is of such a magnitude that the resultant load falls outside the column flange. As a result, there will be uplift on one side, and the anchor bolt will have to furnish the needed tensile force to provide equilibrium.

In this design, the anchor bolts are assumed to have no significant tension due to tightening. As a result, they are assumed not to affect the force system. As the moment is applied to the column, the pressure shifts toward the flange on the compression side. It is assumed that the resultant of this compression is located at the center of the flange.

Example D-2

Repeat Example D-1 with the same column and design stresses, but with the moment increased from 225 ft-k to 460 ft-k. Refer to Fig. D.3.

Solution. Using a W14 × 120(d = 14.5 in, t_w = 0.590 in, b_f = 14.7 in, t_f = 0.940 in)

$$e = \frac{(12)(460)}{620} = 8.90 \text{ in} > \frac{d}{2} = 7.25 \text{ in}$$

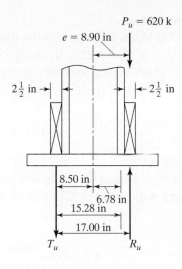

FIGURE D.3

∴. The resultant falls outside the column flange. Taking moments about the center of the right flange, we have

$$(620)(8.90 - 6.78) - 15.28\, T_u = 0$$

$$T_u = 86.02 \text{ k}$$

$$\text{Anchor bolt } A_{\text{reqd.}} = \frac{T_u}{\phi_t 0.75 F_u} \qquad \text{(AISC Table J3.2)}$$

$$= \frac{86.02}{\phi 0.75 F_u} = \frac{86.02}{(0.75)(0.75)(58)} = 2.64 \text{ in}^2$$

Use two $1\frac{3}{8}$-in-diameter bolts each side. ($A_s = 2.97 \text{ in}^2 > 2.64 \text{ in}^2$)

Approximate plate size, assuming a triangular pressure distribution

$$R_u = P_u + T_u = 620 + 86.02 = 706.02 \text{ k}$$

A of plate reqd. $\geq \dfrac{R_u}{\text{Avg}\phi_c F_p} = \dfrac{R_u}{\phi_c F_p/2}$ with $\phi_c F_p$ given $= 3.32$ ksi in statement of Example D-1.

$$A \text{ reqd.} \geq \frac{706.02}{3.32/2} = 425.31 \text{ in}^2$$

Try a 24-in-long plate.

The load is located $\frac{24}{2}$ in minus the distance from the column c.g. to the column flange c.g. $= \frac{24}{2} - 6.78 = 5.22$ in from edge of the plate. Thus, the pressure triangle will be $3 \times 5.22 = 15.66$ in long, and the required plate length B will equal

$$B = \frac{706.02}{\frac{1}{2} \times 3.32 \times 15.66} = 27.16 \text{ in}$$

Try a 28-in-long plate.

The load R_u is located $\frac{28}{2} - 6.78 = 7.22$ in from the edge of the plate. The triangular pressure length will be $(3)(7.22) = 21.66$ in long, and the required plate width will be

$$B = \frac{706.02}{\left(\frac{1}{2}\right)(3.32)(21.66)} = 19.64 \text{ in}$$

If the plate is made 20 in wide, the pressure zone will have an area $= 20 \times 21.66 = 433.2 \text{ in}^2$, and the maximum pressure will be twice the average pressure, or

$$\frac{706.02}{433.2} \times 2 = 3.26 \text{ ksi} < 3.32 \text{ ksi} \qquad \text{(OK)}$$

Taking moments to right at center of right column flange (see Fig. D.4)

$$M_u = 2.16(7.22)\left(\frac{7.22}{2}\right) + \frac{1}{2}(1.10)(7.22)\left(\frac{2}{3}\right)(7.22) = 75.41 \text{ in-k}$$

$$t = \sqrt{\frac{(6)(75.41)}{(0.9)(36)}} = 3.74 \text{ in}$$

Use PL$3\frac{3}{4} \times 20 \times 2$ ft 4 in A36

Design of weld from column to base PL (see Fig. D.5)

Total length of fillet weld each flange

$$= (2)(14.7) - 0.590 = 28.81 \text{ in}$$

$$C = T = \frac{M_u}{d - t_f} = \frac{(12)(460)}{13.56} = 407.08 \text{ k}$$

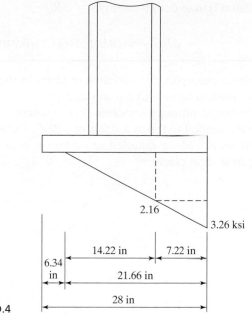

FIGURE D.4

Strength of 1-in-long 1-in fillet weld, using E70 electrodes

$$\phi R_{nw} = \phi(0.60F_{EXX})(0.707)(a) = (0.75)(0.60 \times 70)(0.707)(1.0)$$
$$= 22.3 \text{ k/in}$$

Weld size required $= \dfrac{407.08}{(28.41)(22.3)} = 0.643$ in

Use $\frac{11}{16}$-in fillet welds, E70 electrode, SMAW

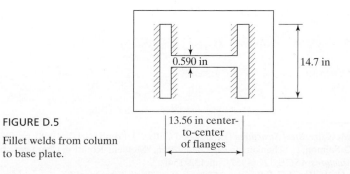

FIGURE D.5

Fillet welds from column
to base plate.

Note: Tension flange design strength $= \phi R_n$

$$= \phi F_y A_f = (0.9)(36)(14.7)(0.940) = 448 \text{ k} > T_u \qquad \text{(OK)}$$

In these examples, the variation of stress in the concrete supporting the columns has been assumed to vary in a triangular or straight-line fashion. It is possible to work with an assumed ultimate concrete theory where the concrete in compression under the plate is assumed to fail at a stress of $0.85 f_c'$. Examples of such designs are available in several texts.[2] More detailed information concerning moment-resisting bases is available in several places.[3,4]

[2]W. McGuire, *Steel Structures* (Englewood Cliffs, NJ: Prentice-Hall, 1968), pp. 987–1004.

[3]C. G. Salmon, L. Schenker, and B. G. Johnston, "Moment-Rotation Characteristics of Column Anchorages," *Transactions ASCE*, 122 (1957), pp. 132–154.

[4]J. T. DeWolf and E. F. Sarisley, "Column Base Plates with Axial Loads and Moments," *Journal of Structural Division*, ASCE, 106, ST11 (November 1980), pp. 2167–2184.

APPENDIX E

Ponding

It has been claimed that almost 50 percent of the lawsuits faced by building designers are concerned with roofing systems.[1] Ponding, a problem with many flat roofs, is one of the most common subjects of such litigation. If water accumulates more rapidly on a roof than it runs off, ponding results because the increased load causes the roof to deflect into a dish shape that can hold more water, which causes greater deflections, and so on. This process continues until equilibrium is reached, or until collapse occurs. Ponding can be caused by increasing deflections, clogged roof drains, settlement of footings, warped roof slabs, and so on.

The best way to prevent ponding is to use appreciable roof slopes (1/4 in per ft or more), together with good drainage facilities. Supposedly, more than two-thirds of the flat roofs in the United States have slopes less than 1/4 in per ft. The construction of roofs with slopes this large will increase roof building costs by only a few percent compared with perfectly flat roofs. The supporting girders for flat roofs with long spans should definitely be cambered to reduce the possibility of ponding (as well as the sagging that is so disturbing to the people occupying a building).

Appendix 2 of the AISC Specification states that, unless roof surfaces have sufficient slopes to areas of free drainage or sufficient individual drains to prevent water accumulation, the strength and stability of the roof systems during ponding conditions must be investigated. The very detailed work of Marino[2] forms the basis of the ponding provisions of the AISC Specification. Many other useful references also are available.[3-5]

---

[1]Gary Van Ryzin, "Roof Design: Avoid Ponding by Sloping to Drain," *Civil Engineering* (New York: ASCE, January 1980), pp. 77–81.

[2]F. J. Marino, "Ponding of Two-Way Roof System," *Engineering Journal*, AISC, vol. 3, no. 3 (3rd Quarter, 1966), pp. 93–100.

[3]L. B. Burgett, "Fast Check for Ponding," *Engineering Journal*, AISC, vol. 10, no. 1 (1st Quarter, 1973), pp. 26–28.

[4]J. Chinn, "Failure of Simply-Supported Flat Roofs by Ponding of Rain," *Engineering Journal*, AISC, no. 2 (2nd Quarter, 1965), pp. 38–41.

[5]J. L. Ruddy, "Ponding of Concrete Deck Floors," *Engineering Journal*, AISC, vol. 23, no. 2 (3rd Quarter, 1986), pp. 107–115.

The amount of water that can be retained on a roof depends on the flexibility of the framing. The specifications state that a roof system can be considered stable and not in need of further investigation if we satisfy the expressions

$$C_p + 0.9C_s \leq 0.25 \qquad \text{(AISC Equation A-2-1)}$$

$$I_d \geq 25(S^4)10^{-6} \qquad \text{(AISC Equation A-2-2)}$$

where

$$C_p = \frac{32L_s L_p^4}{10^7 I_p} \qquad \text{(AISC Equation A-2-3)}$$

$$C_s = \frac{32SL_s^4}{10^7 I_s} \qquad \text{(AISC Equation A-2-4)}$$

L_p = column spacing in direction of girder (primary member length), ft

L_s = column spacing perpendicular to girder direction (secondary member length), ft

S = spacing of secondary members, ft

I_p = moment of inertia of primary members, in^4

I_s = moment of inertia of secondary members, in^4

I_d = moment of inertia of the steel deck (if one is used) supported on the secondary members, in^4 per ft

Should steel roof decks be used, their I_d must at least equal the value given by Equation A-2-2. If the roof decking is the secondary system (i.e., no secondary beams, joists, etc.), it should be handled with Equation A-2-1.

Some other AISC requirements in applying these expressions follow:

1. The moment of inertia I_s must be decreased by 15 percent for trusses and steel joists.
2. Steel decking is considered to be a secondary member supported directly by the primary members.
3. Stresses caused by wind or seismic forces do not have to be considered in the ponding calculations.

Should moments of inertia be needed for open-web joists, they can be computed from the member cross sections or, perhaps more easily, backfigured from the resisting moments and allowable stresses given in the joist tables. (Since $M_R = FI/c$, we can compute $I = M_R c/F$.)

In effect, these equations reflect *stress indexes*, or percentage stress increases. For instance, here we are considering the percentage increase in stress in the steel members caused by ponding. If the stress in a member increases from $0.60F_y$ to $0.80F_y$, we say that the stress index is given by

$$U = \frac{0.80F_y - 0.60F_y}{0.60F_y} = 0.33$$

The terms C_p and C_s are, respectively, the approximate stiffnesses of the primary and secondary support systems. AISC Equation A-2-1 ($C_p + 0.9C_s \leq 0.25$), which gives us an approximate stress index during ponding, is limited to a maximum value of 0.25. Should we substitute into this equation and obtain a value no greater than 0.25, ponding supposedly will not be a problem. Should the index be larger than 0.25, however, it will be necessary to conduct a further investigation. One method of doing this is presented in Appendix 2 of the AISC Specification and will be described later in this section.

Example E-1 presents the application of AISC Equation A-2-1 to a roofing system.

Example E-1

Check the roof system shown in Fig. E.1 for ponding, using the AISC Specification and A36 steel.

Solution

$$C_p = \frac{32L_sL_p^4}{10^7I_p} = \frac{(32)(48)(36)^4}{(10^7)(1830)} = 0.141$$

$$C_s = \frac{32SL_s^4}{10^7I_s} = \frac{(32)(6)(48)^4}{(10^7)(518)} = 0.197$$

$$C_p + 0.9C_s = 0.141 + (0.9)(0.197) = 0.318 > 0.25$$

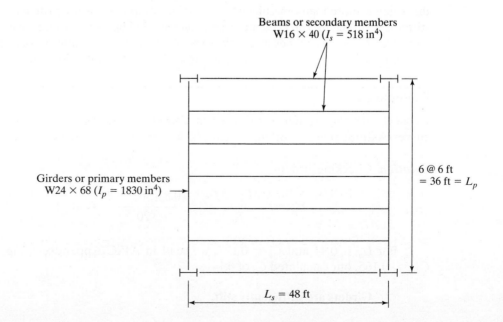

Beams or secondary members
W16 × 40 (I_s = 518 in⁴)

Girders or primary members
W24 × 68 (I_p = 1830 in⁴)

6 @ 6 ft
= 36 ft = L_p

L_s = 48 ft

FIGURE E.1

This value indicates insufficient strength and stability, and thus a more precise method of checking should be used.

The curves in Appendix 2 of the AISC Specification provide a design aid for use when we need to compute a more accurate flat-roof framing stiffness than is given by the Specification provision that $C_p + 0.9C_s \leq 0.25$.

The following stress indexes are computed for the primary and secondary members:

For primary members,

$$U_p = \left(\frac{0.8F_y - f_o}{f_o} \right)_p \qquad \text{(AISC Appendix Equation A-2-5)}$$

For secondary members,

$$U_s = \left(\frac{0.8F_y - f_o}{f_o} \right) \qquad \text{(AISC Appendix Equation A-2-6)}$$

In these expressions, f_o represents the stress due to $D + R$ (where D is nominal dead load and R is nominal load due to rainwater or ice, exclusive of ponding contribution). These loads should include any snow that is present, although most ponding failures have occurred during torrential summer rains.

We enter Fig. A-2-1 in Appendix 2 of the AISC Specification with our computed U_p and move horizontally to the calculated C_s value of the secondary members. Then we go vertically to the abscissa scale and read the upper limit for the flexibility constant C_p. If this value is more than our calculated C_p value computed for the primary members, the stiffness is sufficient. The same process is followed in Fig. A-2-2, where we enter with our computed U_s and C_p values and pick from the abscissa the flexibility constant C_s, which should be no less than our C_s value. This procedure is illustrated in Example E-2.

Example E-2

Recheck the roof system considered in Example E-1, using the AISC Appendix 2 curves. Assume that, as ponding begins, f_o is 20 ksi in both girders and beams.

Solution. Checking girders:

$$U_p = \frac{0.8F_y - f_0}{f_0} = \frac{(0.8)(36) - 20}{20} = 0.44$$

For $U_p = 0.44$ and $C_s = 0.197$, we read in AISC Appendix 2, Fig. A-2-1, that $C_p = 0.165$, our calculated C_p of 0.141.

∴ **Girders are sufficiently stiff.**   (OK)

Checking beams:

$$U_s = \frac{0.8F_y - f_o}{f_o} = \frac{(0.8)(36) - 20}{20} = 0.44$$

For $U_s = 0.44$ and $C_p = 0.141$, we read from AISC Appendix 2, Figure A-2-2, that $C_s = 0.150$. This value is $<$ our calculated value of $C_s = 0.197$.

∴ **Stiffer secondary members are required.**

Glossary

Allowable Strength   The nominal strength of a member divided by the safety factor, R_n/Ω.

Amplification Factor   A multiplier used to increase the computed moment or deflection in a member to account for the eccentricity of the load.

Annealing   A process in which steel is heated to an intermediate temperature range, held at that temperature for several hours, and then allowed to slowly cool off to room temperature. The resulting steel has less hardness and brittleness, but more ductility.

ASD (Allowable Strength Design)   Method of sizing structural members such that the allowable strength equals or is greater than the required strength of the member using service loads.

Aspect Ratio   The ratio of the lengths of the sides of a rectangular panel to each other.

Available Strength   The design strength or allowable strength depending on the design method (ASD or LRFD) used.

Bar Joist   See *Open Web-Joist*.

Bays   The areas between columns in a building.

Beam   A member that supports loads transverse to its longitudinal axis.

Beam–Column   A column that is subjected to axial compression loads as well as bending moments.

Bearing–Type Connection   Bolted connection where shear forces are transmitted by the bolt bearing against the connection elements.

Bearing Wall Construction   Building construction where all the loads are transferred to the walls and thence down to the foundations.

Block Shear   A fracture type shear where fracture may occur on either the tension plane or the shear plane, followed by yielding on the other plane (see Fig. 3.16).

Braced Frame   A frame that has resistance to lateral loads supplied by some type of auxiliary bracing.

Brittle Fracture   Abrupt fracture with little or no prior ductile deformation.

Buckling Load   The load at which a straight compression member assumes a deflected position.

Built-Up Member   A member made up of two or more steel elements bolted or welded together to form a single member.

Camber   The construction of a member bent or arched in one direction so that its deflection will not be so noticeable when the service loads bend it in the opposite direction.

Cast Iron   An iron with a very high carbon content (2% or more)

Charpy V-Notch Test   A test used for measuring the fracture toughness of steel by fracturing it with a pendulum swung from a certain height.

Cladding   The exterior covering of the structural parts of a building.

Cold-Formed Light-Gage Steel Shapes   Shapes made by cold bending thin sheets of carbon or low-alloy steels into desired cross sections.

Column   A structural member whose primary function is to support compressive loads.

Compact Section   A section that has a sufficiently stocky profile so that it is capable of developing a fully plastic stress distribution before buckling.

Composite Beam   A steel beam made composite with a concrete slab by shear transfer between the two (see Fig. 16.1).

Composite Column   A column constructed with rolled or built-up steel shapes, encased in concrete or with concrete placed inside steel pipes or other hollow steel sections (see Fig. 17.1).

Connection   The joining of structural members and joints used to transmit forces between two or more members.

Coping   The cutting back of the flanges of a beam to facilitate its connection to another beam (see Fig. 10.6).

Cover Plate   A plate welded or bolted to the flange of a member to increase cross-sectional area, moment of inertia or section modulus.

Dead Loads   Loads of constant magnitude that remain in one position. Examples are weights of walls, floors, roofs, fixtures, structural frames, and so on.

Design Strength   The resistance factor times the nominal strength, ΦR_n.

Diagonal Bracing   Inclined structural member typically carrying only axial force in a braced frame.

Direct Analysis Method   A design method for stability that includes the effects of residual stresses and initial out-of-straightness of frames by reducing member stiffness and applying notional loads in a second-order analysis.

Drift   Lateral deflection of a building.

Drift Index   The ratio of lateral deflection of a building to its height.

Ductility   The property of a material by which it can withstand extensive deformation without failure under high tensile stress.

Effective Length   The distance between points of zero moment in a column; that is, the distance between its inflection points.

Effective Length Factor *K*   A factor that, when multiplied by the length of a column, will provide its effective length.

Elastic Design   A method of design that is based on certain allowable stresses.

Elastic Limit   The largest stress that a material can withstand without being permanently deformed.

Elasticity   The ability of a material to return to its original shape after it has been loaded and then unloaded.

Elastic Strain   Strain that occurs in a member under load before its yield stress is reached.

Endurance Limit   The maximum fatigue-type stress in a material for which the material seems to have an infinite life.

Euler Load   The compression load at which a long and slender member will buckle elastically.

Eyebar   A pin-connected tension member whose ends are enlarged with respect to the rest of the member so as to make the strength of the ends approximately equal to the strength of the rest of the member.

Factored Load   A nominal load multiplied by a load factor.

Fasteners   A term representing bolts, welds, rivets, or other connecting devices.

Fatigue   A fracture situation caused by changing stresses.

Faying Surface   The contact or shear area of members being connected.

Fillet Weld   A weld placed in the corner formed by two overlapping parts in contact with each other (see Fig. 14.2).

First-Order Analysis   Analysis of a structure in which equilibrium equations are written based on an assumed nondeformed structure.

Flexural Buckling   A buckling mode in which a compression member deflects laterally without twist or change in cross-sectional shape.

Flexural-Torsional Buckling   A buckling mode in which a compression member bends and twists simultaneously without change in cross-sectional shape.

Floor Beams   The larger beams in many bridge floors that are perpendicular to the roadway of the bridge and that are used to transfer the floor loads from the stringers to the supporting girders or trusses.

Fracture Toughness   The ability of a material to absorb energy in large amounts. For instance, steel members can be subjected to large deformations during erection and fabrication, without failure, thus allowing them to be bent, hammered, and sheared, and to have holes punched in them.

Gage   Transverse spacing of bolts measured perpendicular to the long direction of the member (see Fig. 12.4).

Girder   A rather loosely used term that usually indicates a large beam and, perhaps, one into which smaller beams are framed.

Girts   Horizontal members running along the sides of industrial buildings, used primarily to resist bending due to wind. They often are used to support corrugated siding.

Government Anchors   Bent steel bars used when ends of steel beam are enclosed by concrete or masonry walls. The bars pass through the beam webs parallel to the walls and are enclosed in the walls. They keep beams from moving longitudinally with respect to the walls.

Gravity Load   A load, such as dead load or live load, acting in a downward direction.

Groove Welds   Welds made in grooves between members that are being joined. They may extend for the full thickness of the parts (complete-penetration groove welds), or they may extend for only a part of the member thickness (partial-penetration groove welds) (see Fig. 14.2).

Hybrid Member   A structural steel member made from parts that have different yield stresses.

Impact Loads   The difference between the magnitude of live loads actually caused and the magnitude of those loads had they been applied as dead loads.

Inelastic Action   The deformation (of a member) that does not disappear when the loads are removed.

Influence Line   A diagram whose ordinates show the magnitude and character of some function of a structure (shear, moment, etc.) as a unit load moves across the structure.

Instability   A situation occurring in a member where increased deformation of that member causes a reduction in its load-carrying ability.

Intermediate Columns   Columns that fail both by yielding and by buckling. Their behavior is said to be inelastic. Most columns fall in this range, where some of the fibers reach the yield stress and some do not.

Ironworker   A person performing steel erection (a name carried over from the days when iron structural members were used).

Joists   The closely spaced beams supporting the floors and roofs of buildings.

Jumbo Sections   Very heavy steel W sections (and structural tees cut from those sections). Serious cracking problems sometimes occur in these sections when welding or thermal cutting is involved.

Killed Steel   Steel that has been deoxidized to prevent gas bubbles and to reduce its nitrogen content.

Lamellar Tearing   A separation in the layers of a highly restrained welded joint, caused by "through-the-thickness" strains produced by shrinking of the weld metal.

Lateral Load   A load, such as wind load or earthquake load, acting in a lateral or horizontal direction.

Leaning Column   A column designed to carry only gravity loads, having connections that do not provide lateral load resistance.

Limit State   A condition at which a structure or some point of the structure ceases to perform its intended function either as to strength or as to serviceability.

Lintels   Beams over openings in masonry walls, such as windows and doors.

Live Loads   Loads that change position and magnitude. They move or are moved. Examples are trucks, people, wind, rain, earthquakes, temperature changes, and so on.

Load Factor   A number almost always larger than 1.0, used to increase the estimated loads a structure has to support, to account for the uncertainties involved in estimating loads.

Local Buckling   The buckling of a part of a larger member that precipitates failure of the whole member.

Long Columns   Columns that buckle elastically and whose buckling loads can be predicted accurately with the Euler formula if the axial buckling stress is below the proportional limit.

LRFD (Load and Resistance Factor Design)   A method of sizing structural members such that the design strength equals or is greater than the required strength of the member using factored loads.

Malleability   The property of some metals by which they may be hammered, pounded, or rolled into various shapes—particularly, thin sheets.

Mild Steel A ductile low-carbon steel.

Milled Surfaces Those surfaces that have been accurately sawed or finished to a smooth or true plane.

Mill Scale An iron oxide that forms on the surface of steel when the steel is reheated for rolling.

Modulus of Elasticity, or **Young's Modulus** The ratio of stress to strain in a member under load. It is a measure of the stiffness of the material.

Moment Connection A connection that transmits bending moment between connected structural members.

Moment Frame A frame that has resistance to lateral loads supplied by the shear and flexure of the members and their connections.

Net Area Gross cross-sectional area of a member minus any holes, notches, or other indentations.

Nominal Loads The magnitudes of loads specified by a particular code.

Nominal Strength The theoretical ultimate strength of a member or connection.

Noncompact Section A section that cannot be stressed to a fully plastic situation before buckling occurs. The yield stress can be reached in some, but not all of the compression elements before buckling occurs.

Notional Load Virtual load applied in a structural analysis to account for destabilizing effects that are not otherwise accounted for in the design provisions.

Open-Web Joist A small parallel chord truss whose members are often made from bars (hence the common name *bar joist*) or small angles or other rolled shapes. These joists are very commonly used to support floor and roof slabs (see Fig. 19.5).

P-Delta Effect Changes in column moments and deflections due to lateral deflections.

Partially Composite Section A section whose flexural strength is governed by the strength of its shear connectors.

Pitch The longitudinal spacing of bolts measured parallel to the long direction of a member (see Fig. 12.4).

Plane Frame A frame that, for purposes of analysis and design, is assumed to lie in a single (or two-dimensional) plane.

Plastic Design A method of design that is based on a consideration of failure conditions.

Plastic Modulus The statical moment of the tension and compression areas of a section taken about the plastic neutral axis.

Plastic Moment The yield stress of a section times its plastic modulus. The nominal moment that the section can theoretically resist if it is braced laterally.

Plastic Strain The strain that occurs in a member after its yield stress is reached with no increase in stress.

Plate Girder A built-up steel beam (see Fig. 18.3).

Poisson's Ratio The ratio of lateral strain to axial or longitudinal strain in a member under load.

Ponding A situation on a flat roof where water accumulates faster than it runs off.

Post-Buckling Strength The load a member or frame can support after buckling occurs.

Proportional Limit Largest strain for which Hooke's law applies, or the highest point on the straight-line portion of the stress–strain diagram.

Purlins Roof beams that span between trusses (see Fig. 4.4).

Quenching Rapid cooling of steel with water or oil.

Required Strength The forces, stresses, and deformations produced in a structural member determined from a structural analysis using factor or service loads.

Residual Stresses The stresses that exist in an unloaded member after it's manufactured.

Resistance Factor ϕ A number, almost always less than 1.0, multiplied by the ultimate or nominal strength of a member or connection to take into account the uncertainties in material strengths, dimensions, and workmanship. Also called *overcapacity factor*.

Rigid Frame A structure whose connectors keep substantially the same angles between members before and after loading.

Safety Factor A number, typically greater than 1.0, divided into the nominal strength to take into account the uncertainties of the load and the manner and consequences of failure.

Sag Rods   Steel rods that are used to provide lateral support for roof purlins. They also may be used for the same purpose for girts on the sides of buildings (see Fig. 4.5).

St. Venant Torsion   The part of the torsion in a member that produces only shear stresses in the member.

Scuppers   Large holes or tubes in walls or parapets that enable water above a certain depth to quickly drain from roofs.

Second-Order Analysis   Analysis of a structure for which equilibrium equations are written that include the effect of the deformations of the structure.

Section Modulus   The ratio of the moment of inertia, taken about a particular axis of a section, to the distance to the extreme fiber of the section, measured perpendicular to the axis in question.

Seismic   Of or having to do with an earth-quake.

Serviceability   The ability of a structure to maintain its appearance, comfort, durability, and function under normal loading conditions.

Serviceability Limit State   A limiting condition affecting the ability of a structure to maintain its appearance, maintainability, durability or comfort of its occupants or function of machinery, under normal usage.

Service Loads   The loads that are assumed to be applied to a structure when it is in service (also called *working loads*).

Shape Factor   The ratio of the plastic moment of a section to its yield moment.

Shear Center   The point in the cross section of a beam through which the resultant of the transverse loads must pass so that no torsion will occur.

Shear Lag   A nonuniformity of stress in the parts of rolled or built-up sections occurring when a tensile load is not applied uniformly.

Shear Wall   A wall in a structure that is specially designed to resist shears caused by lateral forces such as wind or earthquake in the plane of the wall.

Shims   Thin strips of steel that are used to adjust the fitting of connections. Finger shims are installed after the bolts already are in place.

Short Columns   Columns whose failure stress will equal the yield stress, and for which no buckling will occur. For a column to fall into this class, it would have to be so short as to have no practical application.

Sidesway   The lateral movement of a structure caused by unsymmetrical loads or by an unsymmetrical arrangement of building members.

Simple Connection   A connection that transmits negligible bending moment between connected members.

Skeleton Construction   Building construction in which the loads are transferred for each floor by beams to the columns and thence to the foundations.

Slenderness Ratio   The ratio of the effective length of a column to its radius of gyration, both pertaining to the same axis of bending.

Slender Section   A member that will buckle locally while the stress still is in the elastic range.

Slip-Critical Joint   A bolted joint that is designed to have resistance to slipping.

Space Frame   A three-dimensional structural frame.

Spandrel Beams   Beams that support the exterior walls of buildings and perhaps part of the floor and hallway loads (see Fig. 19.3).

Steel   An alloy consisting almost entirely of iron (usually, over 98 percent). It also contains small quantities of carbon, silicon, manganese, sulfur, phosphorus, and other materials.

Stiffened Element   A projecting piece of steel whose two edges parallel to the direction of a compression force are supported (see Fig. 5.6).

Stiffener   A plate or an angle usually connected to the web of a beam or girder to prevent failure of the web (see Figs. 18.10 and 18.11).

Story Drift   The difference in horizontal deflection at the top and bottom of a particular story.

Strain-Hardening   Range beyond plastic strain in which additional stress is necessary to produce additional strain.

Strength Limit State A limiting condition affecting the safety of the structure, in which the ultimate load-carrying capacity is reached.

Stringers The beams in bridge floors that run parallel to the roadway.

Tangent Modulus The ratio of stress to strain for a material that has been stressed into the inelastic range.

Tension Field Action The behavior of a plate girder panel which, after the girder initially buckles, acts much like a truss. Diagonal strips of the web act similarly to the diagonals of a parallel chord truss. The stiffeners keep the flanges from coming together, and the flanges keep the stiffeners from coming together (see Fig. 18.8).

Toughness The ability of a material to absorb energy in large amounts. As an illustration, steel members can be subjected to large deformations during fabrication and erection without fracture, thus allowing them to be bent, hammered, and sheared, and to have holes punched in them without visible damage.

Unbraced Frame A frame whose resistance to lateral forces is provided by its members and their connections.

Unbraced Length The distance in a member between points that are braced.

Unstiffened Element A projecting piece of steel having one free edge parallel to the direction of a compression force, with the other edge in that direction unsupported (see Fig. 5.6).

Upset Rods Rods whose ends are made larger than the regular bodies of the rods. Threads are cut into the upset ends, but the area at the root of the thread in each rod is larger than that of the regular part of the bar (see Fig. 4.3).

Warping Torsion The part of the resistance of a member to torsion that is provided by the warping resistance of the member cross section.

Weathering Steel A high-strength low-alloy steel whose surface, when exposed to the atmosphere (not a marine one), oxidizes and forms a tightly adherent film that prevents further oxidation and thus eliminates the need for painting.

Web Buckling The buckling of the web of a member (see Fig. 10.9).

Web Crippling The failure of the web of a member near a concentrated force (see Fig. 10.9).

Working Loads See *Service Loads*.

Wrought Iron An iron with a very low carbon content ($\leq 0.15\%$)

Yield Moment The moment that will just produce the yield stress in the outermost fiber of a section.

Yield Stress The stress at which there is a decided increase in the elongation or strain in a member without a corresponding increase in stress.

Index